W9-CAH-327

PRINCIPLES OF EARTH SCIENCE

Chapters 14–16

PRINCIPLES OF EARTH SCIENCE

ARTHUR N. STRAHLER

HARPER & ROW, PUBLISHERS
NEW YORK EVANSTON SAN FRANCISCO LONDON

Sponsoring Editor: Dale Tharp
Project Editor: Elizabeth Dilernia
Designer: T. R. Funderburk
Production Supervisor: Stefania J. Taflinska
Compositor: Ruttle, Shaw & Wetherill, Inc.
Printer and Binder: Halliday Lithograph Corporation

PRINCIPLES OF EARTH SCIENCE

Library of Congress Cataloging in Publication Data

Strahler, Arthur Newell, 1918-
Principles of earth science.

Includes index.
1. Earth sciences. I. Title.
QE28.S83 550 75-26638
ISBN 0-06-046451-8

CONTENTS

PREFACE

In shaping this new work, *Principles of Earth Science,* special attention has been given to the needs of those students who have only one quarter or one semester available for an overview of earth science. Few of them have developed a strong basic background in science, yet they have matured in a time of great scientific achievement in the earth sciences. One aim of this book is to give today's student a nontechnical but meaningful survey of the advances in astrogeology and astronomy achieved through the Apollo and Mariner space missions. Another aim is to review the steps leading up to the recent revolution in geology—plate tectonics—and to summarize the major features of this remarkable earth science breakthrough

Today's student is aware that many environmental phenomena are dangerous and destructive. Earthquakes, volcanic eruptions, seismic sea waves, landslides, floods, tornadoes, hailstorms, hurricanes, and storm surges repeatedly make the front pages of our newspapers. The need to understand how and where these environmental hazards originate is met throughout this book.

Many of today's youth are deeply concerned with man's impact upon the environment. Environmental topics are explained in the text, including air pollution, urban climate change, global climate change, and man-induced erosion.

Young persons today are also aware that we face a prolonged energy shortage and shortages in vital mineral resources. They realize that a knowledge of earth science will greatly increase their understanding of the nature, origin, and occurrence of natural resources.

However, although these special aspects of earth science are included, the main thrust of *Principles of Earth Science* is toward increased understanding of a wide range of planetary phenomena spanning the lithosphere, hydrosphere, and atmosphere. Today's students are wide-ranging in their travels. Some probe the shallow ocean floor as scuba divers; other climb high above the timberline as backpackers and skiers. Their travels in jet aircraft present them with a wide range of landscapes and a diversity of weather and climate patterns. Curiosity about the origin of rock formations, landforms, clouds, storms, planets, and stars needs to be satisfied by reasonable and straightforward explanations.

Over the last two decades earth science textbooks have grown in bulk and complexity in desperate attempts to include all the knowledge accumulated. In constructing a text scaled down to fit a brief college survey course, not only must a wide range of topics of secondary importance be deleted, but the vocabulary of science must be severely and selectively pruned as well. Special attention has been given to limiting the number of geoscience terms and to making sure that each is adequately defined and explained as it is introduced. A checklist of important terms is included with each chapter, along with a group of self-testing questions on the content of the chapter. The Glossary at the end of the book defines all terms in alphabetical order.

A special feature of this new book is the brief human-interest essays that open each chapter. Some give a taste of the gropings and controversies that attended the emergence of natural science; others are news reports detailing the impact of natural phenomena on mankind. A few describe modern scientific research and technology. Collectively the essays are intended as a small advance reward to make easier a confrontation with the serious content of the chapter.

Illustrations have been culled, leaving only those with direct and clear impact. Many new line drawings and graphs have also been structured to this end. Tables have been reduced to the fewest possible number. It has become fashionable today to make lavish use of large photographs and art works in full color, even though in most instances these give little information and serve only to raise the price of the book. We have chosen instead to be generous with photographs carefully selected to illustrate precisely the subjects under discussion. We hope that our careful attention to design and content has generated a product in which educational values are optimized at a time when economic values are a paramount concern.

Arthur N. Strahler

PRINCIPLES OF EARTH SCIENCE

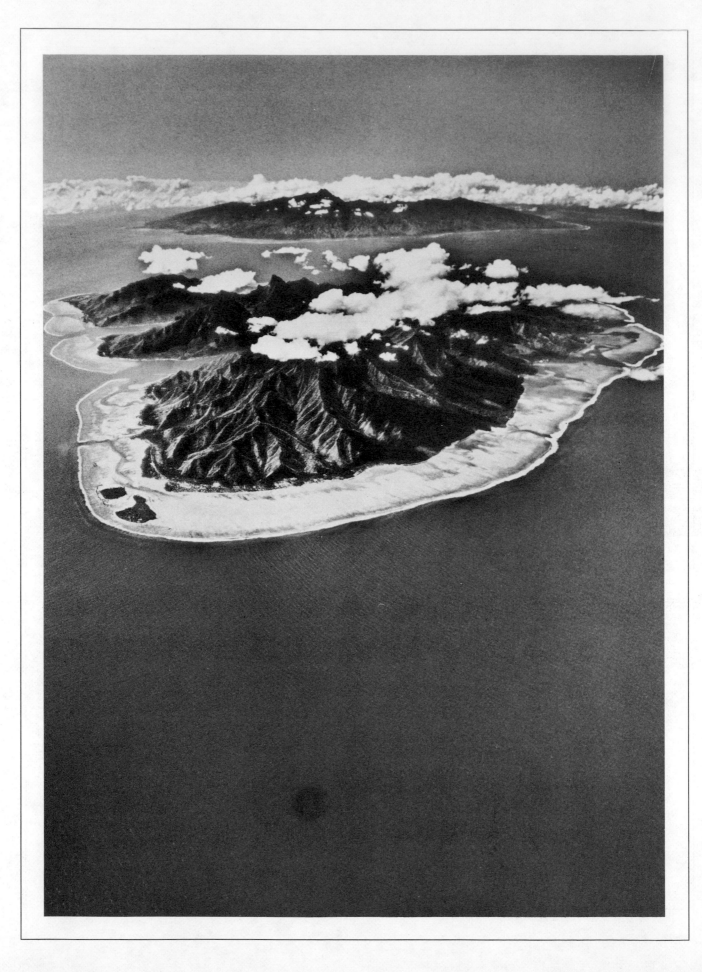

INTRODUCTION TO EARTH SCIENCE

As your sleek airways jet makes its approach to the airport at Papeete, a scene of striking beauty unfolds below you. Near at hand is Moorea—"the perfect South Sea island." Your travel brochure was not exaggerating things that much, after all. Moorea's lush green mountain slopes culminate in sharp peaks touched by snow-white puffy clouds. Ringing her shores is a broad belt of reef—pale blue-green shoal water separated from the deep blue ocean and the green land by narrow strips of dazzling white coral sand. A strong surf breaks over the outer edge of the reef. Beyond Moorea rises your destination, the island of Tahiti. Paul Gaugin painted there and you know his canvases well.

You look forward to a memorable stay on Tahiti. There will be traditional native dances; there will be skin-diving in fabulous coral grottoes, some wild outrigger rides, and in the evening that famous Tahitian feast. You look forward to a side trip to Moorea, the Bali Hai island in whose bays the movie *South Pacific* was filmed. You can imagine yourself wading in cascading mountain streams and submerging in deep pools beneath high waterfalls almost hidden in tropical rainforest.

You begin to take a closer look at Moorea. In the scene before you three great earth realms meet and interact. A solid mass of rock, representing the lithosphere, rises in stately fashion over a boundless ocean, representing the global hydrosphere. In the zone of interaction between ocean and land, reef-building organisms have thrived for millenia. The skeletal parts of corals and the limy secretions of algae have accumulated into a barrier of white rock. Looking higher, you notice that both the ocean and the land are interacting with the third great earth realm, the atmosphere. Moisture evaporated from the vast surface of the Pacific Ocean is condensing in clouds over the island summit. Why are clouds concentrated over the island and not over the surrounding ocean?

Now you notice that the island of Tahiti, in the background, is quite different in appearance from Moorea. Moorea is cut up into deep valleys and knife-edged crests, while Tahiti seems to have smoother sides. Your knowledge of earth science leads you to suspect that Tahiti is a much younger island, perhaps an extinct volcano still largely preserving its original shape. In contrast, Moorea, if it too originated as a volcano, has been extinct for a much longer time and has been eroded by rain and streams.

Another feature also arouses your curiosity. Moorea has a broad coral reef encircling it, whereas Tahiti has only a narrow reef. Yet Tahiti is the larger island, so why shouldn't it have an even wider reef than Moorea? Then you realize that if Moorea is in fact a much older island, there has been a much longer time for the reef to grow. Young Tahiti has been built so recently that it possesses only a narrow reef. These are speculations, to be sure, but they are consistent with what you have learned about earth forms and earth processes.

Earth science deals with the earth's crust, oceans, and atmosphere. But earth science also considers the earth as a whole planet. To understand our own planet we must study the sister planets and the entire solar system. At the hub of the solar system lies our sun—an average star. The steady flow of radiant energy from the sun's incandescent surface powers atmospheric storms and those activities which erode the lands and transport sediment. Our survey of earth science is completed by an overview of the sun as a star in its galaxy, and of our galaxy in the universe.

(*Above*) Molten lava glows brightly on the floor of a pit crater in Kilauea Volcano, Hawaii. (Mike Roberts Color Productions.)

(*Below*) This fast-moving tongue of lava from Mauna Ulu ran down a paved road, solidifying as a wrinkled flow with a smooth, glassy surface. (Dr. Willard H. Parsons.)

1 THE IGNEOUS MINERALS AND ROCKS

INSTANT ROCK

Not long ago, I took my family to Hawaii, where we visited an actively erupting volcano named Mauna Ulu. (*Mauna Ulu* means "growing mountain.") We parked at the end of a highway that abruptly disappeared beneath a great mass of black lava rock; here molten rock had engulfed the highway only a couple of years earlier. Following a rough, steep trail, barely traceable over the contorted lava surface, we made our way to the crater rim. Facing heat like that from an open-hearth steel furnace, we looked down some 200 feet to see a great churning pool of molten rock, as free as the waters below Niagara Falls. Along one side of the huge pit molten rock was welling up in incredible volume. It then moved slowly across the pit to the opposite side, where it disappeared into a great subterranean cavity. Over much of the pool, a thin crust of solidified magma had formed quickly, like ice on a pond. However, the moving crust was continually being fractured and we could see the magma glowing red between the solid plates. Solid rock was being formed before our eyes—"instant rock" it surely was. Created from a fiery melt, this was *igneous* rock (in Latin the word *ignis* means "fire").

Six thousand miles from Hawaii, citizens of a small Icelandic island recently fought a bitter battle with another "growing mountain." The island, called Heimaey, is occupied by a volcano, but it had been dormant for thousands of years. Heimaey had become the site of an important village, for it has a fine harbor to protect a fishing fleet. Then, suddenly, in January 1973 a fiery fissure opened up across the dormant volcano, spewing forth sheets of flame and tossing up solid volcanic fragments. Soon lava flows emerged from the old volcano and began to spread toward the town and into the sea. The black lava fragments, called *tephra,* accumulated on rooftops and in the streets like a heavy fall of snow. Bravely the villagers swept the tephra from their roofs and trucked it away from the streets, but they were powerless to stop the lava flow creeping into the edge of town.

Then Iceland's scientists put their minds to work and came up with a plan. The lava flow could be stopped with massive applications of cold sea water, taking out heat from the lava and causing the forward edge to solidify. Soon pumps were brought in and the experiment began. At the peak of the operation 47 pumps were in action delivering water at the rate of nearly a million gallons an hour. Altogether over five million tons of sea water were sprayed on the lava flow, and it came to a halt. Molten lava continued to flow from the volcano to the sea, in a direction away from the town. Five months and five days after it started, the eruption was officially declared to be over. Thanks to the new lava flows the harbor is better protected than ever. Altogether, the area of the island was increased by about 20% by new lava flows. Damaged buildings were dug out and most were repaired and reoccupied. Some of the tephra has been used to extend the runways of the island's airport.

In this chapter our subject is igneous rock, its origin and composition. We shall first examine igneous rock in detail to learn what minerals it contains and to determine the chemical elements of which these minerals are composed.

ROCKS AND MINERALS

Geology is the science of the solid earth, or **lithosphere. Geoscience** is often used as a synonym for geology. Geology deals with rocks and minerals, with the fossilized remains of animals and plants contained in rocks, and with the physical history of the solid earth starting from the time of the earliest-formed rocks. The geologist studies rocks and minerals as old as 3½ billion years, as well as rocks solidifying before his eyes at the present moment. Rocks and minerals will comprise the subject matter of our first eleven chapters.

What is rock? In a broad sense, rock is the solid substance comprising the earth's outer shell. Rock is composed of mineral matter in the solid state. Most rock consists of several minerals in combination. Commonly the minerals occur as individual grains, so that the rock is a physical mixture of such mineral grains.

What, then, is a mineral? First, a mineral consists of inorganic matter in the solid state. The living tissues of plants or animals and their partly decomposed remains are excluded from the mineral category. Substances that exist as liquids and gases at ordinary atmospheric temperatures are also excluded. Liquid water is not strictly a mineral, but water qualifies as a mineral in the form of ice. Second, a particular mineral in its pure state has a definite chemical composition which can usually be stated by a fairly simple chemical formula. For example, one form of rock salt—the mineral halite—has the composition sodium chloride. Within such a mineral the atoms are arranged in a specific geometric pattern in space, and this arrangement persists uniformly throughout the mineral. In other words, the composition is homogeneous throughout. A substance with such solid structure is described as a **crystalline solid.** (There are exceptions to the crystalline state in minerals.) Strictly speaking, a mineral is a substance of natural occurrence, ruling out such man-made materials as glass or furnace slag.

To summarize these points within a formal definition, a **mineral** is a homogeneous, naturally occurring, inorganic substance, usually having a definite chemical composition and a characteristic atomic structure.

The various rock groups and individual rock types are distinguished in terms of the mineral varieties present and the proportions in which they occur. Then there are further distinctions among rock types based upon the size of the individual crystal grains, or the lack of any observable crystalline structure. Hundreds of mineral varieties have been identified. Fortunately, however, the bulk of the common rocks consists of combinations of only a few important minerals. We shall limit our mineral list to those important common minerals, and our list of rocks to the most common combination of those minerals.

Our first concern is with a class of rocks designated as **igneous rock;** it is rock that has solidified from a high-temperature molten condition. In other words, molten mineral matter, called **magma,** has undergone a transformation from liquid to solid state.

CHEMICAL COMPOSITION OF THE EARTH'S CRUST

Before beginning a study of the principal minerals comprising the igneous rocks, it will be helpful to examine figures on the abundance of elements in the earth's crust. The crust is the outermost of the earth's solid shells and has an average thickness of about 10 mi (17 km). Although the rocks of the crust constitute only about one-half of 1% of the total mass of the earth, they are the only rocks available to the geologist for direct examination and chemical analysis. The bulk of the crust (about 95%) is composed of igneous rock.

Table 1.1 lists the eight most abundant chemical elements in the earth's crust. The order of listing is according to percentage by weight. Several points are of interest in the figures in Table 1.1. Notice, first, that the eight elements constitute between 98% and 99% of the crust by weight and that almost half of this weight is oxygen. Measured in other ways, the importance of oxygen is even greater—in numbers of atoms it makes up over 60% of the total. Because oxygen is an atom of comparatively large radius, it represents almost 94% by volume. Notice that silicon is in second place with about 28%, or roughly half the value for oxygen. Aluminum and iron occupy intermediate positions, while the last four elements —calcium, sodium, potassium, and magnesium—are in the range of 2% to 4%.

Table 1.1 **The most abundant elements in the earth's crust**

Element	Percentage by weight
Oxygen	46.6
Silicon	27.7
Aluminum	8.1
Iron	5.0
Calcium	3.6
Sodium	2.8
Potassium	2.6
Magnesium	2.1
Total	98.5

The ninth most abundant element is titanium, followed in order by hydrogen, phosphorus, barium, and strontium. It is interesting that the metals copper, lead, zinc, nickel, and tin, which play such an important role in our modern technology, are present only in very small proportions. These metals are indeed scarce elements. Fortunately for mankind, these and other rare but important elements have been concentrated locally into ores. Metals occurring as ores can be profitably extracted in useful quantities.

None of the eight elements on our list exists by itself in any important quantities in nature as a solid substance. (Very small amounts of metallic iron do occur naturally; it is called native iron.) Instead, these elements are found in combinations known as **chemical compounds.** Both oxygen and silicon combine freely with the remaining six elements to form abundant compounds. From a chemist's point of view, oxygen and silicon are both described as **nonmetallic elements,** in contrast to the remaining six, which are **metallic elements.**

THE SILICATE MINERALS

The vast bulk of all igneous rock consists of mineral compounds containing the elements silicon and oxygen. Collectively, these minerals are known as **silicates.** In a **silicate mineral** both silicon and oxygen are combined with one or more of the metallic elements listed in Table 1.1.

We can gain a good appreciation of the nature of igneous rocks as a class by noting the proportions of only seven silicate minerals, or mineral groups. These are shown in Figure 1.1. The mineral list begins with **quartz,** containing only silicon and oxygen. The next five silicate compounds all contain aluminum and can be designated **aluminosilicates.**

The first aluminosilicate is **potash feldspar.** Potassium is the dominant metallic element in potash feldspar. The mineral name for a common kind of potash feldspar is **orthoclase.**

Next come the **plagioclase feldspars.** They span a continuous range from the **sodic plagioclase** end of the series, with sodium making up most of the metallic element, to **calcic plagioclase** at the other end, with calcium making up most of the metallic element. Plagioclase of intermediate composition contains about equal proportions of sodium and calcium. Quartz and the feldspars are light in color.

Biotite is the dark-colored representative of the **mica group** of silicate minerals. Biotite is a complex aluminosilicate of potassium, magnesium, and iron, with some water. Continuing down the list, we come to two more mineral groups. In each group are several closely related minerals, each with its own name and distinctive chemical composition. The **amphibole group** is represented by the mineral **hornblende;** the **pyroxene group** by **augite.** Both of these groups are complex aluminosilicates of calcium, magnesium, and iron. Minerals of this group are usually very dark in color. Finally, there is **olivine,** a dense greenish mineral which is a silicate of magnesium and iron, but without aluminum.

Figure 1.2 shows specimens of some of the minerals we have listed. These are showpiece specimens of the pure minerals. Notice that pure quartz is glasslike; the feldspars are white and porcelain-like; biotite is black and opaque to light; olivine is greenish and glassy.

Figure 1.1 Common silicate minerals and igneous rocks.

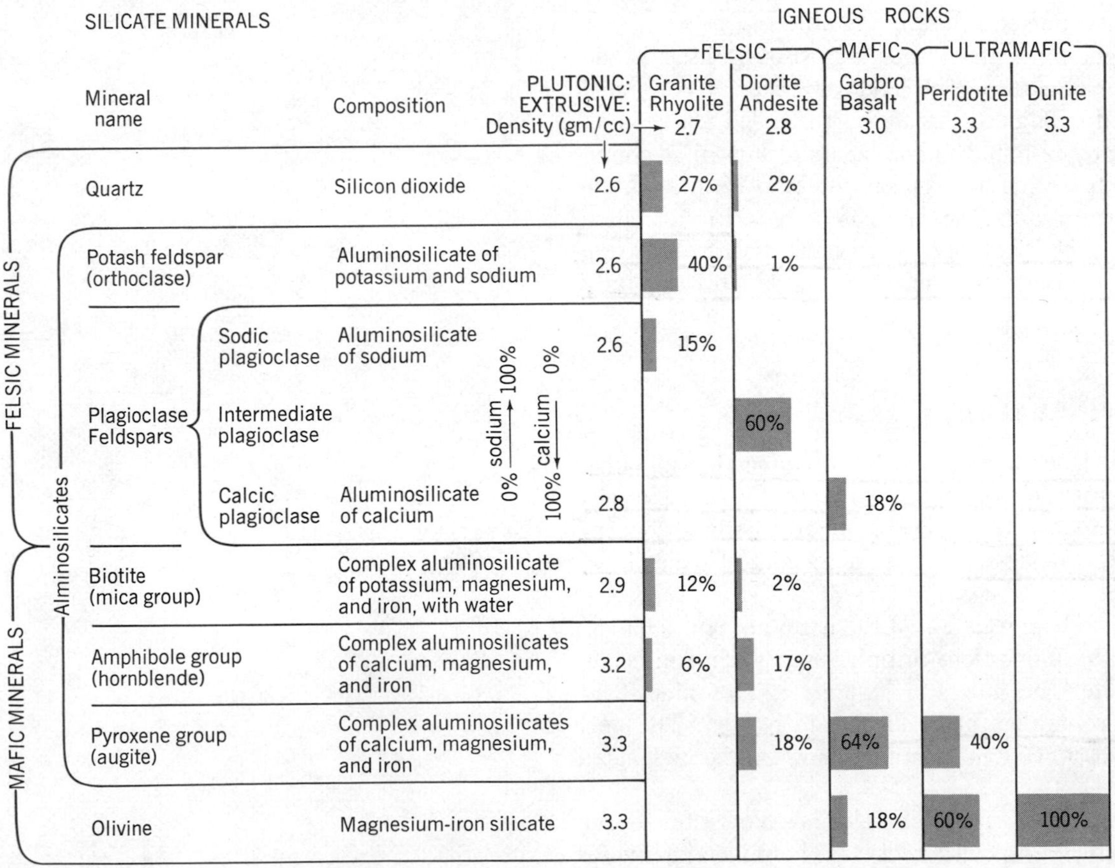

A. Crystals of pure quartz, about one-half natural size. The crystals are hexagonal (six-sided), with pyramidal free ends. (Ward's Natural Science Establishment, Inc., Rochester, N.Y.)

B. A mass of clear quartz of a variety often called rock crystal. It is enclosed by fracture surfaces and resembles a chunk of clear glass. The needlelike objects inside are crystals of tourmaline. (Ward's Natural Science Establishment, Inc., Rochester, N.Y.)

C. Microcline, a potash feldspar, with well-developed cleavage surface. (Ward's Natural Science Establishment, Inc., Rochester, N.Y.)

D. Albite, a sodic plagioclase feldspar, with well-developed cleavage surfaces. (American Museum of Natural History.)

E. A cleavage piece of biotite mica. (Ward's Natural Science Establishment, Inc., Rochester, N.Y.)

F. Olivine, showing a glassy fracture surface. (Ward's Natural Science Establishment, Inc., Rochester, N.Y.)

Figure 1.2 Silicate mineral specimens.

MINERAL DENSITIES

A very important concept in understanding how minerals and rocks are arranged in the earth's crust is that of **mineral density.** In general, density refers to the degree of compaction of matter into a given volume of space. A gas, such as air, has very low density, because the molecules of matter comprising the gas are widely separated. Iron, on the other hand, has a high density—its molecules are very closely packed and the individual molecules have great weight compared to the oxygen and nitrogen molecules of the air.

Figure 1.3 illustrates the principle of mineral density. We imagine that cubes of the same dimensions of each of several substances are hung in turn from a coil spring. The **mass,** or quantity of matter, in each cube differs from one substance to the next. The greater the mass in the cube, the stronger is the force of gravity pulling upon the mass, and the greater is the amount of stretching of the spring. Pure water is used as the reference standard in this case. Each cubic centimeter (cc) of water contains one gram (1 gm) of matter. The density of water is therefore 1 gm/cc. By comparison, a gram of quartz contains 2.6 grams of matter; its density is 2.6 gm/cc. Olivine has a density of 3.3 gm/cc, while pure iron has a density of about 8 gm/cc.

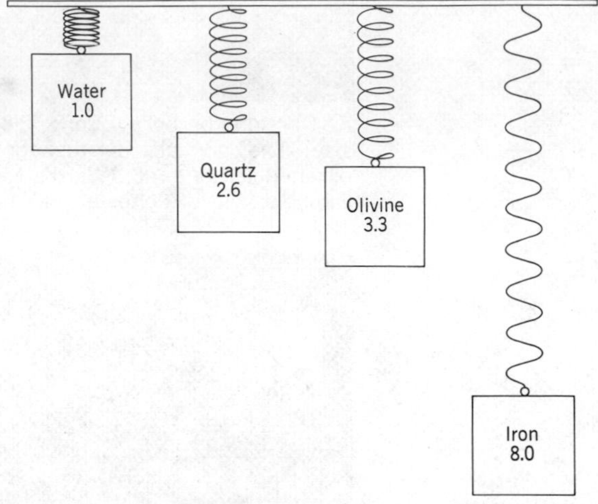

Figure 1.3 Substances of differing densities have different weights, as shown when cubes of equal volume are compared.

FELSIC AND MAFIC MINERAL GROUPS

Now refer back to Figure 1.1, where densities are listed opposite each mineral. Notice that density increases from top to bottom of the list. Quartz and the feldspars range from 2.6 to 2.8 gm/cc; they comprise the **felsic group.** "Felsic" is a coined word; it is derived from "fel" in "feldspar" and "si" in "silica." Not only do the felsic minerals have comparatively lower density; they are typically light in color. The remaining minerals range in density from 2.9 for biotite to 3.3 for augite and olivine. These denser minerals comprise the **mafic group.** "Mafic" is a word coined from the syllable "ma" in "magnesium" and "fic," a contraction of "ferric" (an adjective describing iron).

The presence of iron, a high-density element, largely accounts for the greater densities of the mafic minerals. We see this effect strongly in two black, iron-rich mafic minerals, neither of which is a silicate. The first is **magnetite,** an iron oxide, with a density of 4.5 g/cc. Second is **ilmenite,** an iron-titanium oxide, with a density of 5.5 g/cc. Both magnetite and ilmenite occur in minor proportions in a wide variety of igneous rocks.

Looking ahead, we can reason that an igneous rock composed of felsic minerals will have lower density than an igneous rock composed of mafic minerals. This difference must influence the rearrangement of rocks into layers in response to gravity.

MINERAL CLEAVAGE

Certain silicate minerals—and a wide range of other minerals as well—exhibit a physical property known as **cleavage.** Such minerals, when struck a sharp blow or crushed, break apart along planelike partings. These planes tend to be oriented parallel with one another to form a set. A mineral of this type separates (cleaves) into sheets. There may also be another set of parting planes at a right angle to the first, or at some intermediate angle. In such a case the mineral breaks apart into prisms. Even a third set of planes may exist, yielding cubes or rhombohedrons when the mineral is broken. Calcite is a mineral illustrating rhombohedral cleavage (Figure 1.4). A distinctive cleavage is often a help in identifying a mineral. Then there are some minerals, such as quartz or olivine, having no cleavage—they break irregularly, like a chunk of glass. The micas, such as the muscovite shown in Figure 1.5, cleave so perfectly in parallel planes that sheet after sheet of paper-thin mica can be stripped away. (Muscovite sheets are used in electrical insulation and as viewing windows in furnace doors.)

Figure 1.4 These cleavage rhombs of pure calcite are of a variety known as Iceland spar. (Ward's Natural Science Establishment, Inc., Rochester, N.Y.)

SILICATE MAGMA

Igneous rocks are composed almost entirely of the seven silicate minerals or mineral groups we have studied: It is estimated that these seven make up as much as 99% of all igneous rocks.

A **silicate magma** is a mass of molten mineral matter capable of yielding silicate minerals as it solidifies. Naturally, we don't know much about the nature of silicate magma as it exists deep in the earth—there's simply no way to examine it directly. But we can infer a number of facts from studies of magmas as they emerge at the surface in molten form, as **lava,** from volcanoes. Other observations are made in the laboratory, where silicate minerals are melted and allowed to cool and solidify under various controlled conditions.

We infer that silicate magma deep in the earth—say 25 mi (40 km) down—ranges in temperatures from 900° to 2200° F (500° to 1200° C). Here, confining pressures are 6000 to 12,000 times as great as the normal pressures of our atmosphere at sea level. As magma rises in the crust, it passes through zones of progressively lower temperature and lower pressure. In some cases the magma solidifies within the surrounding rock; in other cases it is able to reach the surface, emerging as lava.

All silicate magmas contain chemical substances other than the elements that yield silicate minerals. These substances are termed **volatiles,** because they remain in a liquid or gaseous state at much lower temperatures than the mineral-forming compounds. Volatiles are separated from the magma as temperatures drop and the silicate minerals crystallize into the solid state.

Much has been learned about volatiles in magmas by sampling substances emitted from volcanoes along with lava. Water (as steam) is by far the most important constituent of the volatiles emitted as gases from volcanoes. In fact, about 90% by volume of the gas contained in an average magma is water. (Some of this water may be rainwater that has percolated down from the surface to reach the magma.) Estimates of the proportion of water present in magmas range from 0.5% to 8%. Fresh igneous rocks commonly contain about 1% of water entrapped within the minerals during their crystallization.

The importance of even a small amount of water in magma is

Figure 1.5 A sheet of muscovite mica. The vertical surfaces facing the observer are cleavage planes; surfaces at top and sides are crystal faces. (American Museum of Natural History.)

very great, because the magma can remain molten at considerably lower temperatures than if it consisted only of silicate compounds without water. As a result, intrusions of magmas can reach closer to the earths surface before solidifying. Magma can also pour out upon the surface in greater amounts than would otherwise be possible.

CRYSTALLIZATION OF MAGMA

Crystallization is the process of change from liquid state to solid state. Crystallization begins to take place in a silicate magma at a certain critical combination of temperature and pressure. However, all minerals do not begin to crystallize at the same time. Moreover, a mineral, once formed, does not necessarily remain intact and unchanged from that point on. Instead, the early-formed minerals may subsequently be changed gradually in composition. Also, certain silicate minerals are dissolved and reformed as temperatures continue to fall. This process of change is referred to as **reaction.**

Next, we must recognize a further complexity in the crystallization process. Certain of the crystallized minerals may be removed from the silicate melt, a process termed **fractionation.** For example, early-formed crystals may be left behind as the magma migrates upward, or the crystals may simply settle to the base of the magma body and accumulate there. Removal of crystallized minerals changes the average chemical composition of the remaining fluid magma and modifies the series of reactions that can follow.

Geologists recognize a definite order in which silicate minerals are crystallized from a magma, as temperature gradually decreases. The order is as follows:

olivine
calcic plagioclase
pyroxene group (augite)
intermediate plagioclase
amphibole group (hornblende)
sodic plagioclase
biotite
potash feldspar (orthoclase)
quartz

Perhaps it occurs to you that the minerals are named in just the reverse of the order we presented them earlier. In general, the mafic minerals (olivine, pyroxene, amphibole) crystallize ahead of certain of the felsic minerals (quartz, potash feldspars). The plagioclase feldspars also crystallize early, beginning with the calcic types and ending with the sodic types.

Well, what is the significance of an order of crystallization? It means that two or more varieties of igneous rocks can be produced from a single silicate magma. This is how it works: Suppose that fractionation takes place after olivine and pyroxene have been formed. These dense minerals settle out in a layer at the base of the magma body. Here they form a distinctive type of rock known as peridotite. Crystallization then continues to a stage in which pyroxene (augite) and intermediate plagioclase feldspar are formed. At this point fractionation could produce a rock known as gabbro.

The remaining components of the magma are now comparatively richer in silicon, aluminum, and potassium because most of the calcium, iron, and magnesium have been used up. This remaining magma may crystallize after migrating to a different physical location from the earlier-formed minerals. The result is an igneous rock predominantly composed of quartz and potash feldspar. Granite is such a rock.

The final residual matter of the magma, remaining fluid at comparatively lower temperatures, is a watery solution rich in silica. From this solution are deposited rock veins, of a type known as **pegmatite**, which consist of a large crystals of quartz, potash feldspar, and mica (Figure 1.6). Ultimately, the water and other minor volatile constituents may reach the earth's surface, to escape in **fumaroles** (vents emitting hot gases) or in hot springs.

Geologists use the expression **magmatic differentiation** for the process we have just described. "Differentiation," in this case, simply means "to break up into parts." Geologists think magmatic differentiation can be responsible for an arrangement of igneous rocks into a series ranging from a mafic group rich in iron, magnesium, calcium, and silica to a felsic group rich in aluminum, sodium, and potassium, and with excess silica. So we find that the classification of igneous rocks which we are about to present is based upon concepts of mineral fractionation and magmatic differentiation.

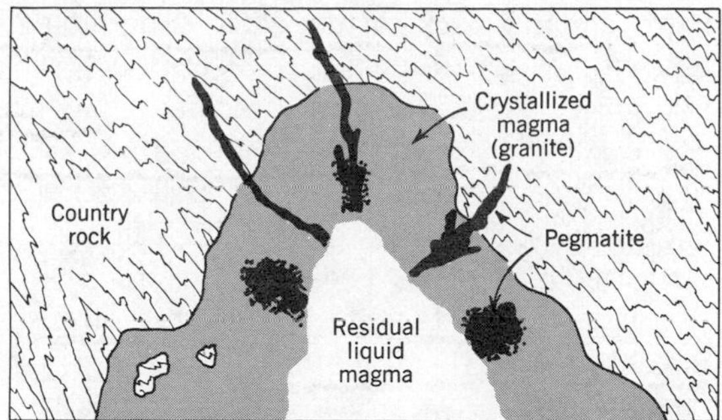

Figure 1.6 (*Upper*) Occurrence of pegmatite bodies in relation to an intrusive igneous body and the country rock. (*Lower*) Enormous crystals from the Etta pegmatite deposit, Pennington County, South Dakota. The hammer rests upon a single large spodumene crystal. (J. J. Norton, U.S. Geological Survey.)

INTRUSIVE AND EXTRUSIVE IGNEOUS ROCKS—ROCK TEXTURE

Now that we can approach the problem of classifying igneous rocks on a meaningful basis, we will limit ourselves to only a few important rock varieties, those that figure in later discussions of geologic processes and structures.

Consider first, however, that igneous rocks are classified not only by mineral composition, but also in terms of the sizes of individual crystals that make up the rock. The word **texture** covers crystal sizes and arrangements.

Crystal size is largely dependent upon the rate of cooling of the magma through the stages of crystallization. As a general rule, rapid cooling results in very small crystals, while extremely sudden cooling produces a natural glass. Very slow cooling, on the other hand, tends to produce large crystals.

From this principle we can deduce that igneous rocks cooling in huge masses at great depths where escape of heat is extremely slow, will tend to develop a texture consisting of large crystals. The specimen of granite shown in Figure 1.7 illustrates such texture. Individual mineral grains can easily be distinguished with the unaided eye. We say that this granite is **coarse-grained** in texture. Large bodies of coarse-grained igneous rock, crystallized slowly at great depth below the surface are described as **plutons.** Plutonic igneous rocks belong to a general class of igneous rocks called **intrusive igneous rocks.** Rocks of this class solidify beneath the surface, completely enclosed in preexisting solid rock. We give the

Figure 1.7 Close-up of a coarse-grained granite. The light-colored grains are quartz and feldspar; the dark grains are mostly biotite mica. (A. N. Strahler.)

name **country rock** to this older, surrounding rock. In Figure 1.6, illustrating pegmatite, the large central rock body is an intrusive igneous rock; specifically, it is a pluton. The narrow veins of pegmatite, extending out into the country rock, are also intrusive igneous rocks. Pegmatites have extremely coarse texture but are quite different in structure than the pluton. We will take up forms of the various intrusive igneous rock bodies in Chapter 3.

Igneous rock that emerges at the earth's surface, cooling rapidly in contact with the atmosphere or ocean, is referred to as **extrusive igneous rock.** Lava, which is fluid magma pouring out from a vent and spreading over the surface, is one expression of extrusive action. Because the lava cools rapidly, mineral crystals are extremely small—most cannot be distinguished even with a good magnifying lens. These rocks are said to be **fine-grained** in texture. In some instances, cooling is so rapid that the magma solidifies into a **volcanic glass,** or **obsidian,** illustrated in Figure 1.8. Next to the obsidian is a specimen of a rock full of spherical cavities; this is **scoria.** It is a type of lava in which expanding gases produced countless bubble-holes in the rock. A related type of igneous rock is **pumice,** in which the bubble holes are very tiny. Pumice may have such a low density that chunks of it will actually float on water. Pumice is often used as decorative boulders in landscaping lawns and gardens.

Figure 1.8 Volcanic glass, or obsidian (*right*); scoria (*left*). (A. N. Strahler.)

THE GRANITE-GABBRO ROCK SERIES

In terms of bulk composition, most igneous rock of the earth's crust belongs to the **granite-gabbro series.** Customarily, these rocks are presented in sequence from the felsic end toward the mafic end. We shall follow this practice, although with the knowledge that it is the reverse of the order required by the reaction series and magmatic differentiation.

The right-hand part of Figure 1.1 lists rocks of the granite-gabbro series. In each column, under the rock name, are bars showing the proportions of minerals making up the rock in a typical example. Notice that each plutonic (intrusive) rock has an equivalent extrusive (lava) type. Although the names of the extrusive rock and the equivalent intrusive rock differ, their compositions are alike.

Granite is dominated in composition by the feldspars and quartz. Potash feldspar of the orthoclase variety is the most important mineral, while sodic plagioclase may be present in moderate amounts or absent. Quartz, which accounts for perhaps a quarter of the rock, reaches its most abundant proportions in granite. Biotite and hornblende are common accessory minerals. Magnetite, not shown on the chart, is also a common accessory.

Granite is a light colored igneous rock and is grayish to pinkish, depending upon the variety of potash feldspar present. Its density, about 2.7 gm/cc, is comparatively low among the igneous rocks. Most granites are sufficiently coarse in texture for the component minerals to be identified with the unaided eye (Figure 1.7). The grayish cast of the quartz grains, with their glassy luster, sets them apart from the milky white or pink feldspars. Black grains of biotite or hornblende contrast with the light minerals. The extrusive equivalent of granite is **rhyolite,** a light gray to pink form of lava.

Geologists examine igneous rocks by mounting a very thin rock slice on a glass slide and placing it under a microscope. Figure 1.9 is a sketch showing how the rock slice looks when illuminated by a strong beam of polarized light entering from below. A specimen of granite is at the left. Quartz is quite clear, with grayish tones. Mineral cleavage structure gives cross-hatched patterns to the potash feldspar and hornblende. The plagioclase feldspar shows strong light and dark banding. Each mineral grain has a sharp boundary in contact with its neighbors.

The granite-gabbro series progresses through transitional rocks, not named here. During this transition potash feldspar and quartz decrease in proportion while plagioclase feldspar increases and moves from the sodic end toward the intermediate varieties.

Diorite is the next important plutonic rock on our list. Its extrusive equivalent, **andesite,** occurs very widely in lavas associated with volcanoes. Looking at the bars in Figure 1.1, we see that diorite is dominated by plagioclase feldspar of intermediate composition, while quartz is a very minor constituent. At this point in the granite-gabbro series, pyroxene of the augite variety makes its appearance. Amphibole, largely hornblende, is also important, and some biotite is present.

Gabbro is an important though not abundant plutonic rock, but it is greatly overshadowed in importance by its extrusive equivalent, **basalt.** We find that basalt makes up huge areas of lava flows and is the predominant igneous rock underlying the floors of the ocean basins. Basalt is also a major rock type at the surface of the moon. Gabbro and basalt are composed largely of pyroxene and calcic plagioclase feldspar with varying amounts of olivine. (Some types lack olivine.) Gabbro and basalt are dark-colored rocks—dark gray, dark green, to almost black—and of relatively high density. Figure 1.9 shows a slice of olivine-rich gabbro viewed under the microscope.

We can now apply the adjectives "felsic" and "mafic" to igneous rocks as well as to silicate minerals. **Felsic igneous rocks** are those rocks dominantly composed of felsic minerals; they include granite and diorite. **Mafic igneous rocks** are those rocks composed dominantly of mafic minerals. However, the mafic rocks include some types, such as gabbro and basalt, containing substantial amounts of calcic plagioclase feldspars. As shown in Figure 1.1, the felsic rocks have densities less than 3.0 gm/cc. The mafic rocks have densities of 3.0 gm/cc or higher.

Continuing the igneous rock series depicted in Figure 1.1, we arrive at **peridotite,** a rock composed almost entirely of olivine and pyroxene. Although widespread in occurrence, peridotite occurs in relatively small plutonic bodies. Peridotite is a dark-colored rock of high density, 3.3 gm/cc, and belongs to a group designated as **ultramafic igneous rocks.**

Finally, we include a variety of igneous rock consisting almost entirely of the mineral olivine. Thus ultramafic rock is called **dunite.** Like peridotite, dunite has a density of 3.3 gm/cc. Although dunite is a rare rock at the earth's surface its geologic importance is very great: There is evidence to show that dunite comprises much of the earth beneath the outer crust.

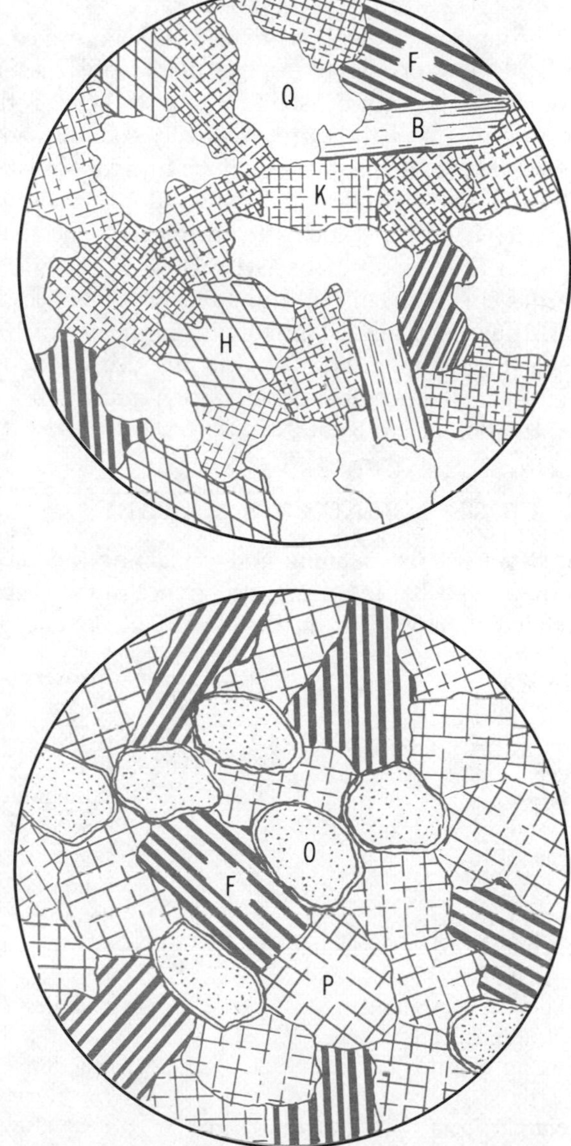

Figure 1.9 Sketch of mineral grains as seen in thin section under the polarizing microscope, enlarged about five times natural size. (*Upper*) Granite. (*Lower*) Olivine gabbro. Q—quartz, K—potash feldspar, F—plagioclase feldspar, B—biotite, H—hornblende, P—pyroxene, O—olivine.

IGNEOUS ROCKS IN REVIEW

Our sampling of silicate minerals and igneous rocks has focused attention upon the way in which elements form into minerals, and minerals form into rocks. The concept that igneous rocks fall into major composition groups—felsic, mafic, and ultramafic—is perhaps the most important single concept to carry forward into the following chapters. Remember that these rock groups differ in density. In looking at the structure of our planet, we shall find that the principal planetary rock layers are arranged in order from the least dense (felsic) rocks at the surface, to the most dense (ultramafic) at the base. How did such a layered structure come to exist? Perhaps the answer lies in a vast process of magmatic differentiation that operated in early stages of our planet's history.

Many other questions remain to be answered. For example, at what depth do bodies of magma form? Is there a single magma layer everywhere beneath the solid rock crust? What is the source of heat in magma? Will the earth eventually cool off and igneous activity cease? These are questions we hope to clarify in later chapters.

Our study of silicate minerals and magmas will prove helpful in a later chapter in which we investigate mineral deposits of value to man in his industrial society. How do the rarer elements in the earth's crust—metallic elements such as copper, lead, zinc, and silver—become concentrated into ore bodies rich enough to be mined and refined into pure metals? We will find that basic concepts of igneous extrusion and magma crystallization provide a large part of the explanation of ore bodies.

YOUR GEOSCIENCE VOCABULARY

Test yourself by defining and explaining the significance of each term listed below in order of occurrence in the chapter. All terms are defined alphabetically in the Glossary at the end of the book.

geology
lithosphere
geoscience
rock
mineral
crystalline solid
igneous rock
magma
earth's crust
chemical element: metallic, nonmetallic
chemical compound
silicates, silicate minerals
aluminosilicates
quartz
feldspar: potash, plagioclase, calcic, sodic, intermediate
orthoclase
mica group: biotite, muscovite
amphibole group: hornblende
pyroxene group: augite
olivine
density
mass
felsic mineral, felsic group
mafic mineral, mafic group
magnetite
ilmenite
cleavage
silicate magma
lava
volatiles
crystallization
reaction
fractionation
pegmatite
fumarole
magmatic differentiation
texture of rock: coarse-grained, fine-grained
pluton
intrusive igneous rock
country rock
extrusive igneous rock
volcanic glass
obsidian
scoria
pumice
granite-gabbro series
granite
rhyolite
diorite
andesite
gabbro
basalt
felsic igneous rocks
mafic igneous rocks
peridotite
ultramafic igneous rocks
dunite

SELF-TESTING QUESTIONS

1. Name in order the eight most abundant elements in rocks of the earth's crust, giving approximate percentage by volume for each. Name several essential metals of industry missing from this list.

2. What are the silicate minerals? Name them and describe their chemical compositions in terms of elements present. In what way do these minerals form a series? In what way do the terms *felsic* and *mafic* apply to the silicate minerals?

3. Explain the concept of mineral density. In what way is mineral density an important influence in determining the arrangements of rock layers in the earth?

4. Describe natural cleavage in minerals, using calcite and muscovite as examples. Name two minerals lacking in cleavage.

5. Describe a silicate magma. What are volatiles, and what is their importance in a magma?

6. In what order do the silicate minerals crystallize from a magma? How does this order relate to the formation of varieties of igneous rocks? How can magmatic differentiation take place?

7. What controls the size of crystals in an igneous rock? Distinguish between intrusive rocks and extrusive rocks (lavas) in terms of physical properties. How do pegmatites form?

8. Describe the granite-gabbro series of igneous rocks. Describe each of the principal members of the granite-gabbro series and give the mineral components of each. Which rocks are felsic, which are mafic? What are the ultramafic rocks? Name two examples.

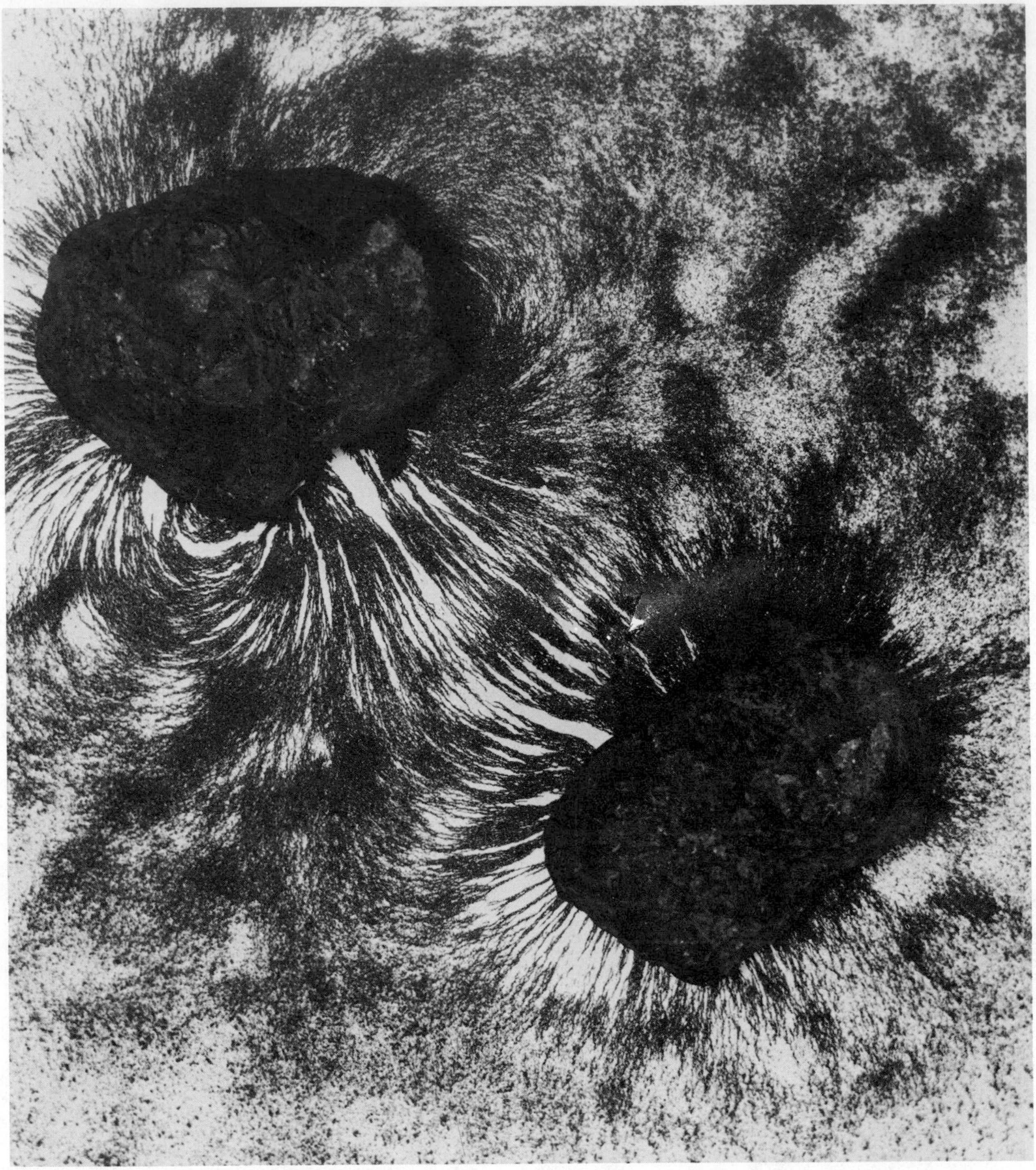

(*Above*) Acting like two bar magnets, these two pieces of magnetite, of a variety known as lodestone, attract and hold iron filings (Ward's Natural Science Establishment, Inc., Rochester, N.Y.)

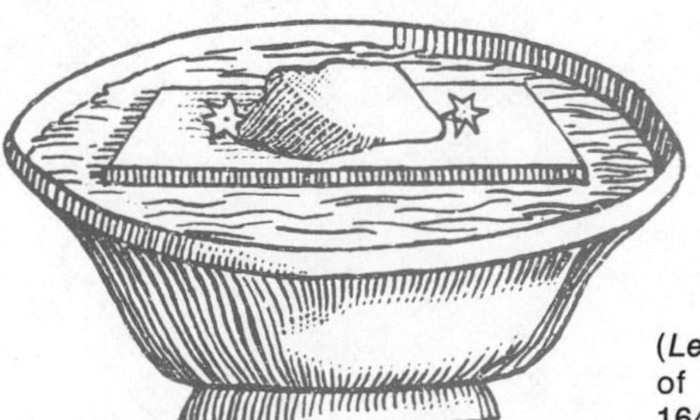

(*Left*) A medieval floating compass. Stars mark the poles of the piece of lodestone. (From Athanasius Kircher, 1643.)

2 THE EARTH'S INTERIOR AND CRUST

SAGA OF THE LODESTONE

Every field geologist worth his salt carries a magnetic compass on his belt. You will see him take out his compass from time to time, cradle it lovingly in his hands, and peer into its face, mumbling strange incantations like "strike, north–twenty-three–east." The geologist's hammer, a mean thing with a long, sharp pick, gives him a sense of power, but his compass gives him a sense of security. Together they make him invincible as he tramps the wilds in search of fossils and minerals.

The magnetic compass began with the lodestone, a black, naturally magnetic iron mineral that spontaneously attracts iron to itself. Iron filings cling to lodestone in long stringy masses (photograph on opposite page). In Western civilization, the Greeks first found lodestone in the hills of Magnesia, a Macedonian province on the Aegean coast.* Thales, a Greek living between 640 and 546 B.C., mentions lodestone in his writings. Lodestone intrigued people of the Mediterranean lands. It was believed to have curative properties, relieving such complaints as toothache, gout, dropsy, hemorrhage, and convulsions.

About the tenth century A.D., it was discovered that a piece of lodestone could be floated on a chip of wood in a dish of water, and that a mark on the stone would always come to rest pointing to the same place on the horizon. Thus the magnetic compass came into existence.

In the Orient the compass may have been known far earlier than in the West, but this can't be proved. The first clear record in Chinese writings dates from late in the eleventh century A.D. It was known that when a needle is rubbed on a piece of lodestone and suspended on a thread, the needle will always point south.

* This historical account is based on data of David G. Knapp, 1962, "Origins of Geomagnetic Science", Chapter VI of *Magnetism of the Earth,* Publ. 40-1, Coast and Geodetic Survey, U.S. Govt. Printing Office, Washington, D.C.)

Alexander Neckham, an English monk, gives us the first good European account of the magnetic compass. In his book *De Naturis Rerum* (circa A.D. 1200) he wrote: "Mariners at sea, when through cloudy weather in the day, which hides the sun, or through the darkness of night they lose knowledge of the quarter of the world to which they are sailing, touch a needle with a magnet which will turn around until, on its own motion ceasing, its point will be directed toward the north." A drawing of a medieval floating compass is reproduced on the opposite page.

In the thirteenth century the compass was greatly improved by Petrus Peregrinus, a Frenchman. He replaced the floating lodestone with a needle pivoted on a vertical shaft between two bearings. He added a sighting device and graduated rim. After Peregrinus came many other improvements in the compass as a tool of navigation, which ushered in the great age of global discovery.

In this chapter we will investigate the phenomenon of earth magnetism, in which the entire globe acts as a single sphere of lodestone. In recent years earth magnetism has proved to be the key to understanding how continents drift apart or collide.

EXPLORING THE EARTH'S INTERIOR

How can scientists possibly find out what material exists deep in the earth's interior, down to the very center? Certainly this is one of the most difficult problems faced by earth scientists. The deepest mines allow us to collect rock samples down to depths of only about 10,000 ft (3000 m). This depth is only about $\frac{1}{2000}$ of the distance to the earth's center. The deepest bore holes put down in search of petroleum now penetrate to depths somewhat greater than 20,000 ft, or about 4 mi (6 km). Even this depth is trivial in comparison with the earth's radius.

Without having samples of materials in the earth's interior, how can the geologist determine the earth's mineral composition? How can he estimate the conditions of temperature, pressure, density, and rock strength prevailing at various depths? Of course, some information can be had by study of magmas. But there is little reason to believe that magmas originate deeper than about 100 mi (160 km) below the surface.

Obviously, the earth scientist must use indirect methods to probe the depths of our planet. One investigator who uses such indirect methods, applying the principles and instruments of physics, is the geophysicist. He uses earthquake waves to furnish information about the earth's interior. In a later chapter, we shall study earthquake waves in some detail. For the moment, we shall simply summarize basic information supplied by **geophysics,** the branch of earth science that applies the principles of physics to the study of the earth.

THE EARTH'S CORE AND MANTLE

Figure 2.1 is a cutaway diagram of the earth showing its principal interior subdivisions. The crust ranges from 10 to 25 mi (16 to 40 km) in thickness—a rock skin too thin to show to correct scale on the diagram. Beneath the crust lies the **mantle,** a solid rock shell some 1800 mi (2895 km) thick. It is generally believed that the mantle consists of the mineral olivine, that is, silicate of iron and magnesium. As we found in Chapter 1, a silicate rock of olivine composition is dunite; it is classed as an ultramafic rock.

Beneath the mantle lies the earth's **core,** a sphere 2160 mi (3475 km) in diameter. There is good evidence to show that the core is composed of iron, mixed with some nickel. In other words, the core is believed to be metallic, in contrast to the silicate rock mantle that surrounds it. From data of earthquake waves it is known that the outer portion of the core is in a liquid state. However, the inner portion of the core, with a radius of 780 mi (1255 km) is in a solid state. We will present evidence for this assertion in Chapter 6.

We can make the inference that the earth must have a very dense core even without evidence from earthquake waves. The physicist can measure the earth's mass with considerable accuracy. He can also calculate its volume. Recall that density is a measure of the quantity of mass within a given volume. Dividing the earth's volume by its mass, we arrive at an average earth density of about 5.5 gm/cc.

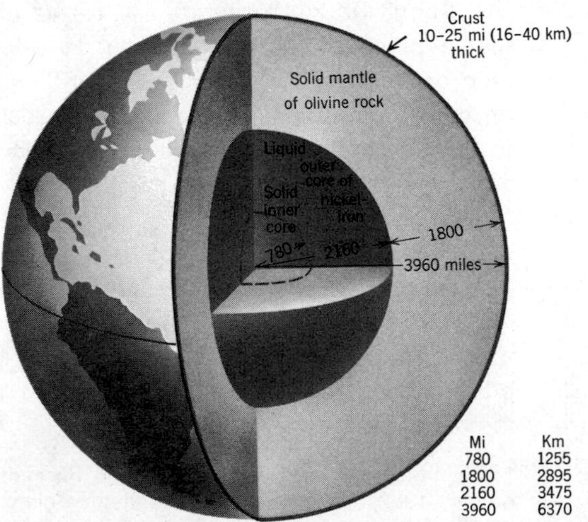

Figure 2.1 Zones of the earth's interior. (© 1973, John Wiley & Sons, New York.)

Now the density of the ultramafic rocks at the earth's surface is only about 3.3 gm/cc. If the earth has a thick, rocky mantle of that density, it must have a very dense core to bring the average density up to 5.5 gm/cc. This line of reasoning is supported by studies of meteorites, solid fragments of matter from outer space reaching the earth's surface. Many meteorites are of nickel-iron composition; they are believed to be core fragments of an early planet (or planets), formed at about the same time as the earth.

PHYSICAL CONDITIONS IN THE EARTH'S INTERIOR

Let us trace the changes in physical conditions of matter inward to the earth's center. We will take up in order (a) pressure, (b) temperature, and (c) density, using a separate graph for each property (Figure 2.2). The depth scale runs from left to right across the bottom of each graph. The uppermost graph shows how confining pressure increases from surface to core. The pressure scale uses units of millions of atmosphere. One "atmosphere" is the pressure of the earth's atmosphere at sea level, about 15 lb/sq in. (1 kg/sq cm). Pressure increases rapidly through the mantle, reaching a value of about 1½ million atmospheres at the core boundary. Pressure increases rapidly through the outer core, then levels off somewhat in the inner core. At the very center of the earth, pressure is about 3½ million atmospheres.

The middle graph shows internal earth temperatures, using the Kelvin scale (Fahrenheit temperatures are shown at the right). Temperature increases very rapidly in the crust and upper mantle, then rather abruptly changes to a more gradual increase. At a depth of 600 mi (1000 km) the temperature is about 2000° K (3000° F). This would be well over the melting point of silicate rock under conditions prevailing at the earth's surface, but under the great confining pressure, the mantle rock is well under its melting point. At the core boundary the temperature is about 2700° K, somewhat greater than the melting point of iron at the prevailing pressure. Consequently, at this point we enter the liquid portion of the core. Temperature in the solid part of the core levels off at just under 3000° K (4500° F).

The bottom graph shows earth density from surface to center. Notice that density rises steadily through the mantle because confining pressure increases, causing the silicate rock to be compressed into smaller volume. However, at the core boundary, where the material changes to iron, there is an abrupt jump to nearly 10 gm/cc. Although iron has a density of about 8 gm/cc at the earth's surface, it is compressed to smaller volume at this depth.

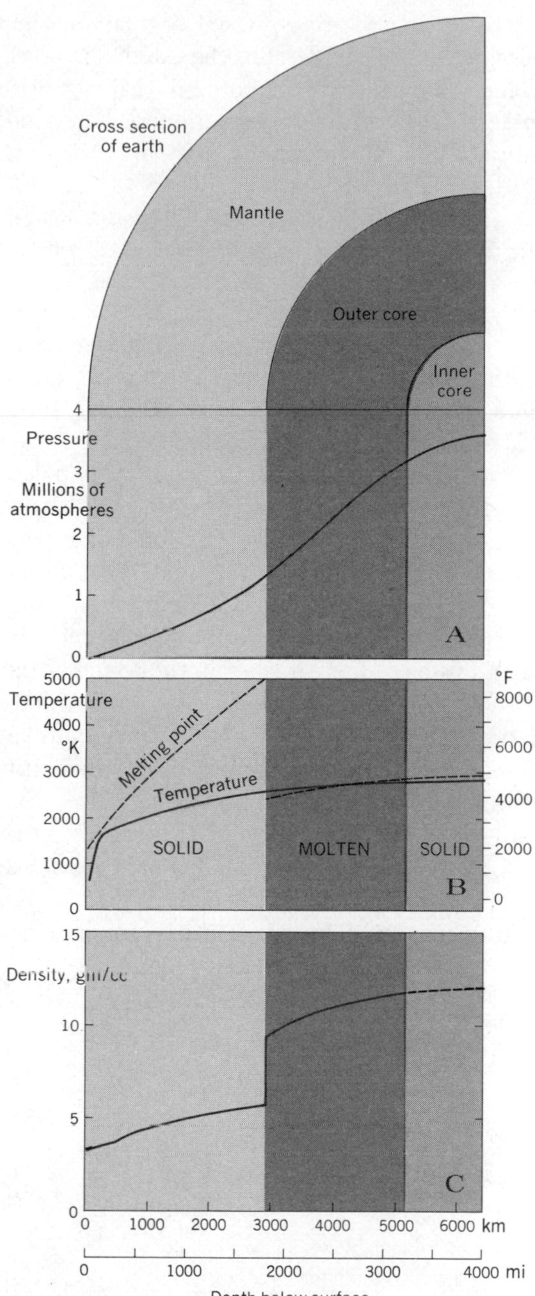

Figure 2.2 Physical conditions in the earth's interior. (*A*) Increase in pressure with depth in the earth. (*B*) Increase in temperature with depth. (*C*) Increase in density with depth.

THE EARTH AS A MAGNET

It is to Sir William Gilbert, physician to Queen Elizabeth, that credit is due for important advances in the scientific study of **earth magnetism.** In his treatise *De Magnete*, published in 1600, Gilbert described his experiments with a 5-in. sphere of lodestone, known as a *terrella.* Using a tiny pivoted compass needle the size of a barleycorn, he was able to describe the external magnetic field (Figure 2.3). Gilbert proposed the hypothesis that the earth is formed of an interior sphere of lodestone surrounded by a nonmagnetic shell. He assumed that the magnetic axis of the sphere coincides with the earth's pole of rotation.

In its most simple aspect, the earth's magnetic field resembles that of a bar magnet located at the earth's center (Figure 2.4). The axis of the imaginary bar magnet is situated approximately coincident with the earth's geographic axis. At the points where the projected line of the magnetic axis, or **geomagnetic axis,** emerges from the earth's surface are the **north magnetic pole** and **south magnetic pole.** Note that the earth's magnetic axis forms an angle of about 20° with respect to the geographic axis. As a result, the magnetic poles do not coincide with the geographic poles.

Figure 2.4 shows lines of force of the earth's magnetic field in relation to the earth's core. The force lines pass through a common point close to the earth's center. The magnetic axis is oriented vertically in this diagram. There exists a **magnetic equator,** lying in a plane at right angles to the geomagnetic axis and encircling the earth's surface approximately in the region of the geographic equator. Visualized in three dimensions, the lines of force of the earth's magnetic field form a succession of doughnutlike rings, suggested in Figure 2.4. The small arrows show the attitude that would be assumed by a small compass needle, free to orient itself parallel with the force lines close to the earth's surface. (Compare with Gilbert's terrella, Figure 2.3.) Force lines extend out into space surrounding the earth to distances as great as 100,000 mi. This entire field of magnetic effect is called the **magnetosphere.**

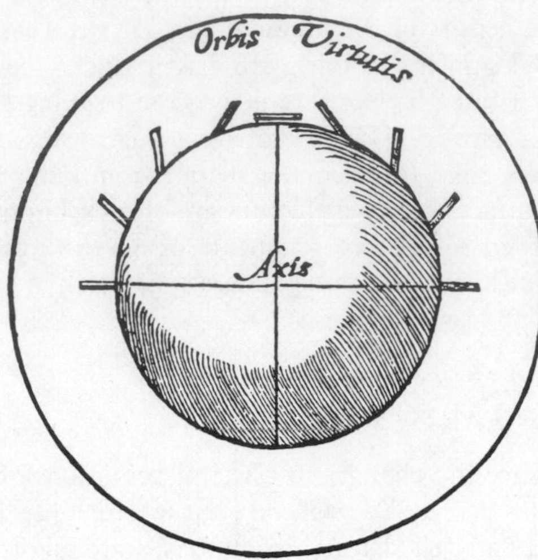

Figure 2.3 Gilbert's diagram of his terrella includes small magnets to show inclination. (From W. Gilbert, 1600, *De Magnete.*)

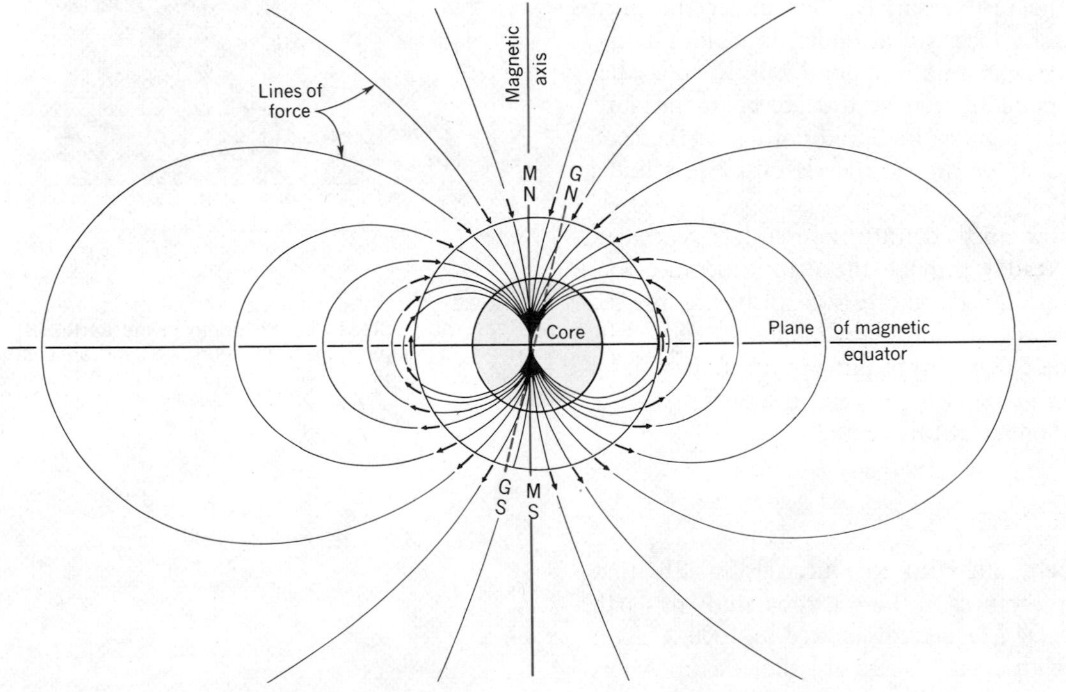

Figure 2.4 Lines of force in the earth's magnetic field are shown here in a cross section passing through the magnetic axis. Letter *M* designates *magnetic,* and *G, geographic.* Arrows at the surface of the earth show the orientation of a magnetized needle.

Earth magnetism is explained by the **dynamo theory.** This theory postulates that the liquid iron of the core is in slow rotary motion with respect to the solid mantle that surrounds it. It can be shown that such motion will cause the core to act as a great dynamo, generating electrical currents. These currents at the same time set up a magnetic field (Figure 2.5). A single, symmetrical current system can thus explain the magnetic field as essentially resembling a simple bar magnet.

One of the most remarkable scientific discoveries of recent decades has been that the earth's magnetic field has undergone repeated changes in polarity. In other words, the magnetic north pole and south pole have switched places, but with the axis unchanged in position. The record of such changes can be read from basalt rock, which preserves a record of magnetic polarity at the time it solidifies. Yet another remarkable discovery has been that the north and south magnetic poles have wandered widely over the globe throughout a large segment of geologic time. We will take up these topics in more detail in later chapters on earth history.

We turn our attention next to some details of the earth's crust and the mantle layer immediately beneath it.

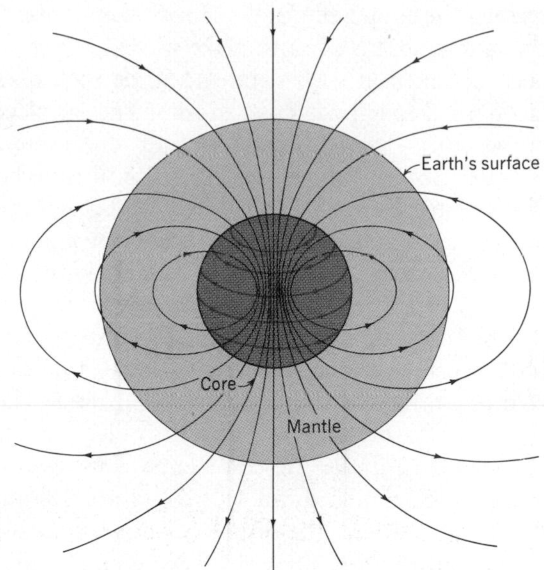

Figure 2.5 Electric currents, shown as heavy lines on the earth's core, are believed capable of producing the earth's magnetic field.

COMPOSITION OF THE EARTH'S CRUST

It is essential to understand that the crust beneath the continents is very different in both composition and thickness from the crust beneath the ocean floors. Roughly one-third of the globe is covered by **continental crust,** about two-thirds by **oceanic crust.** How these two kinds of crust differ is easy to state; why they differ is a story which will unfold in later chapters.

Simple reasoning would lead us to conclude that if the silicate rock-forming minerals were permitted to assemble freely under the attractive force of the earth's gravity, we would find felsic rocks in a surface layer, mafic rocks next below, and ultramafic at the bottom. Actually, this general arrangement is accepted as the most reasonable model for the earth's crust and mantle.

Figure 2.6 shows this arrangement in a very rough way for continents and ocean basins. Felsic rock forms the upper part of the continental crust in a layer with an average thickness of perhaps 10 mi (16 km). This felsic rock is largely of the composition of granite, and can also be described as **granitic rock.**

The lower part of the crust is probably largely of mafic rock, down to an average depth of 25 mi (40 km). This mafic rock is largely of the composition of basalt, and can also be described as **basaltic rock.** At the base of the basaltic layer there is an abrupt change to a denser mantle rock, which we interpret to be ultramafic rock with a composition resembling dunite.

The surface of abrupt change from mafic to ultramafic rock is known as the **Moho.** This word is the first part of the name of a Yugoslav scientist, *A. Mohorovičić,* who discovered the discontinuity on the basis of earthquake studies. Actually, the only evidence con-

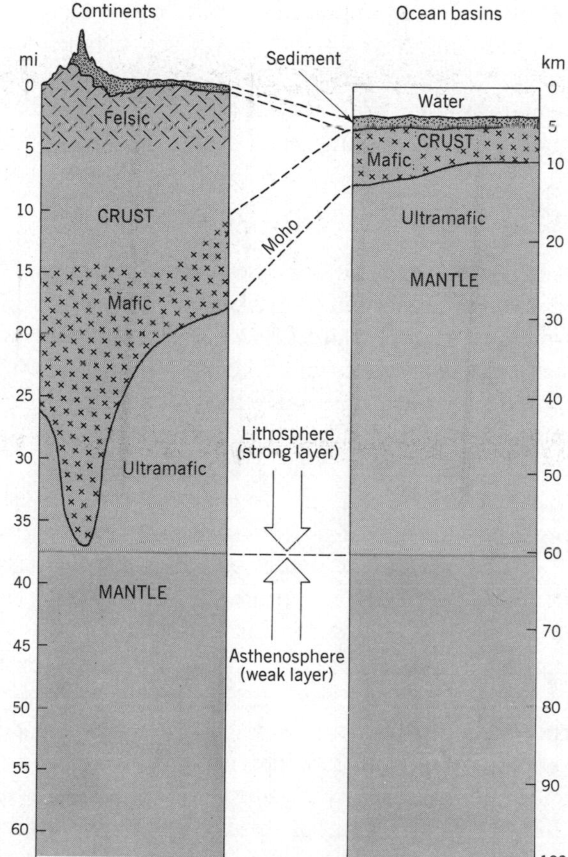

Figure 2.6 Comparison of the crust and mantle under continents and ocean basins.

cerning the supposed change of rock composition comes from abrupt changes in speeds of earthquake waves. As yet no rock samples have been obtained at such depths. Mantle rock does make its way surfaceward through the crust in some special places of the globe, but in the process it is changed physically and even chemically.

Perhaps the most striking point driven home by the crustal diagram in Figure 2.6 is the difference between continental crust and oceanic crust. First, the continental crust is much the thicker—ranging mostly from 20 to 40 mi (30 to 60 km) in thickness. The thickest portions lie beneath the highest mountain ranges, as the diagram shows. In other words, the mountains have **crustal roots.** In contrast, oceanic crust averages only some 4 to 5 mi (7 to 8 km) thick. As a result, the Moho is encountered at much shallower depths under the ocean basins—about 7 to 8 mi (11 to 13 km) below the ocean surface—than under the continents. This means that if we are ever to put a bore hole down to take samples of the mantle rock, we should do it under the oceans. Confining pressure and temperature will be considerably lower at the Moho under the oceanic crust. Rock at such depths tends to close in upon any drilling tools we use, so the procedure will be extremely difficult at best.

A second striking point of difference between the two types of crust is that the felsic, or granitic, upper layer of the continental crust is missing from the oceanic crust. Instead, mafic (basaltic) rock comprises the entire igneous portion of the oceanic crust. There is a layer of sediment and sedimentary rocks over much of the basaltic layer.

Leaving for the time being the detailed examination of the earth's crust, we return to broader questions about the earth as a planet. What is the source of the earth's internal heat? Has the earth gradually been cooling off?

RADIOACTIVITY AND HEAT

Answers to many of our questions about the earth's internal heat, igneous activity, and the evolution of the earth's interior can be found in the phenomenon of radioactivity. **Radioactivity** is the spontaneous breakdown of certain elements, leading to permanent changes in the atoms involved. Radioactivity is accompanied by the emission of energetic atomic particles and the production of heat. Another way of describing radioactivity is that it is a natural process of conversion of matter into energy. Let us investigate some of the principles of nuclear physics involved in radioactivity.

The **nucleus,** or dense core, of an atom consists of two types of particles, **neutrons** and **protons.** For a given element the number of neutrons is only approximately constant, whereas the number of protons is fixed. Take, for example, an important radioactive form of the element uranium. In the nucleus of this form of uranium there are 146 neutrons and 92 protons. The total of neutrons and protons is therefore 238. This quantity is known as the **mass number** and is designated by a superscript after the symbol for uranium, thus: U^{238}. This form of the element is also written as uranium-238, or simply U-238. Although in the case of uranium-238, the number of neutrons is 146, there exists another form of uranium with 143 neutrons. The latter form thus has a mass number of 235, and it is designated as uranium-235. These differing varieties of the same elements are referred to as **isotopes.**

A key to the understanding of radioactivity is that certain isotopes are unstable. This instability can result in the flying off of a small part of the nucleus. The original element is thus transformed into a different element, having a different name. In this spontaneous breakdown, mass is converted into energy, released into the surrounding matter and finally transformed into sensible heat. The term **radioactive decay** covers the entire process. The total quantity of heat produced per unit of time in the radioactive process can be exactly calculated for a given quantity of an unstable isotope.

The radioactive disintegration of one parent isotope may lead to the production of another unstable isotope, known as a **daughter product.** This product, in turn, may produce yet another unstable isotope, and so forth, until ultimately a stable isotope results and no further radioactivity occurs.

Take as an example the system of uranium-238. This isotope is transformed through a dozen or so daughter products. Decay ends in a stable lead isotope, lead-206. In this series each gram of uranium produces 0.71 calorie of heat per year. Other important heat-producing decay sequences in the rocks of the earth are those of uranium-235 and thorium-232. Uranium-235 produces 4.3 calories of heat per gram per year; thorium-232 produces 0.20 calorie.

Now we can get back to the subject of the earth's internal heat. We have solved the problem of origin of the heat itself—it is heat generated by radioactivity, or simply **radiogenic heat.**

RADIOGENIC HEAT INSIDE THE EARTH

Heat flows continuously upward from the depths of the earth toward the surface. The rate of increase in temperature with depth is called the **geothermal gradient.** Observations in deep mines and in bore holes show that the geothermal gradient has a value of about 1 F° per 50 ft (3 C° per 100 m).

The total upward heat flow at the earth's surface in one year is about enough to melt an ice layer ⅕ in. (6 mm) thick. This quantity of heat is extremely small compared with that received by the earth's surface from solar radiation. The earth's heat flow from its depth is of no significance in heating the earth's surface or in powering the atmospheric and oceanic circulation systems.

Referring back to the temperature-depth graph in Figure 2.2, notice the flattening of the temperature curve in the lower mantle and core. This flattening means that there is a very low thermal gradient within the deep interior. A logical interpretation of this fact is that the rate of production of radiogenic heat is greatest near the earth's surface and decreases rapidly with depth. We must conclude that the concentration of radioactive isotopes is greatest in the rocks of the crust. Concentration falls off rapidly in the mantle rocks and is very small in the lower mantle and core. In short, the main source of the earth's internal heat is concentrated near the earth's surface, not at the center.

Scientists have analyzed both the concentrations and rates of heat production of the radioactive isotopes of uranium and thorium in each of the three classes of igneous rocks. Felsic rocks with the composition of granite produce radiogenic heat at a rate about twice that of mafic rocks of basalt composition. Basalt rock produces heat at a rate over 17 times as great as ultramafic rock, such as dunite or peridotite. In other words, the most rapid production of radiogenic heat is by granitic rocks of the upper zone of the continental crust. It has been estimated that about one-half of all radiogenic heat is produced above a depth of 22 mi (35 km) in the continental crust.

As we find conditions today, the rate of surfaceward flow of heat from the upper mantle and crust closely balances the rate of heat production. The mantle remains for the most part at a temperature lower than its melting point. There exists, however, a shallow layer of the mantle in which melting on a large scale is a likely occurrence.

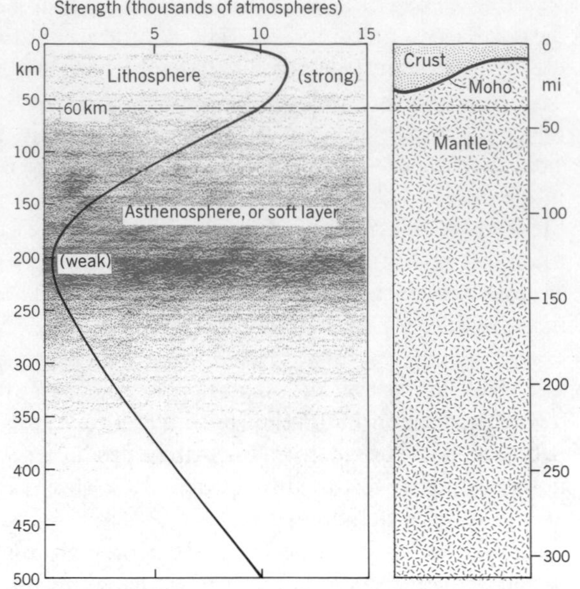

Figure 2.7 Strength of rock in the lithosphere and asthenosphere. The curve represents a rough estimate only.

A SOFT LAYER IN THE MANTLE

A critical layer is found in the upper mantle in the depth range of 40 to 125 mi (60 to 200 km); it has very indefinite upper and lower boundaries. Here the mantle rock is at a temperature very close to its melting point and is, therefore, in a condition of reduced strength. In other words, the rock is soft, just as white-hot iron is soft, compared with cold iron. The term **soft layer** has been applied to this part of the mantle.

Figure 2.7 shows the estimated strength of the soft layer. Notice the marked drop to low strength, with a minimum at about the 125 mi (200 km) depth. Apparently, while remaining solid and in a crystalline form, the mantle rock at this depth develops a plastic quality, because here it is close to its melting point. A plastic zone here was suspected as far back as 1926, when the seismologist Beno Gutenberg presented evidence that earthquake waves are slowed in velocity in this zone. Reduction of earthquake wave velocity is interpreted by seismologists as caused by a reduction in strength of the rock.

The plastic quality of rock in the soft layer makes possible very slow flowage movements when this rock is subjected to unequal stresses. As we will see, flowage of mantle rock plays a vital part in modern concepts of crustal changes and mountain-making. At the same time, it is clear that the mantle also behaves as a crystalline solid in response to stresses that are applied suddenly and over short periods of time. Earthquakes, which are generated by sudden slip movements of rock masses, can originate in the plastic zone.

The mantle thus exhibits properties of both an elastic solid and a plastic substance, depending upon the nature of the stresses applied. Certain common substances, among them ordinary tar, show such properties. A piece of cold tar will shatter into fragments when struck, but if left unsupported for long periods of time it will flow very slowly in a thick stream. We must remember that the extremely high pressures and temperatures existing in the mantle impart to rock qualities that we cannot observe or duplicate at the earth's surface.

Geologists have given the name **asthenosphere** to the soft layer of the mantle and have restricted the term **lithosphere** to the strong, rigid overlying zone (including the crust and part of the upper mantle). As shown in Figure 2.7, the base of the lithosphere is placed at about 40 mi (60 km) as a rough average figure.

Identification of a plastic layer of reduced strength in the mantle leads to a most important hypothesis: The strong, rigid lithospheric shell may be capable of rotating independently of the deeper mantle, the motion between the two bodies being accommodated by yielding within the soft asthenosphere. Under such a hypothesis the continents and ocean basins might shift widely in latitude and longitude.

Discovery of the asthenosphere was a major step forward in understanding of global geologic processes. Yielding in the soft layer takes place much like the gliding of playing cards over one another when a card deck is pushed from left to right, as in Figure 2.8. Such gliding in parallel layers is referred to as **shearing.** Shearing will be most rapid in the soft layer, but entirely missing in the rigid lithosphere. (Imagine that the uppermost cards in the deck are glued together to make a solid block.) Of course, the rock layers we refer to as gliding over one another are extremely thin, approaching the thickness of layers of atoms in the minerals. Ice deep within glaciers experiences such shearing motion.

If the lithosphere is a brittle layer, it is conceivable that it might break up into numerous plates, and that these rigid plates might move over the asthenosphere independently. We shall leave this topic for further exploration in later chapters.

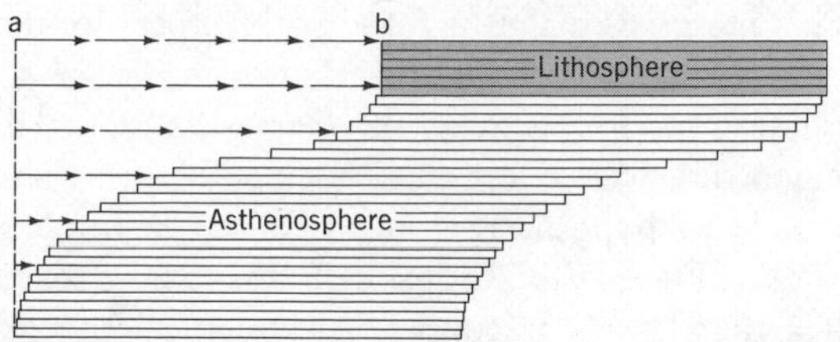

Figure 2.8 Shearing motion of rock in the asthenosphere resembles slip of cards in a deck in which cards in the upper part are glued together and move as a solid plate.

ISOSTASY

Earlier, we referred to the fact that the continental crust is thicker under mountains than under low plains. The crust beneath mountains projects down in the form of deep crustal roots, or mountain roots. The felsic rock zone is also thicker beneath the mountains, as shown in Figure 2.6. The question is: How can we account for the existence of mountain roots?

Existence of a plastic layer in the mantle allows for the possibility that in certain places the lithosphere may be free to rise and in other places to sink. Sinking of the lithosphere will require that the mantle material of the asthenosphere be displaced sidewise by slow flowage. Rising of the crust will require that mantle material be brought in from surrounding zones of the asthenosphere. As this material rises it is cooled somewhat and becomes part of the lithosphere.

Ability of the lithosphere to sink or rise is explained by the principle of **isostasy,** one of the fundamental concepts of geology. The word isostasy comes from Greek words meaning "equal" and "stand." The concept is that lithospheric masses seek an equilibrium level of stability, just as icebergs float at rest in the ocean.

A crustal model of isostasy was proposed over a century ago by Sir George Airy, Astronomer Royal of England.

Airy supposed that the relatively light material of which mountains are composed extends far down into the earth to form roots.

This material, which is felsic rock, protrudes downward into a location normally occupied by denser mafic or ultramafic rock. The higher a mountain mass, the deeper its roots will be. Under a plains region the layer of less dense rock will be very shallow.

Airy's hypothesis is illustrated by a simple model using floating blocks (Figure 2.9). Suppose that we take several blocks, or prisms, of a metal such as copper. Although all the prisms have the same dimensions of cross section, they are cut to varying lengths. Because copper is less dense than mercury, the prisms will float in a dish of that liquid metal. All blocks are floated side by side in the same orientation. The longest block floats with the greatest amount rising above the level of the mercury surface, and the shortest block has its upper surface lowest. With all blocks now floating at rest, it is obvious that the block rising highest also extends to greatest depth.

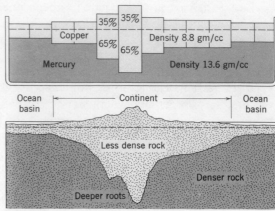

Figure 2.9 The Airy hypothesis of mountain roots is illustrated by some blocks of copper floating in a pan of mercury.

THE EARTH'S INTERIOR IN REVIEW

This chapter has dealt with the structure and composition of our earth as a whole planet. A major concept has been that the earth, early in its history, developed a layered arrangement, with the densest material nearest the center. A stable earth resulted, free from great upheavals that would have rendered the evolution of life impossible.

Two important physical phenomena have been reviewed: earth magnetism and radioactivity. Magnetism was traced to a liquid core in slow motion, acting as a dynamo. Although earth magnetism is a very weak force, it sets up a most important environmental shield around the earth. Radioactivity is the great heat-producing process within the earth and maintains the upper mantle in a condition close to melting. Here the rock is weakened to the point of permitting slow movement of the rigid lithosphere over the soft asthenosphere. Discovery of these facts sets the stage for our understanding the evolution of continents and ocean basins.

YOUR GEOSCIENCE VOCABULARY

geophysics
mantle
core
earth magnetism
geomagnetic axis
magnetic poles; north, south
magnetic equator
magnetosphere
dynamo theory
continental crust

oceanic crust
granitic rock
basaltic rock
Moho
crustal roots
radioactivity
nucleus of atom
neutron
proton
mass number

isotope
radioactive decay
daughter product
radiogenic heat
geothermal gradient
soft layer of mantle
asthenosphere
lithosphere
shearing
isostasy

SELF-TESTING QUESTIONS

1. Describe the earth's core and mantle, giving approximate dimensions. Of what substances are the core and mantle composed, and what is their physical state?

2. Describe the changes in pressure, temperature, and density from the earth's surface to its center. How does pressure affect density?

3. Give a brief historical review of early studies of the earth's magnetism. What was the motivation behind much of this investigation?

4. Describe the earth's magnetic field. Define the magnetic poles and equator. What theory has been proposed to explain the earth's magnetism?

5. Describe the earth's crust, giving thickness and rock composition. At what depth is the mantle encountered under continents and ocean basins? What is the Moho? What is the composition of the mantle?

6. Describe the structure of the atomic nucleus. What is the mass number? Explain what an isotope is.

7. Describe the process of radioactive decay. How is heat produced?

8. Describe the geothermal gradient. How does the quantity of heat reaching the surface compare with heat derived from solar radiation? How does the thermal gradient change with increasing depth? What is the significance of this change?

9. In what earth zones is the production of radiogenic heat concentrated? Which types of rock produce most of the earth's radiogenic heat?

10. What is the soft layer of the earth's mantle? Describe the physical conditions within this layer. At what depth does it lie? What is the distinction between the asthenosphere and the lithosphere?

11. Explain the concept of isostasy and illustrate with a simple physical model. According to this concept, what is the relationship between depth of the Moho and surface elevation of the continents? Does isostasy explain the shallow depth of the Moho beneath the oceanic crust?

This cloud of heated gases and dust is the type of cloud which destroyed St. Pierre. The top of the cloud rises to 13,000 feet above sea level. (A. Lacroix.)

3 IGNEOUS ACTIVITY AND VOLCANOES

THE GLOWING CLOUD OF DEATH

St. Pierre, Martinique Island, West Indies. May 8, 1902. Assistant Purser Thompson was aboard his ship, the *Roraima.* Shortly after mooring was completed, Thompson surveyed the colorful city of 25,000 inhabitants. Behind the city rose the conical slopes of Mount Pelée, an active volcano reaching a summit elevation of nearly 4000 feet. For some days, ashes had rained down upon the city, and the smell of sulfur was heavy in the streets. Muffled explosions had been heard from time to time. There had been some panic, and many inhabitants had fled the city, but they were replaced by refugees crowding in from the countryside. Then, at 7:50 A.M. it happened. Four deafening explosions were followed by the rise of a black cloud from the crater of Mount Pelée. Another black cloud shot horizontally outward, speeding down the mountain toward the city. Here is how Thompson described the events that swiftly followed:

The remains of St. Pierre, photographed not long after its destruction by the death cloud. Mount Pelée can be seen in the distance. (A. Lacroix.)

I saw St. Pierre destroyed. It was blotted out by one great flash of fire. Nearly 40,000 people were killed at once. Of eighteen vessels lying in the Roads only one, the British steamship *Roddam* escaped and she, I hear, lost more than half on board. It was a dying crew that took her out. . . . The mountain was blown to pieces. There was no warning. The side of the volcano was ripped out, and there hurled straight toward us a solid wall of flame. It sounded like a thousand cannon. The wave of fire was on us and over us like a lightning flash. It was like a hurricane of fire, which rolled in mass straight down on St. Pierre and the shipping. The town vanished before our eyes, and then the air grew stifling hot and we were in the thick of it. Wherever the mass of fire struck the sea, the water boiled and sent up great clouds of steam. I saved my life by running to my stateroom and burying myself in the bedding. The blast of fire from the volcano lasted only for a few minutes. It shriveled and set fire to everything it touched. Burning rum ran in streams down every street and out into the sea. Before the volcano burst, the landings at St. Pierre were crowded with people. After the explosion, not one living being was seen on land. . . . The fire swept off the ship's mast and smoke stack as if they had been cut by a knife.*

Only two persons are known to have survived the catastrophe at St. Pierre. One of these was a prisoner in an underground dungeon at the time. Even so, he was badly burned by the hot gas which penetrated into every open space. The type of eruption that destroyed St. Pierre has since been given the name of nuée ardente, or "glowing cloud." A similar event occurred in December of the same year and was photographed by a French scientist who had come to study the earlier eruption (photograph on p. 28). He also took the accompanying photograph of the devastated city.

Volcanic activity is one of the subjects we will investigate in this chapter. Volcanoes have posed a threat to man since the earliest recorded history. The buildup of heat in rock beneath the surface is going on today in many places. There will be many more disasters before science can predict volcanic events accurately and people can learn to cope with this form of environmental hazard.

* From Thompson's account, as quoted by L. Don Leet in *Causes of Catastrophe,* Whittlesey House, New York, 1948.

INTRUSION AND GRANITE BATHOLITHS

The geologist identifies many bodies of coarse-grained intrusive igneous rock, exposed at the earth's surface. Much of this is felsic rock—granite or closely related types. A good example is found in the Sawtooth Mountains of Idaho (see Figure 15.2), where a great mass of granite is exposed. While some of these bodies are as extensive as a large state, their individual size is very small in terms of the entire earth's surface.

No known igneous rock on earth exceeds an age of about 3¾ billion years. It is generally agreed that the earth achieved its identity as a planet from 4½ to 5 billion years ago. Note that this leaves about one billion years or more of earth history unaccounted for by rock records. One must conclude that all known igneous rocks are formed of magma that invaded and completely replaced whatever older crustal rock previously existed.

The earth's earliest history may have been one of repeated melting and solidification, whether as a whole globe, or in parts, until mineral and rock layering was completed and the radiogenic isotopes were concentrated near the surface. In the final stages of this segregation process, it is likely that many small magma pockets were formed and invaded the overlying crust. The magma replaced preexisting rock. This process seems to have continued into relatively recent geologic periods and is probably active today, although we cannot observe what magmas are doing at depth at the present time.

Geologists visualize upward movements of magma bodies in the crust and refer to this process as **intrusion.** It is understandable that if a rock body is heated until it melts, the magma, being of lower density than its solid equivalent at the same level, will tend to rise. In exerting upward pressure upon the overlying rock layers, rising magma can incur mechanical changes, such as the lifting and rupturing of the overlying mass.

Blocks of the enclosing solid rock (the country rock) can break off and sink into the magma, allowing the magma to rise and occupy the cavity. Such a process of magma rise is termed **stoping**, from the miner's term for mining upward into a ceiling. The finding of unmelted blocks of country rock enclosed by igneous rock attests to stoping as a real process. These strange angular fragments are called **xenoliths** (Figure 3.1). (The prefix *xeno* comes from the Greek for "foreigner.")

There is also the possibility that magma can rise by melting the overlying rock and incorporating the molten material into the original magma. The magma is said to **assimilate** the country rock as melting proceeds, and the magma composition may be changed by addition of different mineral components. For example, if a mafic magma of basaltic composition were to assimilate rock rich in free silica and aluminosilicates of potassium and sodium (feldspars) the average composition for the magma would be changed in the direction of the felsic rocks.

Whatever the details of the process of magma rise and intrusion, the end result can be studied after the process of denudation (erosional removal of rock) has exposed to view plutonic rocks that formerly were many miles below the earth's surface. In such exposures we recognize an igneous body known as a **batholith.** Batholiths

Figure 3.1 The surface of this boulder shows xenoliths of various kinds of igneous and metamorphic rocks enclosed in granite. Prescott, Arizona. (A. N. Strahler.)

are mostly of felsic rock in a rather coarse crystalline state. Granite as a particular rock is a major constituent of batholiths. Batholiths seem to be "bottomless" because their great extent downward does not permit the observation of a lower boundary. Actually, there is evidence that the bulk of most batholiths extends down no more than about 6 mi (10 km). They may have rootlike appendages extending much deeper. Figure 3.2 shows how the upper part of a batholith is related to the country rock. A small, domelike projection of a batholith is known as a **stock.**

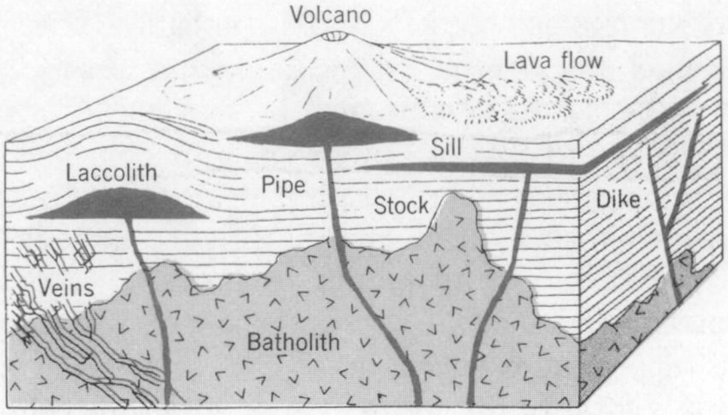

Figure 3.2 shows other forms of intrusive rock bodies. The country rock is depicted as having a layered structure, as would be the case for sedimentary strata. Magma intruding these layers may spread out into a thin sheet of relatively great horizontal extent named a **sill** (Figure 3.3). Where magma pressure lifts the overlying layers into a dome, a **laccolith** results. In thick sills and large laccoliths the rock texture is that of a pluton.

Fractures in previously formed solid rock may be invaded by magma, which forces the enclosing rock mass apart. There result more or less vertical **dikes,** wall-like igneous rock bodies (Figure 3.4). Dikes are typically of small thickness—from a few inches to a few yards—and have a fine-grained crystalline texture. Shrinkage in cooling of thin sills and dikes results in a system of joint fractures that produce long rock columns of prismatic form with four, five, or six sides. This structure is termed **columnar jointing** (see Figure 3.5). The sill shown in Figure 3.3 has good columnar jointing. The columns are oriented with the long dimension at right angles to the enclosing country-rock surfaces. Consequently, columns are typically vertical in sills but horizontal in dikes.

As explained in Chapter 1, watery, silica-rich solutions that remain after a magma has largely crystallized are forced to penetrate fractures in either the newly formed igneous body or the adjacent country rock. Minerals deposited from such solutions take the form

Figure 3.2 Forms of occurrence of the igneous rocks. (© 1973, John Wiley & Sons, New York.)

Figure 3.3 The prominent cliff with columnar jointing is a basalt sill intruded into sedimentary beds. Yellowstone River, Yellowstone National Park. (George A. Grant, U.S. National Park Service.)

of **veins** (Figure 3.2). One important class of veins consists of pegmatite, having a rock texture of unusually large crystals. Other types of veins contain concentrations of uncommon minerals, among them the ores of various metals.

Figure 3.4 A dike of basalt cutting granite. Cohasset, Massachusetts. (John A. Shimer.)

THE GRANITE CONTROVERSY

Granite bodies may actually have more than one origin. We have already considered the possibility that the magma of a granite batholith has come from a deeper parent body of igneous rock. The parent magma rose by stoping and assimilation, to recrystallize higher in the crust. This concept of the origin of granite is referred to as the **magmatic theory.**

The magmatic concept has been strongly challenged for several decades. The challengers point out that the magmatic theory requires enormous volumes of country rock to be assimilated in order to make room for the invading magma. They propose that granite has been formed in place by gradual chemical and physical change of the older country rock. It is supposed that the country rock was already rich in most of the ingredients of a felsic igneous rock. Then, under the influence of heat, pressure, and infusion of certain chemical solutions, the earlier rock underwent slow recrystallization, becoming granite. However, melting did not occur, so that at no time was there a liquid magma body. **Granitization** is the word applied to this total process. We shall return to this process in our investigation of another class of rocks (metamorphic rocks) in Chapter 5.

OUTPOURING OF BASALTS ON THE CONTINENTS

Next to the granites in importance as igneous rocks within the continents are some vast accumulations of basalt. These extrusive rocks represent outpourings of basaltic magma upon the surface of the continental crust. The molten basalt is extremely fluid at high temperatures. This magma has issued from near-vertical cracks, known as **fissures,** and has spread in thin sheets to solidify rapidly in the form of **lava flows.** Basaltic lavas often exhibit the same type of columnar jointing seen in thin sills and dikes (Figure 3.5).

Figure 3.5 As a result of shrinkage upon cooling, this basaltic lava flow shows columnar jointing. The columns are about 15 ft (5 m) long. Palisades of the Columbia River, Washington. (U.S. Geological Survey.)

In certain regions, outpouring of basalts occurred in enormous quantities within fairly short spans of geologic time. The accumulation of basalt layers may total several thousands of feet thick, while the areas covered run from 50,000 to 100,000 sq mi (130,000 to 260,000 sq km). These accumulations are known as **flood basalts.** Two notable examples are shown in the maps of Figure 3.6. One is the Columbia Plateau region of Washington, Oregon, and Idaho. A second is the Deccan Plateau of peninsular India. Figure 3.7 is an air view of the Columbia Plateau basalts, now deeply eroded.

Flood basalts are believed to originate from very great depths in the crust. Figure 3.8 shows the source of basalt as a pocket of molten basalt from the basaltic rock layer of the lower part of the continental crust. The depth of such a source might be between 20 and 30 mi (30 and 50 km).

The rise of basaltic magma from a deep source requires that the overlying crust subside to occupy the space vacated by the magma. Fracturing of the subsiding rigid crust into cracks provides additional planes of weakness through which more magma is able to rise.

Recall that the crust beneath the ocean basins is basalt. The supposition was long held that this crust represents a single ancient rock layer that has undergone little subsequent change. This concept has been completely destroyed in recent years: Oceanic basalt is very much younger, generally, than most rocks of the continental crust. The new concepts are discussed in Chapter 9.

VOLCANIC EXTRUSION

Volcanism is a term applied generally to the formation of extrusive igneous rocks. Volcanism includes both the outpourings of flood basalt lavas and the more localized accumulations of magma in the form of individual volcanoes and groups of volcanoes. By **volcano,** we mean a massive structure built by emission of magma and its contained gases from a pipelike conduit or from fissures. As eruption

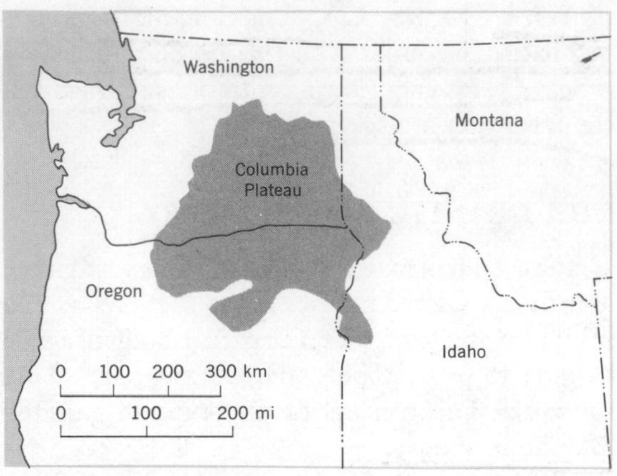

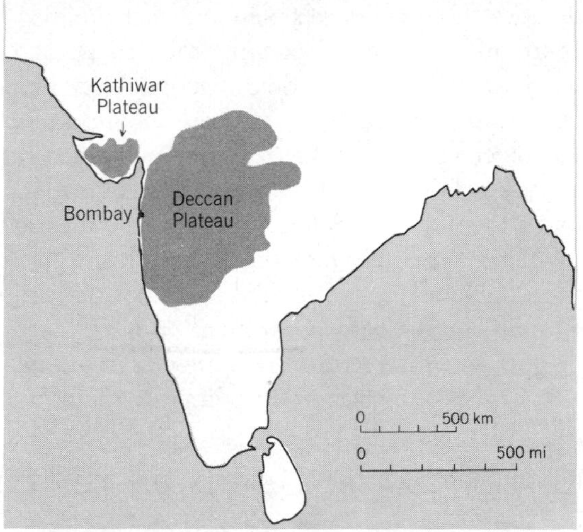

Figure 3.6 Approximate present surface extent of the Columbia Plateau basalts (*upper*) and the Deccan Plateau basalts of India (*lower*).

Figure 3.7 Flood basalts of the Columbia Plateau region. The basalt layers have been eroded to produce steep cliffs rimming broad, flat-topped mesas. Dry Falls, Grand Coulee, central Washington. (John S. Shelton.)

continues through time, the accumulated igneous rock must form a more or less conical mountain mass, a **volcanic cone.** The cone surrounds the **vent,** or point of emergence of the conduit.

Magmas of both felsic and mafic composition, as well as intermediate types, can erupt to produce volcanoes. Important differences in form and distribution of volcanoes can be traced to differences in the magma composition. The principal lava groups are as follows:

Parent magma type	Name of lava	Classification of lava
Felsic	Rhyolite	Acidic
Intermediate	Andesite	Intermediate
Mafic	Basalt	Basic

Acidic lavas and **intermediate lavas** are highly viscous and retain large amounts of gas under pressure. As a result, these lavas tend to give explosive eruptions. **Basic lavas,** as a group, are of low viscosity, and the contained gases readily escape. Consequently, large basalt emissions are typically quiet, as in the case of the flood basalts.

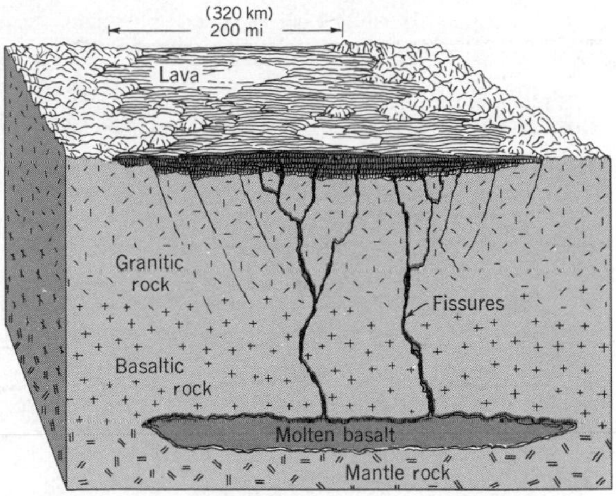

Figure 3.8 This schematic drawing suggests the relation of flood basalts to crustal zones beneath.

COMPOSITE VOLCANOES

Eruption of acidic and intermediate lava typically produces a tall, steep-sided cone. The cone characteristically steepens to the summit. At the summit is a depression, the **crater,** marking the position of the vent (Figure 3.9). Familiar to all are the graceful conical profiles of such volcanoes as Mt. Fuji in Japan and Mt. Hood in the Cascade range. Cones of this type are called **composite volcanoes,** because they consist in part of lava flows and in part of ash. The internal structure of such a volcano is shown in Figure 3.10. By **volcanic ash** we mean finely divided igneous rock that results from the explosive emission of magma heavily charged with gases under high pressure. Collectively, all material blown out of a volcanic vent is called **tephra.** Besides fine ash, tephra includes many

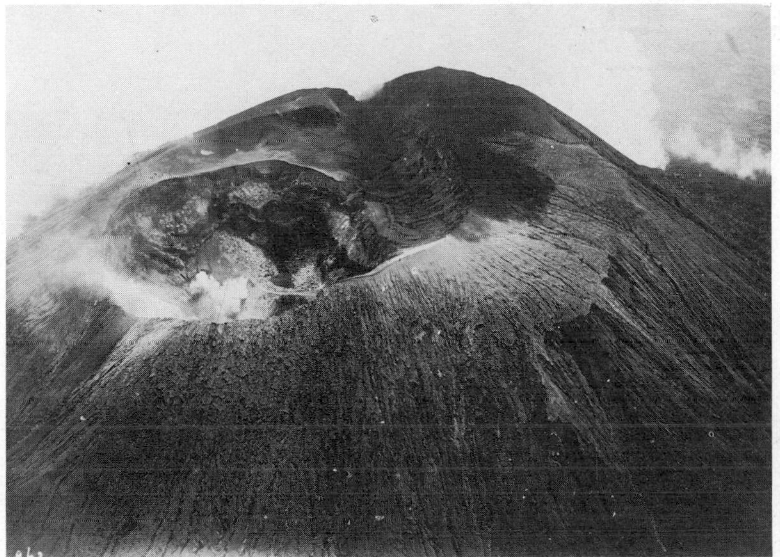

Figure 3.9 (*Left*) Smoke and gases issue from the crater of a great composite volcano in central Java. Torrentrial rains have scored the soft slope of ash with long, narrow gullies. (Luchtvaart-Afdeeling, Ned. Ind. Leger., Bandoeng.)

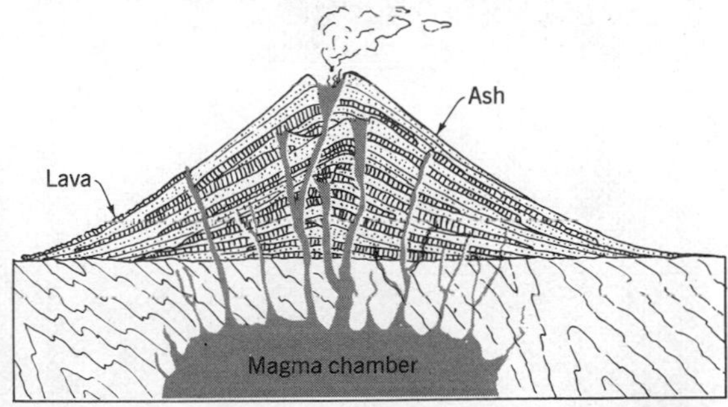

Figure 3.10 (*Above*) This idealized cross section of a composite volcanic cone shows feeders rising from a magma chamber beneath.

larger fragments of pebble and cobble sizes. Some fragments are of boulder size (**volcanic bombs**); these can be hurled only a short distance from the vent (Figure 3.11). Particles come to rest at various distances from the vent, depending upon their size. Very fine volcanic dust can travel many miles from the vent (Figure 3.12).

Flows of lava are also emitted from the volcano, typically emerging from vents on the flanks of the cone. Thus the cone is constructed of both tephra and flows.

Another important form of explosive emission is a cloud of incandescent gases and fine ash, known as a **nuée ardente** (French for "glowing cloud"). This cloud moves rapidly down the side slopes of the cone, as described in the opening paragraphs of this chapter.

Figure 3.11 During the 1914 eruption of the Japanese volcano Sakurajima, a blocky lava flow advanced slowly over a ground surface littered with volcanic blocks and bombs. (T. Nakasa.)

BASALTIC CINDER CONES

We turn now from the largest volcanoes to the smallest—the **basaltic cinder cones** (Figure 3.13). These small cones are usually but a few hundred feet high and less than 1 mi (1.6 km) in basal diameter. They are formed entirely of tephra, composed of scoriaceous basalt. The larger tephra fragments accumulate close to the vent, building up a broadly rounded cone, whereas the finer particles of ash are carried in the wind to fall in a surrounding apron. An ash layer up to several inches deep may be found within a radius of a few miles. In some cases a basaltic lava flow emerges from the same vent, spreading in a tonguelike stream away from the cone and continuing for several miles down the nearest stream valley (Figure 3.13).

Cinder cones commonly occur in groups of as many as several dozen. A fine example of a cinder cone field is the one surrounding the San Francisco Peaks in northern Arizona. Other groups lie near Mt. Lassen in northern California and in Craters of the Moon National Monument, Idaho.

SHIELD VOLCANOES

The continued outpouring of great quantities of highly fluid basaltic lavas from a radiating series of fissures produces the **shield volcano.** Unquestionably the greatest assemblage of shield volcanoes is the

Figure 3.12 Volcanic ash almost buried this village during the 1914 eruption of Sakurajima. (T. Nakasa.)

Hawaiian Islands. Each island is formed of one or more such volcanoes (Figure 3.14). Another locality famous for its shield volcanoes is Iceland.

The Hawaiian volcanoes were built upward from the floor of the Pacific Ocean basin, averaging about 16,000 ft (4900 m) below sea level in this region. The highest volcano, Mauna Loa, rises to an elevation over 13,000 ft (4000 m) above sea level. Thus, measuring from their bases on the ocean floor, these volcanoes are on the order of 5 mi (8 km) high. This is vastly greater than the height of most other forms of volcanoes. Side slopes of the Hawaiian volcanoes are usually quite gentle—not more than 4° to 5° in the freshly built condition. Lava flows emerge from fissures on the flanks of the shield and travel long distances before solidifying (Figure 3.15). There is comparatively little explosive activity and little accumulation of ash.

A characteristic feature of the Hawaiian volcanoes is the broad, steep-walled **central depression,** up to 2 mi (3.2 km) or more wide and several hundred feet deep (Figure 3.14). The central depression is produced by a subsidence that follows withdrawal of basaltic magma from below. Upon the floor of the depression are smaller **pit craters,** 0.5 mi (0.8 km) across or less (Figure 3.16). Molten basalt is often exposed in the floors of the pit craters.

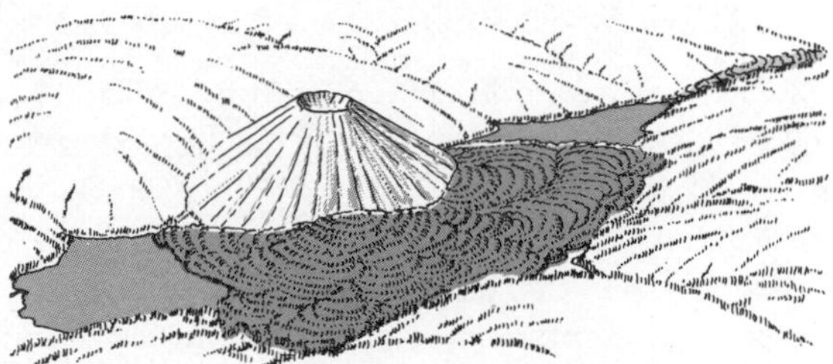

Figure 3.13 A fresh cinder cone and its basaltic lava flow (*color*) have blocked a valley, forming a small lake. (© 1973, John Wiley & Sons, New York.)

Figure 3.14 The summit of Mauna Loa, seen from the air, consists of a chain of pit craters leading to the central depression in the distance. The summit elevation is 13,680 ft (4170 m). On the distant horizon you can see the snow-capped summit of Mauna Kea, an extinct shield volcano. (U.S. Army Air Corps.)

CALDERAS—EXPLODED VOLCANOES

The explosive eruption of a composite volcano occasionally blows out an enormous mass of previously solidified lava, as well as magma from a considerable depth. This event may be accompanied by a collapse or subsidence of the central part of the cone. A deep, steep-sided crater is the result. Explosion craters of large composite volcanoes are commonly less than 1 mi (1.5 km) in diameter and represent only a small proportion of the diameter of the cone at its base.

A much larger explosion depression, the **caldera**, may be from 3 to 10 mi (5 to 16 km) or more in diameter. It represents a large proportion of the total cone diameter. Formation of a caldera is one of the most violent of natural catastrophes. Perhaps the best-known event of this kind was the explosive destruction in 1883 of the Indonesian volcano, Krakatoa. Some 18 cu mi (80 cu km) of rock are estimated to have disappeared from the volcano, demolishing the cone and leaving a caldera about 4 mi (6 km) across. Much of this lost material is believed to have disappeared by subsidence into a cavity left by the loss of gases and magma. In addition, enormous quantities of volcanic dust and pumice spread outward. The explosion produced a great seismic sea wave, or tsunami, that caused the death of many thousands of coastal inhabitants of the islands of Java and Sumatra.

In 1912 another such explosion demolished the volcano Katmai, on the Alaskan Peninsula, producing a caldera 3 mi (5 km) wide and 2000 to 3700 ft (600 to 1130 m) deep. As far away as Kodiak, 100 mi (160 km) distant, the ashfall from this explosion totaled 10 in. (25 cm), and the sound of the explosion was heard at Juneau, 75 mi (120 km) away.

Of the older calderas, those produced in prehistoric time, perhaps the best known is the basin of Crater Lake, Oregon (Figure 3.17). The caldera is about 5.5 mi (9 km) in diameter and surrounded by steep cliffs, rising to heights of 500 to 2000 ft (150 to

Figure 3.15 A basaltic lava flow from Mauna Loa moves toward the village of Hoopuloa, Hawaii, April 1926. (U.S. Air Force.)

Figure 3.16 Halemaumau, an active pit crater on Mauna Loa, seen in 1952. Hawaii Volcanoes National Park. (National Park Service, U.S. Department of the Interior.)

Figure 3.17 Crater Lake, Oregon, occupies a great caldera. Wizard Island, a cinder cone surrounded by its lava flows, is seen at the lower right. (U.S. Army Air Service.)

600 m) above the lake. The lake is up to 2000 ft (600 m) deep and covers 20 sq mi (52 sq km). The original volcano, given the name Mt. Mazama, probably rose 4000 ft (1200 m) higher than the present caldera rim; it was an imposing composite volcano resembling Mt. Hood and other volcanoes of the Cascade Range. More recently, a small cinder cone, Wizard Island, and its associated lava flow were built up in the floor of the caldera.

THE RING OF FIRE

Most of the world's active and recently active volcanoes tend to be concentrated in chain-like fashion in long, narrow belts. Others are clustered, and some are geographically quite isolated, seemingly not part of any recognizable belt. The volcanoes within each chain or region are alike in terms of their lava type—acidic, intermediate, or basic.

Andesite lavas, of the intermediate classification, form much of the greatest of all volcano chains—the **circum-Pacific belt**, sometimes called "The Ring of Fire" (Figure 3.18). This ring extends from the Andes range of South America (note that the word andesite is derived from *Andes*) through the West Indian archipelago and Mexico, along the Cascade range of the American Cordillera, then along the Alaskan coast and out along the Aleutian Islands chain. Continuing through Kamchatka, the Kuriles, and Japan, the andesite volcano ring runs southward in the form of island chains of the western Pacific and passes through New Zealand.

Distinct in location from the andesite lavas are the oceanic basalt lavas that comprise volcanoes of the Pacific Ocean basin and the

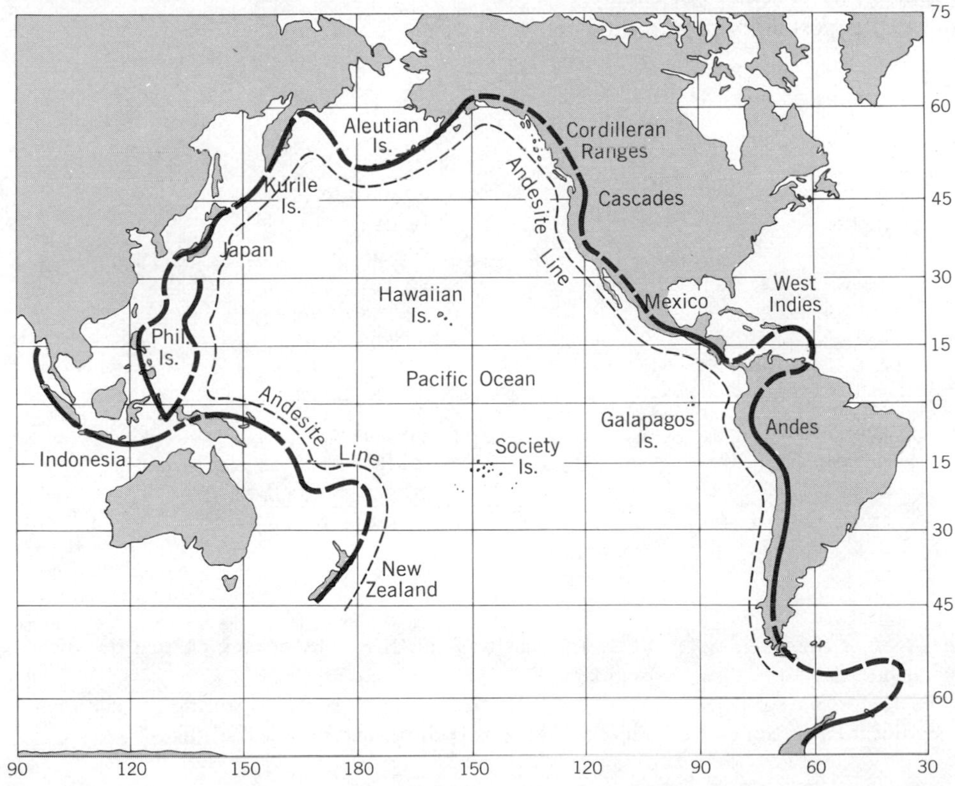

Figure 3.18 The circum-Pacific Ring of Fire, a belt of recent volcanic activity. The andesite line marks the limit of Pacific Ocean basin basaltic lavas.

middle zone of the Atlantic Ocean. These basalts are rich in the mineral olivine. In the Pacific basin, geologists have identified the **andesite line**, separating andesite lavas from basaltic lavas. The position of the andesite line is drawn on the map, Figure 3.18.

A particularly important chain of andesitic volcanoes forms a great arc in the East Indies. We shall see that this Indonesian chain is closely related to the circum-Pacific chain.

Of secondary importance on a global basis are the Mediterranean volcanoes, among them Etna, Stromboli, and Vesuvius. Important minor groups include the volcanoes of east central Africa.

Interpretation of the great volcano belts and their origin is closely tied in with the breaking and bending of the lithosphere. We will develop this topic further in Chapter 5.

IGNEOUS ACTIVITY IN REVIEW

Although the earth as a planet has provided a generally stable environment for life for at least two billion years, there are many signs of local instability of the crust. Local buildup of heat, most likely of radiogenic origin to begin with, causes melting of pockets of silicate rock at depths of many miles in the lower crust and upper mantle. From these sources, magma rises to form new masses of intrusive and extrusive igneous rock. We know that the locations of such activity are not scattered at random over the globe. Instead they occur in distinct belts. We must pursue this subject in later chapters to find a coherent explanation of how volcanism and the breaking and bending of the lithosphere operate in a unified manner.

Igneous activity represents a great amount of heat dissipation. The reservoir of heat contained in igneous bodies at shallow depths beneath the surface is one of the potentially useful sources of energy for industrial purposes. We will evaluate this possibility in our later review of energy resources.

YOUR GEOSCIENCE VOCABULARY

intrusion
stoping
xenolith
assimilate
batholith
stock
sill
laccolith
dike
columnar jointing
vein
magmatic theory
granitization
fissure
lava flow
flood basalts
volcanism
volcano
volcanic cone
vent
acidic lava
intermediate lava
basic lava
crater
composite volcano
volcanic ash
tephra
volcanic bomb
nuée ardente
basaltic cinder cone
shield volcano
central depression
pit crater
caldera
circum-Pacific belt
andesite line

SELF-TESTING QUESTIONS

1. Is it possible to find the original rock of the earth's crust? Defend your answer with geologic evidence.

2. Describe the intrusion of magma to form a batholith. What is stoping? What is the significance of xenoliths? In what way can the country rock change the invading magma?

3. What structures are produced during the cooling of sheetlike igneous bodies, such as sills and dikes?

4. What is "the granite controversy"? Describe two divergent opinions as to the origin of granite.

5. Describe the rise and extrusion of flood basalts in large volumes. What may be the source region of this magma? Name two regions of great flood basalt accumulations.

6. What is the process of volcanism? Name three groups of lavas and classify them as to mineral affinities. How does lava composition influence viscosity of the magma? How is explosiveness related to composition?

7. Describe the form and structure of a composite volcano. Name two prominent composite cones. Contrast the composite volcano with the basaltic cinder cone in terms of size, form, and composition.

8. How and where do shield volcanoes originate? Contrast shield volcanoes with composite volcanoes in terms of form, structure, and composition.

9. Describe the events typically involved in formation of a caldera. Account for the material which disappears during caldera formation.

10. Name and describe the earth's major belts and groups of volcanoes. What is the andesite line?

(*Above*) On this nineteenth century French map, the extinct volcanoes of the Auvergne district appears as craters not unlike those of the moon. In the dialect of the Auvergne, a puy is a conical hill. (*Right*) A sketch, by the distinguished American geologist-artist, Armin K. Lobeck, shows the Rock of Saint Michel; it is a volcanic neck consisting of volcanic breccia. (From *Geomorphology,* by A. K. Lobeck. Copyright 1939 by the McGraw-Hill Book Company. Used by permission of the publisher.)

4 SEDIMENTS AND SEDIMENTARY ROCKS

THE MAN WHO ERRED

How wrong can a scientist be? The gold medal for the greatest single error in the history of geology would surely go to Abraham Gottlob Werner, a German mineralogist who lived from 1750 to 1817. Werner was an excellent mineralogist. But perhaps his love of precision and detail made him so inflexible of mind that he could not be receptive to scientific evidence, even when it stared him in the face. What was Werner's great mistake? He stoutly maintained that basalt was laid down in ocean water as a sediment, that it was a rock precipitated from an aqueous solution. To thousands of tourists who each year watch basalt solidify from glowing hot magma on the volcanoes of Hawaii, the igneous origin of basalt is too obvious to be stated. How could Werner have gone so far wrong?

Rocks exposed in the hills of Saxony were visited and studied by Werner and his students. Interbedded with sedimentary strata are a few prominent layers of basalt. (The word basalt had been applied to these dark volcanic rocks long before their origin was understood.) Volcanic rocks were well known to Werner through their occurrences in the active Mediterranean volcanoes—Etna, Vesuvius, and Stromboli. We can even assume that he had specimens of recent volcanic rock from those sources in his collections. However, after studying the Saxony basalts Werner wrote that there was ". . . not a trace of volcanic action, nor the smallest proof of volcanic origin. . . . After further more-matured research and consideration, I hold that no basalt is volcanic but that all these rocks . . . are of aqueous origin."

Werner's insistence on the aqueous origin of basalt and other igneous rocks led to his being dubbed a Neptunist. Disagreeing with him were the Plutonists, led by Nicholas Desmarest, a French student of geology. Desmarest might better be called a dilettante, for he was a government official, and not an academic geologist. Desmarest studied geology in the Auvergne district of France. Here there are many volcanic cones and lava flows of recent age, although there has been no volcanic activity in historic time.

Desmarest observed that where basalt rests upon soil, the soil is definitely scorched and baked from the heat of the lava. He also identified a frothy zone in the lava, with the texture of scoria, which lent itself to the interpretation that bubbles of expanding gas had formed cavities in the magma before it cooled. Desmarest made detailed maps of lava flows of the Auvergne field and assembled a mountain of evidence showing the volcanic origin of basalt.

After the controversy between Neptunists and Plutonists had raged throughout Europe for some time, two of Werner's students—D'Aubuisson and von Buch—traveled to Auvergne to examine Desmarest's field evidence. Despite their initial bias in favor of their teacher, these men became convinced of the volcanic origin of basalt. Their frank reports went far toward settling the controversy. Later, the same baking of mineral matter beneath the basalt flows and the same scoriaceous texture were found in the basalt exposures of Saxony. Werner had simply not observed the field evidence, so determined was he that his theory was right.

In this chapter we turn to a study of rocks that are genuinely sedimentary in origin, composed of materials deposited in layers under water.

EXTERNAL EARTH PROCESSES

So far, we have investigated only **internal earth processes**—activities involving magmas and crustal heat of radiogenic origin. Rock produced by internal processes comes in contact with water and air at the earth's surface. Here, an entirely different set of processes—**external earth processes**—comes into play. External processes attack igneous rock, altering the silicate minerals and allowing their component elements to be released into the earth's surface waters. External processes create soils, on which all terrestrial plant life depends as a medium of growth.

Besides the lithosphere there are three "spheres" of importance to man. One is the **atmosphere,** or envelope of gases surrounding the earth; a second is the **hydrosphere,** or earth's liquid water in streams, lakes, and oceans. Third is the **biosphere,** or organic realm. All three spheres are important in geologic processes, including the formation of minerals and rocks.

One can think of the external earth processes as acting at a distinctive **interface,** or contact layer. This layer, the surface zone of the continents, brings both atmosphere and hydrosphere into direct contact with the lithosphere. This interface is a zone of intensive physical and chemical activity. Here energy is expended and transformed, while materials, both inorganic and organic, are constantly synthesized or decomposed. The sun is the sole source of energy for the external earth processes. Recall from Chapter 2 that heat conducted upward to the earth's surface through rock—geothermal heat—is a trivial fraction of the heat made available by the sun's rays. We can say that the external earth processes are solar powered processes.

SEDIMENT AND THE ROCK CYCLE

The solar powered, external processes of the atmosphere and hydrosphere attack and alter any exposed rock mass. As a result, rock is changed physically and chemically to produce **sediment,** broadly defined as any finely divided mineral and organic matter derived directly or indirectly from the disintegration, decomposition, and reprocessing of preexisting rock, and from life processes.

External earth processes also transport and redistribute sediment, producing accumulations that constitute a second rock class—the **sedimentary rocks.** These in turn can be acted upon by internal processes of crustal change and heating to be transformed into a third rock class—the metamorphic rocks. It is also possible for sedimentary rocks to be melted to form successive generations of igneous rocks.

From this analysis of cause and effect we recognize a **rock transformation cycle,** in which the mineral matter of the earth's crust is continually reprocessed (Figure 4.1). These changes are accompanied by enormous expenditures of energy, some of which is supplied by the sun and some of which is supplied by radiogenic heat inherited from early in the earth's history.

This chapter concentrates upon the part of the rock cycle in which solid rock is transformed into sediment, in turn becoming sedimentary rock.

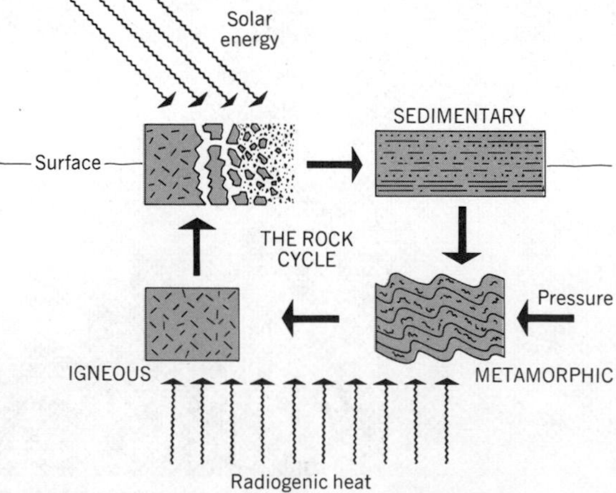

Figure 4.1 The rock cycle in its simplest form.

WEATHERING AND THE SURFACE ENVIRONMENT

The interface between lithosphere and atmosphere represents a specialized environment as far as minerals and rocks are concerned. This surface environment is one of relatively low temperature and low confining pressure. Such conditions are in contrast to the high-temperature and high-pressure environment in which plutonic igneous rocks are formed deep within the earth's crust.

The surface environment is one of instability for the silicate minerals, for they succumb readily to the presence of free oxygen, carbon dioxide, and water. We like to think that no substance is more enduring than the granite we use in tombstones, yet in fact granite is one of the most decay-susceptible mineral assemblages in the face of the "elements." Most varieties of metamorphic rocks, too, were produced in an environment of high pressures and temperatures and are not well adapted to endure exposure to atmospheric conditions.

The alteration of minerals in the presence of water, oxygen, and various natural acids is greatly aided by a group of physical forces of disintegration. These forces can break apart hard igneous and metamorphic rocks. Breakup into very small particles increases the mineral surface area exposed to chemical solutions. Were it not for the forces of physical disintegration, rock alteration would have proceeded very slowly throughout the geological past, and the course of earth history would have been quite different.

Geologists use the term **weathering** for the total of all processes acting at or near the earth's surface to cause physical disintegration and chemical decomposition in rocks. We will have more to say about the physical processes of weathering in Chapter 12. Our attention in this chapter will be focused on chemical changes in rock. We shall be particularly interested in the chemical changes affecting the silicate minerals of igneous rocks, because we are tracing steps in the cycle of rock transformation leading to formation of sediments and sedimentary rocks.

CHEMICAL WEATHERING

Chemical weathering consists of several important chemical reactions causing change in rock composition. All of these reactions may occur more or less simultaneously. Consider, first, that all surface water—whether it is in the form of raindrops, soil water, ground water, or water in streams, lakes, and the oceans—contains in solution the gases of the atmosphere. We can disregard nitrogen—the major atmospheric component—because it is a comparatively inactive element. The principal gases of interest in rock weathering are oxygen and carbon dioxide.

Molecules of oxygen, when dissolved in water, separate into individual oxygen atoms. Each atom takes on an electrical charge, becoming an **ion.** Oxygen ions in water are readily available for the process of **oxidation,** in which the oxygen ion combines with such metallic ions as may be available.

Carbon dioxide in solution in water forms a weak acid capable of reacting with certain susceptible minerals. Other acids, those of organic origin, are also active in the water found in soil and rock.

Water itself is capable of dissolving certain minerals directly, a process we see daily in the solution of table salt (the mineral halite).

All of these chemical processes require water, and the earth's surface is abundantly endowed with water. There is no truly dry environment on the earth's surface, not even in the most nearly rainless deserts. If there is any surface environment in which rocks can escape the chemical processes of decay, it is in the perpetually frozen layer found below the surface in arctic and antarctic lands.

Chemical union of water with mineral compounds is termed **hydrolysis.** The process is a true chemical change and is not reversible in the surface environment. It is not merely a form of water absorption, which can be followed by drying out. Hydrolysis produces a new mineral compound from the original mineral.

THE CLAY MINERALS

The chemistry of mineral alteration is very complex. We offer here only a few examples of alteration of common silicate minerals into other minerals. Those which are soft and plastic when moist are collectively referred to as **clay minerals.**

Potash feldspar, an abundant constituent of granite, takes up water to yield a clay mineral, **kaolinite.** This soft, white mineral becomes plastic and exudes a distinctive clay odor when moistened. It is widely used in the manufacture of chinaware, porcelain, and tile.

Feldspars and muscovite mica are altered to another clay mineral, **illite,** an abundant constituent of sediments. Figure 4.2 shows illite particles highly magnified under the electron microscope. A third important clay mineral is **montmorillonite,** also shown in Figure 4.2; it is derived from the alteration of igneous rocks.

Through hydrolysis, plagioclase feldspar is altered to **bauxite,** a mixture of several clay minerals composed largely of aluminum oxide and water. Bauxite occurs in abundance in warm, wet climates and forms important deposits of aluminum ore. The mafic silicate minerals contain iron in abundance, and this element, when released by chemical weathering, combines with oxygen to form very stable oxides in combination with water. An example is **limonite,** an earthy substance of brown to yellowish color, abundant in soil of warm, wet climates.

An important point about chemical weathering is that it changes very hard igneous and metamorphic rocks into soft substances. In this way weathering weakens the bedrock and enables it to be broken up by mechanical processes. The clay minerals consist of extremely small particles down to the size of **colloids.** Colloids are particles so small they remain suspended indefinitely in water. Colloids are thus easily transported long distances in streams. In contrast, the hydrous oxides of iron and aluminum (limonite, bauxite) tend to accumulate as hard masses and layers in and beneath the soil. Being almost immune to further change, they resist removal and transportation.

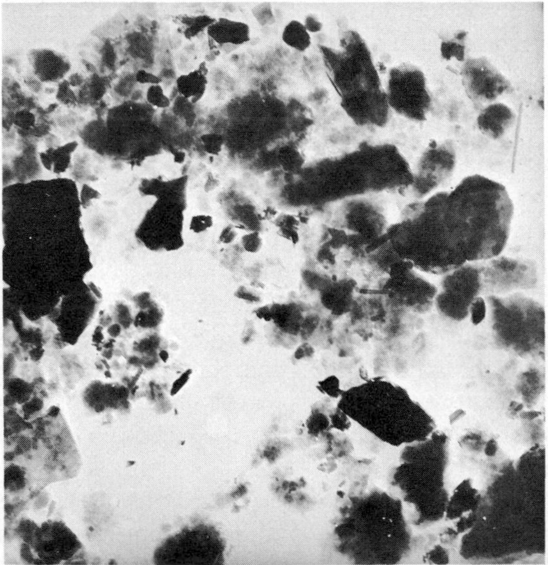

Figure 4.2 Fragments of the clay minerals illite (*sharp outlines*) and montmorillonite (*fuzzy outlines*) which have settled from suspension in tidal waters of San Francisco Bay. Enlargement about 20,000 times. (Corps of Engineers, U.S. Army.)

In warm humid climates, chemical decay of igneous and metamorphic rocks extends to depths as great as 300 ft (90 m) and has produced a thick layer of soft clay-rich rock known as **saprolite** (means "rotten rock"). Examples may be found throughout the Appalachian region of the southeastern United States.

Generally speaking, the felsic rocks, with abundant quartz and potash feldspar, are more resistant to chemical decay than the mafic rocks, rich in plagioclase feldspar, pyroxene, amphibole, and olivine. Olivine in particular decomposes very readily. As a result, where felsic and mafic igneous rock masses are side by side (as where a dike of felsic rock cuts through a mass of mafic rock, or vice versa), the felsic rock usually stands out sharply (Figure 4.3).

Chemical weathering produces the parent matter of the soil. Silicate minerals as they occur in igneous rock have no nutrient value to plants. Only when chemical weathering occurs can the nutrient elements, such as calcium, magnesium, and potassium, be released in the form of ions usable by plants.

Figure 4.3 A felsic dike (angular blocks at center) in deeply altered mafic rock, Sangre de Cristo Mountains, New Mexico. (A. N. Strahler.)

THE SEDIMENTARY ROCKS

Weathering transforms rock into sediment, and this in turn may become sedimentary rock. Usually, the sediment must be transported and then redeposited in protected places to produce significant accumulations. In such protected places major deposits of sedimentary rock can form.

We can classify sedimentary rocks in terms of the possible origins of the sediment constituting the rock (Figure 4.4). The first order of classification is into clastic and nonclastic divisions. **Clastic sediment** consists of particles broken away individually from a parent rock source. The clastic rocks are in turn subdivided into two

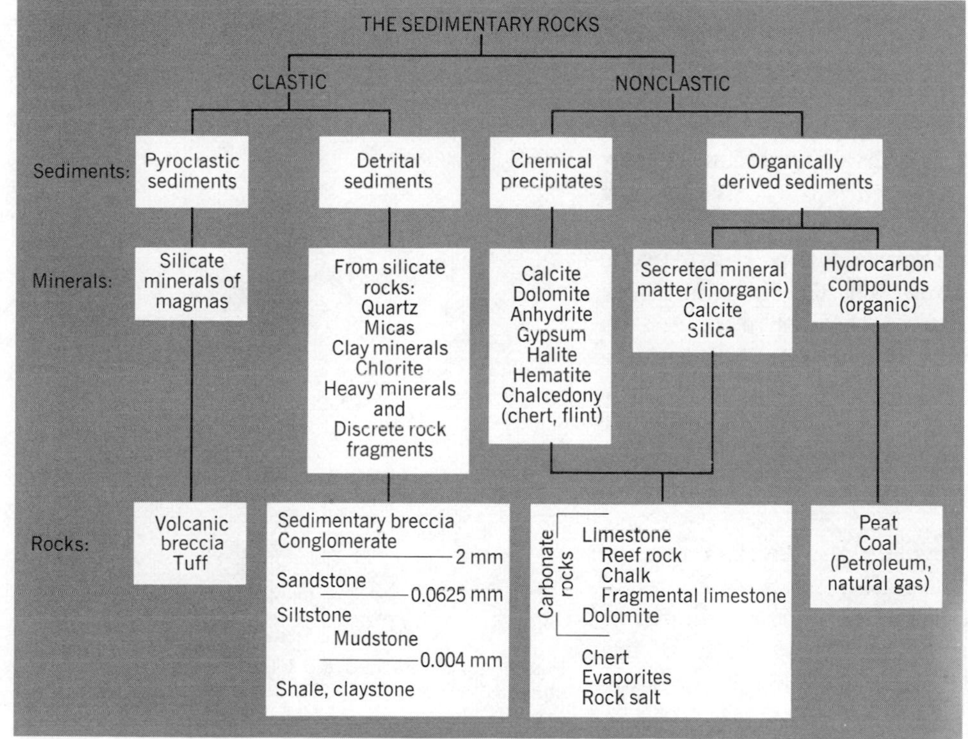

Figure 4.4 Composition and classification of the sedimentary rocks.

groups. One rock is made up of **pyroclastic sediments.** This material was called tephra in Chapter 3. A second group consists of rocks made up of **detrital sediments,** mineral fragments derived by the weathering of preexisting rocks of any classification.

The nonclastic division also includes two basic subdivisions, **chemical precipitates** and **organically derived sediments.** Chemical precipitates are inorganic compounds representing solid mineral matter precipitated from a water solution in which that matter has been transported. The organically derived sediment consists of both the remains of plants or animals and mineral matter produced by the activities of plants and animals. This includes, for example, the shell matter secreted by animals, which is a true mineral and constitutes an inorganic sediment. On the other hand, accumulating plant remains, consisting of hydrocarbon compounds, form a truly organic sediment. We shall need to be careful to distinguish between organically derived mineral matter and organic sediment (hydrocarbon compounds).

Now that we have recognized the broad classes of sediments, we can list the component minerals. Minerals that have not been described in previous chapters will be included in our discussion of the appropriate sediment group. The naming of a particular sedimentary rock depends not only upon its mineral composition but also upon the size of the component mineral grains. Clastic sediments are named primarily on the basis of the sizes of the component fragments.

STRATA

Sedimentary rocks are usually recognizable through the presence of distinct layers resulting from changes in particle size and composition during the period of deposition. These layers are termed **strata,** or simply beds. The planes of separation between layers are **stratification planes,** or bedding planes. The rock is described as being stratified, or bedded (see Figure 4.8). Bedding planes in their original condition are nearly horizontal, but they may have become steeply tilted (see Figure 5.1) or otherwise distorted into wavelike folds by subsequent movements of the earth's crust.

When reporting upon geologic relations, we need to describe the attitude of various natural rock planes. The **dip** of a natural rock plane is the acute angle formed between the rock plane and an imaginary horizontal plane of reference (Figure 4.5). Dip is stated in degrees, and ranges from zero for a horizontal plane to 90° for a vertical plane. Instruments for measurement of dip use a level bubble for determination of the horizontal. The direction of dip may also be stated, using compass directions. In Figure 4.5, the direction of dip is toward the west.

The **strike** of an inclined rock plane is the direction assumed by the line of intersection between the rock plane and a horizontal plane of reference. In Figure 4.5 the horizontal plane is indicated by the surface of a lake. Geologists conventionally give the strike as a compass direction. In the illustration, the strike is north (or north-south).

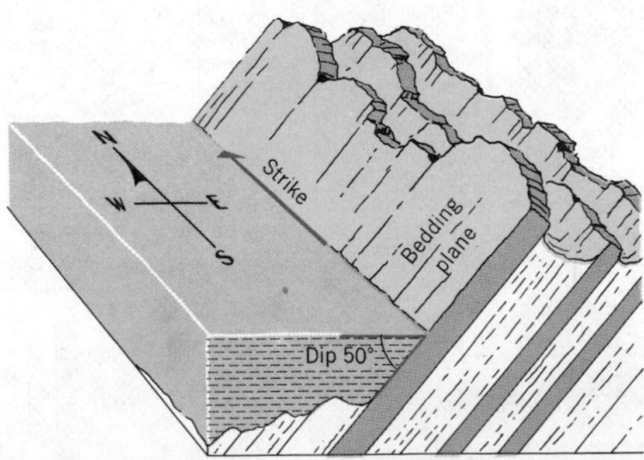

Figure 4.5 Strike and dip illustrated by strata at the shore of a lake. (© 1973, John Wiley & Sons, New York.)

THE DETRITAL SEDIMENTS

The most abundant particles of detrital sedimentary rocks consist of quartz, rock fragments, feldspar, and clay minerals. Fragments of unaltered fine-grained parent rocks can easily be identified in coarse sandstones by microscopic examination. Such fragments are typically second in abundance to quartz grains and are the chief component in the coarser grades of detritus. Mica and other minerals generally make up less than 3% of coarse-grained detrital rock. Clay minerals may be abundant in the finer-grained detrital sediments.

There are, in addition to quartz, feldspar, and muscovite, a number of minor minerals, found in the igneous and metamorphic rocks, which are highly resistant to physical abrasion and chemical alteration. These durable minerals remain intact during transportation. Because of their relatively greater density, as compared with that of quartz and other felsic minerals, these detrital minerals are referred to as the **heavy minerals.** An example is magnetite, an oxide of iron.

The heavy detrital minerals are easily separated from the less dense quartz and mica by processes of water transportation or by winds. Consequently, these minerals form local concentrations as dark layers in many types of sand accumulations. A magnet dragged through such dark sand will usually emerge heavily coated with magnetite grains.

Because the naming of clastic rocks depends in large part upon the sizes of component mineral grains, it is important to establish a system of size grades. Among geologists the **Wentworth scale** is widely accepted (Table 4.1). The units of length are millimeters.

A sediment accumulation that has become hardened into rock is described as lithified. The noun **lithification** refers to the hardening process. Usually, cementation by mineral matter, or compaction, or both, are responsible for lithification.

Table 4.1 **Grade sizes of sediment particles**

Grade name	Diameter, mm
Boulders	Over 256
Cobbles	64–256
Pebbles	2–64
Sand	0.06–2
Silt	0.004–0.06
Clay	Under 0.004

THE DETRITAL SEDIMENTARY ROCKS

Coarsest of the detrital sedimentary rocks is **sedimentary breccia,** consisting of large angular blocks in a matrix of finer fragments. These rocks often represent ancient submarine landslides, or terrestrial flows of mud, and are comparatively rare rocks. **Volcanic breccia** is the equivalent rock in the pyroclastic group.

A **conglomerate** consists of pebbles or cobbles, usually quite well rounded in shape, embedded in a fine-grained matrix of sand or silt (Figure 4.6). The main distinction between a conglomerate and a breccia is that the large fragments in the breccia are angular. Rounding of the conglomerate pebbles is a result of abrasion (wearing action) during transportation in stream beds or along beaches. Essentially, then, conglomerates represent lithified stream gravel bars and gravel beaches.

Sandstone is composed of grains in the range from 2 mm to 0.06 mm (1/16 mm). Perhaps the most abundant and familiar form is quartz sandstone, in which quartz is the predominant constituent. Beautifully rounded quartz grains extracted from a sandstone are

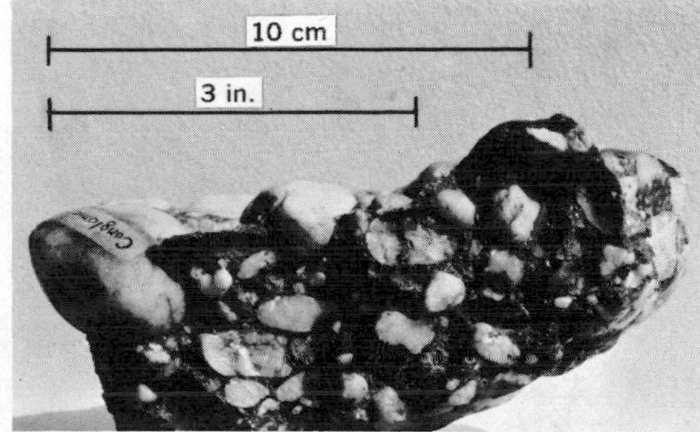

Figure 4.6 A conglomerate, consisting of well-rounded quartzite pebbles in a matrix of fine sand and silt. (A. N. Strahler.)

pictured in Figure 4.7. In this example, rounding was perfected by wind transport in ancient sand dunes. Quartz sandstones contain minor amounts of the heavy detrital minerals and frequently small flakes of muscovite mica, some grains of feldspar, and rock fragments.

The quartz sandstones commonly represent lithified sediment deposits of the shallow oceans bordering a continent or of shallow inland seas. The quartz grains have survived a long distance of travel. Finer particles have been sorted out and removed during the transportation process. As we implied above, certain quartz sandstones were formed from large deposits of dune sands in ancient deserts on the continents. Also, some quartz sandstones are formed largely of recycled grains derived from preexisting sandstones.

Lithification of quartz sands requires **cementation** by mineral deposition in the interstices (open spaces) between grains. This cementation is accomplished by slowly moving ground water importing the cementing matter as ions in solution. The cementing mineral may be silica (silicon dioxide). In this case, the sandstone is an extremely hard rock with great resistance to weathering and erosion. If the cementing material consists of calcium carbonate, a less durable rock results.

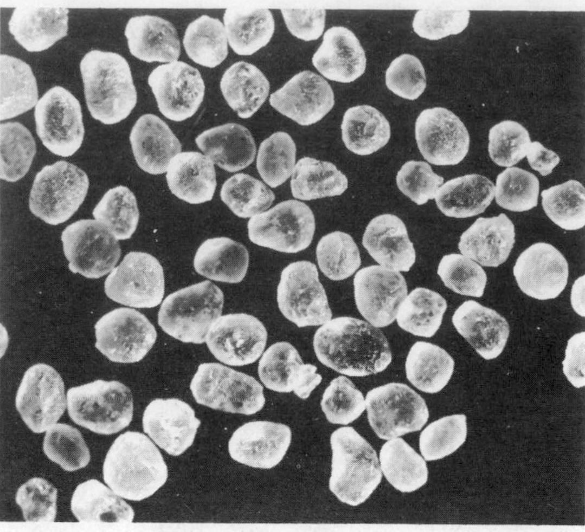

Figure 4.7 Well-rounded quartz grains extracted from sandstone. The grains average about 1 mm (0.04 in.) in diameter. (A. McIntyre, Columbia University.)

The compaction and cementation of layers of silt gives a compact fine-grained rock known as **siltstone.** It has the feel of very fine sandpaper and is closely related to fine-grained sandstone, with which there is a complete intergradation.

A mixture of silt and clay with water is termed a **mud,** and the sedimentary rock derived from such a mixture is a **mudstone** (Figure 4.8). The compaction and consolidation of clay layers forms **claystone.**

Many sedimentary rocks of mud and clay composition are laminated in such a way that they break up easily into small flakes and plates. A rock that breaks apart in this way is described as **fissile** and is generally called a **shale.** Shale is fissile because clay particles lie in parallel orientation with the bedding to form natural surfaces of parting.

The bulk of claystone and clay shale consists of the clay minerals derived from the alteration of the silicate minerals. Kaolinite, illite, and montmorillonite are the most common of these minerals.

Figure 4.8 The Big Badlands of South Dakota have been eroded to show horizontal strata—largely soft muds and clays. (Douglas Johnson.)

Compaction of clay sediments into rock is largely a process of exclusion of water under pressure from the overlying sediments. Because the clay minerals consist of minute flakes and scales, the proportion of water held in the initial sediment is very large. Once thoroughly compacted, claystone and clay shale do not soften appreciably when exposed to water. However, shale breaks apart easily upon impact.

Shales of mud and clay composition make up the largest proportion of all sedimentary rocks. They can be subdivided by color. The red shales owe their color to finely disseminated oxide of iron. Red shales are associated with red siltstones and red sandstones in enormously thick accumulations. Collectively known as **red beds,** these strata are interpreted as having been deposited in an environment of abundantly available oxygen. Favorable conditions are found on river floodplains and deltas of arid climates.

The gray and black shales, also found in great thicknesses, are interpreted as deposited in a marine environment in which oxygen is deficient. The dark color is due to disseminated carbon compounds of organic nature.

Referring back to the pyroclastic sediments, layers of fine volcanic ash and dust become lithified into **tuff,** a fine-grained sedimentary rock. The mineral composition of tuff is usually the same as for rhyolite and andesite, which are the lavas associated with explosive volcanoes.

THE NONCLASTIC SEDIMENTARY MINERALS

Perhaps the most important minerals of the nonclastic sediment class are the **carbonates.** These are compounds of the calcium ion or magnesium ion, or both, with the carbonate ion. Calcium carbonate is the composition of one of the most abundant and widespread of minerals, **calcite** (see Figure 1.4). An important chemical precipitate is **dolomite,** a carbonate of both magnesium and calcium.

Evaporites form a major class of chemical precipitates. These are highly soluble salts deposited from salt water bodies when evaporation is sustained under an arid climate. We are all familiar with **halite,** or rock salt, one of the commonest of the evaporites; it consists of sodium chloride. Two sulfate compounds of calcium, **anhydrite** and **gypsum,** are also among the common evaporites.

Two other common minerals important in nonclastic sedimentary rocks are **hematite,** an oxide of iron, and **chalcedony,** a form of silica lacking obvious crystalline structure. Chalcedony occurs abundantly as nodules and layers in sedimentary rocks and is then referred to as **chert.**

THE CARBONATE ROCKS

Most nonclastic sedimentary rocks are either carbonate rocks or evaporites (Figure 4.4). Chert forms a class by itself.

Of the carbonate rocks, the most important is **limestone,** a sedimentary rock in which calcite is the predominant mineral. Because either clay minerals or silica (as quartz grains or chert) may be present in considerable proportions, limestones show a wide variation in chemical and physical properties. Limestones range in color

Figure 4.9 Layers of chalk make up this marine cliff on the Normandy coast of France. (Photographer not known.)

from white through gray to black, in texture from obviously granular to very dense.

The most abundant limestones are of marine origin. Some of these are formed by inorganic precipitation, others are by-products of organic activity. The marine limestones show well-developed bedding and may contain abundant fossils. Dark color may be due to finely divided carbon. Many limestones have abundant nodules and inclusions of chert and are described as cherty limestones. An interesting variety of limestone is **chalk,** a soft, pure-white rock of low density (Figure 4.9). It is composed of the hard parts of minute algae.

Important accumulations of limestone consist of the densely compacted skeletons of corals and the secretions of associated algae —they are seen forming today as coral reefs along the coasts of warm oceans. Rocks formed of these deposits are referred to as **reef limestones.** These limestones are in part fragmental, since the action of waves breaks up the coral formations into small fragments that accumulate among the coral masses or in nearby locations. Limestones composed of broken carbonate particles are recognized in Figure 4.4 as fragmental limestone.

Dolomite is a rock composed largely of the mineral of the same name. Dolomite rock poses a problem of origin, since the mineral is not excreted by organisms as shell material. Direct precipitation from solution in sea water is not considered adequate to explain the great thicknesses of dolomite rock that are found in the geologic record. The most widely held explanation of the formation of dolomite rock is that it has resulted from the alteration of limestone, by the substitution of magnesium ions of sea water for part of the calcium ions.

THE EVAPORITE ROCKS

The great bulk of evaporite rocks consists of the minerals gypsum, anhydrite, or halite. Evaporite minerals occur in association with one another in sedimentary strata, usually with marine sandstones and shales, but in some instances with limestones and dolomites.

Most hypotheses of the origin of thick sequences of the evaporites in the geologic record require a special set of environmental conditions. First, there must exist an arid climate in which evaporation on the average exceeds precipitation. Such climates are widespread in tropical latitudes today and can be presumed to have been present in the geologic past. Second, a shallow evaporating basin is required, and this may have been a large shallow bay or lagoon cut off from the open sea by a bar. A narrow inlet, through which ocean water could enter to replace water lost by evaporation, is necessary to account for thick beds of evaporites. A slow subsidence of the area of deposition is required to accommodate the accumulating beds.

Huge accumulations of halite have formed at several points in geologic time. A good example is the salt beds and associated layers of red shales and sandstones, gypsum, and anhydrite of the Permian basin of Kansas, Oklahoma, and parts of northern Texas and southeastern New Mexico. In the Permian Period of geologic time, dozens of halite beds were deposited here. A few individual salt beds over 300 ft (90 m) in thickness are known. Thick anhydrite beds also occur in this series.

HYDROCARBON COMPOUNDS IN SEDIMENTARY ROCKS

Hydrocarbon compounds, consisting of carbon, hydrogen, and oxygen of organic origin, make up the last class of sedimentary rocks shown in Figure 4.4. Only **coal** qualifies for designation as a rock. The solid forms of hydrocarbon compounds—peat and coal—remain in the place of original accumulation. On the other hand, the liquid and gaseous forms—petroleum and natural gas—can migrate far from the places of origin to become concentrated in distant rock reservoirs.

In a swamp or bog environment, plant remains accumulate faster than they can be destroyed by bacterial activity, because oxygen is deficient in the stagnant water. Also, the organic acids released by the decay process inhibit further bacterial activity. The product of this environment is **peat,** a soft fibrous material ranging in color from brown to black.

Under the load of accumulating sediments, layers of peat have become compacted into **lignite,** or "brown coal," a low-grade fuel intermediate between peat and coal. Lignite has a woody texture.

Upon further compaction, lignite is transformed into **bituminous coal,** or "soft coal," and this in turn may be transformed into **anthracite,** or "hard coal." Anthracite is found in strata that were subjected to intense pressures of folding in the mountain-making process. The changes from peat to bituminous coal and anthracite required millions of years.

Coal occurs in seams, interbedded with sedimentary strata, which are usually thinly bedded shales, sandstones, and limestones (Figure 4.10). Collectively, such accumulations are known as **coal measures.** Individual coal seams range in thickness from a fraction of an inch to several tens of feet. Coal measures and coal mining are described in more detail in Chapter 11.

Figure 4.10 Outcrop of an 8-ft (2.4-m) coal seam, Dawson County, Montana. Large blocks of coal have slumped to the base of the cliff (foreground). (M. R. Campbell, U.S. Geological Survey.)

Petroleum, or crude oil, is a mixture of several fluid hydrocarbon compounds of organic origin. Petroleum is found in localized concentrations in certain sedimentary strata. In composition, a typical crude oil might run as follows: carbon, 82%; hydrogen, 15%; oxygen and nitrogen, 3%. Petroleum is neither a mineral nor a rock, but its close association with the clastic sedimentary rocks justifies our discussing it here. **Natural gas,** which is closely linked in origin and occurrence with petroleum, is a mixture of gases, principally methane (marsh gas), and small amounts of other hydrocarbons.

Geologists agree that petroleum and natural gas are of organic origin. A major hypothesis of oil origin attributes the oil to microscopic plant forms living in vast numbers in the seas. Upon death of the organism a minute particle of oil was released on the ocean floor, becoming incorporated into accumulating sediment. Where this was a dark mud the sediment eventually became a shale formation. Today we find petroleum disseminated through **oil shale,** which can be processed to derive petroleum. Eventually, the petroleum must have been forced to migrate from the shale to the porous reservoir rock.

Coal, petroleum, and natural gas are referred to collectively as **fossil fuels,** and are modern civilization's major source of energy. The limited world supplies of fossil fuels are discussed in Chapter 11.

CONTINENTAL SEDIMENTS IN REVIEW

The rock transformation cycle encompasses all known varieties of rock. This cycle involves changes in mineral matter brought about by both internal and external earth processes. If we choose to start the rock cycle with igneous rocks, we find that silicate minerals formed under internal environments of high temperature and high pressure are later decomposed in the weathering process because of exposure to air and water. Weathering produces sediment, which may be transported, deposited, and hardened into sedimentary rock. Organisms make significant additions to sediment accumulations, and it is here that the biosphere makes its contribution to geologic processes and materials.

Production of sediment and sedimentary rocks has gone on throughout at least 3½ billion years of geologic time. It is thought that much of the enormous bulk of sedimentary rock produced during this time has been remelted to form new igneous rock. We shall inquire further into the rock cycle in the next chapter.

YOUR GEOSCIENCE VOCABULARY

internal, external earth processes
atmosphere
hydrosphere
biosphere
interface
sediment
sedimentary rock
rock transformation cycle
weathering
chemical weathering
ion
oxidation

hydrolysis
clay minerals
kaolinite
illite
montmorillonite
bauxite
limonite
colloids
saprolite
clastic sediments, rocks
pyroclastic sediment
detrital sediment

chemical precipitate
organically derived sediment
strata
stratification planes
dip
strike
heavy minerals
Wentworth scale
lithification
sedimentary breccia
volcanic breccia
conglomerate

sandstone
cementation
siltstone
mud
mudstone
claystone
fissile
shale
red beds
tuff
carbonates
calcite
dolomite: mineral, rock
evaporites
halite
anhydrite
gypsum
hematite
chalcedony
chert
limestone
chalk
reef limestone
hydrocarbon compounds
coal
peat
lignite
bituminous coal
anthracite
coal measures
petroleum
natural gas
oil shale
fossil fuels

SELF-TESTING QUESTIONS

1. What geologic role is played by the internal earth processes? by the external processes? Use the concept of an interface in organizing your answer.

2. Outline the major features of the rock transformation cycle. What principal kinds of rocks are involved in the cycle?

3. Evaluate the earth's surface environment in terms of chemical stability of mineral and rocks. What is mineral alteration? Define weathering. Comment on the geological importance of weathering.

4. What are the principal chemical reactions involved in chemical weathering? Describe hydrolysis of feldspars and name the principal alteration products. Where does saprolite occur?

5. Describe a classification of sedimentary rocks based upon origins of sediments. What are the subdivisions within the clastic and nonclastic groups? How do chemical precipitates differ from the organically derived sediments?

6. Describe the appearance of sedimentary strata as seen in exposures. What terms are applied to the various layered structures? Explain how strike and dip are used to describe the attitude of strata.

7. What are the most abundant constituents of the detrital sedimentary rocks? Describe the heavy minerals occurring in sedimentary rocks. What grade scale is in general use to describe dimensions of sedimentary particles?

8. Name the common detrital sedimentary rocks in order of texture from coarse to fine. Describe the appearance and physical properties of each type. How are sediments lithified? What clay minerals are common in claystone and shale? What is the climatic significance of red beds?

9. Describe the minerals making up carbonates and evaporites. Describe limestone and dolomite. Where do these rocks originate? How do coral reefs produce limestone?

10. Under what conditions do evaporites accumulate in large quantities? Where are abundant evaporites found in the geologic record?

11. What are hydrocarbon compounds? In what forms do hydrocarbons occur in sedimentary rocks? What is the composition of lignite and coal? Under what conditions is anthracite formed?

12. What is the composition of petroleum? of natural gas? In what rocks do petroleum and natural gas accumulate? What is the origin of petroleum?

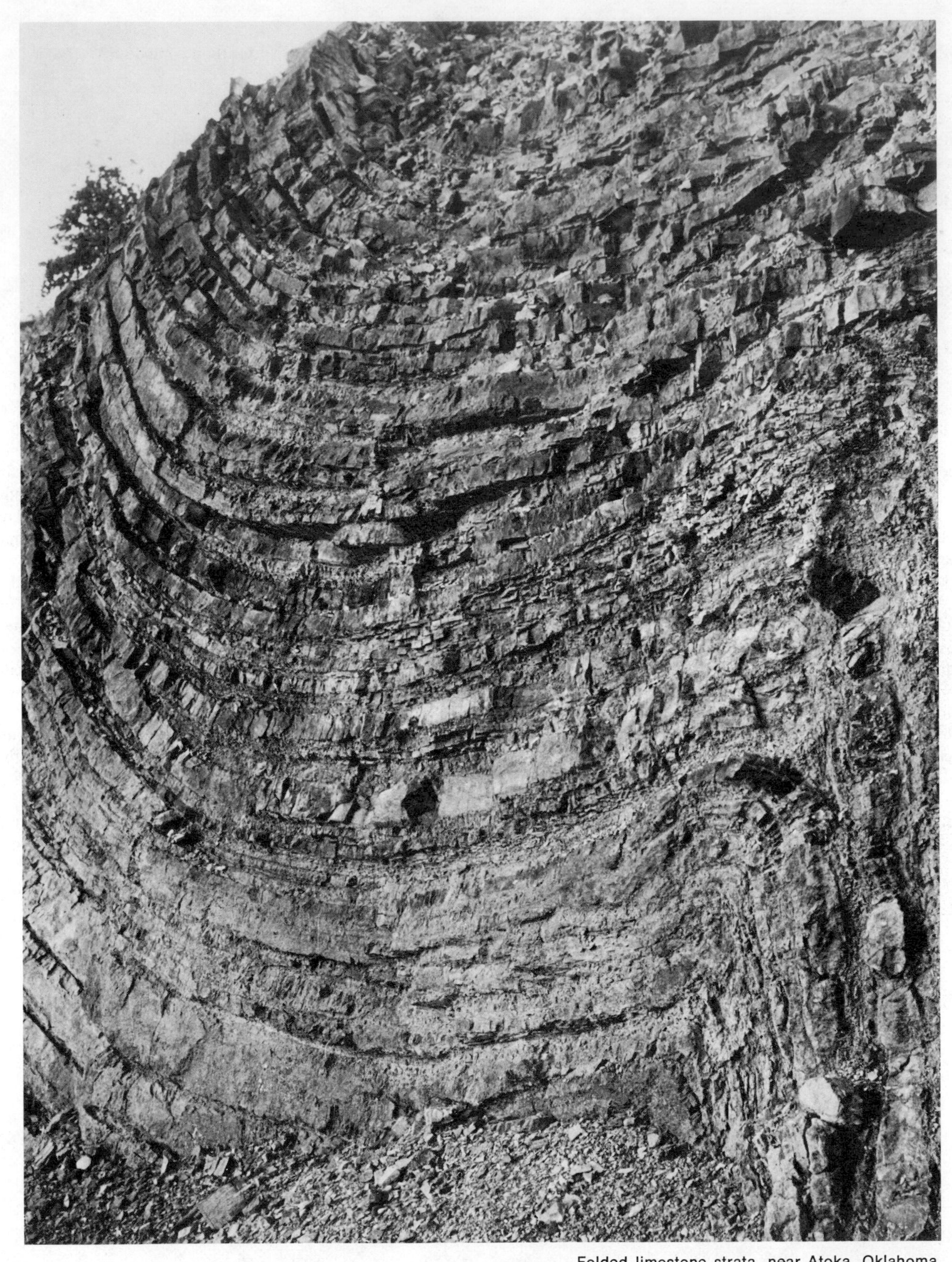

Folded limestone strata, near Atoka, Oklahoma.
(Photograph by Lofman, courtesy of Exxon Corporation.)

5 TECTONIC ACTIVITY AND THE METAMORPHIC ROCKS

CATASTROPHISM AND UNIFORMITARIANISM

As the geologist examines exposed rocks, he finds unmistakable evidence that certain parts of the earth's crust have been crumpled and fractured on an enormous scale. What he sees is difficult to explain satisfactorily in terms of natural forces and rock properties as he observes them at the earth's surface. In certain localities rock is sedimentary, but instead of lying horizontally the strata are wrinkled into folds, as if the rock had been plastic like soft clay, at the time crumpling took place.

Late in the eighteenth century, geological science was not yet established along the lines that it is today. One group of naturalists of that period explained folding and breakage of strata as the result of one or more sudden catastrophes. Leader of this school was an able French student of fossil life forms, Baron Cuvier, who wrote:

> The dislocation and overturning of older strata show without any doubt that the causes which brought them into the position which they now occupy, were sudden and violent. . . . The evidences of those great and terrible events are everywhere to be clearly seen by anyone who knows how to read the record of the rocks.

Contorted strata on the coast of Berwickshire, Scotland. (After Sir James Hall, Bart., 1812.)

Those who supported this view were known as **catastrophists.** They held that not only were our present mountains, cliffs, and canyons formed by violent catastrophe, but all the animals whose shells and bones we now find as fossils in the strata were suddenly killed in the cataclysm. This theory was tenable only as long as evidence consisted of a few observations in only one region. Later, it gradually became clear that no single worldwide catastrophe can explain all known relations among strata and their fossil content. Furthermore, the length of time needed for the various events of geologic history was grossly underestimated by the catastrophists.

A sound basis for reconstructing geologic events then began to emerge. It is the principle that processes acting in the past have been essentially the same as those seen in action over the face of the earth today. This principle was strongly maintained by a Scottish geologist, James Hutton (1726–1797). He termed it **uniformitarianism.**

The geologist believes that if he can watch a volcano in eruption he will learn the explanation for lava forms now found enclosed in ancient rocks. If he studies the manner in which sediments are being laid down around the mouth of the Mississippi River today he can learn how to interpret similar layers of shale and siltstone which he finds exposed in the walls of canyons far inland from present-day shores. The concept of uniformitarianism is often summarized in the statement that "the present holds the key to the past."

FOLDED AND FAULTED STRATA

Sir James Hall, a Scot, observed folded strata along the Berwickshire coast of his native land. His sketch of folds, published in 1812, is reproduced on the opening page of this chapter. The idea that these beds had once been flat-lying was, in Hall's time, a new concept to be greeted with skepticism. To Hall, an experienced geologist, it seemed reasonable that the strata must have been soft, like clay and loose sand, at the time the crumpling took place.

Evidence that such contorted strata were once flat-lying and that they were deposited beneath the sea is often undeniable. For example, the surfaces of the sandstone layers often show ripple markings identical to those seen in sand under shallow water near a beach (Figure 5.1). Shale and limestone strata commonly contain great numbers of fossil shells. These could not have clung to a steeply sloping sea bottom, for many are half shells or broken shell fragments like those one sees lying loose on beaches.

Figure 5.1 These ripple marks were formed on a nearly horizontal sea floor, but have since been tilted by mountain-making movements to an almost vertical attitude. Precambrian quartzite strata in the Baraboo Range, Wisconsin. (A. N. Strahler.)

Some 75 years after Hall made his observations in Scotland, a Swiss geologist and alpinist, Albert Heim, was busy unraveling the geological intricacies of the Swiss Alps. Heim observed that folded strata are often sharply broken across, or faulted, providing evidence that the rock behaved as a brittle solid (Figure 5.2). Obviously the rock yielded with a sudden movement when its elastic limit was exceeded. The strata appear sheared off cleanly along the break, and if a particularly distinctive layer is present on one side, its continuation may perhaps be found by searching higher or lower among the strata on the other side of the break.

A most important question facing the early geologists was to explain how these strata could be crumpled into folds, like a soft clay, but also have sharp breaks, like a brittle solid. Was it possible that the rock was plastic at one time and later became brittle? This might seem to be the answer, but in some localities Heim found indications that both the folding and faulting occurred at about the same time, with the rock in the same physical state. Heim realized that the strata were deeply buried and under tremendous confining pressure during their deformation. Such extreme pressure is now recognized as the cause of the seemingly strange behavior of rock in mountain zones.

Other questions present themselves. At what depth did this deformation occur? What was the nature of the forces involved? How long did the process take? Why did deformation occur at this place, but not affect similar sedimentary strata of the same geologic age 200 miles distant?

Figure 5.2 Folded and faulted strata, Glacier National Park, Montana. (Douglas Johnson.)

MOUNTAIN ARCS, ISLAND ARCS, AND TRENCHES

If, as Hutton proposed, the present is the key to the past, we may learn much about ancient geologic history by investigating the active crustal zones of the present. Present-day crustal activity in the form of earthquakes, volcanism, uplift of high mountain chains, and downsinking in deep trenches is principally concentrated in long, narrow, broadly curving zones known as the **primary arcs.** Examine the location of these arcs on a world map (Figure 5.3). The arcs fall into two chains. First is the circum-Pacific belt, already referred to

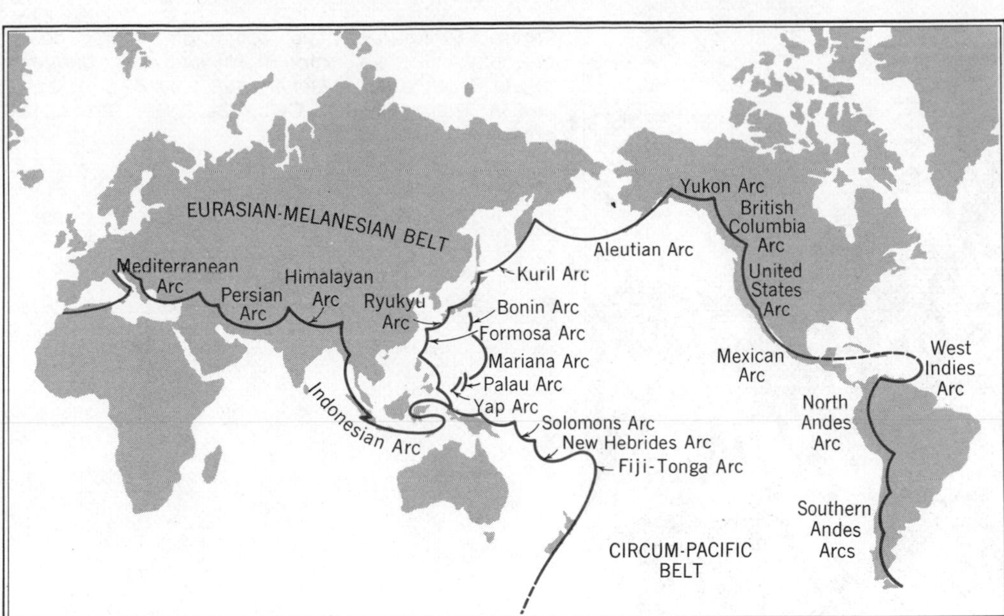

Figure 5.3 The primary arcs.

in Chapter 3 as the locus of much recent volcanic activity. Throughout North and South America, arcs of the circum-Pacific belt lie along the western continental margins, except for the West Indies Arc, which is oceanic in location.

Starting with the Aleutian Arc, and continuing south along the western side of the Pacific, the primary arcs are in oceanic positions at some distance from the Asiatic shoreline. In large part, these arcs are represented by chains of volcanic islands.

Second of the chains of primary arcs is the **Eurasian-Melanesian belt,** extending from the Mediterranean region, eastward through southern Asia. This belt terminates in the Indonesian Arc, which appears to intersect the circum-Pacific belt in a T-junction. Notice that the primary arcs are convexly bowed outward from the continental centers, a form that must be highly significant in terms of the mechanics of their origin.

The primary arcs contain the world's great mountain ranges; they are commonly described as belonging to the **alpine system.** Fine examples are the Alps of Europe, the Himalayas of southern Asia, the Andes of South America, and the Cordilleran ranges of western North America. Comparative recency of the uplift of these ranges is well established by identification and dating of fossils of marine origin among the summit rocks.

Mountain-making in the primary arcs occurs in two ways: volcanism and tectonic activity. **Tectonic activity** consists of the bending and breaking of crustal rock through the action of internal earth forces. Some sections of the primary arcs are dominated by volcanism; other sections are dominated by tectonic activity and consist of masses of severely deformed strata.

Of equal interest to the alpine ranges are deep **oceanic trenches,** typically located adjacent to the mountain chains or to chains of volcanic islands, called **island arcs.** Trenches of the western Pacific Ocean are particularly striking (Figure 5.4). These oceanic trenches represent narrow zones of extreme crustal sinking. They have bottom

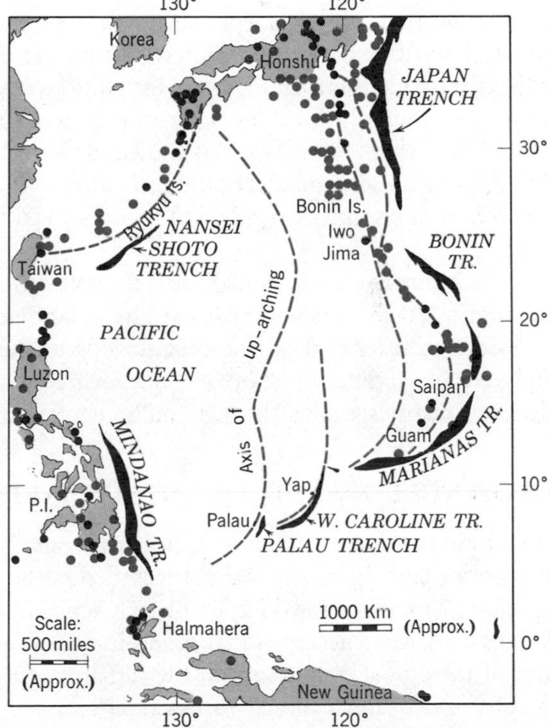

Figure 5.4 This map of the western Pacific Ocean shows trenches (*black*), island arcs (*dashed lines*), active volcanoes (*black dots*), and epicenters of deep-focus earthquakes (*color dots*). (© 1973, John Wiley & Sons, New York.)

Figure 5.5 The Peru-Chile Trench, off the west coast of South America. (Portion of *Physiographic Diagram of the South Atlantic Ocean,* 1961, by B. C. Heezen and M. Tharp, Boulder, Colo., Geol. Soc. Amer., reproduced by permission.)

depths on the order of 24,000 to 30,000 ft (7.5 to 10 km) below sea level. Equally impressive is the Peru-Chile Trench (Figure 5.5).

Because sources of sediment are quite limited in the vicinity of an oceanic trench, these depressions are only partly filled with sediment. The situation is quite different for belts of crustal depression adjacent to mountain ranges on the continents. Here the sinking zone is continually filled by sediment derived from stream erosion of the mountains. An example is the Indo-Gangetic plain of northern India, Bangladesh, and Pakistan. This sinking belt has been filled to a depth of thousands of feet by sediment derived from the Himalayan range.

In summary, each primary arc consists basically of an elevated mountain chain or island volcano chain bordered by a deeply depressed trench located on the oceanic side of the arc. Such tectonic and volcanic activity must involve deep-seated crustal processes. How do such features evolve through millions of years of geologic time?

GEOSYNCLINES AND MOUNTAIN-MAKING

Mountain-making belts have appeared at many times throughout the earth's past 3½ billion years of recorded geologic history. A typical episode of mountain-making involves a series of evolutionary stages. We can identify the individual stages in various parts of the primary arcs of the alpine system and in the remains of older belts long since deeply eroded and exposed to examination.

The earliest stages of mountain-making are represented by volcanic island arcs and trenches, such as those of the western Pacific. The upper block diagram of Figure 5.6 shows an island arc and its adjacent trench.

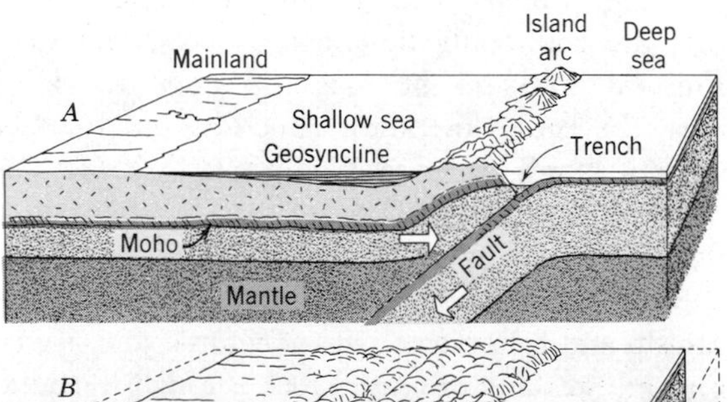

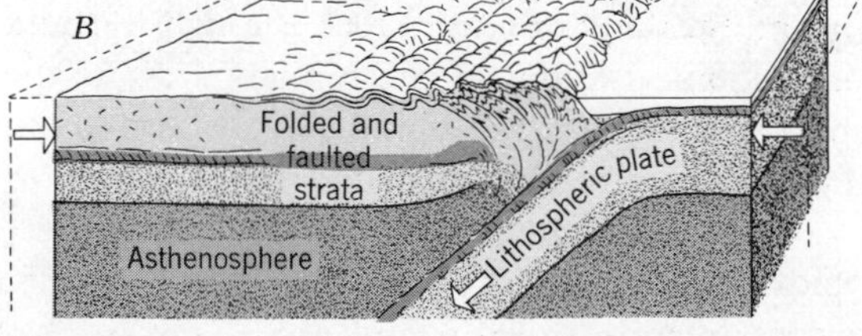

Figure 5.6 These block diagrams suggest the evolution of a volcanic island arc into a belt of intensely deformed rocks of the continental crust.

Notice that the rigid lithosphere has been fractured beneath the island arc. As a result, there exist two **lithospheric plates.** The plate at the right has been forced to move toward the adjacent plate on the left. The edge of right-hand plate has bent down sharply and is "diving" steeply into the soft asthenosphere. Note that the right-hand plate has the oceanic type of crust, which is thin and consists of basalt. The left-hand plate, in contrast, has thick continental crust with a granitic (felsic) upper zone. Thus, the fracture zone separates oceanic crust from continental crust.

As the downbent lithospheric plate slides into the asthenosphere it is intensely heated. We infer that the basalt is remelted into new magma bodies and these begin to rise. Let us suppose that magma differentiation then takes place (Chapter 1). The more mafic materials perhaps remain behind, while the more felsic materials continue to rise. The result is a magma of intermediate composition. When this magma reaches the surface, it produces andesite volcanoes and lava flows. As the andesite lavas accumulate, they first form a submarine mountain chain. Eventually, the volcanoes are built to a great height above sea level, becoming islands. Between the mainland and the island arc is a shallow sea underlain by continental crust.

Erosion of the rapidly forming volcanic islands produces large quantities of clastic sediment. Much of this sediment is clay, derived by weathering of the volcanic rock. This sediment is carried by streams into the adjacent coastal areas, then spread over the floor of the shallow sea by tidal and bottom currents. The sediment accumulates in extensive layers, later becoming consolidated into sedimentary rock strata.

The accumulated sedimentary strata of the shallow sea constitute a **geosyncline,** labeled in the upper block diagram of Figure 5.6. As sediment layers continue to accumulate, the geosyncline thickens and may reach a total accumulation of as much as 50,000 ft (15 km) in the central part. Under the weight of the accumulating sediment, the crust beneath the geosyncline sags into the shape of a shallow trough.

The period of geosynclinal sediment deposition comes to an end after several tens of millions of years. In a change of pace, the downbent lithospheric plate now butts strongly against the continental plate. Intense crustal compression sets in, crumpling the geosyncline and throwing the rock layers into wavelike corrugations, called **folds.** This event is suggested in the lower block diagram of Figure 5.6.

Some further details of the ensuing changes are shown in Figure 5.7. The belt of volcanic rocks remains active adjacent to the trench. Between this belt and the undisturbed rocks of the mainland the sedimentary strata of the geosyncline are intensely crumpled.

Under high pressures and temperatures, the shearing action of compression taking place deep in the zone of intense folding has altered much of the sedimentary rock to produce metamorphic rock. We will investigate this process in later paragraphs. As shown in Figure 5.7, the deeply buried metamorphic rock in the zone of maximum compression has melted. The resulting body of magma makes its way upward by melting, dissolving, and stoping the overlying rock. Upon cooling, this magma body has become a granite batholith.

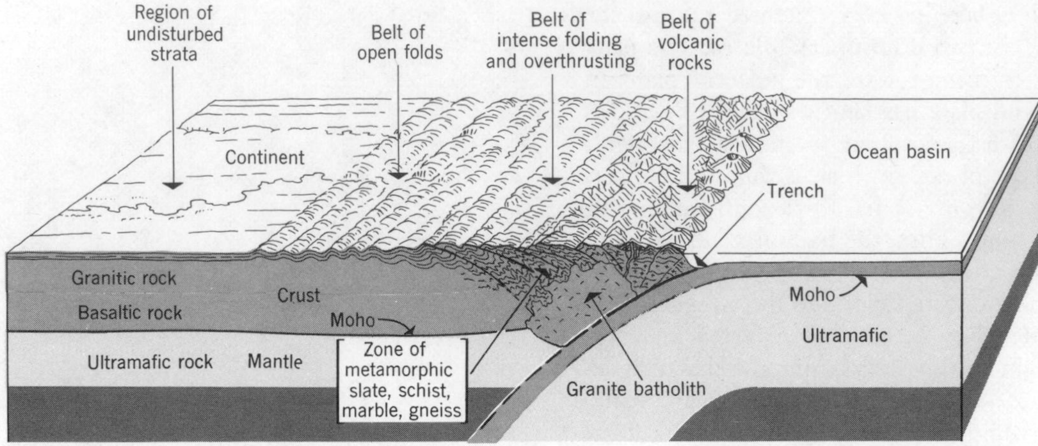

Figure 5.7 A mountain belt after folding, metamorphism, and intrusion have been completed.

OPEN FOLDS OF SEDIMENTARY STRATA

Closer to the continent, the thin layer of sedimentary strata has been thrown by compressional forces into a belt of open, wavelike folds. This is a shallow effect compared with the metamorphism and intrusion taking place deep within the belt of intense tectonic activity. Fold belts are typically located in positions on the continental-interior side of belts of more intensive tectonic activity.

The geologist refers to a troughlike downfold as a **syncline,** and to a crestlike upfold as an **anticline.** Figure 5.8 is a block diagram of a geologically famous region of simple open folds, the Jura Mountains, bordering the European Alps along the northwest side. Notice that anticlines of limestone form ridge crests, while synclines form the intervening valleys. A river has cut a winding gorge across the mountains, exposing the strata to view in each anticline. Figure 5.9 is a ground view of an anticline cut through by a river.

ALPINE-TYPE FOLDING OF STRATA

Referring to Figure 5.7, notice that there is a belt of intense folding next to the belt of open folds. Here the strata are not only intensely

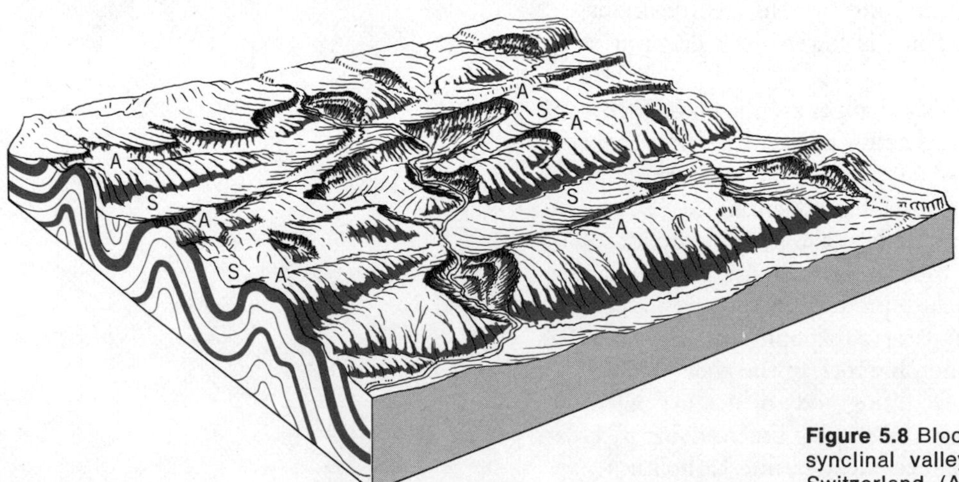

Figure 5.8 Block diagram of anticlinal ridges (A) and synclinal valleys (S). Jura Mountains, France and Switzerland. (After Erwin Raisz.)

Figure 5.9 This great anticline of sandstone strata has been cut through by a river, revealing the arched strata on both sides of the gap. The locality is just north of Cape Town, South Africa. (Douglas Johnson.)

crumpled, but also fractured into slices. The term **fault** is applied by geologists to a sharp rock break along which there has been a slippage movement. The process of breakage and slippage is referred to as **faulting.** Figure 5.7 shows that faulting occurred in a series of slices, and that each slice rode over the slice beneath. This kind of movement is referred to as **overthrust faulting.** The fault itself is known as an **overthrust fault,** or simply a thrust fault.

Figure 5.10 shows how strata are folded in an alpine mountain region, and how the process of overthrusting is also involved. Marine sedimentary strata of a geosyncline have been crumpled into tightly compressed folds. First, the anticlines were rotated forward, becoming **overturned folds.** Overturning of anticlines continued, producing **recumbent folds.** The recumbent folds then developed low-angle overthrust faults upon which they slid forward as **thrust sheets.** A single thrust sheet may have a width of many miles in the direction of movement.

The European Alps have long been the object of intensive geologic study because of the extreme complexity of their folded and thrust-faulted structures, revealed in striking exposures in the steep mountain walls. Thin thrust slices, resting one upon the other, are illustrated in Figure 5.11. A thrust fault may escape notice unless a detailed examination is made of the strata above and below the thrust plane. If fossils in the overlying rock are of an older geologic period than those below, an overthrust is clearly indicated. A fine example

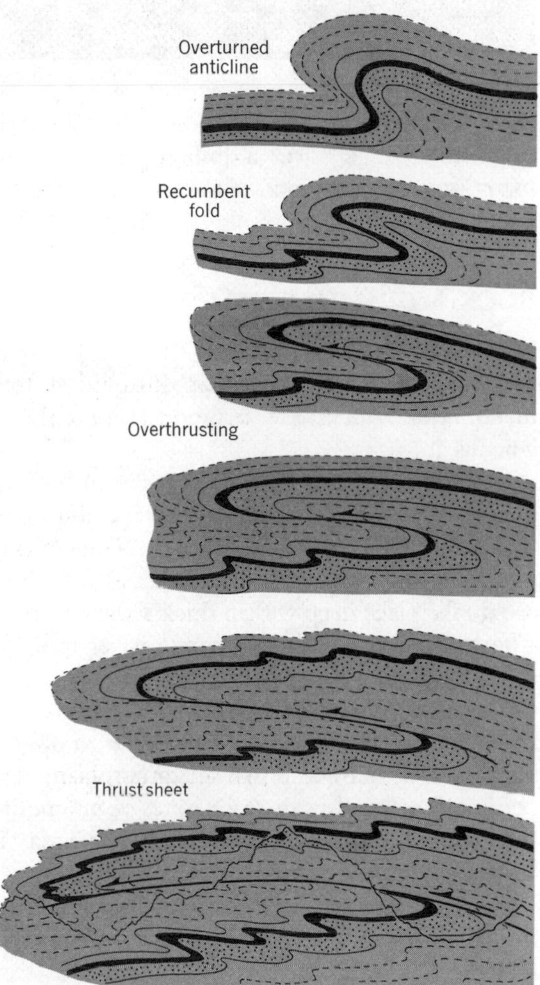

Figure 5.10 Development of overturned and recumbent folds and thrust sheets.

Figure 5.11 A cross section through a portion of the Helvetian Alps, Switzerland, shows thrust slices. Horizontal and vertical scales are the same.

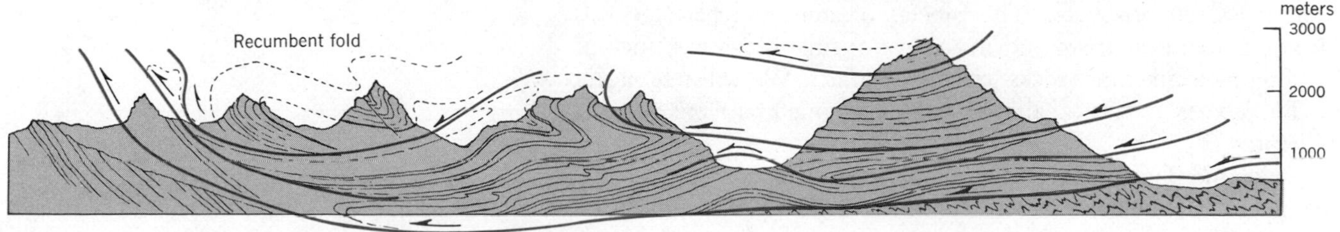

Figure 5.12 The Lewis overthrust fault is marked by the light-colored slanting line about midway up the mountainside (*arrow*). The rock mass above the thrust moved from left to right. Northern Rocky Mountains, Montana. (Douglas Johnson.)

is the Lewis overthrust in Montana (Figure 5.12). Here rocks of Precambrian age (over a billion years old) were thrust eastward many miles over much younger weak shales of Cretaceous age (about 70 million years old).

ROCK METAMORPHISM

We have delayed introducing the third major class of rocks—the **metamorphic rocks**—until we introduced tectonic activity, since metamorphic rocks are byproducts of either tectonic activity or igneous intrusion.

Metamorphism may affect rocks of both igneous and sedimentary origins. The changes can occur during mountain-making as unequal stresses are applied to rock. These changes occur under conditions of high confining pressure and high temperature; they most often take place deep within thick sedimentary strata of geosynclines. The total process of rock change is referred to as **dynamothermal metamorphism;** it is felt in two ways. First, original minerals recrystallize and new minerals are formed. Second, a new set of structures is imposed on the rock and may replace or obliterate original bedding structures. Dynamothermal metamorphism has affected enormous bodies of rock within the root zones of mountain chains of the alpine type. Consequently, the effects are seen today in surface rocks over large areas.

Application of high temperatures alone can also cause metamorphism. Such effects are often conspicuous in country rock close to an igneous intrusion. The changes are essentially those of baking in a high-temperature oven. A shale rock close to an igneous contact may experience a hardening and color change not unlike that caused by baking of brick or tile. However, most large igneous intrusions cause **contact metamorphism,** which consists of changes caused by emanations of hot watery solutions containing many different kinds of ions. These are highly active solutions and cause mineral alteration of the country rock. The process of mineral replacement is known as **metasomatism** and may leave original structures, such as bedding in sedimentary rocks, essentially intact. We will refer again to this process as one capable of producing important ore deposits (Chapter 11).

Metamorphism is a change of mineral state in response to a change in environment. Many common igneous and sedimentary minerals are unsuited to an environment of deformation under stress at high pressures and/or high temperatures. They will be altered to form minerals capable of attaining equilibrium under those environments. There will often be changes in grain size and shape as well.

Many of the most common and abundant minerals of metamorphic rocks are also abundant in igneous and sedimentary rocks. These have been described in Chapters 1 and 4. Quartz, one of the most abundant of minerals in rocks of all classes, persists unaltered in metamorphic rocks. Quartz may also be recrystallized or introduced by invading solutions. Feldspars are commonly produced by contact metamorphism. Hornblende, olivine, biotite, and muscovite are commonly formed during metamorphism from the constituents of clay-rich sediments. Calcite usually persists from the original carbonate rock but easily recrystallizes under pressure. Dolomite also tends to persist from original sedimentary rocks.

Figure 5.13 Garnet crystals in schist. The larger crystal is nearly 1 in. (2 cm) across. (A. N. Strahler.)

Among the more important new minerals distinctive of dynamothermal metamorphism, many are aluminosilicates. As a group, they are hard minerals with densities comparable to those of the mafic minerals. Crystals of distinctive form grow in the metamorphic rock and are often easily recognized.

An example of a metamorphic mineral of aluminosilicate composition is garnet. Most of us know garnet as a semiprecious gemstone of red color. Figure 5.13 shows beautifully formed crystals of garnet embedded in a metamorphic rock rich in mica. Very large garnet crystals of poor quality, and of no value as gems, were once extensively mined for use as an abrasive powder on sandpaper. This use illustrates the extreme hardness of garnet.

METAMORPHIC ROCKS

It has been common practice to subdivide the metamorphic rocks into two groups. One group consists of rocks with obvious parallel structures which appear as lines on the rock surface. One expression of linear structure is **foliation,** a crude layering along which the rock easily separates. A second structure is **banding,** a layered arrangement of strongly knit crystals forming a massive rock. Another group of metamorphic rocks lacks obvious parallel structures, but is characterized by granular texture.

Slate is a very fine-grained rock that splits readily into smooth-surfaced sheets along cleavage surfaces. Slate is largely derived from fine-grained, clay-rich marine clastic sedimentary rocks (shale, claystone). The cleavage of slate is a new structure imposed by metamorphism and usually cuts across the original bedding. Slate colors range from gray to green to red.

Schist is a foliated rock and comes in many varieties. Foliation results from the parallel alignment of easily cleavable minerals such as mica. The reflecting surface of these minerals give a characteristic glistening sheen to the foliation surfaces (Figure 5.14). Schists have undergone a high degree of metamorphism and their origin is not always clear. Most schists are interpreted as altered clastic sedimentary strata rich in aluminosilicate minerals. It is commonly inferred that slates represent an intermediate grade of metamorphism between shale and schist. This sequence is actually demonstrated

Figure 5.14 This fragment of mica schist, about 6 in. (15 cm) long, shows a glistening, undulating surface of natural parting (*upper*). An edgewise view (*lower*) shows the thin foliation planes. (A. N. Strahler.)

in some localities by tracing the changes continuously from shale, through slate, to schist.

Basaltic lava flows subjected to dynamothermal metamorphism yield a foliated rock of dark greenish color that has long been known to geologists as **greenstone.**

Gneiss, a metamorphic rock showing banding, requires subdivision into gneisses of different origins. Certain granite bodies show an elongation of crystals into streaklike or pencil-like forms, suggestive of flowage of the granite in its final stages of solidification. Such a rock is often termed a **granite gneiss.**

Certain banded rocks consist of alternate layers of foliated rock and granular rock. The granular layers may be granitic or composed of quartz and feldspar (Figure 5.15). Where the rock clearly consists both of schist layers and igneouslike layers, the rock is presumed to have resulted from the injection of igneous components by solutions penetrating the schist layers. Such rocks are described as **injection gneisses.** The bands may be contorted into small folds (Figure 5.16). Injection gneisses can often be traced into masses of pure granite, suggesting that the invading granite magma has assimilated the country rock.

Figure 5.15 Close-up view of a granular banded gneiss. The lighter bands are rich in quartz and feldspar; the darker bands are rich in hornblende and biotite. (A. N. Strahler.)

Of the granular metamorphic rocks, the most widespread are metamorphosed from sedimentary rocks. Pure quartz sandstone undergoes minor physical change and virtually no chemical change when subjected to the same tectonic process that produces slates and schists. Under extreme pressure, the quartz grains are crushed and forced into closer contact. Strongly cemented by silica, this process results in a **metaquartzite,** one of the hardest and most durable rocks known.

Limestone and dolomite are metamorphosed into **marble,** which is typically a light-colored granular rock exhibiting a sugary texture on a freshly broken surface. Although white when pure, marbles come in many colors, depending upon the presence of impurities.

Metamorphic rocks are found today over wide areas of the continental crust. They represent the root structures of intensely folded geosynclines from which many thousands of feet of overlying rock have been uncovered. Banding and foliation of these rocks, along with the orientation of folds within them, can be interpreted to identify ancient tectonic belts. In this way it is possible to reconstruct stages in the growth of the continents.

THE ROCK CYCLE IN REVIEW

This is a good place to summarize the cycle of rock transformation, or, simply, the rock cycle. Figure 5.17 is a triangular diagram showing that any one of the three major rock classes—igneous, sedimentary, and metamorphic—can be derived from either of the other two classes. Sequences of changes are labeled on the sides of the triangle.

Let us relate the rock cycle to the contrasting physical-chemical environments found at depth within the earth and at the earth's surface (Figure 5.18). The completed diagram now represents a schematic vertical cross section of the crust, say to a depth of about 20 mi (30 km). Throughout the rock cycle, large masses must be moved from the **deep environment** of high temperatures and pres-

Figure 5.16 This banded gneiss, over 2 billion years old, is exposed on the east coast of Hudson Bay, south of Povungnituk, Quebec. (Photograph G.S.C. No. 125221 by F. C. Taylor, Geological Survey of Canada, Ottawa.)

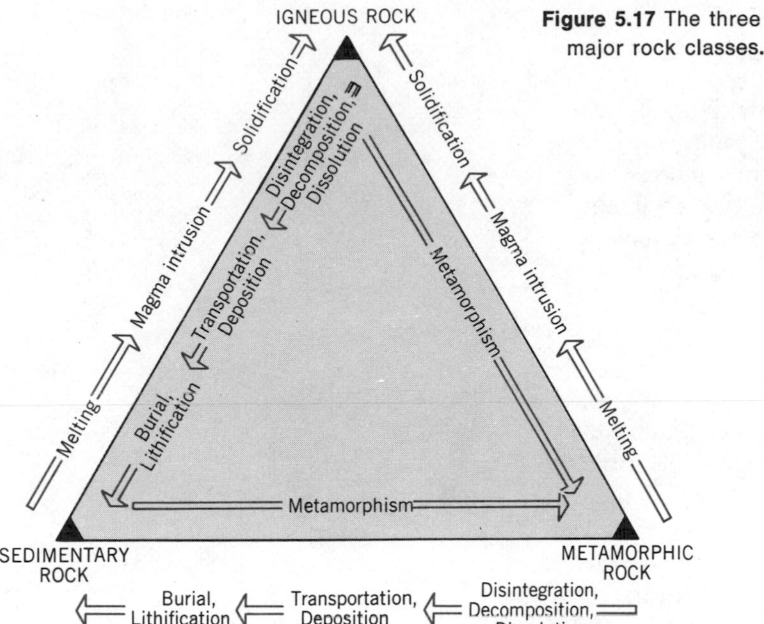

Figure 5.17 The three major rock classes.

sures to the **surface environment** of low temperatures and pressures. We see that this change of environment can be accomplished by two processes: (1) Rising magma may bring igneous material to various intermediate positions, where it solidifies into plutonic rock bodies, or by extrusion it may reach the surface to form volcanic rocks. (2) Rock formed at depth may appear at the surface of the earth by uncovering, as a result of the combined processes of crustal uplift and denudation. Thus large bodies of igneous, metamorphic, or sedimentary rock can migrate upward from the deep environment to the surface environment.

Transition from the surface environment to the deep environment can be accomplished by burial and sinking of the earth's crust.

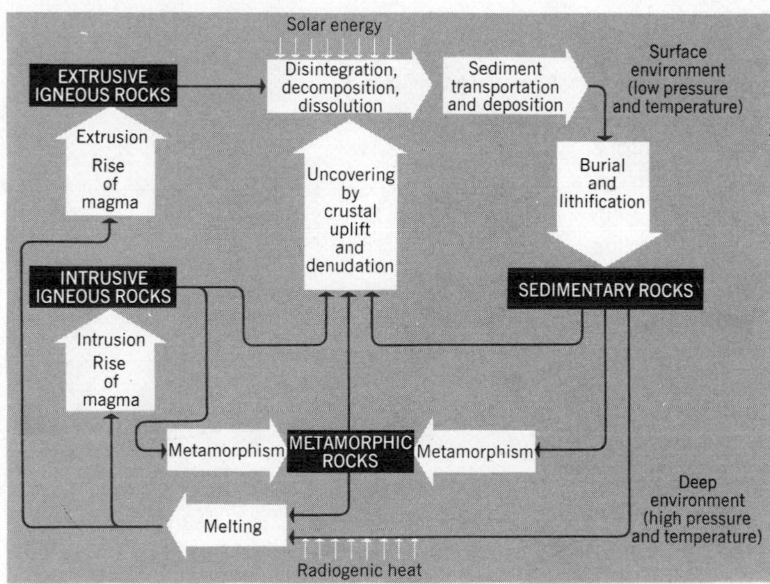

Figure 5.18 The rock transformation cycle.

Both sedimentary strata and extrusive volcanic rocks can eventually reach the deep environment, where either metamorphism or remelting can take place.

The large quantities of energy required to operate the rock cycle are derived from both the internal geologic system of radiogenic heat and the system of solar energy acting through external processes of the atmosphere and hydrosphere. This linkage of two great and unlike systems is the key to understanding of the earth's unique surface environment.

TECTONIC EVENTS IN THE SCALE OF GEOLOGIC TIME

It is necessary to refer the events of intrusion, extrusion, sedimentation, and tectonic activity to an established geologic time scale. Without a time scale we cannot synchronize the history of these events from place to place over the earth.

The search for a means by which to establish the age in years of an event in the earth's past history was for decades frustrating and misleading. The dilemma over the age of the earth and the duration of periods of geologic time was solved with the discovery of radioactivity.

RADIOMETRIC AGE DETERMINATION

Basic principles of spontaneous decay of radioactive isotopes, explained in Chapter 2, provide us with a method of determining the age of an igneous rock, a procedure of science known as **geochronometry.** Ages thus determined are referred to as **radiometric ages.**

At the time of solidification of an igneous rock from its liquid state, minute amounts of minerals containing radioactive isotopes are trapped within the crystal lattices of the common rock-forming minerals, in some cases forming distinctive radioactive minerals. At this initial point in time there are present none of the stable daughter products that constitute the end of the decay series. However, as time passes the stable end member of each series is produced at a constant rate and accumulates in place.

Each radioactive isotope has its own rate of decay, which is absolutely constant. Suppose we start with a given quantity of the isotope. After the passage of a certain interval of time, the isotope will be reduced to one-half the initial quantity, as shown in Figure 5.19. This time interval is called the **half-life.** For example, the half-life of uranium-238 is 4½ b.y. In that period of time the starting quantity of uranium will have been reduced by one-half, and an equal quantity of the stable daughter product, lead-206, will have been produced. (By quantity we mean the number of atoms in the sample.)

Knowing the half-life of the decay system, we can estimate closely the time elapsed since mineral crystallization occurred. An accurate chemical determination of the ratio between the radioactive isotope and the stable daughter product must be made. A fairly simple mathematical equation is used to derive the age in years of the mineral under analysis.

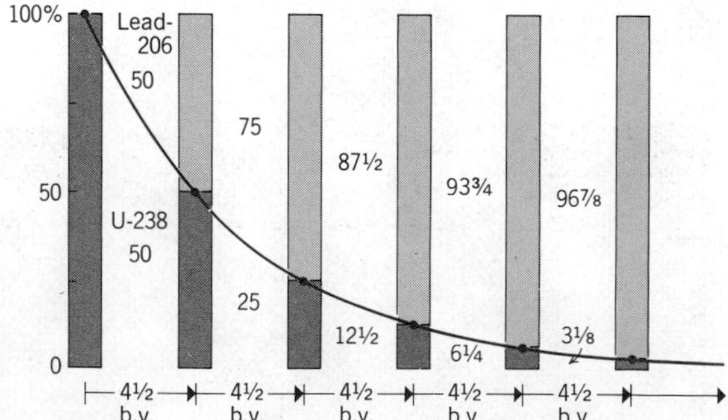

Figure 5.19 As the amount of U-238 decreases with time, the amount of lead-206 increases. The smooth line is the decay curve.

For example, in the uranium-lead series, quantities of both uranium-238 and lead-206 are measured in a mineral sample. The ratio of lead to uranium is entered into the equation, which is then easily solved for age in years.

The radiometric ages given for various events in the timetable of the earth's history are now accepted by geologists as valid within small percentages of error. The radiometric age determination of rocks stands as a strikingly successful application of principles of chemistry to geology.

Radiometric ages are now assigned to the divisions of a scale of geologic time (see Table 7.1). Geologic events older than about 600 million years (m.y.), are referred to as **Precambrian time.** This vast and obscure block of time, about 3 b.y. in duration, holds the history of evolution of the continental crust.

EVOLUTION OF THE CONTINENTAL CRUST

The origin of the continents has long been a major problem of geology. Most modern thought favors the supposition that the continental crust was not present when the accretion of the earth was completed, some 4½ billion years ago. Instead, the continents were formed later of rock of felsic mineral composition gradually segregated from an original crustal rock of mafic composition. Perhaps this original rock was similar to basalt of the present oceanic crust. Some support for this inference lies in the fact that among the oldest known rocks of the continental crust there is found an abundance of greenstone. Recall that greenstone is a metamorphosed volcanic rock of basaltic composition.

Our scenario of continental evolution requires that small masses of granitic rock were added to the crust each time an episode of mountain-making was completed. Thus, the continents grew in size throughout the early part of geologic time.

CONTINENTAL SHIELDS

If the hypothesis of gradual continental accretion is valid we should find that continental interiors are composed largely of metamorphic and intrusive igneous rocks of great geologic age. Such rocks represent the roots of mountain ranges produced in a succession of tectonic events.

The continental interiors are indeed formed of such rocks, and are known as **continental shields.** Figure 5.20 shows the distribution of shields in the Northern Hemisphere. These are areas largely of rock of Precambrian age. Most of the rock is older than 1 b.y. Younger strata form thin covers over large areas of the shields.

Younger mountain belts surround the shields along the continental margins. The marginal position of these younger belts, together with the bordering primary arcs of present-day tectonic activity, suggest that new material has been added to the continental crust by successive mountain-making episodes.

If the continents developed by growth throughout Precambrian time, we should look to the distributions of rock ages in the continental shields. A pattern suggestive of growth stages emerges. The

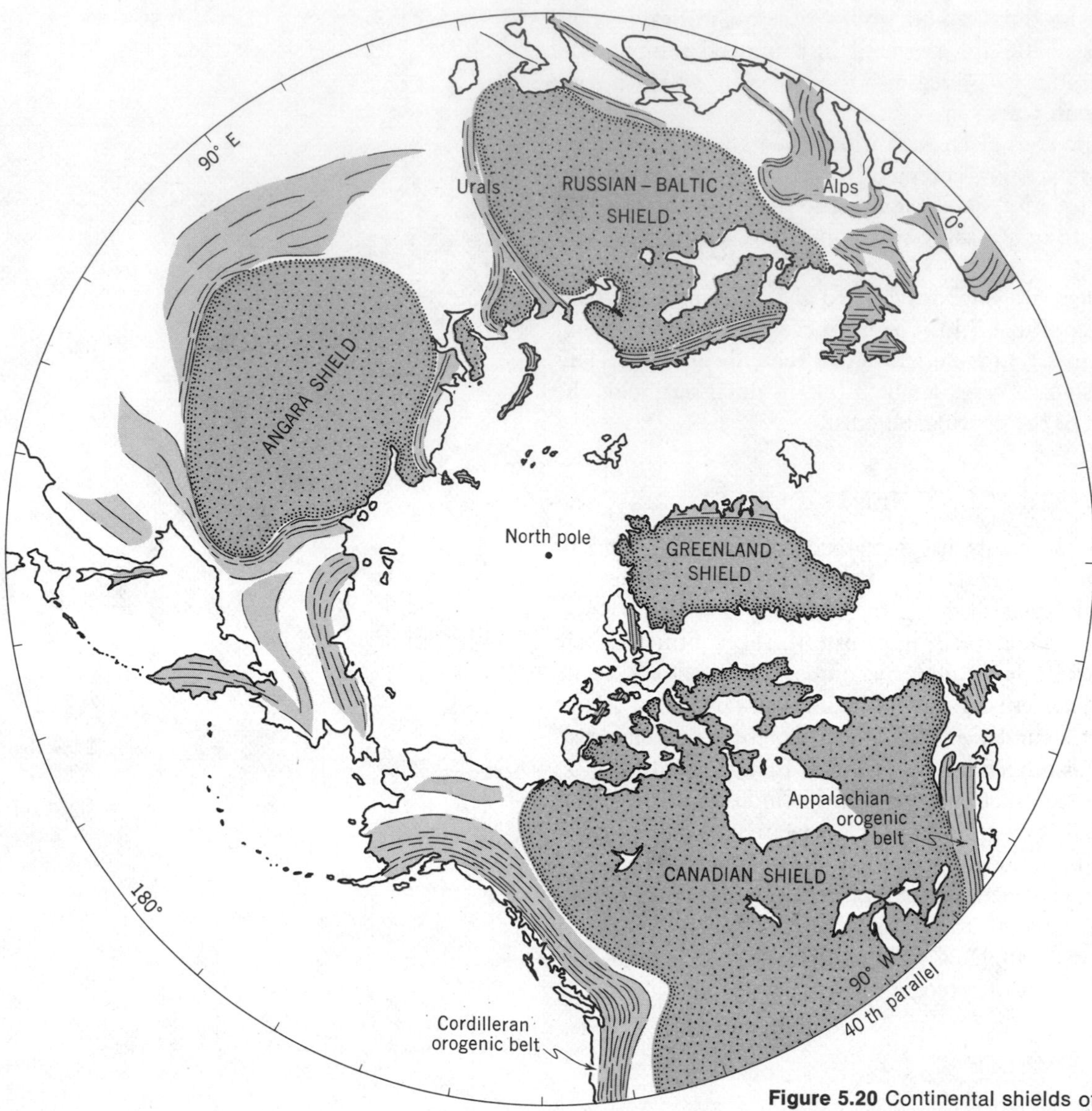

Figure 5.20 Continental shields of the Northern Hemisphere and their bordering mountain arcs. (© 1973, John Wiley & Sons, New York.)

oldest rocks of the shields, older than about 2½ b.y., make up relatively small patches of shield and are designated as **continental nuclei** (see Figure 9.10).

The continental nuclei are surrounded by or are contiguous to larger areas of shield rock with maximum ages falling in middle and upper Precambrian time. This arrangement adds strength to the hypothesis of continental evolution by accretion.

We shall continue the narrative of crustal evolution in Chapter 8. A closer look at the ocean basins will show many features completely different from the continents.

MOUNTAIN-MAKING IN REVIEW

In trying to understand the deformation of strata and the making of mountain ranges geologists have followed the doctrine of uniformitarianism. Mountain-making must have occurred over and over in much the same pattern through nearly 4 b.y. The chain of events typically involves accumulation of sediment in a geosyncline, followed by compression and severe crumpling of those strata. In a pattern of crustal configuration repeated many times in the past, there occurs a major break in the lithosphere. The lithospheric plate on the oceanic side of the break plunges down steeply into the asthenosphere, giving rise first to andesite magma and formation of a volcanic island arc. Renewed movement of this plate against its neighbor continental plate ends the period of geosynclinal sedimentation, deforming the strata into alpine folds and thrust slices. The root zone of the mountains is transformed into metamorphic rock, and there are intrusions of granite batholiths as well.

Radioactivity has made possible the dating of such events and has shown that the continental shields have evolved through geologic time from small nuclei. The rock cycle has been repeated with each period of mountain-making, so that mineral matter of the crust has been recycled many times in the past 4 billion years of earth history.

YOUR GEOSCIENCE VOCABULARY

catastrophists
uniformitarianism
primary arcs
Eurasian-Melanesian belt
alpine system
tectonic activity
oceanic trench
island arc
lithospheric plate
geosyncline
folds
syncline
anticline
fault
faulting
overthrust faulting
overthrust fault
overturned fold
recumbent fold
thrust sheet
metamorphic rock
metamorphism; dynamothermal, contact
metasomatism
foliation
banding
slate
schist
greenstone
gneiss
granite gneiss
injection gneiss
metaquartzite
marble
deep environment
surface environment
geochronometry
radiometric age
half-life
Precambrian time
continental shields
continental nuclei

SELF-TESTING QUESTIONS

1. In what way does the doctrine of catastrophism differ from that of uniformitarianism? Give an example of how uniformitarianism can be applied to a particular geologic problem.

2. Describe the world distribution of the primary arcs. How is volcanism related to these arcs? How are oceanic trenches related to these arcs?

3. Review the stages in development of a mountain belt. In what way are lithospheric plates related to these stages? Include a description of geosynclinal deposition At what stage does alpine-type deformation set in? What pattern of igneous extrusion and intrusion is associated with mountain-making?

4. Describe the development of alpine-type folds and thrust slices. How might the plane of a thrust fault be identified in a rock exposure on a mountainside?

5. Describe rock metamorphism. How does the adjective dynamothermal apply to this process? How does contact metamorphism occur? Compare minerals of metamorphic rocks with those of igneous rocks.

6. What two groups of metamorphic rocks are recognized in terms of structure? Describe slate, schist, metaquartzite, and marble. From what rocks are they derived? What varieties of gneiss are found, and what is the significance of each?

7. Describe the complete cycle of rock transformations and illustrate by means of a triangular diagram. Which of the rock types represents the original crustal rock? Relate rock types to environments of temperature and pressure.

8. How is absolute age of a rock determined? What age in billions of years before the present is spanned by Precambrian time?

9. How and when did the continental crust originate? What are the continental shields? Describe the continental nuclei. How do ages and arrangements of shield rocks suggest the pattern of continental evolution?

(*Above*) A tsunami off the coast of Japan as depicted by the nineteenth century print maker, Hokusai. The crest of a great wave towers over the terrified fishermen in their open boats. Far in the distance is a familiar landmark—Fujiyama. (Metropolitan Museum of Art, bequest of Mrs. H. O. Havemeyer, 1929. The H. O. Havemeyer Collection.)

(*Below*) This raging surf was set up by the arrival of the second wave of the 1946 tsunami at Kawela Bay on the north coast of Oahu, Hawaii. Normally the body of water in this view is a quiet lagoon behind a sheltering coral reef. (F. P. Shepard, Scripps Institution of Oceanography.)

6 FAULTING AND EARTHQUAKES

A DEADLY TSUNAMI

Disaster was on its way to the shores of Hawaii during the early morning hours of April 1, 1946, but no one there knew it. Among its potential victims was a noted geologist, Professor Francis Shepard, an authority on the geological features of the oceans. Shepard was staying at a remote coastal locality on the north shore of the island of Oahu, working on a new textbook of submarine geology. The disaster was the dreaded tsunami, or seismic sea wave, which many times before had brought death and destruction to island and continental shores of the Pacific Ocean. Set off at 2:00 AM by a major earthquake close to the Aleutian Islands, the tsunami was moving southward at a speed of nearly 500 miles an hour, headed directly for the Hawaiian Islands. Five hours later, as the wave came ashore on Oahu, Shepard and his wife had to run for their lives to escape the rising sea water and its angry surf.

The tsunami made itself felt as a rise and fall of sea level, repeated several times at 12-minute intervals. During the first rise, the ocean surface easily overtopped the protective barrier reef of coral lying offshore. Advancing as a surging water mass tossed by breaking surf, the first wave inundated the coastal land up to a height of 13 feet above tide level. The water then retreated, but was followed minutes later by an even higher landward surge, this time reaching 17 feet above tide level. Professor Shepard was able to photograph this second surge from a position of safety. A third, even higher, surge followed the second.

In the city of Hilo, on the northern shore of the island of Hawaii, the rising water inundated a large stretch of low-lying coast, with waves reaching a height of 30 feet above sea level. A large section of the city was destroyed, as houses were floated from their foundations and swept inland. Some buildings were crushed into tangled masses of debris, others floated almost intact to new locations. The tsunami breached the protective breakwater of Hilo Bay, lifting and dropping blocks of rock weighing more than 8 tons. Many small boats were carried inland and damaged. The force of the incoming water was particularly strong in the channel of the Wailuku River, a stream entering Hilo Bay. Here an entire steel span of a railroad bridge was ripped loose and carried 750 feet upstream.

The sudden rise of water and its pounding surf caught many of Hawaii's coastal inhabitants by surprise, and many were drowned. In Hilo alone, 83 persons died, while the total for the islands was over 170 dead. Hundreds of homes were demolished or damaged, and the total loss came to $10 million.

In the words of a U.S. government account of the disaster, "The tsunami of April 1946 is distinguished from the rest: it was the worst natural disaster in Hawaii's history; and the last destructive tsunami to surprise those islands." In 1948, the Seismic Sea-Wave Warning System (SSWWS) was put into operation with headquarters at Honolulu. Because most tsunamis are generated by distant earthquakes, advance warning is given by the earthquake waves sent out at the time the earthquake takes place. These waves travel swiftly through rock of the crust and mantle, and are picked up by sensitive recording instruments at the warning system observatories. Ample time is available to alert people living on low ground close to the shore. How did the system pay off? In 1952, a major tsunami, generated near the Kamchatka Peninsula, reached Hawaii. The damage it inflicted came to $800,000, the death toll was zero. The SSWWS had paid its way. The warning system was not entirely successful in eliminating casualties in later tsunamis. A tsunami in 1960 resulted in 61 deaths in Hawaii.

In this chapter we investigate earthquakes and earthquake waves. Quite apart from their role as major hazards to man and his cities, earthquakes are valuable tools of earth science. Much of what we know of the earth's crust, mantle, and core has been interpreted from the wave records of innumerable earthquakes, both weak and strong.

EARTHQUAKES AND FAULTS

The science of **seismology**, a branch of geophysics, deals with earthquake waves. The earthquake is known to humans directly as a trembling or shaking of the ground. Commonly the motion is barely perceptible to the senses. On occasion the motion is so violent as to collapse strong buildings, sever water and gas mains, open gaping cracks in the ground and thus bring great loss of life and property.

Earthquake waves, or **seismic waves,** are recorded by the **seismograph,** a sensitive instrument that can detect earthquakes thousands of miles distant. A seismograph can measure small vibrations that could not possibly be recognized by the human senses. These records of seismic waves provide the evidence we seek about the earth's interior regions.

To understand how earthquake waves give evidence about the earth's interior requires that we first learn how earthquakes are generated. The vast majority of important earthquakes that can be analyzed at long distances from their sources are produced by faulting in the earth's crust. This phenomenon is simply a sudden slippage between two rock masses separated by a fracture surface. We see small-scale demonstrations of faulting in common materials such as dry soil, concrete, or rock. These brittle materials fracture when compressed or when the underlying support is removed—examples of this phenomenon are cracks in pavements, sidewalks, or masonry walls.

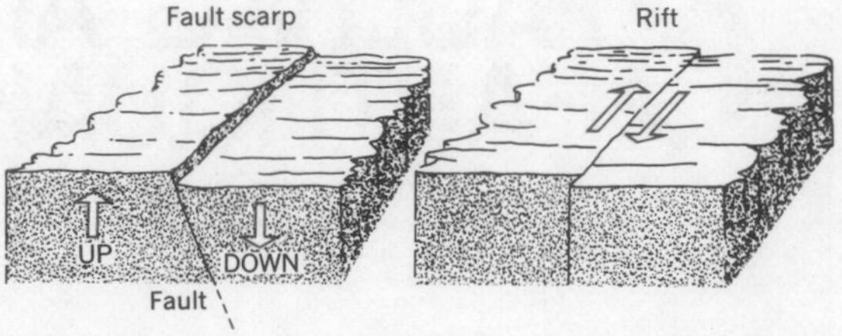

Figure 6.1 A normal fault (*left*) and a transcurrent fault (*right*).

KINDS OF FAULTS

The fracture surface upon which slippage occurs is called the **fault plane;** it may take any orientation with respect to the horizontal. Movement occurs by a series of slips, each involving but a few inches to a few feet of displacement and occurring almost instantaneously.

Figure 6.2 This fault scarp was produced instantaneously during the Hebgen Lake, Montana, earthquake of August 1959. (Irving J. Whitkind, U.S. Geological Survey.)

Each slip generates an earthquake shock. The displacement of corresponding points on two sides of the fault may range from a few inches, in very small faults, to several tens of miles in certain great faults, accumulated over spans of thousands of years. It is by such long-continued faulting that small and large blocks of the earth's crust are displaced with respect to one another, bringing major landscape features into existence.

Figure 6.1 illustrates two basic types of faults. In the **normal fault,** the fault plane is steeply inclined in the direction of the downthrown block. Normal faulting produces a **fault scarp** (Figure 6.2). A scarp is simply a clifflike feature where the ground surface undergoes an abrupt change in level.

The **transcurrent fault,** also shown in Figure 6.1, is characterized by displacement only in the horizontal direction, along a near-vertical fault plane. Thus one block slides past the other. If transcurrent faulting occurs on a plain of very low relief, as the diagram suggests, no topographic scarp will result. In hilly terrain a discontinuous narrow trench, or **rift,** marks the line of the fault (see Figure 6.19).

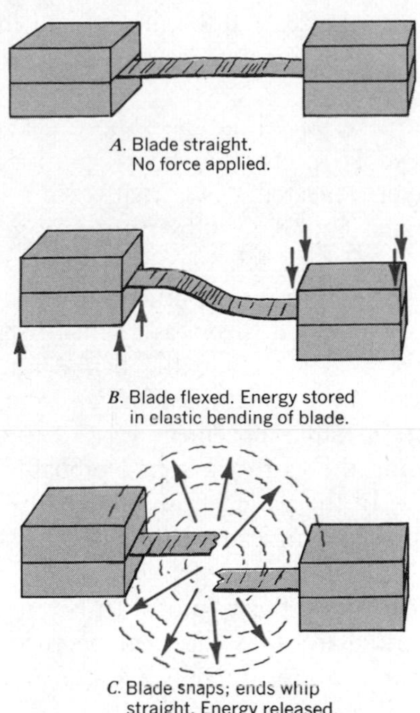

A. Blade straight.
No force applied.
B. Blade flexed. Energy stored in elastic bending of blade.
C. Blade snaps; ends whip straight. Energy released. Sound waves sent out.

Figure 6.3 A steel blade, bent until it snaps, illustrates certain basic features of the earthquake mechanism.

HOW EARTHQUAKES ARE GENERATED

Hard rocks of the earth's crust and upper mantle—the lithosphere—are both strong and brittle. However, this rock is subjected to enormous deforming stresses when one lithospheric plate is forced down beneath another. Under such stress, rock actually bends elastically like steel. The amount of bending is scarcely detectable in small masses. Despite its ability to withstand great stress with only slight bending, or **strain,** a given rock has an **elastic limit.** If it is strained beyond this limit, a fault is formed and the bent rock snaps suddenly back to its normal shape. An earthquake is the disturbance set off by the sudden release of elastic strain. Like a bow slowly bent, the rocks have gradually accumulated energy, only to release it with great suddenness.

A simple model of the mechanism of earthquakes can be made with a strip of tempered steel, such as a coping saw blade, the ends tightly clamped in wooden blocks (Figure 6.3). If the blocks are forced to move parallel with one another to produce an S-bend in the blade, the blade can be bent slowly to the breaking point. When the blade snaps, the broken ends whip back into straight pieces, but the ends are now considerably offset. We are aware of the energy release through a train of sound waves sent out from the broken ends of the blade.

The same model can be applied on a more realistic basis to a common form of earthquake, illustrated in Figure 6.4. Here we are looking down upon a small square of the earth, perhaps one mile on a side. Imagine that in prehistoric time a line (*AB*) has been drawn on the surface. Lateral forces acting in opposite directions along the two sides of the block have gradually bent the rock in such a way as to deform the straight line into an S-shaped bend, as shown in the second block. In modern times a railroad or fence is built straight across the deformed area, the presence of strain being unknown. In the third block, the fault movement has occurred, releasing the elastic

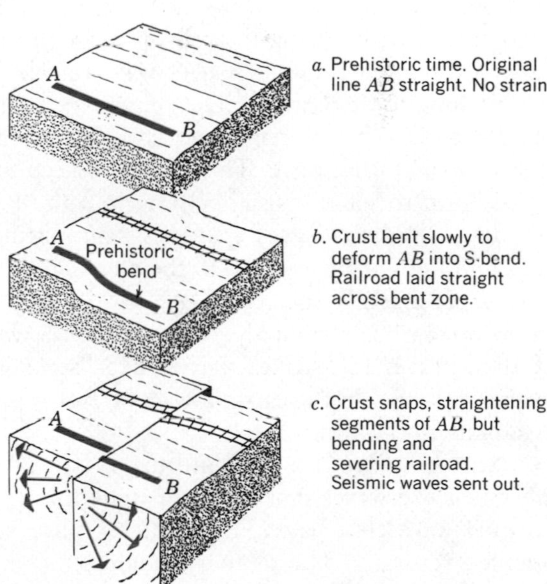

A
B
a. Prehistoric time. Original line *AB* straight. No strain.
A
Prehistoric bend
B
b. Crust bent slowly to deform *AB* into S-bend. Railroad laid straight across bent zone.
A
B
c. Crust snaps, straightening segments of *AB*, but bending and severing railroad. Seismic waves sent out.

Figure 6.4 An earthquake results from the sudden release of elastic strain that has been accumulated in rock over a long period of time.

strain and causing the track to be broken and its ends to assume a bent plan. The original reference line, *AB*, is again restored to straightness.

The earthquake model outlined above is known as the **elastic-rebound theory.** To prove the theory, precision ground measurements were taken repeatedly over many years on both sides of a known earthquake fault line. Observations were also made of the offsetting and bending of reference lines, such as fences and roads (Figure 6.5). The evidence established beyond a doubt the correctness of the elastic-rebound theory as applied to the great San Francisco earthquake of 1906 along the San Andreas Fault, a major transcurrent fault, in northern California. Here it was ascertained that the maximum fault movement was 20 ft (6.4 m). The slow bending preceding the earthquake had probably been going on for a hundred years. In the period from 1851 to 1906, geodetic surveys actually measured a significant deformation of the rock preceding the slippage.

Figure 6.5 This road in the Santa Cruz Mountains of California was offset by fault movement during the San Francisco earthquake of April 18, 1906. Lateral displacement was about 4 ft (1.2 m). (Sketched from a photograph by E. P. Carey.)

The elastic-rebound theory probably does not apply to all major earthquakes, particularly those at great depth, where conditions of pressure and temperature are unlike those at the earth's surface. Nevertheless, the general principle of the sudden release of stored energy of rock strain appears to be behind all major earthquakes.

EARTHQUAKE WAVES

Earthquakes are generated in the earth's crust and outer mantle down to depths as great as 400 mi (640 km). The point of slippage and energy release is known as the **focus,** while the ground-surface point directly above the focus is the **epicenter.** Energy is dispersed from the focus in the form of seismic waves.

There are three basic seismic wave forms: primary waves, secondary waves, and surface waves. **Primary waves** show the same kind of motion as observed in sound waves. As illustrated in Figure 6.6, particles transmitting the primary wave form move forward and backward only in the direction of wave travel. This motion constitutes a **longitudinal wave.** The primary wave is commonly called the **P-wave.** We can remember this relation by thinking of the P-wave as a "push" wave. This is easy to keep in mind because the words "primary" and "push" both begin with "P."

In **secondary waves,** particles transmitting the waves move back and forth at right angles to the direction of wave travel (Figure 6.6). Consequently the secondary wave motion is termed a **transverse wave.** It is commonly designated the **S-wave.** We may think of these waves as "shake" waves; like "secondary" the key word begins with "S." In the earth, P-waves travel approximately 1.7 times as rapidly as S-waves.

Surface waves are the third major type of seismic wave. They are much like waves sent out by a stone thrown into a calm pond. As each wave crest passes, the particles move in a complete circle, oriented vertically. The motion resembles that in ocean waves. Surface waves of earthquake origin are, of course, extremely low in height in comparison with common ocean waves, but are similar in that they die out very rapidly with depth below the ground surface.

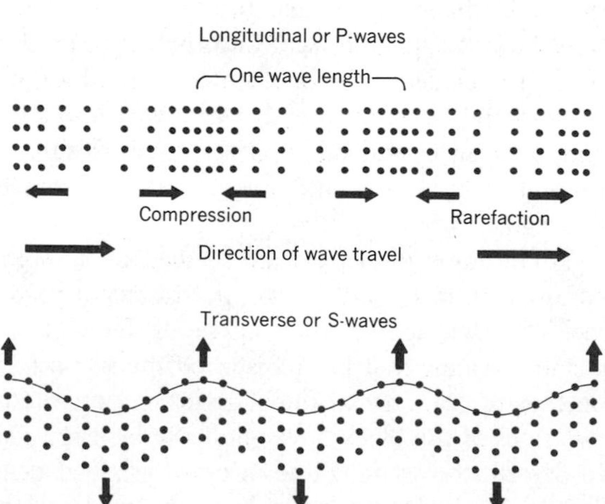

Figure 6.6 Particle motions in longitudinal and transverse seismic waves.

SEISMOGRAPHS

Seismologists must record an earthquake and analyze the directions and amounts of the earth motions involved. The mechanical problem facing the seismologist is that the instrument itself must be resting on the ground and will therefore also move with the ground. Because the instrument cannot be physically separated from the earth, the seismograph designer must make use of the principle of inertia to overcome the effect of the attachment. **Inertia** is the tendency of any mass to resist a change in a state of rest or of uniform motion in a straight line. The greater the mass of the object, the greater its inertia.

To record an earthquake, then, a very heavy mass, such as an iron ball, might be suspended from a very thin wire or from a flexible coil spring, as shown in Figure 6.7. When the earth moves back and forth or up and down in earthquake wave motion, the large mass will stay almost motionless because the supporting wire or spring flexes easily and does not transmit the motion through to the weight. If a pen is now attached to the mass, so that the point is just touching a sheet of paper wrapped around a moving drum, the pen will produce a wavy line on the paper. Strong shocks will give waves of high amplitude. **Amplitude** is the distance of side-to-side or up-and-down swing. Weak shocks will give waves of low amplitude. The number of back-and-forth movements per second is the **frequency.** When frequency is greater, the undulations of the line will be more closely crowded.

Of course, the seismograph as we have described it is too simple to be actually workable. Modern seismographs make use of several magnetic and electronic devices to pick up, amplify, filter, and record the motions of the earth. To analyze earthquakes adequately, a whole battery of seismographs must be operated simultaneously, because each instrument records only the wave motion in one particular line of movement, such as east-west, north-south, or vertically. Then too, earthquake waves include a wide range of frequencies superimposed in a complex way. Like a radio receiver, each seismograph is tuned to receive a particular frequency band; therefore several are needed to register the full range.

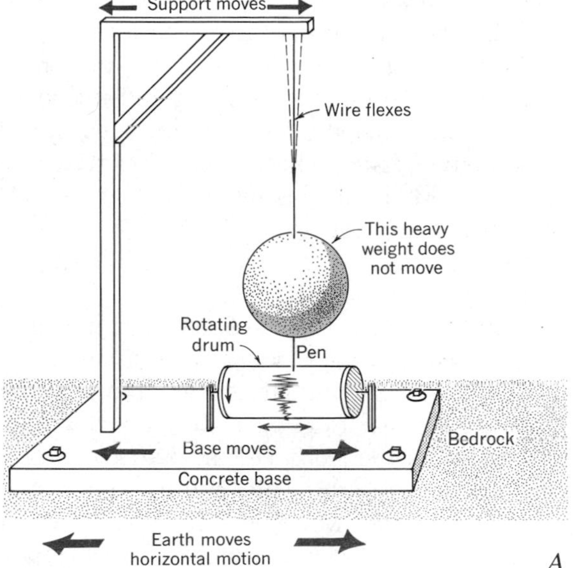

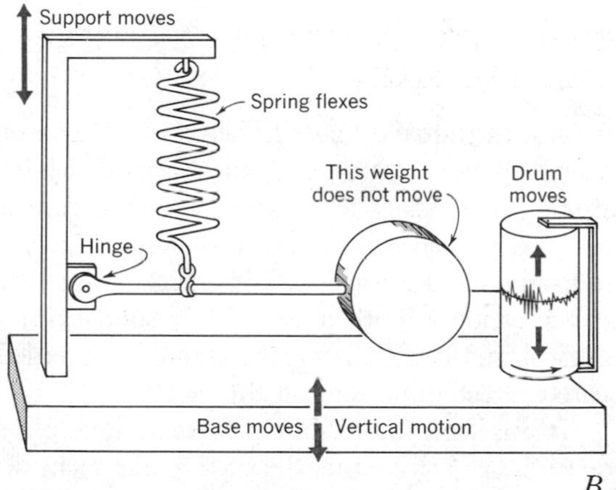

Figure 6.7 Inertia of a large mass provides a means of observing seismic waves. Horizontal motions might be detected by the mechanical arrangement shown in *A*, and vertical motions by that shown in *B*. Neither device would actually be useful unless further refined.

INTERPRETING THE SEISMOGRAM

Figure 6.8 shows a **seismogram,** the record produced by a typical distant earthquake. Figure 6.9 is a cross section of the earth showing how the waves traveled through the mantle and core.

The first indication that a severe earthquake had occurred at a distant point was the sudden beginning of a series of larger-than-average waves; they were the primary waves (P-waves). These waves died down somewhat, then a few minutes later a second burst of activity set in with the beginning of the secondary waves (S-waves). These waves were at first considerably greater in amplitude than the primary waves. There followed smooth waves that increased greatly in amplitude to a maximum and then slowly died down. These last very high-amplitude waves were the surface waves. While the primary and secondary waves had traveled through the earth, the surface waves had traveled along the ground surface (Figure 6.9).

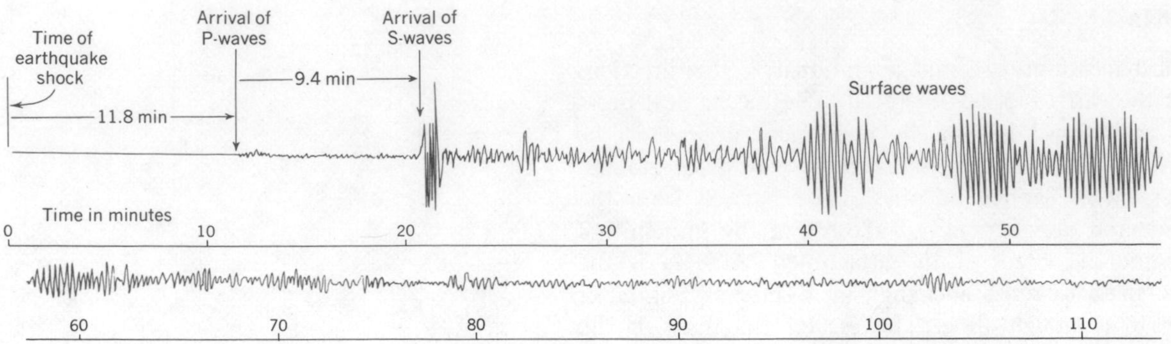

Figure 6.8 This seismogram shows the record of an earthquake whose epicenter was located at a surface distance of 5260 mi (8460 km) from the receiving station, equivalent to 76.4° of arc of the earth's circumference. Figure 6.9 shows the ray paths for this earthquake. (After L. Don Leet, *Earth Waves*, Cambridge, Mass., Harvard Univ. Press. © 1950 by the President and Fellows of Harvard College.)

For an earthquake occurring one-quarter of the globe's circumference away, that is, about 6000 mi (10,000 km), the primary waves will take about 13 min to reach the receiving station, and the secondary waves will begin to arrive about 11 min later.

It was soon apparent to the first students of seismograms that the farther away the earthquake focus, the longer the spread of time between the arrival of the primary and secondary wave groups. Both groups start from the focus at the same instant, but the primary group travels faster. The surface waves travel even more slowly and come in last. From this discovery, about 1900, came the obvious conclusion that the spread of time between arrival of the wave groups can be used to measure the distance from the focus to the seismograph station.

Based on many years of seismogram analysis, tables have been prepared to show the relationship of distance to timespread of P- and S-wave arrivals. These tables allow the seismologist to calculate at once the distance from his observatory to the earthquake epicenter. As shown on the globe (Figure 6.10), a circle can be drawn with the observatory as the center and the known distance as radius. The epicenter lies somewhere on this circle. When three such circles are drawn from three widely separated observing stations, the earthquake epicenter can be located within the limits of a small triangle of error.

SEISMIC WAVES AND THE EARTH'S CORE AND MANTLE

Study of earthquake waves has confirmed the existence of the spherical core at the earth's center and has added insight into its physical nature. (Refer back to Chapter 2 and Figure 2.2 for core dimensions and properties.) If the earth were in a solid state throughout, the P-waves and S-waves would travel through the center in all possible directions. With a completely solid earth the various seismic waves of any large earthquake could be recorded by a seismograph located directly opposite on the globe.

It was soon found, however, that there is a large region on the side of the globe opposite the earthquake focus where simple S-waves are not received. Evidently they are prevented from passing through a central region in the earth. Physicists know that transverse waves, or S-waves, cannot be sent through a liquid; so they are agreed that the earth's core is in a liquid state, in contrast to the surrounding mantle, which is in a solid state.

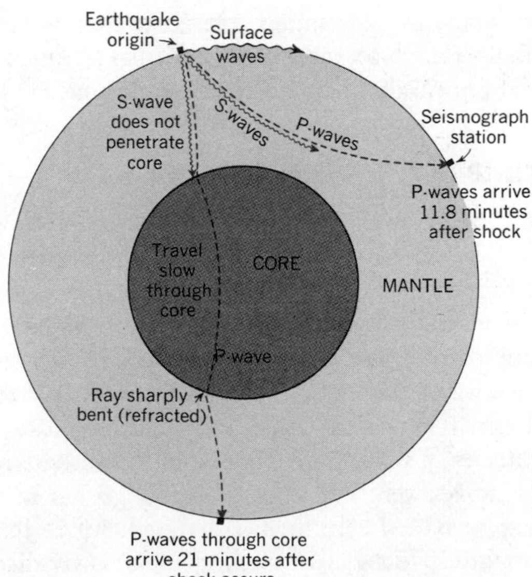

Figure 6.9 Cross section of the earth showing diagrammatically the paths of P-waves, S-waves, and surface waves.

EARTHQUAKE ENERGY AND INTENSITY

Interpretation of seismograms has made possible a calculation of the quantities of energy released as wave motion by earthquakes of various magnitudes. In 1935 a leading seismologist, Charles F. Richter, brought forth a scale of earthquake magnitudes describing the quantity of energy released at the earthquake focus. The **Richter scale** consists of numbers ranging from 0 to 8.6. The scale is logarithmic, which is to say that the energy of the shock increases by powers of 10 in relation to Richter magnitude numbers. Some notes concerning various magnitudes are given below:

Magnitude (Richter scale)	
0	Smallest detectable quake.
2.5–3	Quake can be felt if it is nearby. About 100,000 shallow quakes of this magnitude per year.
4.5	Can cause local damage.
5	Energy release about equal to first atomic bomb, Alamagordo, New Mexico, 1945.
6	Destructive in a limited area. About 100 shallow quakes per year of this magnitude.
7	Rated a major earthquake above this magnitude. Quake can be recorded over whole earth. About 14 per year this great or greater.
7.8	San Francisco earthquake of 1906.
8.4	Close to maximum known. Examples: Honshu, 1933; Assam, 1950; Alaska, 1964.
8.6	Maximum observed between 1900 and 1950. Three million times as much energy released as in first atomic bomb.

Total annual energy release by earthquakes is roughly 50 times the quantity released by a single quake of magnitude 8.4. Most of this annual total is from a few quakes of magnitude greater than 7.

The actual destructiveness of an earthquake also depends upon factors other than the energy release given by Richter magnitude—for example, closeness of inhabited areas to the epicenter is a major factor. Nature of the subsurface earth materials is another. **Intensity scales** are designed to measure observed earth-shaking effects; they are important in engineering aspects of seismology.

An intensity scale used extensively in the United States is the **modified Mercalli scale,** prepared by Richter in 1956. This scale recognizes 12 levels of intensity, designated by Roman numerals I through XII. Each intensity is described in terms of phenomena that any person might experience. For example, at intensity IV hanging objects swing, a vibration like that of a passing truck is felt, standing automobiles rock, and windows and dishes rattle. Damage to various classes of masonry is used to establish criteria in the higher numbers of the scale. At an intensity of XII, damage to man-made structures is nearly total and large masses of rock are displaced (Figure 6.11).

Many of the destructive effects of a severe earthquake are secondary, in the sense that the earthquake movements set off gravity movements of bodies of rock, soil, and clay. An example is the Good Friday earthquake of March 27, 1964, centered about 75 mi (120 km) from the city of Anchorage, Alaska. Magnitude on the Richter scale was 8.4 to 8.6, which is close to the maximum known.

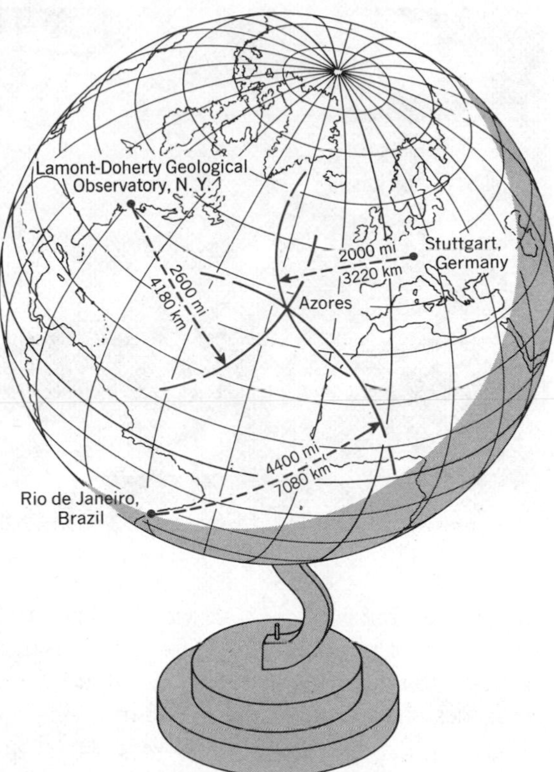

Figure 6.10 Circles drawn from three seismological observatories yield the location of an earthquake epicenter.

Figure 6.11 Severe masonry damage produced by the San Francisco earthquake and fire of 1906. This view is southwestward from the corner of Geary and Mason streets. (W. C. Mendenhall, U.S. Geological Survey.)

Figure 6.12 Slumping and flowage of unconsolidated sediments resulted in severe property destruction at Anchorage, Alaska, during the Good Friday earthquake of March 27, 1964. (U.S. Army Corps of Engineers.)

Intensity on the Mercalli scale was probably VII to VIII in Anchorage, but as most buildings were of frame construction, damage was largely through secondary effects. Of these, the most important were landslides of great masses of gravel overlying layers of unstable clay (Figure 6.12). Major snowslides were set off in the adjacent mountains.

Throughout the region of the Alaskan earthquake sudden changes of land level, both up and down, took place at points as far distant as 300 mi (480 km) from the epicenter and covered a total area of about 80,000 sq mi (200,000 sq km). A belt of uplift reaching a maximum rise of 30 ft (10 m) ran parallel with the coast and largely offshore.

The fault along which slippage occurred to generate the Alaska earthquake is not exposed on land, but lies at a depth of some 25 mi (40 km) in the offshore zone. The entire zone of seismic activity occupies a position between the volcanic Aleutian Arc on the northwest and the deep submarine Aleutian Trench on the southeast.

We are dealing here with a familiar picture: There is a mountain arc along the coast and a deep trench offshore. Evidently the lithospheric plate beneath the Pacific Ocean is being forced to descend beneath the continental lithospheric plate of Alaska and North America, as depicted in Figure 5.6. Numerous earthquakes, many severe, accompany the slippages between these two great lithospheric plates. To follow up this suggestion, we turn to an overview of the global distribution of earthquakes. We will also investigate the depths at which the **foci** (plural of focus) are located.

GLOBAL EARTHQUAKE DISTRIBUTION

A world map shows the plotted epicenters of many of the largest earthquakes (Figure 6.13). The patterns on this map reveal much about tectonic activity. The picture is about the same for earthquakes of all magnitudes and for different spans of time.

Notice particularly the earthquake concentration in the circum-Pacific belt of primary mountain and island arcs and their related

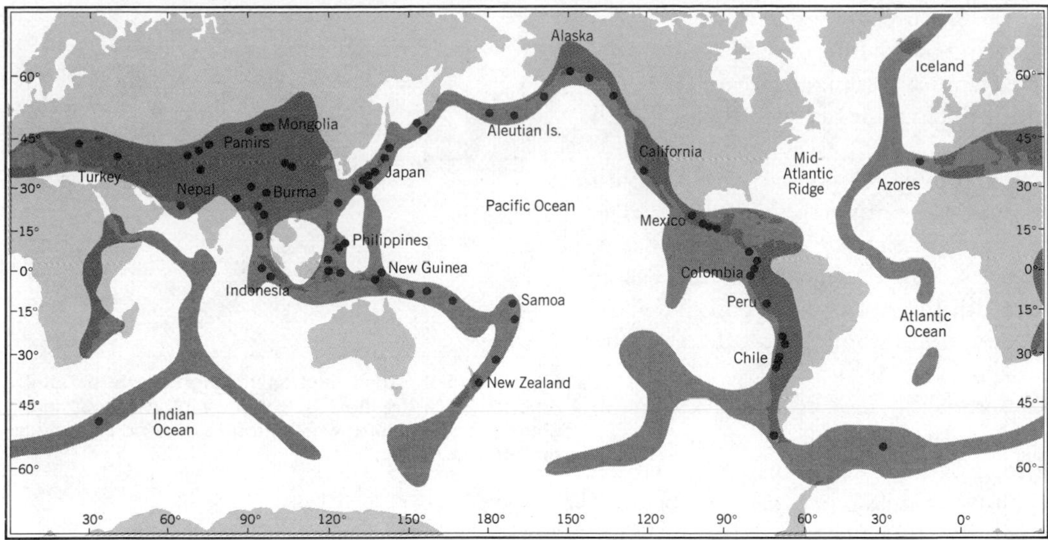

Figure 6.13 World distribution of shallow-focus earthquakes. Dots show epicenters of major earthquakes, measuring 7.9 or over on the Richter scale. The colored zones are principal areas of abundant earthquakes.

trenches. It is estimated that earthquakes of this belt account for about 80% of the total world earthquake energy release. Large earthquakes are shown by black dots; they are abundant around the circum-Pacific belt. Earthquakes also correspond with the Eurasian-Melanesian belt of primary arcs, but are somewhat fewer and more dispersed than in the circum-Pacific belt. This belt accounts for about 15% of the total energy release.

Yet another location of frequent earthquakes is in mid-ocean in the Atlantic Ocean, Indian Ocean, and southern Pacific Ocean, but large earthquakes are rare in these oceanic belts. These seismic zones coincide with a great submarine mountain chain, described in Chapter 8. Although no areas of the world are entirely free of earthquakes, large parts of the continental shields have very few. Faulting is comparatively inactive in these ancient masses of granitic crust.

THE MEANING OF DEEP EARTHQUAKES

From measurement of the elapsed time of arrival of the three seismic wave forms, as well as the amplitude of those waves, the depth to an earthquake focus can be measured. Foci of earthquakes, while not strictly points, have such small dimensions as to be shown as points on a map or cross-sectional diagram.

According to depth, three classes of earthquakes are recognized: (1) **shallow-focus,** centered within 35 mi (55 km) of the surface; (2) **intermediate-focus,** from 35 to 150 mi (55 to 240 km); (3) **deep-focus,** from 185 to 400 mi (300 to 650 km).

Below a depth of about 400 mi (640 km) the properties of the mantle are such that unequal stresses cause continuous slow yielding of the plastic rock, so that sufficient elastic strain cannot be accumulated to produce an earthquake.

Look now at a simplified map of South America (Figure 6.14), showing epicenters of shallow earthquakes (crosses), intermediate earthquakes (open circles), and deep earthquakes (solid dots). The three types lie in three roughly parallel lines. Shallow earthquakes

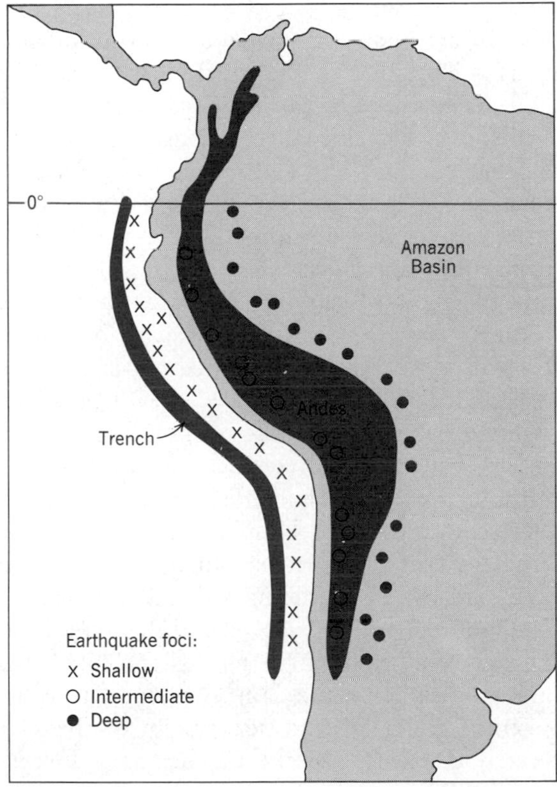

Figure 6.14 This sketch map of South America shows the gross pattern of epicenters of earthquakes of shallow, intermediate, and deep focus.

occur mostly close to the Pacific Ocean shore, near the great Peru Trench and immediately landward of it. Intermediate earthquake foci are located under the great Andes mountain range. The deep foci lie farther eastward, beneath the lowlands of west-central South America.

Figure 6.15, a cross section through this same part of South America, explains the three lines of earthquake epicenters. The quakes are set off along a fault zone of slippage between the downbent plate and the mantle above it. The diving edge of the Pacific plate is extending far down beneath the Andes range and under the continental interior. Consequently, the earthquake foci descend progressively from west to east.

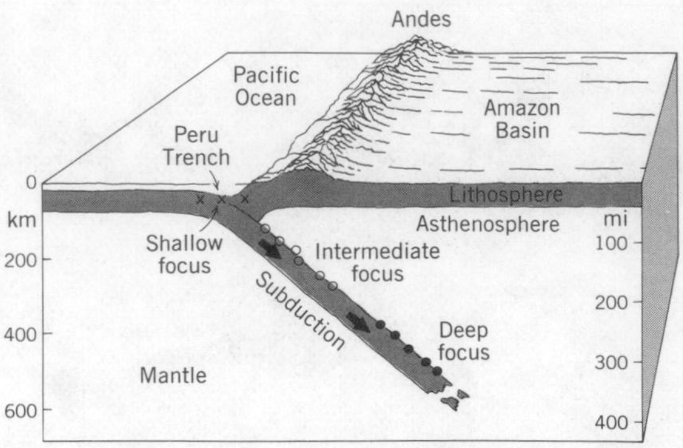

Within the past few years geologists have revived an old term used in the late 1800s. Albert Heim, mentioned in the opening paragraphs of Chapter 5, found evidence that in certain places one crustal mass had been forced by faulting or folding to descend beneath another. The Alpine geologists called this process **subduction.** Subduction as used today in geology means the forcing down of one lithospheric plate beneath another, as illustrated in Figure 6.15.

With this interpretation of depth of earthquake foci, we have fitted one more piece of evidence into the puzzle of crustal deformation and the origin of mountain arcs. The seismic evidence of subduction proved to be an essential building block in the foundation of the revolution in geology.

Figure 6.15 A cross section through western South America showing how subduction of a lithospheric plate is creating a downslanting fault zone extending far into the mantle.

EARTHQUAKES AND CITIES

The impact of an earthquake on a major urban area and its inhabitants is illustrated by the San Fernando earthquake of February 9, 1971. The entire Los Angeles community was shocked into renewed awareness of the wide range of damaging effects an earthquake brings to a modern city. This earthquake was only of moderate magnitude —6.6 on the Richter scale—but in a few places the shaking of the ground was as severe as anything that had been previously measured in an earthquake. Fortunately, the shaking was of brief duration, and only the most susceptible structures were severely damaged. Also fortunate was the fact that the quake occurred at 6:00 AM, when few persons were traveling the freeways or occupying public buildings.

Of the major disastrous effects to buildings, one of the most frightening was the collapse of the Olive View Hospital in Sylmar, a new structure believed to have been built in conformity with earthquake-resistant standards. Severe damage with casualties also affected the Veterans Hospital in Sylmar (Figure 6.16). Hospital buildings that remained standing were weakened and have since been demolished; rebuilding will cost $40 million. The Van Norman Dam was dangerously cracked and the water had to be drained from the reservoir behind it to forestall potential flooding of a densely built-up area below the dam. An important converter station in the electrical power transmission system of the Los Angeles area was severely damaged. On the Golden State Freeway an overpass collapsed, blocking the roadway below, while the freeway pavement itself was badly cracked and dislocated (Figure 6.17).

Figure 6.16 Collapse of buildings of the Veterans Administration Hospital, Sylmar, Los Angeles County, California, caused by the San Fernando earthquake of February 1971. (Wide World Photos.)

Figure 6.17 Broken pavement and collapsed overpass on the Golden State Freeway at the northern end of the San Fernando Valley, California, resulting from the earthquake of February 1971. (Wide World Photos.)

The San Fernando earthquake was generated along a comparatively minor fault some 15 mi (25 km) from the great San Andreas Fault. What would happen if a major earthquake comparable in intensity with the San Francisco earthquake of 1906 or the Alaska earthquake of 1964 were to occur along the San Andreas Fault in this area? A joint panel of experts of the National Academy of Sciences and the National Academy of Engineering found it clear that existing building codes do not provide adequate damage control features. The cost in damage of the San Fernando earthquake of 1971 was estimated as on the order of $500 million. In contrast, experts anticipate that an earthquake as great as the San Francisco earthquake of 1906 would cause damage on the order of $20 billion, should it occur today. In coming decades, the progress of urbanization in the Los Angeles area will have greatly expanded both the population and the building structures subject to devastation.

EARTHQUAKE PREDICTION

One of the most fascinating current activities of seismologists is seeking ways to predict the time of occurrence of an earthquake. Most of the active major faults near large population centers are well known; they have been mapped and there are good records of their past seismic activity. The general approach is to detect physical changes of various sorts that will prove to be reliable forerunners of a potentially severe earthquake. A variety of sensitive instruments can be employed, each capable of measuring some small change in the zone of the fault. One type of instrument can measure the buildup of strain in the rock adjacent to a fault; another instrument measures slight tilting of the ground, a phenomenon known to precede certain earthquakes. Even minute changes in the local strength of the earth's magnetic field are monitored as possible indicators of impending earthquakes.

Changes in speed of P-waves of small tremors near a fault are thought to be tied in with impending larger earthquakes. These waves are generated by frequent, very weak earthquakes and are closely monitored. It has been observed that the P-wave speed becomes less in a period preceding a moderate to severe earthquake, but then increases to greater speed immediately before the event. Reduction in wave speed is attributed to opening of pores in the rock adjacent to the fault as the rock is strained. Then, when ground water fills the pores, wave speed increases. Shortly thereafter, the break occurs. This scenario is debated by seismologists, but on the whole, rapid strides are being made in the field of earthquake prediction.

THE SAN ANDREAS FAULT

The historical approach is also being followed in an attempt to make rather general predictions as to the likelihood of a major earthquake. This approach is particularly well illustrated in the case of the San Andreas Fault of California. As we previously noted, this fault is of the transcurrent type. It runs for a distance of about 600 mi (950 km), starting in the Salton Basin of southern California and trending in a northwesterly direction, about parallel with the Pacific coastline, to the San Francisco Bay region (Figure 6.18). It then follows the northern California coast to Cape Mendocino and passes out to sea. There are many other major active faults in California, as shown in

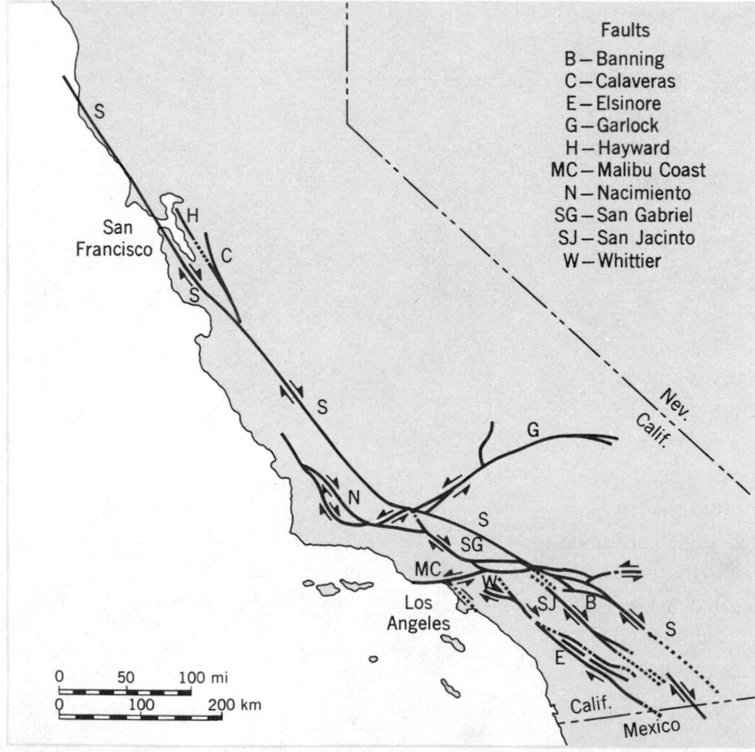

Figure 6.18 The San Andreas Fault and associated major faults of California. (From A. N. and A. H. Strahler, 1973, *Environmental Geoscience*, Santa Barbara, California, Hamilton Publishing Co., Figure 922, p. 222.)

Figure 6.18. Some of these appear as branches of the San Andreas Fault, but others, such as the Garlock Fault, cross the principal trend at near right angles.

Along the San Andreas Fault, motion has consistently been such that the crustal mass on the western (Pacific) side has moved in a northerly direction, relative to the eastern block. With this type of motion the San Andreas Fault is designated as a **right-lateral** transcurrent fault. If you stand close to the fault line, facing across the fault, the block on the opposite side has moved toward your right. (If motion on the opposite side is to your left, it is a **left-lateral** fault.) Examine the rift of the San Andreas Fault shown on the vertical air photo in Figure 6.19. Tracing the channel of a stream as it crosses the rift, you will see that the channel is abruptly offset. The sketch map in Figure 6.20 illustrates the effect. Despite repeated small movements along the fault, each stream has been able to maintain its continuity of flow, but the channel has been gradually "stretched" along the line of the fault.

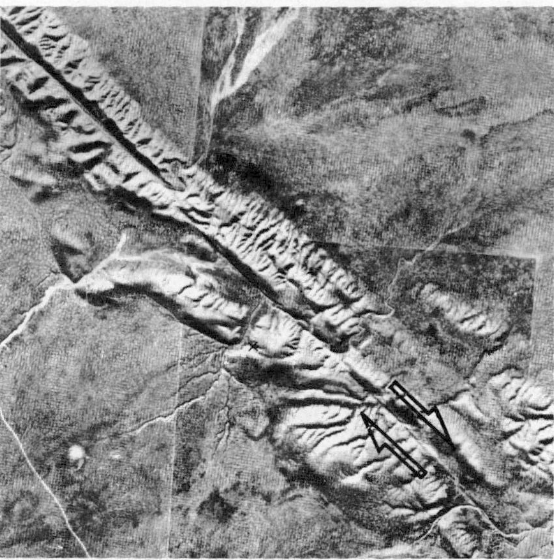

Figure 6.19 In this vertical air view, the nearly straight trace of the San Andreas Fault contrasts sharply with the sinuous lines of stream channels. San Bernardino County, California. (Aero Service Division, Western Geophysical Co. of America.)

Movement along the San Andreas Fault has been going on for many millions of years. Matching of similar rock masses displaced along the two sides of the fault shows a total lateral movement of about 350 mi (560 km) over the past 150 million years. The San Andreas Fault appears to be the contact plane between two major lithospheric plates. That on the Pacific side has been moving northwest relative to the plate on the inland side. We shall identify these great lithospheric plates in Chapter 9.

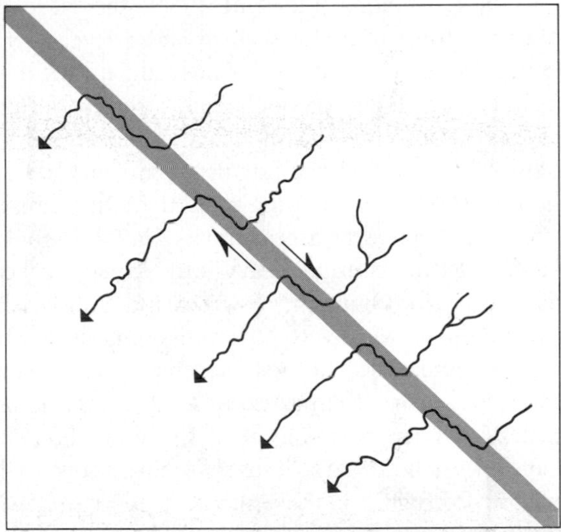

Figure 6.20 Offsetting of streams along an active transcurrent fault.

SEISMIC SEA WAVES, OR TSUNAMIS

Finally, we turn to an important coastal hazard related to earthquakes. This is an extraordinary kind of ocean wave not related to wind or tide. Known as a **seismic sea wave,** or **tsunami,** it is produced by a sudden displacement of the sea floor. The wave may be put in motion by an earthquake setting off a submarine landslide, or causing a sudden rising or sinking of the crust. Another possible cause is a submarine volcanic eruption. The effect of any of these rock movements is very much like that of dropping a stone into a very shallow, quiet pond. A wave is sent outward in an ever widening circle (Figure 6.21).

Seismic sea waves are of enormous length, some 60 to 120 mi (100 to 200 km), whereas the wave height may be only 1 to 2 ft (0.3 to 0.6 m). Such low waves cannot be felt by persons on a ship on the open sea. A typical wave might travel at a speed of 300 mph (500 km/hr).

Seismic sea waves are very long in comparison with the depth of water in which they travel. For example, a 100-mi wavelength is roughly 33 times as great as an ocean depth of 3 mi. In such comparatively shallow water the velocity of travel of the wave varies as the square root of the water depth. This law of wave speed enables the seismologist to issue warnings of a possible destructive seismic sea wave, or tsunami. The information required is a knowledge of the varying depth of the ocean and the known instant of the earthquake. It is then possible to compute the hour at which the first waves will reach a given coast (Figure 6.21).

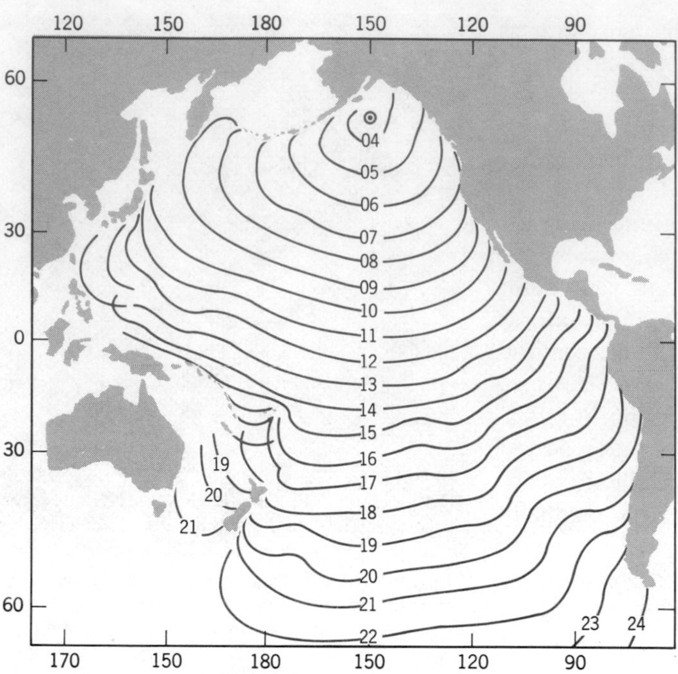

Figure 6.21 This map of the Pacific Ocean shows the location of a tsunami wave front at hourly intervals. The wave originated in the Gulf of Alaska as a result of the Good Friday earthquake of March 27, 1964.

Upon reaching a distant shore, the wave crest of a tsunami takes the form of a slow rise in water level over a period of 10 to 15 min. Superimposed on this are ordinary wind waves. These waves break close to shore, producing a destructive surf. Several great catastrophes in recorded history have been wrought by tsunamis. For example, flooding of the Japanese coast in 1703, with a loss of more than 100,000 lives, may have been of this cause. In May 1960, a tsunami generated by an earthquake in Chile spread across the entire Pacific Ocean, causing heavy surf damage at coastal locations in Hawaii, Japan, Okinawa, New Zealand, and Alaska. One should not confuse the tsunami with coastal flooding caused by storm surges.

In some cases the first evidence of a tsunami is a lowering of water level rather than a rise in level. Lowering of the water surface causes a seaward withdrawal of the water line and exposure of the floors of shallow bays. This is what seems to have happened at Lisbon, Portugal, on November 1, 1755, following an earthquake centered off the Portuguese coast. The sight of the exposed sea floor attracted a large number of townspeople. When the following wave crest arrived, the rapid rise of water level drowned many persons.

During the period 1928 through 1963, a span of 36 years, a world total of 84 tsunamis occurred. About 80% of these were in the Pacific Ocean. The greatest number observed in one year was 6, in 1963. Of the Pacific tsunamis, 5 were widely destructive, 17 caused death or destruction close to the source, and 44 caused no damage.

SEISMIC ACTIVITY IN REVIEW

In this chapter we found that earthquake waves play a varied role in both scientific investigation and the everyday world of man. Fault movements generate most earthquakes, which represent sudden releases of great quantities of energy temporarily stored in crustal rock.

Earthquake waves serve as scientific probes deep into the crust and mantle, and even through the central core. By studying the depths of foci in tectonic belts, we obtain excellent evidence for large-scale subduction of lithospheric plates.

Severe earthquakes are a great natural hazard to man, particularly in urban areas near major active faults. Science seems to be nearing its goal of predicting the time of occurrence of potentially dangerous earthquakes, but urban areas are presently ill equipped to cope with a major earthquake.

YOUR GEOSCIENCE VOCABULARY

seismology
seismic waves
seismograph
fault plane
normal fault
fault scarp
transcurrent fault
rift
strain
elastic limit
elastic-rebound theory
focus, foci
epicenter
primary wave, P-wave
longitudinal wave
secondary wave, S-wave
transverse wave
surface wave
inertia
wave amplitude
wave frequency
seismogram
Richter scale
intensity scale
Mercalli scale
shallow-focus earthquake
intermediate-focus earthquake
deep-focus earthquake
subduction
right-lateral fault
left-lateral fault
seismic sea wave
tsunami

SELF-TESTING QUESTIONS

1. Explain how earthquakes are related to faulting. Describe the two basic types of faults and their surface expression.

2. Describe the elastic strain observed to precede earthquakes. What is the elastic-rebound theory of earthquakes, and how was it demonstrated to apply to the San Francisco earthquake of 1906?

3. Describe the three major types of earthquake waves in terms of the motions involved in each. How do their speeds of travel compare?

4. Explain the principles involved in construction and operation of seismographs.

5. Describe a typical seismogram received from a large, distant earthquake. How is distance to the epicenter calculated? Why is it necessary to have data from at least three observatories to locate the epicenter?

6. Explain how seismic waves provide information on the dimensions and properties of the earth's core.

7. Describe the Richter scale of earthquake magnitudes. How does it express quantity of energy released? From what magnitudes of earthquakes is most of the total annual energy release derived? How does the modified Mercalli scale differ in purpose from the Richter scale?

8. Describe the destructive effects of major earthquakes. What crustal changes of level accompanied the Good Friday earthquake of 1964 in Alaska? How do these changes relate to large features of the crust?

9. Compare the world distribution of earthquake epicenters with that of major crustal features. In what way does the distribution of shallow-focus, intermediate-focus, and deep-focus earthquakes provide evidence of subduction in progress?

10. Using the San Fernando earthquake of 1971 as an example, describe the impact of a major earthquake upon a large urban region. What are the future prospects for earthquake losses upon Los Angeles?

11. What kinds of physical changes in the vicinity of an active fault may prove useful in predicting the time of occurrence of a major earthquake?

12. Describe the seismic sea wave, or tsunami, in terms of origin, dimensions, and manner of propagation. What factor controls speed of this type of wave? What is the environmental impact of a tsunami? How does a tsunami warning system operate?

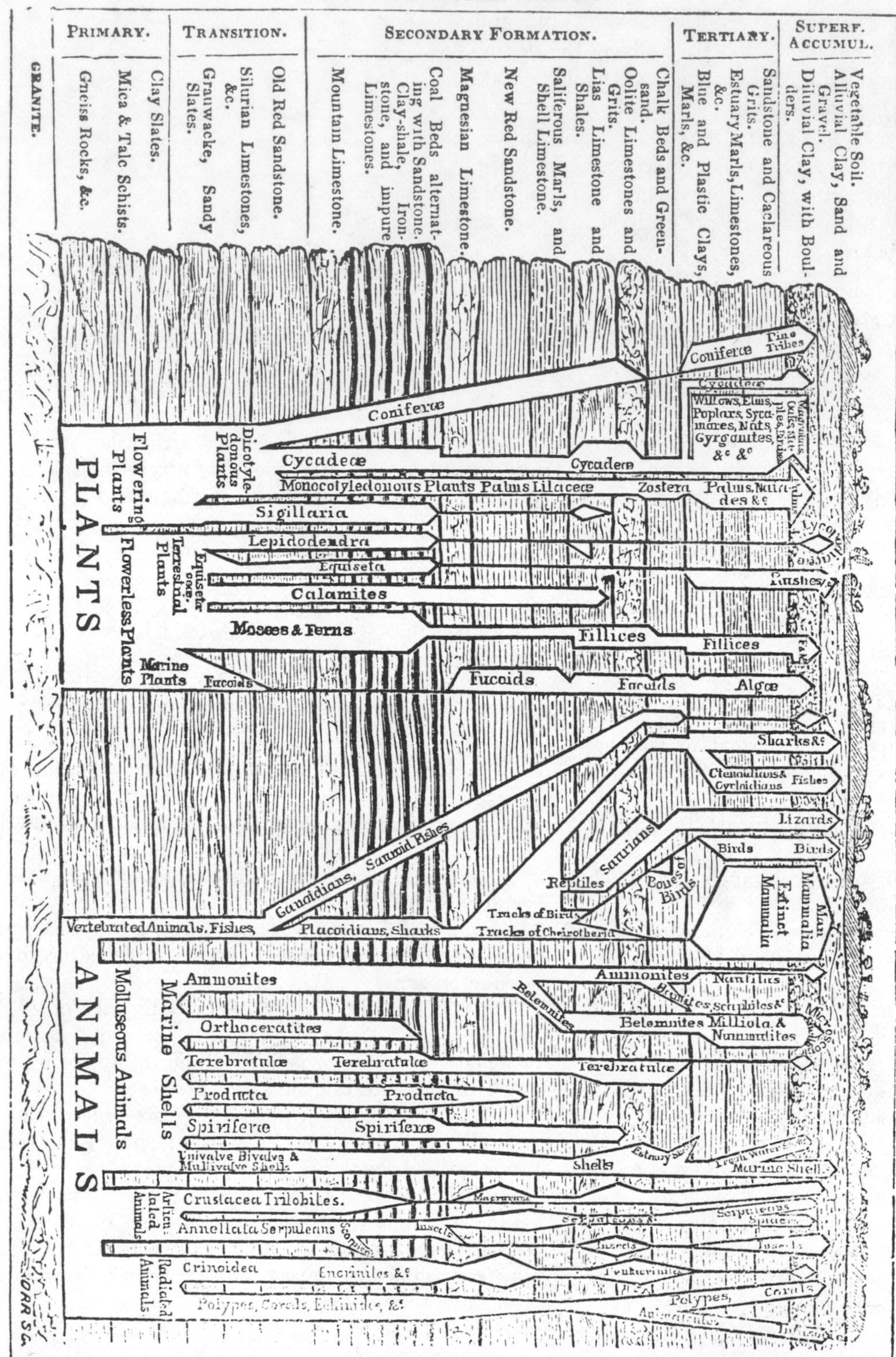

The Geological Distribution of Fossil Organic Remains.

This oryctological chart comes out of Page's geology textbook of 1860. It represents what was known of English stratigraphy and fossils at that time. The accompanying explanation tells us that "the greater or less space occupied by the various tribes of animals and plants on the chart shows their comparative abundance or paucity. The branches designate the species. The whole brings under a glance of the eye the rise, developement, ramification, and extirpation of the different tribes." (From *Elements of Geology*, 1860, Page, Barnes & Burr, New York.)

7 STRATIGRAPHY—THE RECORD OF THE ROCKS

THE FOSSIL CONTROVERSY

Fossils and religion seem not to have mixed very well through the ages. Everyone is familiar with the Scopes trial of 1925 in which Clarence Darrow defended organic evolution against the fundamentalist orator, William Jennings Bryan. Even today, the use of fossils in establishing the evolutionary development of all life forms, including Man, is questioned by a few small groups of religious conservatives.

As early as 450 B.C., Heroditus examined fossil marine shells in Egypt and correctly interpreted them to mean that the Mediterranean Sea had once spread over that area. A century later, Aristotle, a great thinker in other respects, interpreted fossils as organisms which had grown in place in the rock surrounding them. One of his pupils, Theophrastus, explained further that the fossils grew from seeds or eggs planted in the sediment as it accumulated!

Trouble with the forces of organized religion did not really become serious until the Dark Ages, when the Christian church stoutly demanded belief in a six-day creation of the world, taking place only a few thousand years ago. Fossils had no place in this scenario, and the idea that they were once living plants or animals became heresy. However, Man's irrepressible urge to think for himself expressed itself in Italy about A.D. 1500. At this time, canals were being dug in marine strata of Cenozoic age (roughly covering the time span of 10 to 60 million years ago). Marine fossils, strikingly like those living forms seen along the Mediterranean shores today, were uncovered in large numbers from the canal excavations. The fossils attracted the attention of Leonardo da Vinci, who argued forcefully that they were the remains of animals living in those sediments at the time of burial. His pronouncements set off a great fossil controversy between liberal thinkers and the Christian church. Supporters of the church claimed that fossils were not of organic nature, but that they were the product of mystical forces, or even implantations by the Devil to delude honest Christians.

Apparently, the controversy was still going strong in the late 1600s, for at that time a German school teacher, Ernst Tentzel, innocently interpreted some skeletal remains of a mammoth as belonging to a prehistoric monster. To settle the ensuing furor, the medical faculty of his school examined the bones and solemnly declared them to be merely a freak of nature!

When even the most orthodox churchmen could no longer escape the interpretation of fossils as organic remains, a new story was invented: These animals had been killed and buried very suddenly during the Noachian Flood. Thus, when one Johann Jacob Scheuchzer found in some lake beds in Switzerland fossils that resembled the crushed skull and vertebrae of a human, he realized that these were the remains of a victim of the Biblical Flood, and rushed into print with a Latin volume titled *Homo deluvii testis* (Man who is proof of the Flood). In this work, published in 1726, he illustrated and described the strange skull-like fossil. Truth was served, but only much later in the same century, by Baron Cuvier. This competent paleontologist immediately recognized *Homo deluvii testis* as a very large salamander. Appropriately, Cuvier named the new species *Andrias scheuchzeri*.

By the year 1800, Cuvier and his associates had established paleontology on an irrefutable scientific basis, and the great fossil controversy came to an end. The orderly succession of evolutionary changes in fossil forms in strata, arranged one above the next, became the key to unraveling the eras and periods of geologic time. In this chapter we will examine the role of fossils in the interpretation of geologic history.

HISTORICAL GEOLOGY

In more than one sense, this chapter is a step backward into time. Obviously, it is an incursion into vast spans of geologic time. Less obviously, it summarizes progress through early decades in the history of geology as a science. **Historical geology** is a general term for the reconstruction of an orderly sequence of geological events throughout the past 3½ billion years or more of our planet's history.

STRATIGRAPHY

Stratigraphy is a branch of historical geology dealing with the sequence of events in the earth's history as interpreted from evidence found in sedimentary rocks. This is the phase of historical geology we emphasize here. Included in stratigraphy are the records of sediment deposition, of the past geographical distributions of land and sea, and of the past conditions of climate and terrain.

The pursuit of stratigraphy depends on **paleontology,** the study of ancient life based on fossil remains of animals and plants. The **paleonotolgist** identifies, names, and classifies fossils, and determines their evolutionary development. The **stratigrapher** uses fossils both to correlate strata in age of deposition and to establish the physical and chemical environments of deposition.

Stratigraphy uses many of the same methods as the study of civilized man's history. In both fields there must be an intensive search for evidence of what happened. From the actual record, whether it be in rock strata or in a written document, certain interpretations can be made by the historian skilled in guessing at the hidden meanings of seemingly trivial details. In both fields, a major problem is the fragmentary nature of the record. In many cases an event either left no record or the record has since been destroyed. Naturally, the problem of inadequate records generally becomes more serious as older and older periods of history are considered. Then, too, a period well documented in one region is often not documented at all in another.

In this chapter we inquire into the basic principles and methods of stratigraphy in order to obtain an appreciation of the type of problem that the geologist has faced and solved repeatedly. For illustration, we refer to a classic geologic region, familiar to many Americans—the Colorado Plateau region in northern Arizona and southern Utah. Here lie three of our most famous national parks—Grand Canyon, Zion, and Bryce—whose rocks span all the eras of geologic time and are beautifully displayed in cliffs and canyon walls.

THE TABLE OF GEOLOGIC TIME

Before launching into an interpretation of the strata of the Colorado Plateau we will need to examine the table of geologic time and its units. The various time divisions shown in Table 7.1 were worked out during a century or more of patient observation by both paleontologists and stratigraphers. Absolute ages remained only speculative until the early 1900s, when radiometric dating became a reality, as we learned in Chapter 5. That chapter brought us to the close of Precambrian time, about 600 million years before the present (−600 m.y.), when the continental shields were largely completed.

In terms of life forms, the largest time bracket is the **eon.** Precambrian time constitutes the *Cryptozoic Eon. Cryptozoic,* derived from the Greek words *kryptos,* "hidden," and *zoo,* "life," signifies the obscurity and simplicity of life of Precambrian time. The time span of abundant life which followed constitutes the *Phanerozoic Eon,* derived from the Greek word *phaneros,* "visible."

Phanerozoic time is made up of three **geologic eras;** these are the *Paleozoic Era,* the *Mesozoic Era,* and the *Cenozoic Era,* listed in order from earliest to latest. Translating from the Greek roots of these three titles, they can be paraphrased as the eras of ancient (*paleos*), middle (*mesos*), and recent (*kainos*) life, respectively.

The inference of these names is clear. While geologic processes operated in repetitive cycles, each one about the same as the next, life was changing through time in a one-way, irreversible stream. Thus each era is distinct from the next in terms of organic composition. The geologist cannot assign a given stratum of limestone or shale to a given era on the basis of physical rock properties alone, for like rocks were formed in all eras. Instead, it is the distinctiveness of the remains of life forms enclosed in that stratum that permit it to be assigned its place in geologic time.

Table 7.1 **Table of geologic time**

Eon	Era	Period	Duration (m.y.)	Age (m.y.)	Orogenies
PHANEROZOIC EON	CENOZOIC (65)				Cascadian
				65	
	MESOZOIC (160)	Cretaceous	71		Laramian
				136	
		Jurassic	54		Nevadian
				190	
		Triassic	35		
				225	
	PALEOZOIC (345)	Permian	55		Appalachian
				280	(Hercynian)
		Carboniferous	65		
				345	
		Devonian	50		Acadian
				395	(Caledonian)
		Silurian	35		
				430	
		Ordovician	70		Taconian
				500	
		Cambrian	70		
				570	
CRYPTOZOIC EON	PRECAMBRIAN		(b.y.)	(b.y.)	
		Upper Precambrian	0.3–0.4		Kenoran
				0.9–1.0	
		Middle Precambrian (Algonkian)	0.6–0.8		
				1.6–1.7	
			0.7–0.9		
				2.4–2.5	Grenville
		Lower Precambrian (Archean)	0.9–1.0		Hudsonian
		Oldest dated rocks		3.6 ± 0.1	
		Earth accretion completed		4.6–4.7	
		Age of universe		17–18	

(C)

(D)

KAIBAB
KAIBAB
TOROWEAP
COCONINO
HERMIT
SUPAI
COCONINO

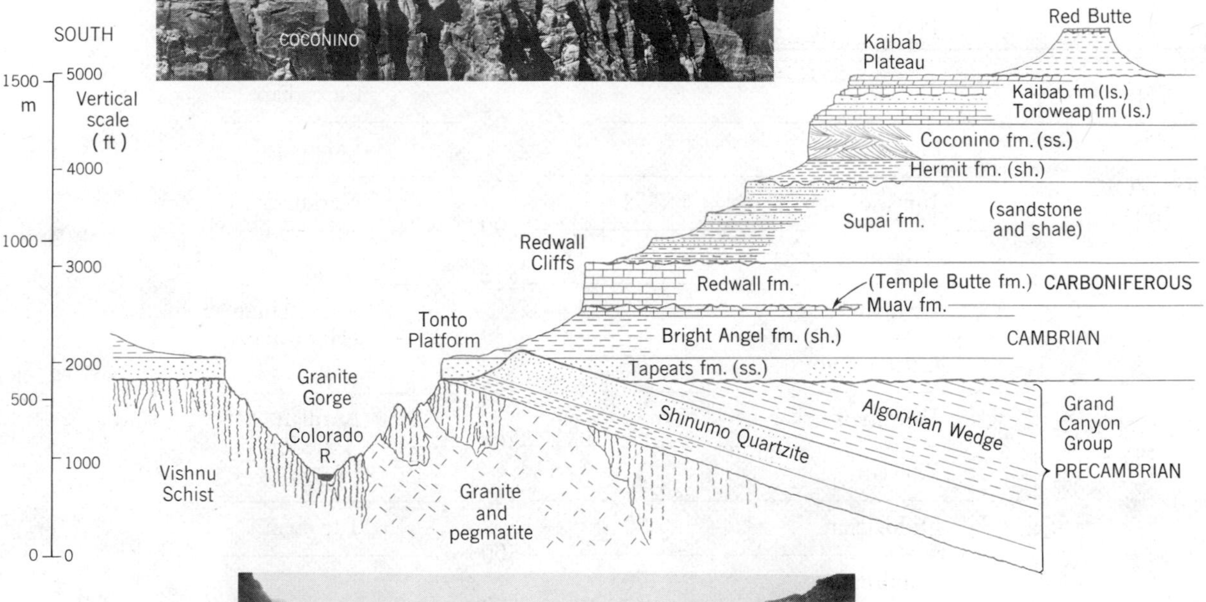

SOUTH

Red Butte
Kaibab Plateau
Kaibab fm (ls.)
Toroweap fm (ls.)
Coconino fm. (ss.)
Hermit fm. (sh.)
Supai fm.
(sandstone and shale)
Redwall Cliffs
Redwall fm.
(Temple Butte fm.)
CARBONIFEROUS
Muav fm.
Tonto Platform
Bright Angel fm. (sh.)
CAMBRIAN
Tapeats fm. (ss.)
Granite Gorge
Colorado R.
Vishnu Schist
Granite and pegmatite
Shinumo Quartzite
Algonkian Wedge
Grand Canyon Group
PRECAMBRIAN

1500 m
1000
500
0
Vertical scale (ft)
5000
4000
3000
2000
1000
0

TAPEATS
VISHNU SCHIST

(A)

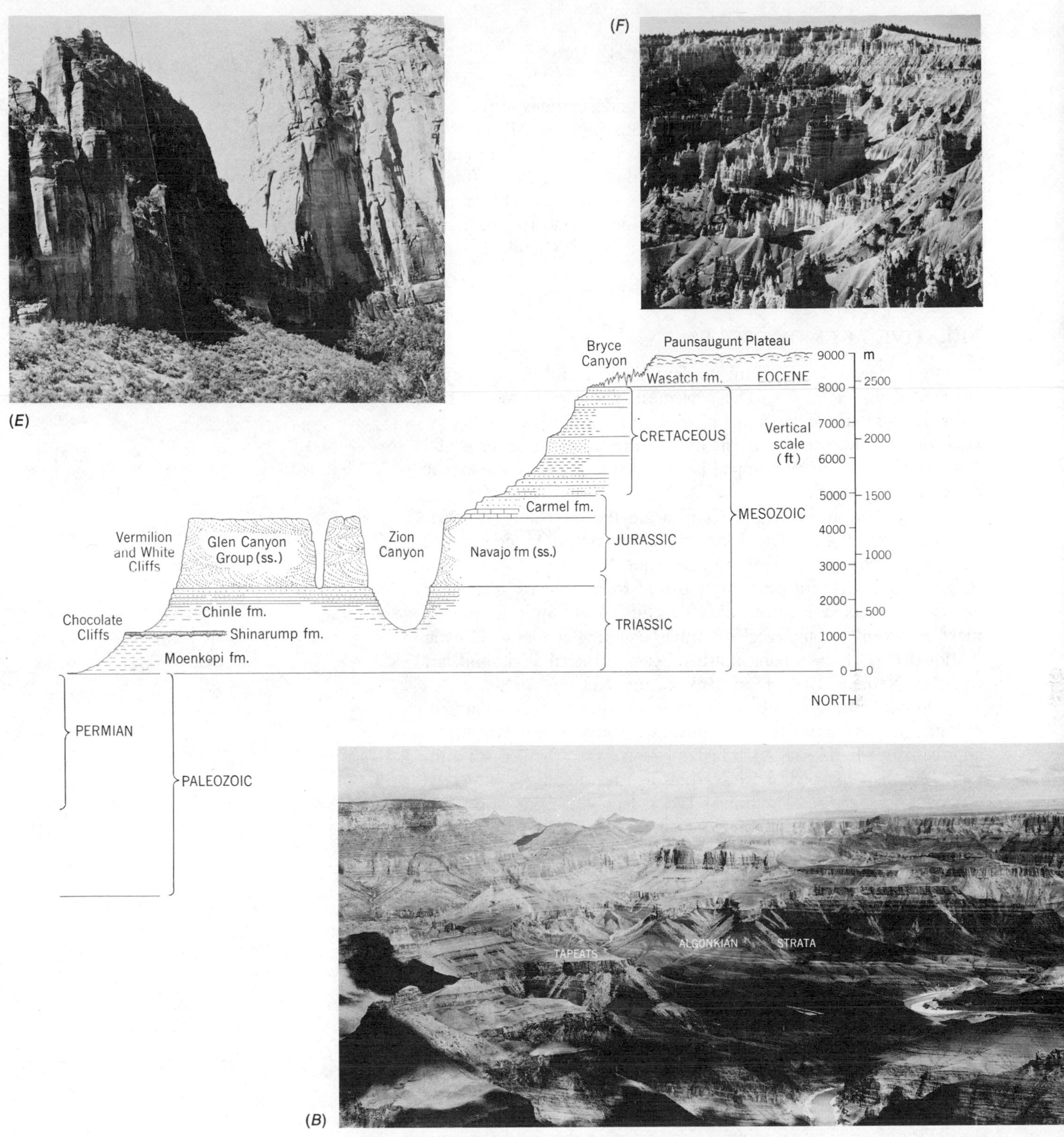

Figure 7.1 The geologic record of rock strata from Grand Canyon, Arizona, through progressively younger formations of Zion Canyon and Bryce Canyon, Utah. (Refer to Figure 7.11 for a map.) (*A*) Archean schists of the Inner Gorge at the foot of the Bright Angel Trail. These metamorphic rocks, of Precambrian age, are the oldest rocks of the entire sequence. (Douglas Johnson.) (*B*) Precambrian sedimentary strata comprising the Algonkian wedge, in the eastern part of Grand Canyon. (U.S. Geological Survey.) (*C*) General view of sedimentary sequence of Grand Canyon from Point Sublime. (A. N. Strahler.) (*D*) Uppermost strata of the rim of Grand Canyon. (Douglas Johnson.) (*E*) The walls of Zion Canyon, Utah. These great cliffs are of Jurassic sandstone. (A. N. Strahler.) (*F*) Bryce Canyon, Utah. Erosional forms in the Wasatch formation of Eocene age. (D. L. Babenroth.)

The table of geologic time gives ages and durations of the three eras, together with their subdivisions into **periods** (Table 7.1). Notice that the Paleozoic Era, with six periods, had a duration of 345 m.y.; the Mesozoic Era, with three periods, lasted only 160 m.y.; while the Cenozoic Era has been too short (65 m.y.) to warrant subdivision by periods; it is only about as long as one average period of the preceding eras. Where information is less abundant, larger blocks of time suffice as subdivisions.

RELATIVE AGES OF STRATA

An overall view of Grand Canyon, Zion Canyon, and Bryce Canyon is revealed in Figure 7.1. The geological profile and cross section shows how the whole panorama of geologic time is exposed to study. Each of the three canyons is cut into a different sequence of strata. Each sequence has been stripped back by erosion processes to expose the sequence beneath.

In the bottom of Grand Canyon are the most ancient rocks, those of Precambrian age. From these we ascend to deposits of successively younger eras of geologic time by a set of great rock stairs. Actually the stair treads are broad platforms; the risers between are clearly marked in great lines of cliffs. You can conveniently make an ascent through geologic time beginning at Grand Canyon National Park, then driving north to Zion National Park, and last to Bryce National Park. (See map, Figure 7.11.)

Our first stratigraphic principle is so simple as to seem self-evident: Among a series of sedimentary strata whose attitude is approximately horizontal, each bed is younger than the bed beneath, but older than the bed above it. This age relationship couldn't be otherwise in the case of sediment layers deposited from suspension in water or air.

So the first inference we make concerning the strata exposed in the walls of Grand Canyon, Zion Canyon, and Bryce Canyon is that they are arranged in order of decreasing age of deposit from bottom to top. We cannot, using this principle alone, say which of the three canyons displays the oldest series of strata, because the three localities are separated by many miles of intervening ground.

The age-layering concept is called the **principle of superposition.** Simple as it seems, there are two possible causes for concern. First, someone might object that the strata have been bodily overturned during mountain-making, as may happen in close Alpine-type folding of strata. In an overturned sequence, the uppermost beds are actually the oldest. The geologist routinely checks against this possibility of error by examining closely certain details of the sedimentary rock. Features such as ripple marking, curvature of fine layers (cross-bedding) in certain sandstones, and orientation of fossil shells give evidence of whether the strata are overturned from their original attitude. A second problem is that the principle of superposition does not tell whether the successive strata differ greatly in age or only by very small intervals of time.

Looking at the upper walls of Grand Canyon in Figure 7.1, your eye spans about 3000 ft (900 m) of thickness of strata in almost perfectly horizontal, parallel arrangement. The entire sequence of strata consists of several major layers, each with a distinctive appearance and composition. Each of these layers is referred to as a geologic **formation** and has been given a name. At the base, forming the edge of the Tonto Platform, is the Tapeats formation, a sandstone layer about 200 ft (60 m) thick. Above this is a soft, gray, sandy shale layer, about 500 ft (150 m) thick, named the Bright Angel formation, which forms smooth, gentle slopes. Above this, forming a great, sheer wall 500 ft (150 m) high, are three formations of limestone: the Muav, Temple Butte, and Redwall formations. Still higher are layers of red sandstone and shale, totaling about 1000 ft (300 m) in thickness, making up the Supai and Hermit formations. These are overlain by a pure creamy white sandstone layer, the Coconino formation, whose sheer 300-ft (90-m) cliff is easily seen in the upper canyon walls. Forming the canyon rim are the Toroweap and Kaibab formations of limestone, together about 500 ft (150 m) thick.

DISCONFORMITIES

Were all the sandstone, shale, and limestone strata of the Grand Canyon walls deposited in quick succession in only a small time span within the 600-m.y. Phanerozoic Eon? Or do they represent widely different periods of geologic time, so that the lowest formation, the Tapeats sandstone, is extremely ancient, but the rim formation, the Kaibab limestone, is very recent? There might conceivably be a difference in age of as much as 600 m.y. in the two formations.

Assuming a great age difference to exist, we are faced with the further possibility that the entire sequence of rocks, 3000 ft (900 m) thick, represents slow, continuous deposition of sediment without interruption of any consequence throughout the entire span of time. A quite different possibility is that each formation was deposited in a short period, but that the records of periods of deposition are themselves separated by long intervals of time, tens of millions of years long, when no deposition took place.

Figure 7.2 shows the interpretation of the history of the Paleozoic strata of the lower walls of Grand Canyon, as worked out by stratigraphers. The bottommost three formations, Tapeats sandstone, Bright Angel shale, and Muav limestone, were deposited in rather

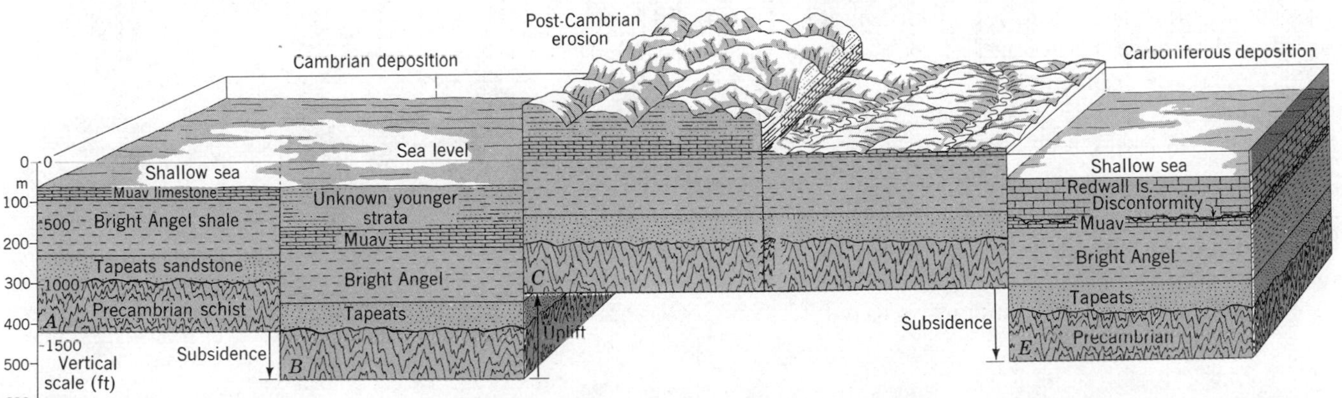

Figure 7.2 Sequence of events leading to the development of a disconformity in the walls of Grand Canyon.

rapid succession in a shallow sea, as shown in diagram A. Additional formations of which we have no record were possibly added to these three (diagram *B*). Then a broad rise of the earth's crust brought these formations above sea level, where stream erosion removed great quantities of rock (diagrams C and *D*). With only the Tapeats, Bright Angel, and Muav formations remaining, a downsinking of the crust occurred, depressing them below sea level and producing a shallow sea in which a new period of deposition began (diagram *E*). This submergence would have allowed deposition of the Redwall limestone formation directly upon the older Muav formation during Carboniferous time.

So the thin line that we now see between the Muav and Redwall formations is the sole indicator of a vast period of lost record. In other words, there existed a time period for which no rock has been retained here. There is no record here of the Ordovician, Silurian, and Devonian periods (see Table 7.1). A surface of separation between two formations, representing a great gap of time, is termed a **disconformity.**

CONTINUITY OF STRATA

Strata of the walls of Zion Canyon do not resemble most of those in Grand Canyon. In Zion the most striking feature is the sheer sandstone wall, 1000 to 2000 ft (300 to 600 m) high, with scarcely a foothold. Steplike forms such as those in Grand Canyon occur only in the lower part of the walls. It seems certain that the strata of the two canyons were deposited under different conditions. Although it is possible that both series were deposited at the same time in unlike environments, this situation is not likely because the two regions are only a few tens of miles apart. It is more likely that the strata of Zion Canyon differ in age from those in Grand Canyon. Which sequence is the younger?

One means of ascertaining whether strata in two localities are of the same or different age is to travel the ground from one locality to the other, observing the strata continuously along the line of march. If one can actually walk upon the same rock layer throughout the entire distance, the similarity of age is proved by the **principle of continuity.** A simple case is shown in Figure 7.3, where the same layer can be followed for miles in the rim of a series of canyons and cliffs.

Where strata have been partly removed from a region by erosion, a combination of continuity and superposition (Figure 7.4) can be used to determine relative ages of rock strata in widely separated places. The combination method is required to prove that the rocks of Grand Canyon are all of older age than those of Zion Canyon and that these in turn are all older than the strata of Bryce Canyon.

The principles of superposition and continuity are attributed to a Danish physician, Nicolaus Steno, who worked out the concepts while serving as physician to the Duke of Florence. Steno studied sedimentary strata exposed along the walls of the Arno Valley, not far southeast of Florence. His findings were published in 1669, comprising the first fruitful effort to work out the geologic history of a region by interpretation of strata.

Figure 7.3 Principle of correlation of strata by direct continuity. The bed at *A* can be traced without interruption to a distant location, *B*.

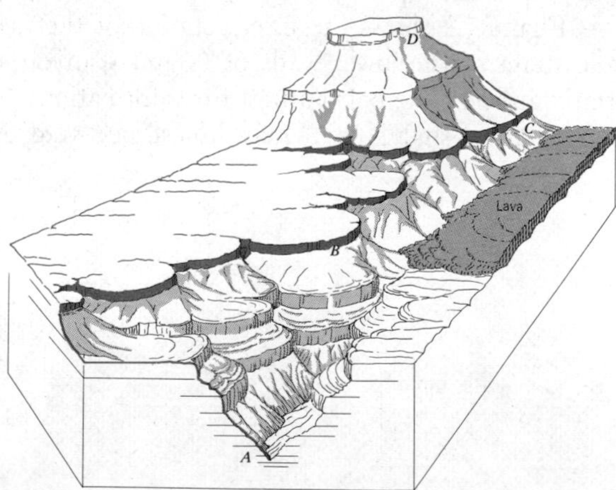

Figure 7.4 Combination of the principles of continuity and superposition in stratigraphic correlation. The order of succession from *A* to *B* is evident from direct superposition, from *B* to *C* by continuity, and from *C* to *D* by superposition.

FOSSILS

Of all sources of information, perhaps the most helpful to the stratigrapher are **fossils,** those ancient plant and animal remains or impressions preserved by burial in sedimentary strata. About the year 1800 an English civil engineer and geologist, William Smith, collected fossils in strata exposed in canal excavations. He found fossils of like species to be present in all parts of a single formation that could be proved to be one and the same bed by direct continuity. Fossil species in strata above or below were found to be distinctively different, but to occur consistently in the same order in widely separate localities.

Once the order of fossils was established by direct observation, the fossils themselves became the evidence for matching like ages of strata elsewhere in the world. For example, the fossils in certain strata in Wales were studied early in the nineteenth century, and these rocks became established as the original standard for the *Cambrian Period* of geologic time (*Cambria* is the Latin name for Wales). One distinctive fossil animal, the trilobite, was abundant in the Cambrian seas, and consequently certain of its many species serve as guide fossils for the Cambrian Period throughout the world. A fossil species particularly well suited to determination of age of strata is known as an **index fossil.**

The value of fossils in telling us the age of rock strata arises from the fact that all forms of plant and animal life have continually and systematically undergone changes with passage of time, a process termed **organic evolution.** If we have before us a complete, or nearly complete, description of past life forms as determined from fossils, and if we know the geologic age to which each fossil form belongs, it is often a simple matter to give the age of any sedimentary layer merely by extracting a few fossils from the rock and comparing them with the reference forms.

Although this practice works well in many cases, there are frequent difficulties. For one thing, many strata contain no fossils, usually because the conditions under which the sediment was being deposited were unsuitable for maintenance or preservation of plant or animal life. A second problem is that some fossil organisms showed such slow changes that almost identical forms survived over a long span of geologic time.

The determination of age is much more convincing when an entire natural assemblage of animal forms, or **fauna,** is studied in a given formation. The principle of **succession of faunas** is simply that each formation has a different fauna (or flora) from that in the formations above it and below it. The principle was proved by use of the principles of superposition and continuity of strata.

In the Bright Angel formation of Grand Canyon many fossil trilobites have been found (Figure 7.5). Specimens can be seen in displays provided by the National Park Service at Grand Canyon. Because these trilobites resemble varieties found in the original Cambrian strata of Wales, the geologist can state that the two sets of strata are of almost the same age. Thus, a jump of thousands of miles with no direct connection between the strata was bridged through the use of index fossils.

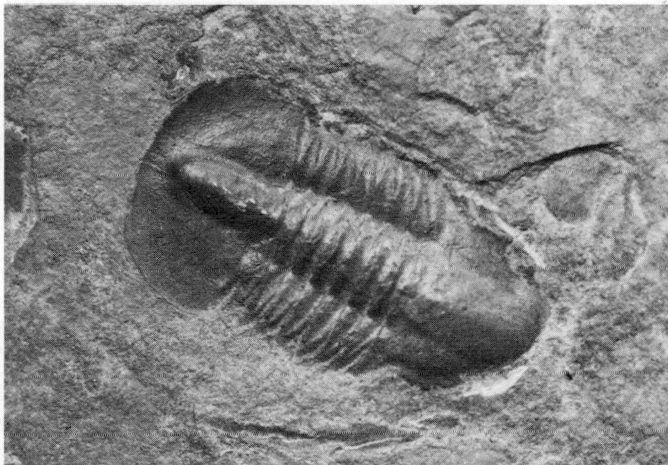

Figure 7.5 This fossil trilobite from the Bright Angel shale of Grand Canyon establishes the formation as being of Cambrian age. The head is to the left. About 1½ times natural size. (Department of the Interior, Grand Canyon National Park.)

Other examples of distinctive fossils of Grand Canyon strata include ancient representatives of the gastropods (snails) and pelecypods (clams) from the Kaibab limestone formation, comprising the canyon rim rocks (Figure 7.6). These fossils are now composed of silica, which replaced the original shells of carbonate matter. They were released from the enclosing rock by use of an acid bath, which dissolved the carbonate matrix and left the silica intact. Extremely minute shell details are preserved in these fossils, even though no original shell matter remains.

Figure 7.7 illustrates a Grand Canyon fossil of a different kind. This object is easily recognizable as a plant leaf, resembling that of a modern fern. What we see here is merely an impression of the leaf on the bedding plane of a fine-grained red shale, the Hermit formation, in the upper walls of the canyon. Many plant fossils consist of a thin layer of carbon representing the altered plant tissue. Some animal fossils consist only of the cavity in which the shell formerly existed, but from which the shell has been removed by action of circulating ground water.

Worldwide studies by stratigraphers and paleontologists over the last 150 years have yielded an extremely detailed and nearly complete reference table of the divisions and subdivisions of geologic time, together with index fossils for all ages from the Cambrian to the recent.

Figure 7.6 Fossils of the Permian Kaibab strata, about 250 m.y. old, in Grand Canyon. (*Upper*) A gastropod, whose shell had a coiled form somewhat resembling that of a modern snail. (*Lower*) A simple clam shell, not greatly unlike certain forms seen today. (N. D. Newell, American Museum of Natural History.)

UNCONFORMITIES IN THE GRAND CANYON

The inner, lower gorge of Grand Canyon provides the geologist with a now-classic example of Precambrian rocks far from the Canadian Shield. Because of their unusual structure and arrangement these rocks allow us to develop further concepts in stratigraphic interpretation.

Precambrian rocks in Grand Canyon lie beneath the Cambrian Tapeats sandstone, which is the rim rock of the Tonto Platform forming the brink of the Inner Gorge (Figure 7.1A). Looking down into the narrow Inner Gorge, one notices that the walls are here completely lacking in horizontal bedding planes, but instead have an extremely rough surface with sets of nearly vertical partings giving a grooved appearance to the rock walls. This rock, the Vishnu schist, is a metamorphosed sedimentary rock rich in quartz, mica, and hornblende. Here and there bands of coarse-grained diorite and granite cut through the schist. No fossils or indications of life have been found in the Vishnu schist, although this is not surprising for a highly metamorphosed rock of this type. The Vishnu schist is dated as of early Precambrian age, older than 2.4 billion years (b.y.) (see Table 7.1). In the Grand Canyon region this lower Precambrian division is named Archean.

Stratigraphers refer to an episode of mountain-making, interpreted from the presence of deformed strata, as an **orogeny.** Folding, faulting, and metamorphism are the usual indications of an orogeny. Names of several major orogenies of worldwide scope appear in Table 7.1. The orogeny in which the Vishnu schist was altered and intruded is perhaps the equivalent of the Kenoran orogeny.

If we follow along the rim of the Tonto Platform, continuing

Figure 7.7 This impression of a fern leaf was found on a bedding plane of fine-grained red shale of the Hermit formation in the walls of Grand Canyon. About natural size. (Department of the Interior, Grand Canyon National Park.)

to study the walls of the Inner Gorge below us, a new geologic feature enters the picture. A sloping wedge of tilted sedimentary strata appears between the Tapeats sandstone and the Vishnu schist (Figure 7.8). The wedge continues to thicken until several thousand feet of strata are exposed. This tilted sedimentary series, consisting of shales and sandstones and belonging to the Grand Canyon Group, includes several individual formations whose names are not important here. In this area the Grand Canyon Group is assigned to the Algonkian time division of the Precambrian, and is correlated with the late Precambrian rocks of the Canadian Shield. The age would perhaps be 1.7 b.y. or younger.

From the principle of superposition, it is evident that the Algonkian sedimentary strata are younger than the Archean Vishnu schist, upon which they rest, but they are older than the Cambrian Tapeats formation, beneath which they lie. Thus, even when we don't know their exact position in geologic time, it is fairly certain that the Algonkian rocks belong to the late or middle Precambrian. Moreover, some hemispherical masses found in limy layers of the Algonkian strata represent deposits made by lime-secreting algae. These are among the earliest known forms of life.

An explanation of the Algonkian rock wedge in Grand Canyon is given in the series of block diagrams of Figure 7.9. Geosynclinal sedimentary strata of early Precambrian age were crumpled and metamorphosed into the Vishnu schist by one of the earliest orogenies of which we have any record. Block A shows a Precambrian mountain range eroded on the ancient schist. Next, these mountains were reduced by erosion to a peneplain (*B*). A **peneplain** is a gently rolling plain, lying close to sea level, produced at the close of a long period of erosion. Crustal sinking followed, and the region became a shallow sea. A great thickness of Algonkian sediments was now deposited in horizontal layers (*C*). These layers were tilted in great fault blocks in the late Precambrian (*D*). Again prolonged erosion removed the mountains, creating a second peneplain above which a few of the harder sandstone masses projected as ridges (*E*).

Figure 7.8 A wedge of Algonkian strata (*A*) lies beneath the Tapeats sandstone (*T*) of the Tonto Platform, but rests upon the Vishnu schist (*V*) of Archean age. (Sketched from a photograph.)

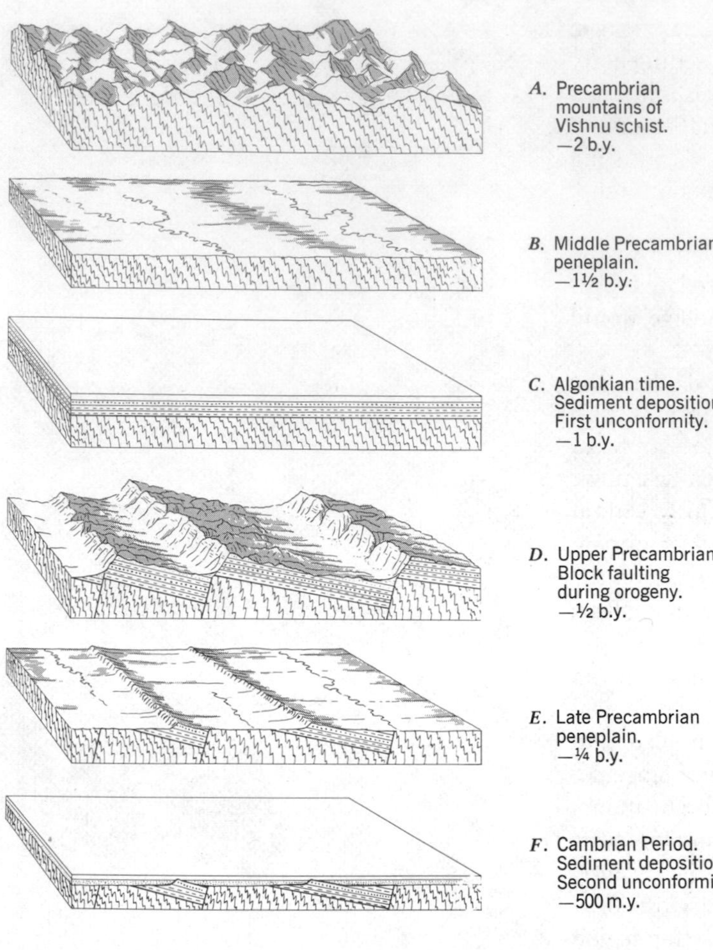

Figure 7.9 This set of block diagrams shows the manner in which the great wedges of Algonkian strata came into existence in lower Grand Canyon.

This topography existed at the close of the Precambrian time. Again crustal sinking took place, causing the region to become a shallow sea, which received the Cambrian sediments (*F*). Consequently, in places, the Cambrian layers rest directly upon the Vishnu schist, but in other places they rest upon a thick wedge of Algonkian strata.

The line of separation seen between the Algonkian beds and the Vishnu schist in the canyon wall is referred to as an **angular unconformity,** since the layers of one group are at an angle to the layers of the other group. The contact line is evidence not only of a vast erosion period that intervened between the formation of the two rock groups, but also of an orogeny that followed the development of the older rock group.

A second unconformity exists in the line of separation between the Cambrian Tapeats sandstone and all of the Precambrian. This unconformity is shown in detail in the geological cross section of Figure 7.1. Notice that a highly resistant formation, the Shinumo Quartzite, stood as a residual hill above the general level of the late Precambrian peneplain. It protrudes through the Tapeats sandstone, which was deposited around the high hill but did not cover it.

In few places on earth are unconformities displayed so clearly and on so grand a scale as in the bottom of Grand Canyon. The term disconformity, explained in connection with the line of separation of the Muav limestone and the Redwall limestone, refers to an erosion interval produced merely by simple vertical rising and sinking of the crust, with no tilting or faulting intervening, only erosion. Thus, in a disconformity the strata above and below the line are horizontal and parallel, whereas in an angular unconformity the rocks are discordant in attitude along the separation line.

Radiometric age determinations on intrusive and extrusive igneous bodies within a complex arrangement of sedimentary strata can be of great assistance in assigning dates to the rock groups lying above and below unconformities. Figure 7.10 shows a hypothetical example as it might be applied to a case resembling that of rocks of the inner Grand Canyon.

Suppose that radiometric ages are found for each of three different igneous rock bodies, labeled 1, 2, and 3 in the diagram. If igneous rock 1 has an age of 2.4 b.y., we can say that the adjacent schist, into which the igneous rock was intruded, is more than 2.4 b.y. old. We can also say that the tilted strata on the right are younger than 2.4 b.y., because they were deposited after igneous body 1 was leveled off by erosion.

Igneous body 2 is a sill which was intruded into the tilted strata, but whether before or after they were tilted can only be pure conjecture from the evidence shown. If the age of igneous rock 2 turned out to be 1.8 b.y., it would mean that the tilted strata are at least that old, perhaps much more, but not exceeding the 2.4 b.y. limit set by igneous rock 1.

If igneous rock 3, a thin vertical dike, yielded an age of 400 m.y., we would know that the Cambrian Period is older than 400 m.y., but that the Carboniferous Period is younger than 400 m.y.

THE RIM ROCKS OF THE GRAND CANYON

Using the principles of stratigraphy, the remaining layers of the Grand Canyon region fall into place in the table of geologic time. Above the Redwall formation with its great limestone cliff are strata of Permian age, and these continue upward to the very rim of Grand Canyon, shown in Figure 7.1*D*. The Permian strata include sandstones and shales (Supai formation), a soft red shale (Hermit

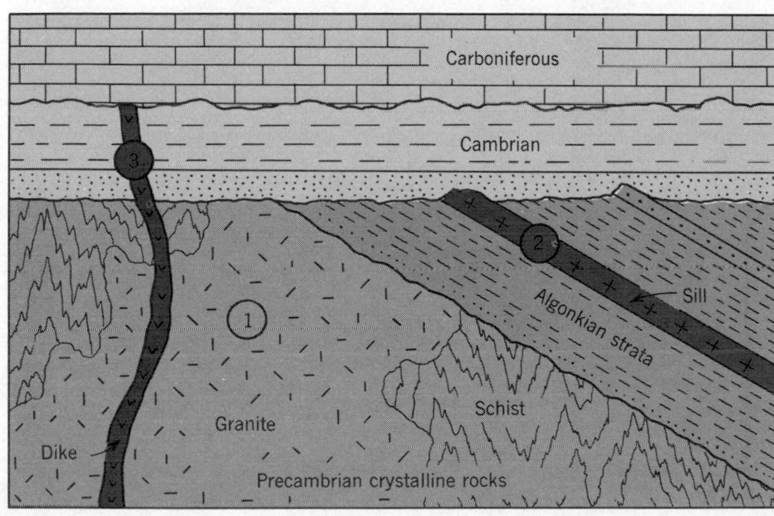

Figure 7.10 Radiometric dating of igneous bodies allows limiting ages to be established for the enclosing and overlying sedimentary strata.

formation), and an almost-white, massive sandstone bed (Coconino formation). Finally, the sequence is capped by hard rimming strata making up the Kaibab and Toroweap formations. These are mostly limestones, and in a semiarid climate, they are highly resistant to erosion.

Figure 7.11 is a **geologic map.** It is simply a map of the extent of surface exposure of each of the various units of strata. At the right is a **structure section,** showing the strata as if cut vertically by a north-south trench, along the line marked by letters A and *B* on the map. The hard Permian strata form a broad platform extending far north and south of the Grand Canyon. This surface is called the Kanab Plateau north of the canyon.

THE WHITE CLIFFS OF ZION

Thousands of feet of strata have been eroded from the Kanab Plateau, and to find these we must continue north almost to the southern border of Utah. Here we encounter the first units of Mesozoic age. Crossing the Chocolate Cliffs and the Vermilion Cliffs, which are the eroded edges of strata of Triassic age, we arrive at the White Cliffs of Zion Canyon.

Zion Canyon derives its character from a sheer cliff of sandstone, rising unbroken through 2000 ft (600 m), as shown in Figure 7.1*E*. This unit is the Navajo formation. Close examination of the weathered surface of this sandstone shows that it consists of sloping

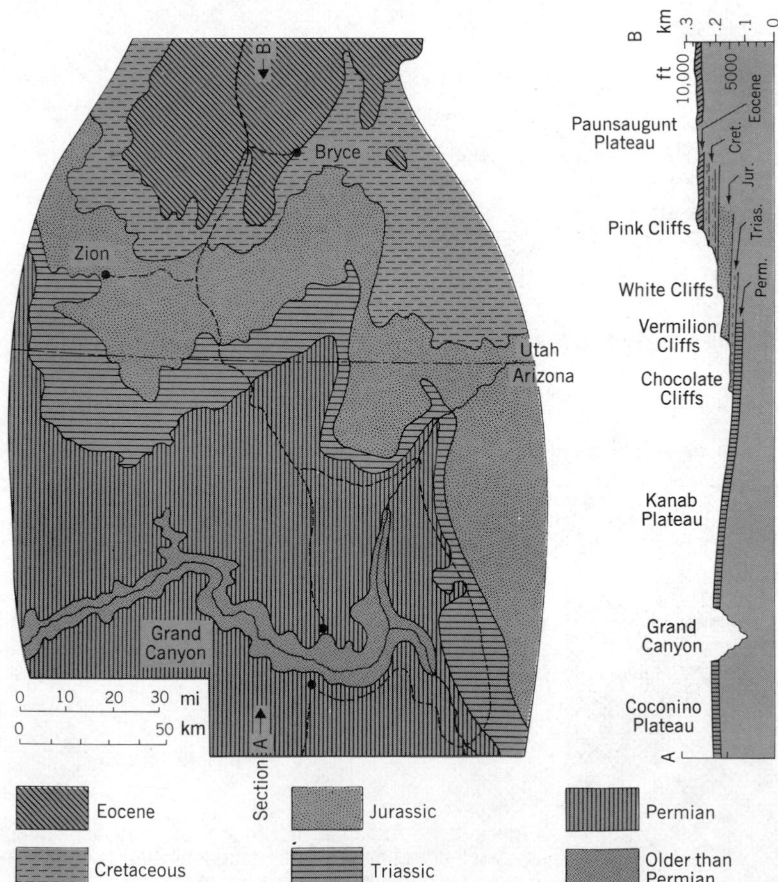

Figure 7.11 Simplified geologic map and structure section of the Grand Canyon–Zion Canyon–Bryce Canyon region of Arizona and Utah.

layers. This structure is called **cross-bedding,** and shows us that the rock originated as dunes of loose sand in a great desert (Figure 7.12).

Leaving Zion behind as we continue north, we cross a great thickness of strata of Cretaceous age on our way to Bryce Canyon.

CENOZOIC STRATA AT BRYCE CANYON

As we approach Bryce Canyon, the highway rises to a high plateau over 8000 ft (2400 m) in elevation. On the eastern rim of this plateau lies Bryce Canyon, eroded by streams that are in the process of undermining the capping formation of the plateau. The plateau is locally named the Paunsaugunt Plateau, and the escarpment which forms its rim is known as the Pink Cliffs. Here we find intricately sculptured forms, including innumerable pinnacles and walls carved in a soft rock of variegated shades of pink, red, and cream (Figure 7.13). Rocks of the plateau comprise the Wasatch formation, about 1000 ft (300 m) thick. These strata are of Cenozoic age.

The Cenozoic Era, or "era of recent life," spans only 65 m.y. of geologic time. Rather than being subdivided into periods, Cenozoic time is broken into seven **epochs.** These are listed in Table 7.2, together with their ages and durations. The Wasatch formation seen at Bryce Canyon belongs to the Eocene Epoch, the longest epoch and second oldest of the Cenozoic Era.

The colorful rock of Bryce Canyon proves, upon examination, to be largely of carbonate composition. In places, there are dense limestone layers; elsewhere the rock is soft calcareous mudstone. Fossils scattered through the formation indicate that the carbonate sediment accumulated in a shallow fresh-water lake.

Other than volcanic rocks, in this part of the Colorado Plateau region no record remains of rocks younger than the Eocene Epoch of the Cenozoic Era. Late in the Cenozoic, the Colorado Plateau region was subjected to renewed faulting, which raised large blocks

Table 7.2 **Epochs of the Cenozoic Era**

Epoch	Duration (m.y.)	Age (m.y.)
		0
Holocene (recent)		
		(11,000 yr)
Pleistocene	2	
		2
Pliocene	11	
		13
Miocene	12	
		25
Oligocene	11	
		36
Eocene	22	
		58
Paleocene	7	
		65
(Cretaceous period)		

Figure 7.12 Cross-lamination reveals ancient dune sands in the Navajo sandstone, Zion Canyon, Utah. (Douglas Johnson.)

Figure 7.13 The sculptured pinnacles of Bryce Canyon, Utah. (D. L. Babenroth.)

to the high elevations we find there today. Erosion was greatly intensified following this uplift. The Colorado River and its tributaries carved deep canyons into the strata.

The Colorado Plateau region has provided us with a succession of strata excellent for historical interpretation. This is a region of spectacular scenery seen by millions of persons each year and explained to many of them by educational programs at three National Parks and a number of National Monuments.

When it comes to unconformities and the Precambrian record, few places on earth can equal the inner Grand Canyon as a display for learning. As downstream excursions on the Colorado River by boat and raft become increasingly popular, more and more Americans will see the Archean and Algonkian rocks at close range. Who is to say whether the mile-thick layered mass of Paleozoic strata is more impressive as we look up from the Inner Gorge, or down from the canyon rim?

YOUR GEOSCIENCE VOCABULARY

historical geology
stratigraphy
stratigrapher
paleontology
paleontologist
eon
era
period
principle of superposition
formation
disconformity
principle of continuity
fossil
index fossil
organic evolution
fauna
succession of faunas
orogeny
peneplain
angular unconformity
geologic map
structure section
cross-bedding
epoch

SELF-TESTING QUESTIONS

1. What is the scope and content of historical geology? What parts do stratigraphy and paleontology play in historical geology?

2. Distinguish between the Cryptozoic Eon and the Phanerozoic Eon. Name the eras of abundant life. How long did each era last? What is the approximate average length of a single period? How many periods fall within the Paleozoic Era? within the Mesozoic Era?

3. Describe the principle of superposition, and explain how it is used. Discuss the question of rates of deposition and durations of periods of nondeposition as interpreted from strata. Use Grand Canyon strata as examples. How are changes of crustal level and sea level, factors in determining the stratigraphic sequence? What is a disconformity? What is the principle of continuity?

4. What is a fossil? How were fossils first used in the study of stratigraphy? How do fossils serve as guides to ages of strata and to correlation of strata? What forms do fossils take and how are they preserved? Give examples.

5. What is the principle of organic evolution as applied to fossils? What is an index fossil? Why is a fossil fauna more useful than a single fossil in stratigraphic research? What is the principle of succession of faunas?

6. What is an unconformity? How does it differ from a disconformity? Describe an angular unconformity exposed in Grand Canyon, and give the geologic history leading to its development.

7. How can radiometric age determinations be used to establish absolute ages for geologic periods and events? Show by means of a simple diagram. How can an orogeny be dated?

8. Interpret the rocks of the great cliffs of Zion Canyon. Interpret the strata of Bryce Canyon in terms of environment of deposition.

9. How is the Cenozoic Era subdivided into time units? What were the final geological events in the history of the Colorado Plateau, leading up to the present time?

(*Above*) The *Glomar Challenger* is 400 feet long and weighs 10,000 tons. Her drilling derrick, which rises 194 feet above waterline, has a capability of raising a weight of one million pounds. The pipe sections used in drilling lie on a rack forward of the derrick Altogether, 24,000 feet of 5-inch pipe is stored here. (Deep Sea Drilling Project.)

(*Left*) Two scientists and the cruise operations manager examine the bottom of a core barrel filled with sediment taken from the bottom of the North Atlantic. The scientific staff changes with each new leg of the cruise. (Deep Sea Drilling Project.)

8 GEOLOGY OF THE OCEAN FLOOR

VOYAGE OF A CHALLENGER

By all odds the strangest ship afloat has wandered the oceans of the globe for a half-dozen years, stopping only at two-month intervals to take on supplies and a fresh staff of scientists. The unusual craft is the *Glomar Challenger;* it bores holes into the sediment and rock of the deep ocean floors, bringing up samples for study. Glomar's namesake, *H.M.S. Challenger,* roamed the oceans a century earlier, but at best her crew could only scrape some mud samples from the ocean bottom, using a dredge and miles of steel cable.

Glomar Challenger looks like an oil-drilling rig, mounted on an ordinary ship's hull. The drilling rig is indeed like other rigs, but the hull is unlike that of any commercial vessel. The difference lies in an elaborate positioning mechanism, by means of which the ship can maintain its exact position while the laborious job of drilling goes on. Four motors stabilize the vessel by directing water jets from tunnels in the hull. They are computer-controlled and respond instantly to the pitching and rolling motions the vessel experiences in heavy seas. The flexible drill string is capable of accommodating a considerable motion of the platform. It is even possible for the drill to be withdrawn, the bit replaced, and the hole again found and penetrated.

Called the most successful scientific expedition in the history of science, the long voyage of the *Glomar Challenger* has enabled scientists to probe the floor of every ocean except the Arctic. Hundreds of holes have been drilled and tens of thousands of feet of core samples recovered. In these cores are found a historical record of the ocean basins undreamed of a few decades ago. We used to think that the dark basaltic rock of the oceanic crust was much older than most rock of the continents. Quite the opposite has proved true. Oceanic basalt is geologically young; the ocean basins themselves are among the most recently formed of the earth's major crustal features.

The voyage of the *Glomar Challenger* is under the supervision of the Deep Sea Drilling Project (DSDP) of the National Science Foundation. The costs run to some $10 million per year, but the scientific results make the outlay seem a bargain. Perhaps the chief gain of the project as a whole has been to reinforce the major concept of the revolution in geology—the concept of continents drifting apart or coming together, opening and closing the ocean basins in the process.

In this chapter we will investigate the strange, dark world of the deep ocean floor. Until the advent of modern methods of sea-floor exploration, developed intensively in World War II, the deep ocean was indeed a world of darkness, scientifically speaking. The fifth decade of the twentieth century ushered in a golden age of discovery of the ocean floors.

CONTINENTS AND OCEAN BASINS

Imagine the waters of the ocean to be completely removed, revealing dry land over the entire globe. When this is done, we see that the natural outlines of the continents are larger and more regular than those appearing on the conventional globe showing merely the shoreline.

A general picture of the earth's solid surface form emerges when we make a graph showing the proportion of surface lying between equal units of vertical distance (Figure 8.1). The unit of vertical distance is the kilometer. The length of each horizontal bar is proportional to the percentage of surface area found within each one-kilometer elevation zone.

The graph shows most of the surface as concentrated in two general zones: (1) on the continents between sea level and 3000 ft (1 km) elevation, and (2) on the floors of the ocean basins from about 10,000 to 20,000 ft (3 to 6 km) below sea level. This graph also tells us that in a general way the continents are broad, tablelike areas whose edges slope away rapidly to the deep ocean floor. Although the floor does not lie entirely within one elevation zone, vast areas lie at approximately the same depth.

You should grasp the idea that the ocean basins are brimful of water, so full that the oceans overlap considerable areas of the continental margins to produce shallow seas bordering the shores. A true picture of the continents in relation to the ocean basins emerges when the ocean level is imagined to be dropped some 500 to 600 ft (150 to 180 m). In so doing we uncover these shallow continental shelves and inland seas. When the area lying above the 600-ft (180-m) submarine contour is included with the lands, it will be found that the continents make up about 35% of the total earth's surface area and the ocean basins about 65%.

RELIEF FEATURES OF THE OCEAN FLOOR

The continental landscape is familiar to all of us, but few persons have seen the deep ocean floor at close range. We now have many undersea photographs taken by a camera lowered to the sea floor from a ship (Figure 8.2), but these give no sense of landscape. We need specialized types of submarine charts and diagrams made from soundings to show the relief forms of the ocean basins.

Figure 8.1 The length of each bar represents the percentage of surface area in each 1 km altitude zone of the solid earth.

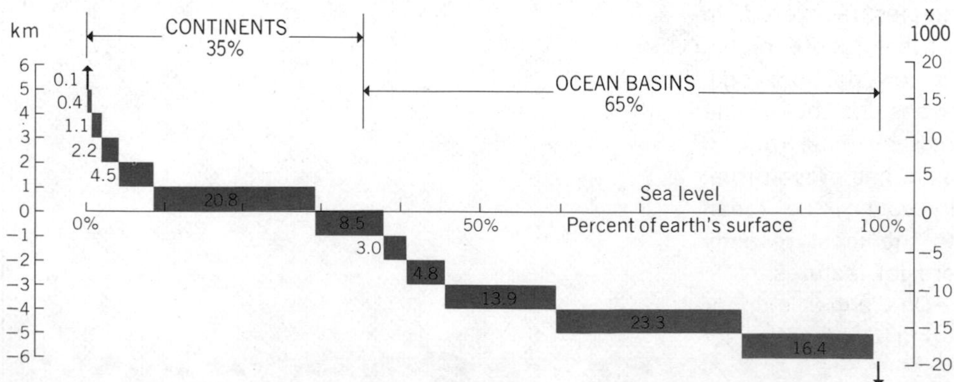

In the early decades of oceanographic research, sounding of the ocean bottom had to be done by lowering a heavy weight on a thin steel cable until the weight reached bottom, allowing the depth to be measured by the length of cable let out. This was a very slow and costly process. Our knowledge of the sea floor was very scanty until the time of World War II, when continuously recording echo-sounding apparatus was put into general use in naval vessels.

The **precision depth recorder** makes use of a sound-emitting device attached to the bottom of the ship. Pulses of sound are sent down through the water from the ship's hull and are reflected from the ocean floor to the ship, where they are picked up by a microphone. An automatic recording device indicates the time required for sound waves to reach the bottom and return. Reflections are plotted continuously by a writing instrument to give a line representing the profile of the ocean bottom (Figure 8.3).

By allowing the precision depth recorder to operate continuously while the ship travels, a profile across the sea floor is obtained. This information can be used to make maps of the sea floor only if the exact position of the ship is known at all times. Fortunately, precise positioning of a vessel is no longer a problem. Echo sounding enormously increased our knowledge of the configuration of the ocean floors within a span of only two decades—the 1940s and 1950s.

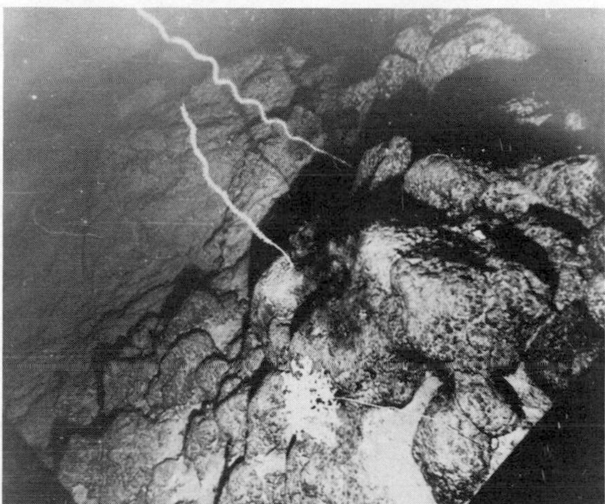

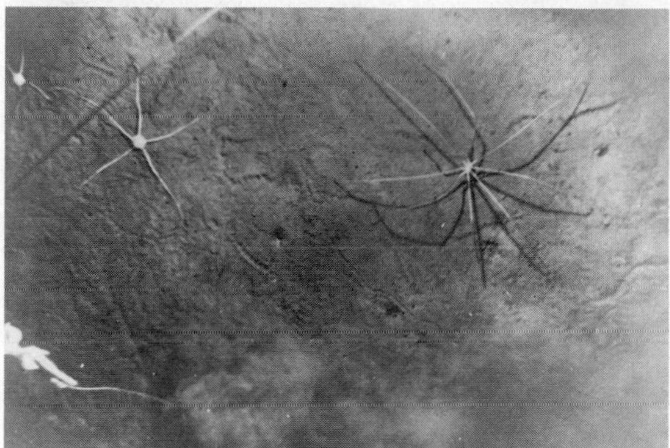

Figure 8.2 Submarine photography. (*Top*) A research scientist of the Lamont-Doherty Geological Observatory staff prepares to lower an undersea camera over the side of the research vessel *Vema.* (National Academy of Sciences, IGY.) (*Center*) This photograph of the ocean floor at a depth of 6600 ft (2000 m) shows an outcrop of bedrock on the side slopes of a seamount. (Lamont-Doherty Geological Observatory of Columbia University.) (*Bottom*) A starfish (*left*) and a sea spider (*right*) seen on a mud bottom at a depth of 6000 ft (1800 m) on the continental slope of the eastern United States. (D. M. Owen, Woods Hole Oceanographic Institution.)

DIVISIONS OF THE OCEAN FLOOR

The topographic features of the ocean basins fall into three major divisions: (1) the **continental margins,** (2) the **ocean-basin floors,** and (3) the **Mid-Oceanic Ridge** (Figure 8.4).

The continental margins lie in belts directly adjacent to the continents, while the Mid-Oceanic Ridge divides the basin roughly in half. Thus, the deep floor of an ocean basin lies in two parts, one on either side of that ridge. These major topographic divisions apply particularly to the North Atlantic basin. Let us consider the various features found in each of the major divisions.

Perhaps the best known and most easily studied of the units within the continental margins are the **continental shelves.** These shelves fringe the continents in widths from a few miles to more than 200 mi (320 km). Generally having very smooth and gently sloping floors, the continental shelves are for the most part less than 600 ft (180 m) deep. A particularly fine example is the continental shelf of the eastern coast of the United States (Figure 8.5). Great thicknesses of sedimentary strata lie beneath the broader continental shelves.

Along their seaward margins the continental shelves give way to the **continental slopes** (Figure 8.5). Although the actual inclination of the slope with respect to the horizontal is only 3° to 6°, this is exceptionally steep for submarine relief features and appears quite precipitous on the highly exaggerated profiles in Figure 8.6. The continental slope drops from the sharply defined brink of the shelf to depths of 4500 to 10,500 ft (1370 to 3200 m). Here the slope lessens rapidly, though not abruptly, and is replaced by the **continental rise,** a surface of much gentler slope, decreasing in steepness toward the ocean-basin floor (Figures 8.5 and 8.6). The continental rise ranges

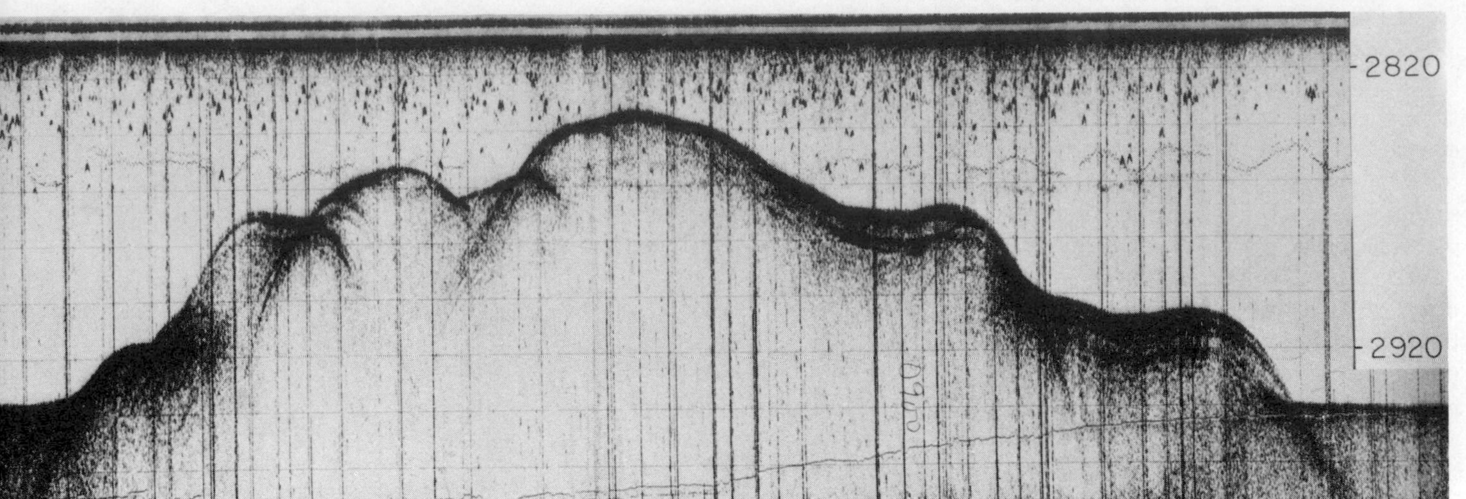

Figure 8.3 Photograph of the actual trace made by a precision depth recorder aboard the research vessel *Vema*. Depth in fathoms is given by figures at the right. The entire profile spans about 10 mi (16 km). An abyssal hill is shown flanked by the very flat surface of the Pernambuco Abyssal Plain. Location is about lat. 13° S, long. 28° W. (Lamont-Doherty Geological Observatory of Columbia University.)

in width from perhaps a hundred to several hundred miles. At its outer margin the continental rise reaches depths of 17,000 ft (5100 m), where it may be in direct contact with the deep floor of the ocean basin.

Notching the continental slope are **submarine canyons,** which may be visualized as resembling gullies cut by water erosion in the side of a hill, but on a huge scale (Figure 8.5). It seems likely that they have been scoured by currents of muddy water that slide in snakelike tongues down the slope to the deep parts of the basins. Such flows of denser muddy water are termed **turbidity currents.** On continental shelves and deltas of large rivers, mud is continually accumulating and may form precariously situated deposits that are easily disturbed and sent sliding by storm waves or earthquake shocks.

Second of the major topographic divisions is the ocean-basin floor, generally lying in the depth range of 15,000 to 18,000 ft (4600 to 5500 m). The ocean-basin floor includes abyssal plains and seamounts.

An **abyssal plain** is an area of the deep ocean floor having a flat bottom with the very faint slope of less than 1 part in 1000 (Figure 8.5). Characteristically situated at the foot of the continental rise, the abyssal plain is represented in all the oceans. Examples are the Hatteras and Nares Abyssal Plains at depths of roughly 18,000 ft (5500 m) (Figure 8.6). The only reasonable explanation

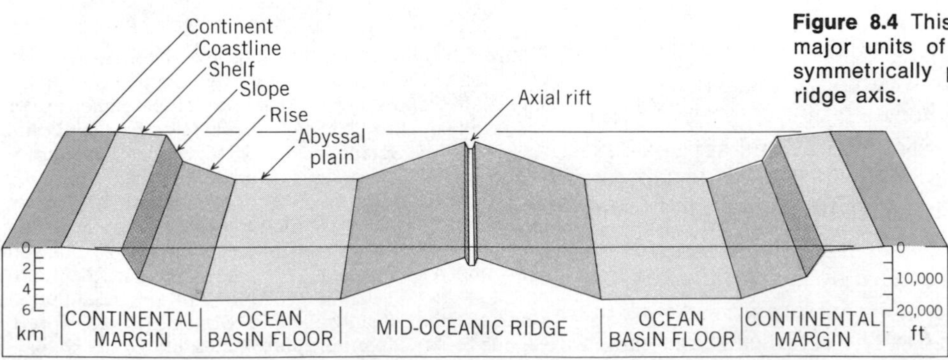

Figure 8.4 This idealized block diagram shows the major units of the North Atlantic Ocean basin as symmetrically placed on both sides of the central ridge axis.

Figure 8.5 The continental margin and ocean-basin floor off the coast of the northeastern United States. Depth in ft; km in parentheses. (Portion of *Physiographic Diagram of the North Atlantic Ocean,* 1968, revised, by B. C. Heezen and M. Tharp, Boulder, Colo., Geol. Soc. of Amer., reproduced by permission.)

for such nearly perfect flatness is that the abyssal plains are surfaces formed by long-continued deposition from turbidity flows. Sediment layers of many successive flows have spread out into thin sheets upon reaching the ocean floor. Previously existing irregularities of the ocean floor have thus been almost entirely buried over large areas.

Perhaps the most fascinating of the strange features of the ocean basins are the **seamounts,** isolated peaks rising 3000 ft (900 m) or more above the sea floor. They are most conspicuous on the ocean-basin floors. A good example from the Atlantic basin is the Kelvin Seamount Group, forming a row of conical peaks extending southeastward across the continental rise and abyssal plain for 600 mi (1000 km) (Figure 8.5). Most seamounts are interpreted as volcanoes of basaltic lava, built upon the sea floor.

We turn now to the third of the major divisions of the ocean basins, the Mid-Oceanic Ridge (Figure 8.4). One of the most remarkable of the major discoveries coming out of oceanographic explorations of the mid-twentieth century has been the charting of a great submarine mountain chain extending for a total length of some

Figure 8.6 Two profiles across parts of the North Atlantic Ocean basin show several major features of the ocean floors. Curvature of the earth has been omitted. (© 1973, John Wiley & Sons, New York. Data by B. C. Heezen, Lamont-Doherty Geological Observatory of Columbia University.)

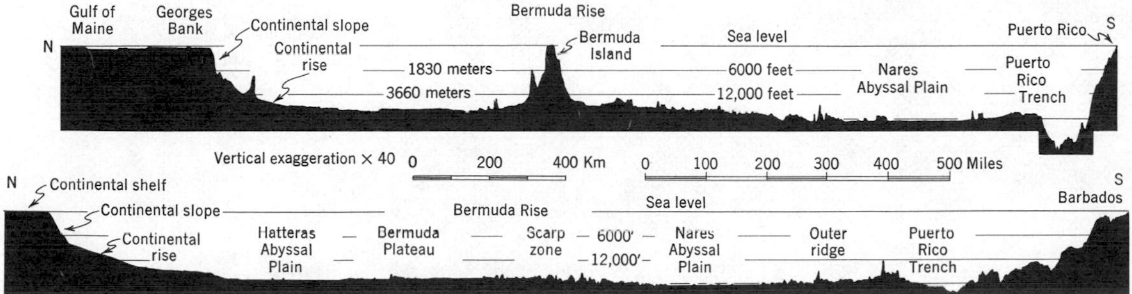

40,000 mi (64,000 km). A world map, Figure 8.7, shows the location and extent of this feature. The ridge runs down the middle of the North and South Atlantic ocean basins, into the Indian Ocean basin, then passes between Australia and Antarctica to enter the South Pacific basin. Turning north along the eastern side of the Pacific basin, where it is named the East Pacific Rise, the ridge contacts the North American continent along the coast of Mexico. The Mid-Oceanic Ridge also extends across the Arctic Ocean basin.

Of the great Mid-Oceanic Ridge, the part first known from detailed studies is the Mid-Atlantic Ridge, seen in detail in Figure 8.8. The ridge in its entirety is a belt, some 1200 to 1500 mi (1900 to 2400 km) wide, in which the surface rises through a series of rough steps from abyssal plains on both sides toward the central axis. Here the ridge assumes mountainous proportions. The higher points lie at depths of 6000 to 9000 ft (1800 to 2700 m). With reference to the adjoining abyssal plains, then, the Mid-Atlantic Ridge has a height of roughly 12,000 ft (3700 m).

A distinctive feature of the principal continuous ridge is the lack of a single high crest line, as many narrow mountain chains of the continents have. Instead, there is a characteristic trenchlike depression, or **axial rift,** running down the midline of the highest part of the ridge. This rift is idealized as a trench in Figure 8.4. Along with other parallel scarps and steplike rises on both sides, the rift strongly suggests that the crust has been pulled apart along the central zone.

A particularly significant feature of the axial rift is that it is broken into many segments, the ends of which appear to be offset along transverse fractures (Figure 8.8). This arrangement of offset segments is particularly striking in the equatorial zone of the Atlantic Ocean, where a single offset displaces the main axial rift by as much as 400 mi (640 km) (see Figure 8.7). It might seem obvious by

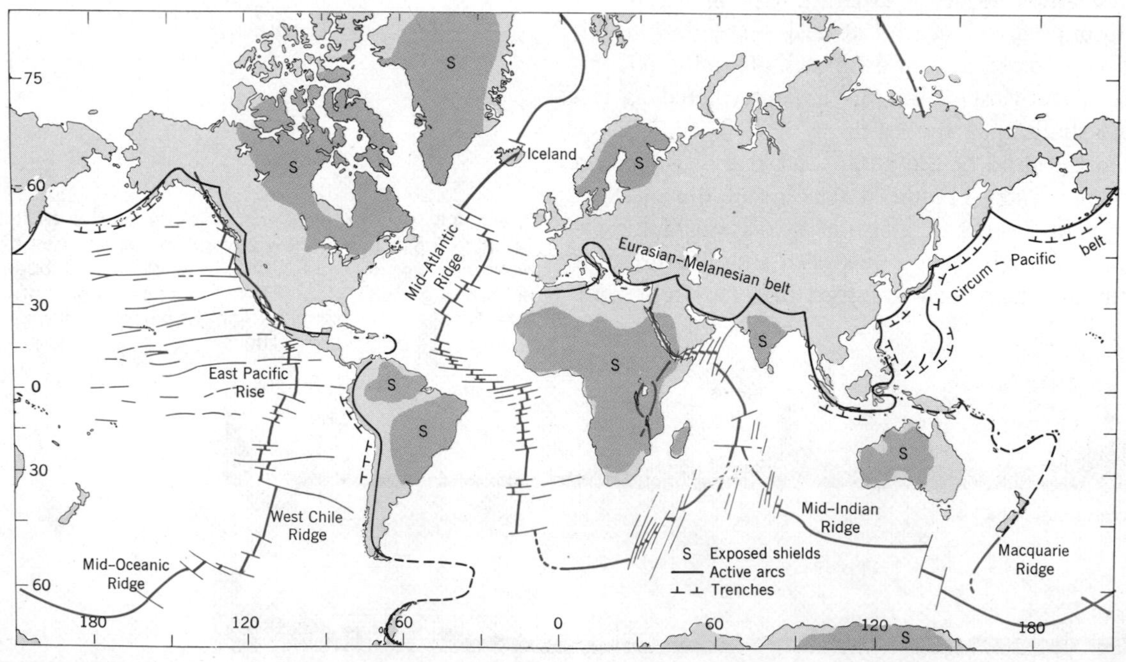

Figure 8.7 The Mid-Oceanic Ridge system and related fracture zones (*color lines*). Island arcs and mountain arcs are shown in black. (© 1973, John Wiley & Sons, New York.)

Figure 8.8 The Mid-Atlantic Ridge offset by transverse fracture zones in the vicinity of the Azores islands. (Portion of *Physiographic Diagram of the North Atlantic Ocean,* 1968, revised, by B. C. Heezen and M. Tharp, Boulder, Colo., Geol. Soc. of Amer., reproduced by permission.)

inspection that the transverse fracture zones are transcurrent faults (Chapter 6). In Chapter 9 we will interpret the transverse fracture zones in terms of lithospheric plate motions.

Before leaving this general description of relief features of the ocean basins, we should include as important elements the deep trenches and their adjoining volcanic island arcs, described in Chapter 5. These crustal features mark the contact of the continents with the ocean basins around much of the Pacific Ocean. As explained in Chapter 5, the trenches occur where an oceanic lithospheric plate is undergoing subduction beneath a continental lithospheric plate. A fine example of a trench, the Puerto Rico Trench, shows clearly at the right-hand end of the profile in Figure 8.6.

CORE SAMPLING OF THE OCEAN FLOOR

Since the 1870s, oceanographers have systematically taken samples of materials of the ocean floors. At first this could be done only by means of dredges that scraped off a thin layer and brought it to the surface for examination.

By the 1930s information about the sediment layer itself began to be obtained by the process of **coring,** which is simply vertical penetration by a long section of pipe that cuts a cylindrical sample, or **core.** Brought to the surface, the core is extruded, giving a complete cross section of the layer (Figure 8.9). Cores over 50 ft (15 m) are readily obtained. The cores are cut in half longitudinally, revealing the bedded structure and permitting small interior samples to be taken for microscopic examination and chemical and physical analysis.

The most recent advance in deep sampling of the sediment of the ocean floor has been through the use of oil-well drilling methods. In the opening paragraphs of this chapter, we described the *Glomar Challenger*, a 10,000-ton vessel built with the capability of drilling into the ocean floor to a depth of 2500 ft (750 m), in water depths as great as 25,000 ft (7600 m). Not only can the drill pass through a sediment layer, but it can also obtain cores of the solid crustal rock beneath.

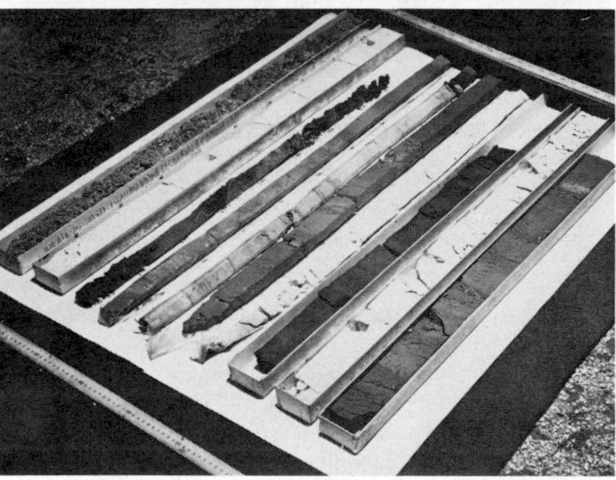

Figure 8.9 Obtaining deep-sea cores aboard the research vessel *Vema*. This scene of the 1950s shows an exciting early phase of ocean floor exploration following World War II. (*Top*) Piston coring tube with its heavy driving weight is prepared for lowering. (*Center*) Members of the ship's company bring a 50-ft (15-m) core aboard. (*Bottom*) Cores that have been extruded into plastic tubes, then sliced through the center to reveal composition and layering for study. Core segments shown here are about 3 ft (1 m) long. (Lamont-Doherty Geological Observatory of Columbia University and National Academy of Sciences, IGY.)

BOTTOM SEDIMENTS TRANSPORTED BY CURRENTS

Very thick accumulations of marine sediments are found beneath the continental shelves. These deposits of sands, silts, and muds are obviously derived from the continents through direct transportation by streams and are spread over the shelves by currents. Geosynclines are large sediment accumulations of this type. As we found in Chapter 5, geosynclines are closely linked with evolution of the continental crust. Our objective in this section is to consider thick sediments of the deep ocean floors farther from land, beneath the continental rise, under abyssal plains, and on the floors of trenches.

A major transport mechanism of the ocean floors is the turbidity current, capable of transporting sediment from relatively high positions of accumulation to the deepest parts of the ocean floor. The turbid mixture originates in soft sediments deposited in fairly shallow water of the continental shelves. The current then flows down the continental slope under the force of gravity because the mixture is denser than the surrounding sea water. Upon reaching the deep ocean floor the tongue spreads into a vast sheet and the sediment settles out in a thin layer.

A turbidity current crossing the continental slope and rise off the Grand Banks of Newfoundland in 1929 was sufficiently powerful to break in succession several trans-Atlantic cables lying across its path. From a knowledge of the exact time of each cable break the velocity of the turbidity current was determined and found to have decreased from 63 mi (100 km) per hour, where the bottom gradient was 1 in 170, to 14 mi (23 km) per hour, where the gradient was only 1 in 2000. In this case the turbidity current was set off by an earthquake shock, causing unconsolidated sediments of the continental slope to slump and to mix with sea water to produce a highly turbid suspension. The turbid tongue ultimately spread out upon the abyssal plain, forming a layer of sediment averaging 3 ft (1 m) thickness over an area of perhaps 80,000 sq mi (200,000 sq km).

Sediments deposited from turbidity currents are called **turbidites.** They are characterized by a most interesting type of layered structure known as **graded bedding.** As shown in Figure 8.10, a single turbidity current produces a single bed, starting at the bottom with coarse sand, grading upward into fine sand, then finally into silt at the top. This arrangement results from the fact that the finer particles settle out from suspension at a slower rate than the coarser particles. Thickness of turbidites can reach many thousands of feet in favorable locations. These are usually basins within reach of the continental shelves, from which the sediment is derived.

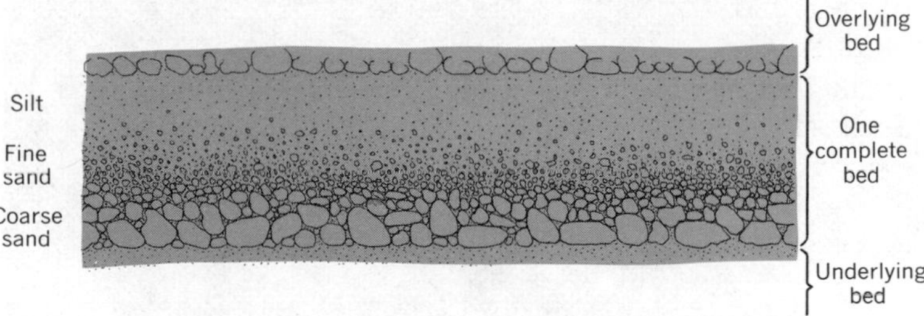

Figure 8.10 Sketch of a cross section through graded bedding. One complete bed represents the deposits of a single turbidity flow.

Besides turbidity currents, which flow only downslope under the influence of gravity, there are other strong currents sweeping the ocean floor in various places. Referred to simply as **bottom currents,** these water movements are capable of moving sands of medium and coarse grades. Ripple marks, shown in Figure 8.11, attest to the strength of bottom currents. These currents can transport sediment on horizontal surfaces, or even up a grade, but they are usually found to flow along the contour of sloping surfaces.

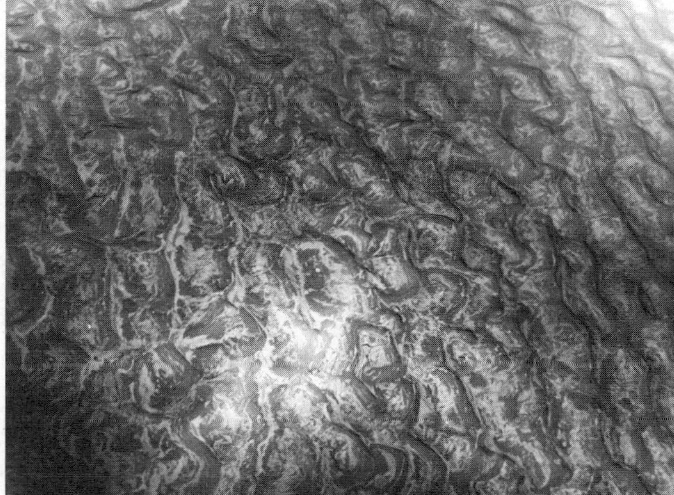

Figure 8.11 Short-crested ripples on the ocean floor attest to action of strong bottom currents. Scotia Sea (lat. 56° S, long. 63° W), depth about 13,000 ft (4000 m). (Charles D. Hollister, Woods Hole Oceanographic Institution.)

The relationship between sediments of the shallow continental shelf and those deposited upon the deep continental rise, at the foot of the continental slope, is shown in Figure 8.12. The locale pictured is the eastern margin of the North American continent. Strata beneath the shelf form a wedge, thickening seaward. This wedge, which attains a thickness of many thousands of feet at its outer limit, consists of clastic sediment brought from the continents. Streams deposit the sediment in broad deltas and it is then spread over the shallow sea floor by currents. Some carbonate strata are usually interbedded with the sands and clays.

At the same time, turbidity flows from the brink of the shelf have been carrying sediment down the slope to the rise, where it has spread out as a series of aprons. This sandy and silty material is also reworked by deep bottom currents. Finer particles have traveled far out over the abyssal plain, but the deposits there are thinner. As a result, there accumulates a thick wedge of turbidites over the oceanic crust, close to its contact with the continental crust. Under the load of these sediments the oceanic crust sinks lower. Altogether, the turbidite wedge can reach a thickness as great as 6 mi (10 km).

The arrangement of sedimentary deposits shown in Figure 8.12 develops where the oceanic-continental contact zone is located within a single lithospheric plate. This zone is not involved in subduction.

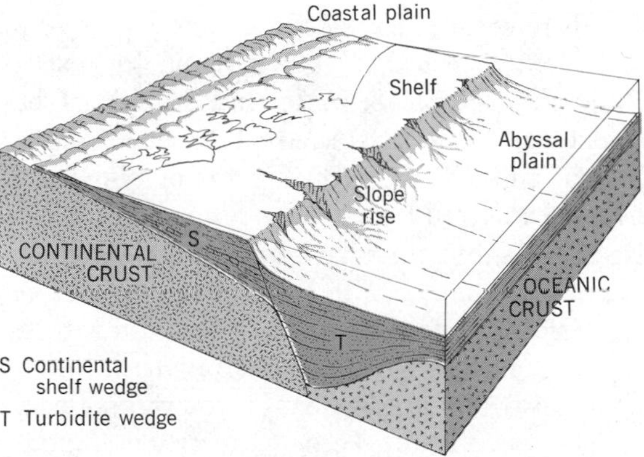

Figure 8.12 Continental-shelf and continental-rise sedimentary accumulations at the contact between continental and oceanic crust.

PELAGIC SEDIMENTS

Sediment settling out from suspension in the surface layer of the oceans is referred to as **pelagic sediment.** One type of pelagic sediment is organic in origin. It consists of the remains of very small plants and animals growing in vast numbers in the shallow, well-oxygenated surface layer. Collectively, such floating organisms are called **plankton.** These organisms secrete hard structures, called **tests.** Upon death of the organisms the organic matter is destroyed, but the tests sink down to great depths. If the tests are not dissolved away as they sink, they will reach the bottom.

Accumulated sediment rich in tests is referred to as **ooze.** Inorganic clays are also present in ooze in substantial quantities. One variety of ooze is calcareous (limy) and is composed of the tests of foraminifera (one-celled animals) and other organisms (Figure 8.13). Another variety is siliceous and is composed of tests of diatoms (one-celled plants) and radiolaria (microscopic animals), which secrete ornate tests of silica. Rate of accumulation of oozes is extremely slow—perhaps from 0.5 to 2 in. (1 to 5 cm) per 1000 years.

A second class of pelagic sediments is detrital and consists of inorganic mineral matter originating on land. Carried by winds or ocean currents, this sediment forms **brown clay,** a soft, plastic material consisting mostly of clay minerals described in Chapter 4: illite, montmorillonite, and kaolinite. Rate of accumulation of brown clay is only from 0.004 to 0.04 in. (0.1 to 1 mm) per 1000 years. Also in the detrital class are sediments dropped into deep water by icebergs. Volcanic ash carried in the atmosphere and settling into the ocean also forms detrital sediment layers on the deep ocean floor.

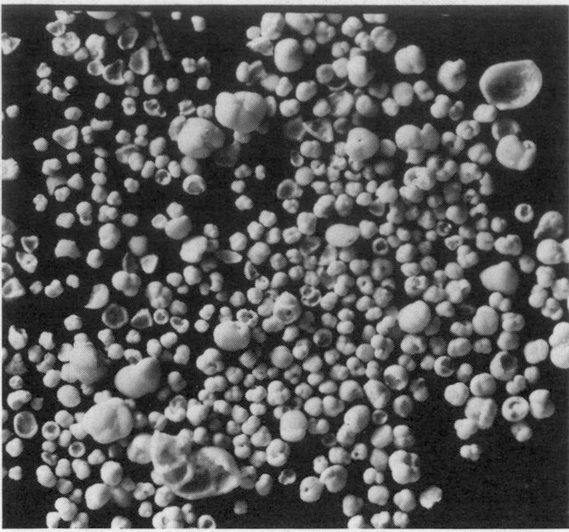

Figure 8.13 These calcareous shells and shell fragments were separated by sieving a sample of Globigerina ooze, the bulk of which is composed of fine clay material. This sample is from a core obtained by scientists of the research vessel *Vema* from a depth of 10,000 ft (3000 m) in the South Atlantic Ocean. Enlargement about 12 times. (A. McIntyre, Lamont-Doherty Geological Observatory of Columbia University.)

MINERAL ALTERATION ON THE OCEAN FLOOR

Finally, there are sediments consisting of minerals formed by alteration in place or re-formed from other minerals. Both basaltic rock and volcanic ash are subject to such alteration. For example, abundant nodules rich in manganese and iron oxides are found exposed on the surface of the deep ocean floors in many places. Referred to as **manganese nodules,** these objects often prove to be thick mineral coatings surrounding nuclei of volcanic rock (see Figure 11.7). Manganese nodules are believed to be formed from manganese and iron derived either from detrital sediments of continental origin or from volcanic rocks of the ocean floor.

IS THE CRUST SPREADING APART ALONG THE MID-OCEANIC RIDGE?

The form of the axial rift of the Mid-Oceanic Ridge strongly suggests that the crust is being pulled apart along the line of the rift. If so, rising basalt magma must be filling the gap that would otherwise be created by crustal spreading. This concept of **sea-floor spreading,** as it is now called, needs to be followed up very closely.

Note that Iceland lies right on the axis of the Mid-Oceanic Ridge (see Figure 8.7). Iceland is built of basalt in the form of shield volcanoes. Extrusion of basalt is active on Iceland at the present time. These facts support our inference as to formation of new oceanic crust along the rift zone of the Mid-Oceanic Ridge. Recently, submarines have brought up extremely recent basalt specimens from the axial rift, confirming our interpretation.

Global patterns of earthquakes lend further support to the concept of sea-floor spreading. Referring back to the world map of earthquake distribution, Figure 6.13, you will notice that the Mid-Oceanic Ridge is an earthquake belt throughout much of its length. These earthquakes are of the shallow-focus type and are not, as a rule, severe in intensity. Shallow earthquakes signify faulting along normal faults and transcurrent faults in the upper part of the crust. Such faulting is just what we would expect in a zone of sea-floor spreading, where crustal fracturing is in progress.

THE GOLDEN AGE OF SEA-FLOOR EXPLORATION

In this chapter we have bridged the gap between the older, classical geology of the continents and the new geology of the ocean basins. Information gathered in the 1940s and 1950s was essential to a new understanding of earth processes. Of the major divisions of the ocean floors, the Mid-Oceanic Ridge proved the most fruitful discovery of all. The finding of a rift at the crest of this ridge quickly led to the idea that the crust is spreading apart along the rift, while new ocean crust is being formed of basalt welling up from the mantle beneath the rift.

Although the concept of sea-floor spreading was developing in this golden age of exploration, direct evidence of crustal separation in progress was lacking. At this stage, geologists could only propose as a hypothesis that lithospheric plates, gliding over a soft asthenosphere, are spreading apart along the Mid-Oceanic Ridge, while at the same time, the leading edges of the plates are nosediving beneath the continental margins in subduction zones. Conclusive evidence was yet to be uncovered—it appeared in the geological revolution of the 1960s.

YOUR GEOSCIENCE VOCABULARY

precision depth recorder
continental margins
ocean-basin floors
Mid-Oceanic Ridge
continental shelf
continental slope
continental rise
submarine canyon
turbidity current
abyssal plain
seamount
axial rift
coring
core
turbidites
graded bedding
bottom currents
pelagic sediment
plankton
tests
ooze
brown clay
manganese nodules
sea-floor spreading

SELF-TESTING QUESTIONS

1. Compare the continents and ocean basins in terms of their extent and the proportion of surface area lying at various elevations. How can the true limits of the continents be defined?

2. Explain the use of the precision depth recorder in making profiles of the ocean floor.

3. Name the three major divisions of the ocean basins. What are the subdivisions of the continental margins? Describe and explain submarine canyons.

4. Why are the abyssal plains so flat? How did seamounts originate? Of what rock are they composed?

5. Describe in detail the form and extent of the Mid-Oceanic Ridge. What is the significance of the axial rift?

6. Describe the procedure for obtaining sediment cores from the deep ocean floor. How is drilling carried out in these same regions?

7. Explain how sediment is transported by currents over the deep ocean floors. What is a turbidity current? Where do such currents act? What kind of sedimentary accumulations do they produce?

8. What is pelagic sediment? What are plankton? What kinds of sediment do plankton produce? What types of detrital sediment accumulate on the deep ocean floors?

9. How are sediments produced by chemical action on the ocean floor? What are manganese nodules? What is the scientific value of the deep-sea sediments in interpretation of global environments of the geologic past?

10. What geologic evidence suggests that sea-floor spreading is taking place along the axis of the Mid-Oceanic Ridge?

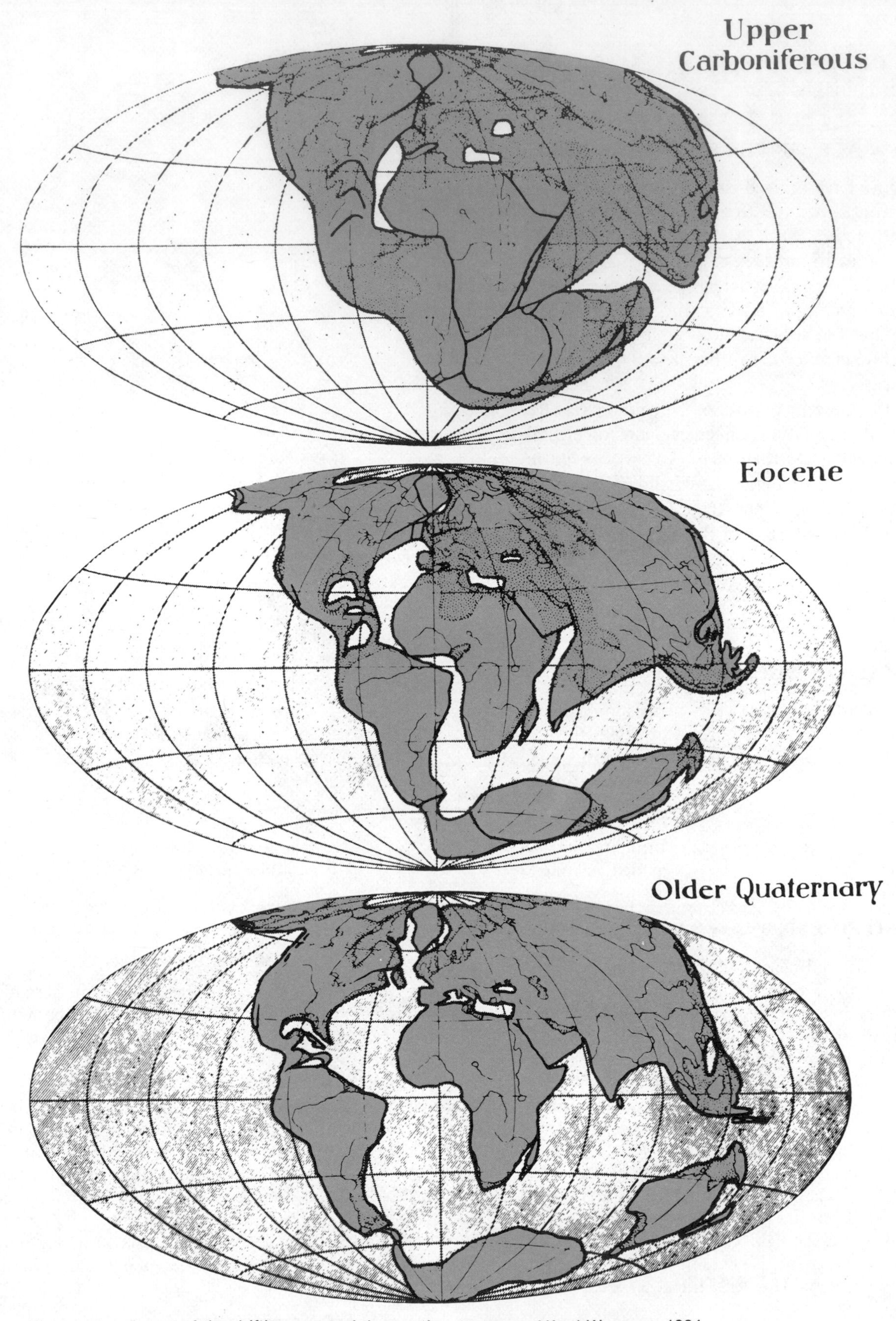

Wegener's concept of the drifting apart of the continents. (From Alfred Wegener, 1924, *The Origin of the Continents and Ocean Basins,* London, Methuen and Co., p. 6, Figure 1.)

9 PLATE TECTONICS AND THE GEOLOGIC REVOLUTION

A PREPOSTEROUS THEORY

The most preposterous notion presented to the geological community in the twentieth century came from a scientist who should have known better than to deal in such nonsense. The scientist was Alfred Wegener, a German meteorologist with impeccable training and sound accomplishments in the field of atmospheric science. Born in Berlin in 1880, the son of an evangelical preacher, Wegener attended the most prestigious universities of his country—Heidelberg, Innsbruck, and Berlin. His interest in geology must have started in 1906, when he traveled to northeast Greenland as a meteorologist with a Danish scientific expedition. An academic life followed and in 1924 Wegener was appointed to a professorship in meteorology and geophysics at the University of Graz in Austria. His career ended tragically in 1930 during his third expedition to Greenland, where he lost his life at a remote field station high on the Greenland Ice Sheet.

Wegener's preposterous idea was that there existed until some 150 million years ago but a single continent on earth; he named it Pangaea. This supercontinent then ruptured into several fragments, which began to drift apart. As the American fragments pulled away from Africa and Eurasia, a narrow ocean basin appeared. It was to become the Atlantic Ocean basin. Wegener visualized Antarctica, Australia, Madagascar, and peninsular India as having been parts of Pangaea, neatly nested together close to the southern tip of Africa. These fragments also drifted apart, opening up the Indian Ocean as they separated. The illustration on the opposite page is Wegener's own concept of the fragmentation of Pangaea; it was published shortly before his death.

To English-speaking geologists, Wegener's scenario was known as **continental drift;** for the most part they had little use for the idea. But Wegener gained a few strong supporters on both sides of the Atlantic. These "drifters" were a source of annoyance, through their perseverance in marshaling evidence that the continents had once been joined together. The evidence was in many ways strong. Wegener visualized each continental fragment as a raft of less-dense, felsic (granitic) rock drifting through a sea of more-dense, mafic (basaltic) rock. Where continents were drifting apart, an ocean basin of basaltic rock opened up between them. The leading (western) edge of the American continental raft was deformed by impact to produce the existing alpine mountain chains that are now the Andes in South America and the Cordilleran ranges of North America.

Geophysicists found Wegener's continental rafting completely unacceptable under known laws of physics; it simply could not happen that way. Of course, these scientists were right; it did not happen that way. Yet today, all but a few earth scientists accept as fact the geographical separation of continents, on a schedule in fairly close agreement with Wegener's early model. A plausible mechanism has been discovered and documented—largely by the dissenting geophysicists themselves.

The recent revolution in geology, based on knowledge about motions of lithospheric plates over a soft asthenosphere, was made possible by discoveries about the nature of the ocean basin floors. This knowledge began to emerge only after World War II. As we found in Chapter 8, the 1950s were a decade of unprecedented oceanic exploration. The 1960s were the decade of the geologic revolution itself. Discoveries regarding sea-floor spreading gathered momentum slowly at first, then reached an exhilarating crescendo late in the decade, as the last odd pieces in the puzzle were quickly fitted into place.

In this chapter we review this final series of discoveries. They deal almost exclusively with the ocean basins, where the final clues to global tectonics lie.

REVERSALS OF THE EARTH'S MAGNETIC FIELD

Continuing our search for evidence about the origin of the ocean basins, we now put to use basic information about the earth's magnetism, developed in Chapter 2. Basaltic lavas contain minor amounts of oxides of iron and titanium. Magnetite, the mineral of which lodestone is a naturally magnetic variety, is an example.

At the high temperatures in a basalt magma, these iron and titanium minerals have no natural magnetism. However, as cooling sets in, each crystallized mineral passes a critical temperature known as the **Curie point**, below which the mineral is magnetized by lines of force of the earth's field. At first, this magnetization is not permanent, but rather of the type known as **soft magnetization**, similar to that acquired by soft iron. With further cooling, however, the soft magnetism abruptly becomes permanent. This final state is known as **hard magnetism**; it resembles the permanent magnetic condition of the alnico magnet. In this way a permanent record of the earth's magnetic field is locked into the solidified lava.

In the study of rock magnetism, a sample of rock is removed from the surrounding bedrock. Orientation of the specimen is carefully documented in terms of geographic north and horizontality. The specimen is then placed in a sensitive instrument, the **magnetometer**, which measures the direction and intensity of the permanent magnetism within the rock. Using a number of samples obtained from a single lava flow, the magnetic values are compared for consistency and averaged, yielding the direction and inclination of the magnetism. The term **paleomagnetism** is used for such locked-in magnetism dating far back into the geologic past. Paleomagnetism can be compared with present conditions and with the magnetic field at other locations and in different times in the geologic past.

As early as 1906, Bernard Brunhes, a French physicist, observed that the magnetic polarity of some samples of lavas is exactly the reverse of present conditions. He concluded that the earth's magnetic polarity must have been in a reversed condition at the time the lava solidified. You might wish to argue that the rock magnetism itself may have undergone a change in polarity. However, in recent years there has been general agreement among earth scientists that the rock magnetism is permanent and is a reliable indicator of the former states of the earth's magnetic field.

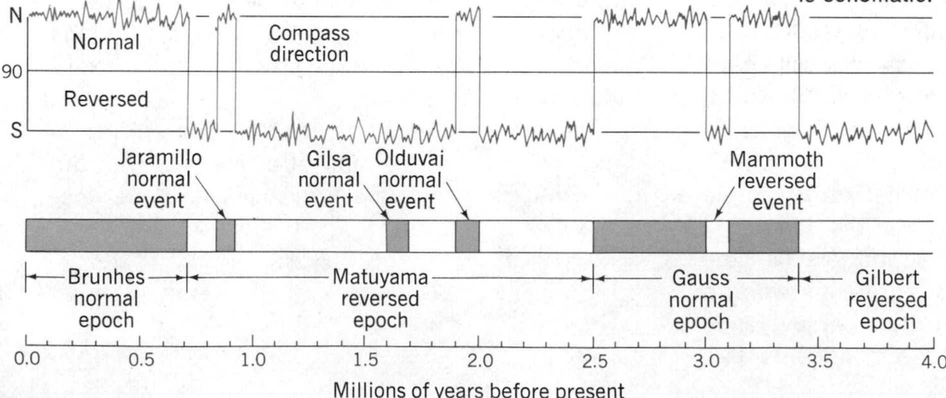

Figure 9.1 The time scale of magnetic polarity reversals. The graph of compass-direction fluctuations is schematic.

In addition to the magnetic data of the lava specimen, we need a radiometric determination of the age of the rock, giving the date of solidification of the magma. Extensive determinations of both magnetic polarity and rock age have revealed that there have been at least nine reversals of the earth's magnetic field in the last 3.5 million years (m.y.) of geologic time.

Figure 9.1 shows the timetable of magnetic polarity changes. Polarity such as that existing today is referred to as a **normal epoch;** opposite polarity, as a **reversed epoch.** Each epoch is named for an individual or a locality. For example, the pioneer work of Bernard Brunhes is recognized in the present normal epoch, which began about 700,000 years ago. An epoch of reversal, named for the Japanese scientist Motonori Matuyama, extends to 2.5 m.y. before present and includes three shorter periods of normal polarity classified as **events.** A still older normal epoch, named in honor of the mathematician Karl Gauss (1777–1855), carries back the paleomagnetic record to about 3.5 m.y. and contains one important reversed event. The oldest of the reversed epochs shown in Figure 9.1 is named after Sir William Gilbert, whose early work on terrestrial magnetism was discussed in Chapter 2.

A large number of polarity reversals have been found throughout the geologic record, going back to nearly 140 m.y., which is through the entire Cretaceous Period (see Table 7.1). As yet, the older reversals in this sequence cannot be dated with great accuracy, but the general pattern in time is known.

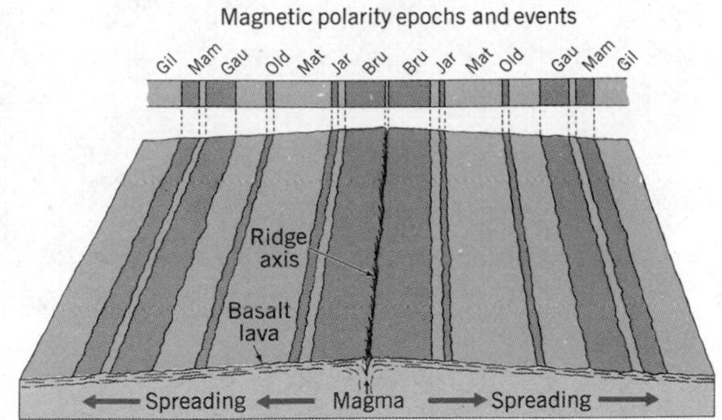

Figure 9.2 Development of symmetrical pattern of magnetic polarity stripes in oceanic basalts during sea-floor spreading. (See Figure 9.1 for time scale.)

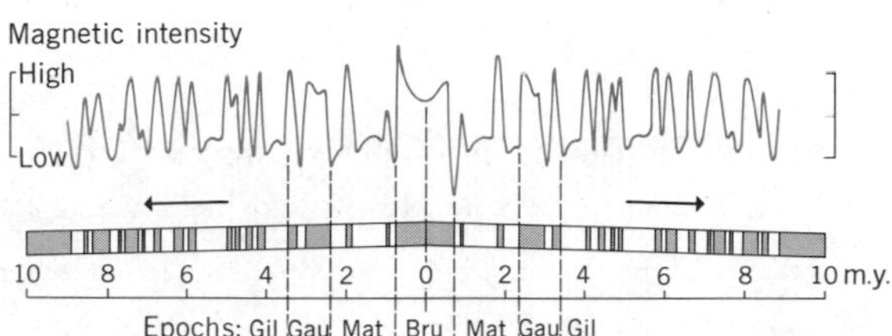

Figure 9.3 Fluctuations in magnetic intensity along a traverse of the Mid-Oceanic Ridge are shown in the upper profile. The corresponding time scale of magnetic polarity reversals is shown below.

ROCK MAGNETISM AND SEA-FLOOR SPREADING

We are now prepared to return to the subject of sea-floor spreading along the Mid-Oceanic Ridge. If the axial rift valley is a line of upwelling of basaltic lavas, and if crustal spreading is a continuing process, the lava flows that have poured out in the vicinity of the axial rift will be slowly moved away from the rift. Lavas of a given geologic age will thus become split into two narrow stripes, one on each side of the rift. As time passes, these stripes will increase in distance of separation, as shown in Figure 9.2. The lavas can be identified and classified in terms of the epochs of normal and reversed magnetic field; these epochs will be represented by symmetrical striped patterns on either side of the rift.

Confirmation of the symmetrical magnetic stripes was gained in the course of oceanographic surveys made during the mid-1960s. We cannot take oriented core samples of lavas from the ocean floors. However, it is possible to operate a sensitive magnetometer during a ship's traverse of the Mid-Oceanic Ridge. When this is done, it is found that there are minute variations in the value of magnetic inclination. (Inclination refers to the downpointing of a magnetic needle balanced on a horizontal bearing.) These departures from a constant normal value are referred to as **magnetic anomalies.**

When several parallel cross lines of magnetometer surveys have been run across the Mid-Oceanic Ridge, the magnetic anomalies can be resolved into a pattern, such as that shown in Figure 9.3. Figure 9.4 is a map showing the striped pattern of magnetic anomalies near Iceland. Notice the mirror symmetry of the striped pattern with re-

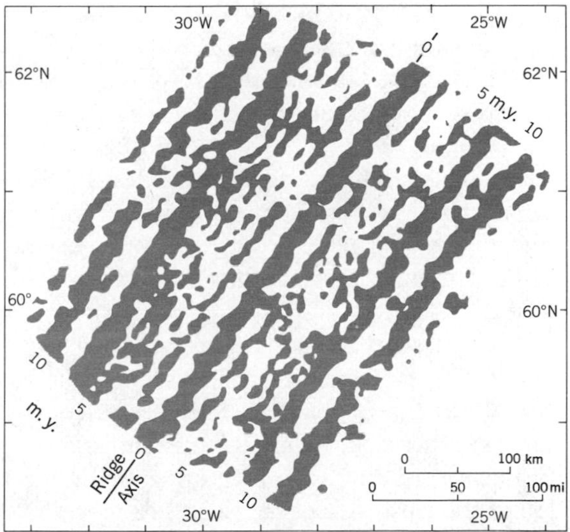

Figure 9.4 Magnetic anomaly pattern for Reykjanes Ridge, located on the Mid-Atlantic Ridge southwest of Iceland, with approximate rock ages in millions of years. (After J. R. Heirtzler, X. Le Pichon, and J. G. Baron, 1966, *Deep-Sea Research*, vol. 13, Figure 1, p. 427.)

spect to the axis of the ridge. From a study of the anomaly pattern, it is possible to identify the normal and reversed epochs, as we have done in Figure 9.2.

It has also been found possible to identify the normal and reversed magnetic epochs in core samples of soft sediment obtained from the ocean floor by piston-coring devices. Here, the epochs are encountered in sequence from top to bottom within the core. In this way numerous older epochs were discovered, extending back through the Cretaceous Period.

Finding the magnetic stripes on the ocean floor proved to be the key to the revolution in geology. There followed rapidly a series of magnetic surveys along many sections of the Mid-Oceanic Ridge, all revealed similar striped patterns in mirror image. Magnetic evidence not only made a virtual certainty of sea-floor spreading, but also allowed the rates and total distances to be estimated as well.

Take, for example, the case of the anomaly pattern shown in Figure 9.4. Here the width of the anomaly zone is 125 mi (200 km), which represents the total distance of crustal separation in about 10 m.y. The average rate of horizontal motion of the crust during this time has been about 0.4 in./yr (1 cm/yr), which means that the rate of separation is double this value, or 0.8 in./yr (2 cm/yr). Elsewhere, the rates of spreading are found to be higher, up to 2 in./yr (5 cm/yr).

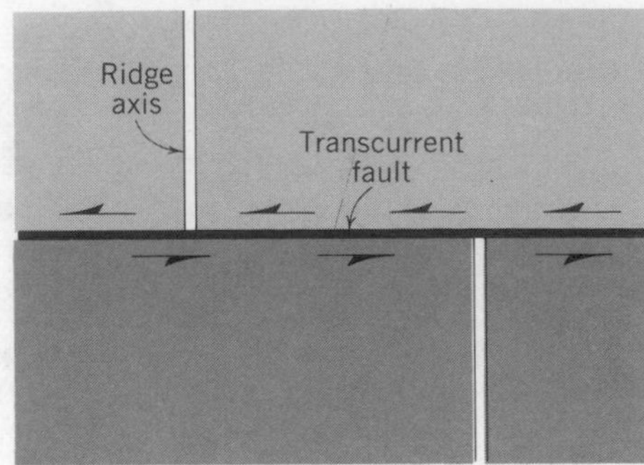

Figure 9.5 Sketch map of the Mid-Oceanic Ridge offset by a transcurrent fault of the left-lateral type.

TRANSFORM FAULTS

A first glance at a map or diagram of the Mid-Oceanic Ridge (Figure 8.7) might lead you to guess that the many transverse fractures are simple transcurrent faults, having horizontal slippage like that along the San Andreas Fault (Chapter 6). It is obvious that segments of the axial rift have been broken across and displaced horizontally along the fractures. But now we have to reckon with sea-floor spreading as a continuing activity, and this places a new twist on the origin of the transverse fractures.

First, let us suppose that an offset of the Mid-Oceanic Ridge is merely a simple transcurrent fault, as shown in Figure 9.5. Assume for the moment that no sea-floor spreading is in progress. In that case, the transcurrent fault illustrated in the figure is of the left-lateral type. (Refer back to Chapter 6 for definitions of left-lateral and right-lateral faults.) Arrows show movement as in a left-lateral fault. The entire crust on the north side moves left; that on the south side moves right.

Next, suppose that transcurrent faulting has ceased, but that sea-floor spreading is occurring, as shown in Figure 9.6. The dashed line shows the trace of the former transcurrent fault, extending far out on either side. Figure 9.6 also shows an offset of the ridge axis. Sea-floor spreading is in progress, as indicated by the broad arrows. All of the crust west of the ridge axis moves westward as a unit (colored area); all of the crust east of the ridge axis moves eastward as a unit (gray pattern). Thus, between *b* and *c* the crust must break on a fault. Horizontal movement between blocks in segment bc will be of right-lateral sense, as the small arrows show. Such a fault has been named a **transform fault.** It is opposite in relative motion to

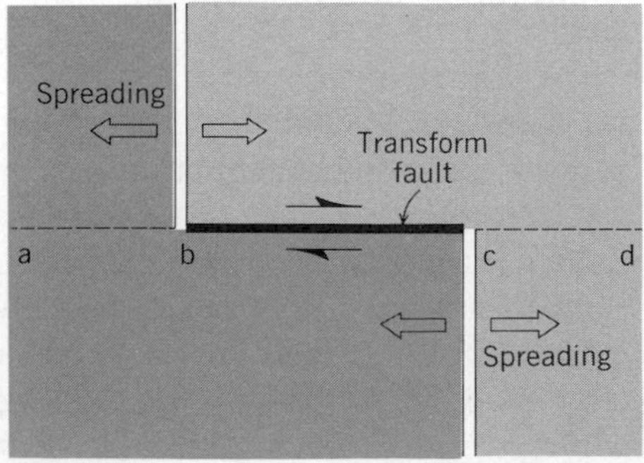

Figure 9.6 Sketch map of a transform fault of the right-lateral type.

the transcurrent fault shown in Figure 9.5, even though the ridge axis has the identical plan of offset in both cases.

Note further that if spreading is at equal rate along the axial rift, there will be no faulting in progress on either of lines ab or cd. Thus the transform fault is limited to the segment between the offset ends. Of course, there may exist older transcurrent faults along the lines ab and cd, and such is apparently the case for much of the Mid-Oceanic Ridge, as a glance at the map will show (Figure 8.7).

Existence of transform faulting was predicted in 1965 as a consequence of crustal spreading. Confirmation came in 1967, when a seismologist examined the records of earthquakes originating along the faults between offset ends of the Mid-Oceanic Ridge. He was able to show that the first motions of the earthquakes along the fracture zones indicated faulting in the transform sense. (Direction of fault motion can now be ascertained by study of the initial wave of the P-wave group.)

Thus seismology gave independent evidence of sea-floor spreading, reinforcing evidence of magnetic anomalies and the ages of sediments and basalts on both sides of the central rift. The scales of evidence suddenly tipped sharply in favor of sea-floor spreading. Almost overnight there were no doubters, only believers.

THE PLATE THEORY OF GLOBAL TECTONICS

Rapid advances in our knowledge of the oceanic crust, and in particular of the wide extent of sea-floor spreading along the Mid-Oceanic Ridge, led in the late 1960s to a general hypothesis of global tectonics. Wide acceptance of this model of earth behavior was the climactic phase in the geological revolution. The noun **tectonics** means the study of structural features of the earth's crust and their origin. Tectonic activity was described in Chapter 5.

The term **plate tectonics** was applied to the hypothesis, which features the largely horizontal movements of platelike elements of the strong, brittle lithosphere over a readily yielding asthenosphere. (Characteristics of the soft layer of the mantle were reviewed in Chapter 2.) Lithospheric plates, described in Chapter 4, consist of both crust and upper mantle and include both oceanic and continental crust.

As first outlined in 1968, the earth's lithosphere was described as divided into six principal rigid plates. Each plate moves horizontally as a unit and may also rotate as it moves over the asthenosphere. Obviously, two major possibilities are that adjacent plates may move apart, creating a widening gap between them, or that they may move together, causing crustal rupture of the edges of the plates. A third possibility is that they may slide alongside each other on transform faults.

Figure 9.7 is a three-dimensional schematic diagram showing relationships among lithospheric plates. Plates spreading apart beneath the oceans produce the Mid-Oceanic Ridge system, with its axial rift and transform faults. Where plates converge, the edge of one plate is bent down and forced into subduction. It thus descends into the asthenosphere, where it is heated and absorbed into the mantle rock at great depth.

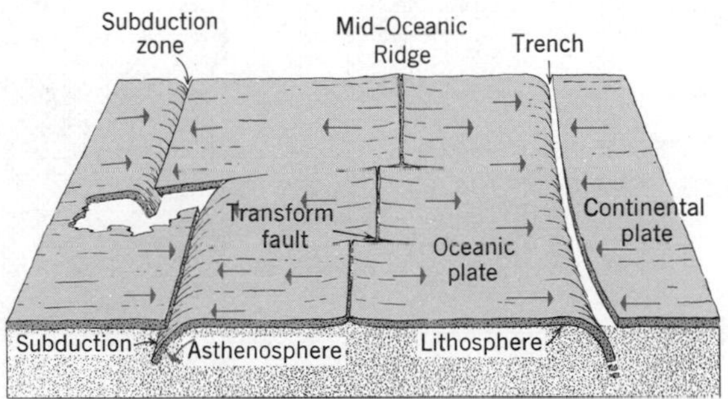

Figure 9.7 A schematic block diagram of plate tectonics. Earth curvature has been removed.

THE GLOBAL SYSTEM OF PLATES

Returning to the global plan of plate tectonics, it is now postulated that there are six major plates and several smaller ones (Figure 9.8). The American plate includes North American and South American continental crust and all of the oceanic crust of the western Atlantic extending eastward to the Mid-Atlantic Ridge. This American plate has a relative westward motion as a single unit, and consequently there is no important tectonic activity along the eastern margins of the American continents. The western edge of the American plate lies along the western continental margins.

The Pacific plate is the only unit bearing only oceanic crust. It occupies all of the Pacific region west of the East Pacific Rise. It undergoes subduction beneath the American plate along the compressional zone of the Alaskan-British Columbia coastal region.

The Antarctic plate occupies the globe south of the Mid-Oceanic Ridge system. The Nazca plate, lying between the East Pacific Rise and South America, moves eastward against the west margin of South America, meeting in the subduction zone of the Peru-Chile Trench and the Andes range. The African plate consists of the African continental crust and a zone of surrounding oceanic crust limited by the Mid-Oceanic Ridge.

A single Eurasian plate, which consists largely of continental crust, is bounded on the east and south by subduction zones of the great alpine mountain chains and island arcs. The Eurasian plate

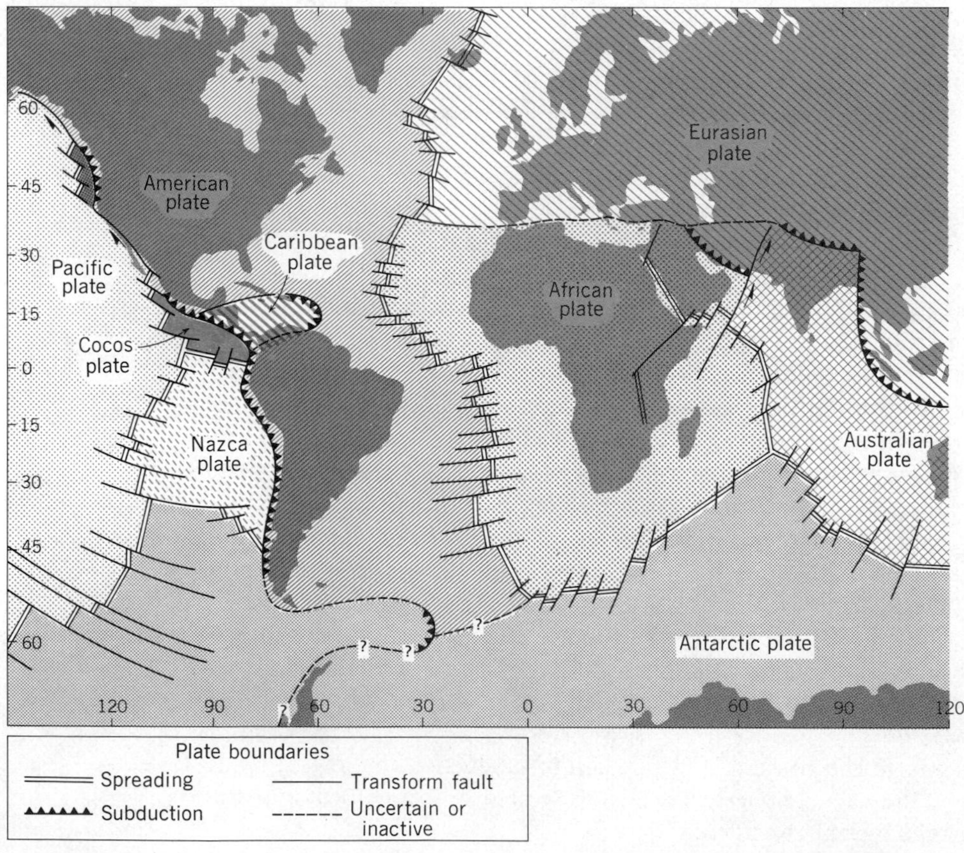

also extends into the North Atlantic as oceanic crust lying east of the Mid-Atlantic ridge. The Australian plate consists of continental crust of India and Australia, as well as of oceanic crust of the Indian Ocean and a part of the southwestern Pacific. It is separated from the Pacific plate by a subduction zone passing through New Zealand.

Because of its vast extent, the Pacific plate can be expected to possess, at its westernmost portion, the oldest oceanic crust on earth. In 1970, deep-sea cores obtained by the *Glomar Challenger* provided confirmation of this inference. Sediments as old as early Cretaceous age or upper Jurassic age, in the range of 125 to 150 million years, were identified in cores.

As might be expected, further study of new data on magnetic stripes and ocean floor topography led to the recognition of small parts of the large units as individual plates. The Caribbean plate was recognized as a separate unit, bounded on the east by the Antillean island arc, a subduction zone. The Philippine plate was recognized as a subdivision of the Pacific plate; it lies between the subduction zones of the Japan-Philippine arc and the Bonin-Marianas arc. West of southern Mexico and Central America is the Cocos plate, located between a spreading plate boundary and a subduction zone. Similar to the Cocos plate, and perhaps formerly joined with it, is the Juan de Fuca plate, lying off the Pacific Northwest coast. For the sake of simplicity we have omitted mention of other small plates. Doubtless, many more will be recognized as research continues.

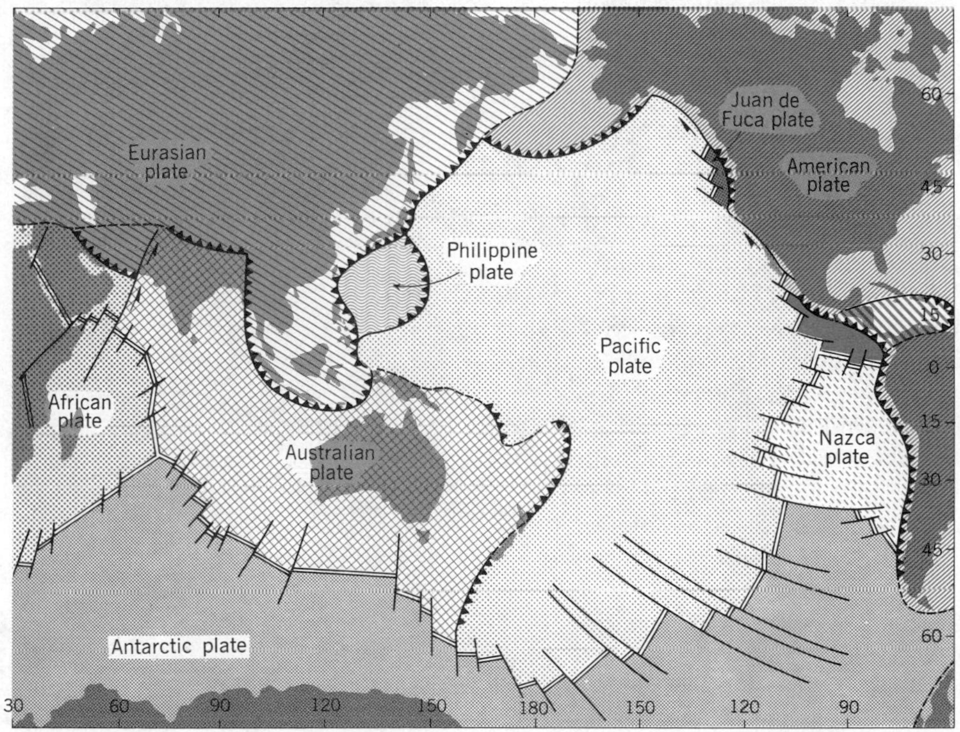

Figure 9.8 A world map of the major lithospheric plates.

CONVECTION CURRENTS IN THE MANTLE

Crustal spreading and plate tectonics require a driving mechanism. Little is known of actual rock movements within the mantle, but most models of global tectonics have been systems of very slow mantle circulation under the general heading of **convection currents.**

Figure 9.9 shows a model of convection involving the entire thickness of the mantle. Rising of less-dense mantle rock under the Mid-Oceanic Ridge and corresponding sinking beneath the subduction zones of the trenches and island arcs are key activities within the convection system. Dominantly horizontal motion under the lithospheric plate would exert a drag, causing the plate to move away from the Mid-Oceanic Ridge and toward the subduction zone.

The energy source for convectional or other large-scale motions of the mantle may lie in the uneven accumulation of radiogenic heat. A completely different hypothesis attributes mantle motion to kinetic energy persisting from a much earlier time in the earth's history. Earth rotation has been invoked as a driving mechanism. Another hypothesis requires energy input from an asteroid impact. Another recently proposed energy source is that of tides.

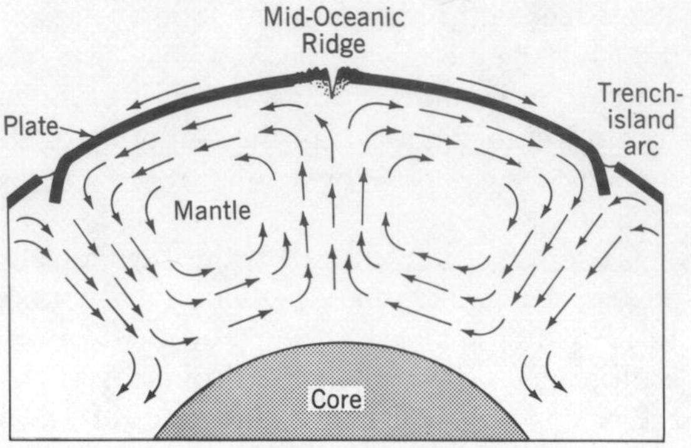

Figure 9.9 The convection hypothesis of lithospheric plate motion.

CONTINENTAL DRIFT—WEGENER VINDICATED

If we accept the evidence of plate separation on a vast scale throughout more than 150 m.y., there is no escape from the conclusion that certain continents were formerly situated much closer together than they are today and may even have been joined together.

This thought leads us back to Alfred Wegener and his supercontinent, **Pangaea.** Far from being rejected as the most preposterous geological theory of the twentieth century, the existence and breakup of Pangaea has been revived in triumph. Once sea-floor spreading was established, fitting the continents back together quickly became an intriguing jigsaw puzzle, preoccupying the time of more than one distinguished geologist.

Figure 9.10 is a map of the united continents based on all available evidence. Shaded areas are the continental nuclei. Although shown nested into one continent, the arrangement of nuclei into two groups suggests that there were originally two centers of continental crust accumulation: **Laurasia** in the Northern Hemisphere and **Gondwana** in the Southern Hemisphere. Notice the absence of Asia in this reconstruction, although it was included in Wegener's Pangaea. Asia consists largely of younger crust. Central America, with its younger mountain arcs, is interpreted as having formed following separation of the continents. Peninsular India, western Australia, Madagascar, and Antarctica are closely clustered beside the African continent.

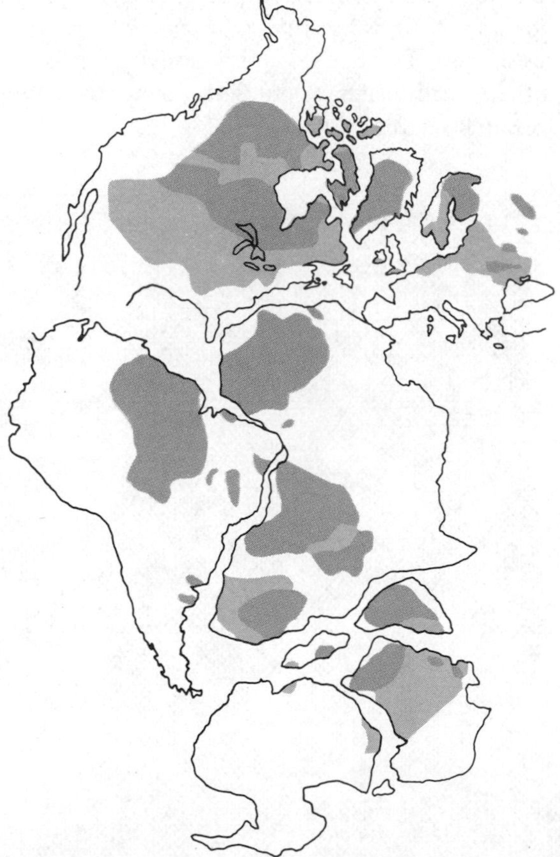

Figure 9.10 Reassembled continents, prior to start of continental drift. *Dark pattern:* areas of oldest shield rocks (older than −1.7 b.y.), *Light pattern:* rocks in the age range −0.8 to −1.7 b.y. (After P. M. Hurley and J. R. Rand, 1969, *Science,* vol. 164, p. 1237, Figure 8.)

The five maps in Figure 9.11 show a reconstruction of the breakup of Pangaea in terms of plate tectonics. Recent studies suggest that continental separation may have begun along the northwestern margin of Africa in mid-Triassic time about 200 m.y. before the present. South America was finally separated from southern Africa in the Cretaceous Period, about −130 m.y. As the Americas drew away from Africa and Europe, new oceanic crust was formed

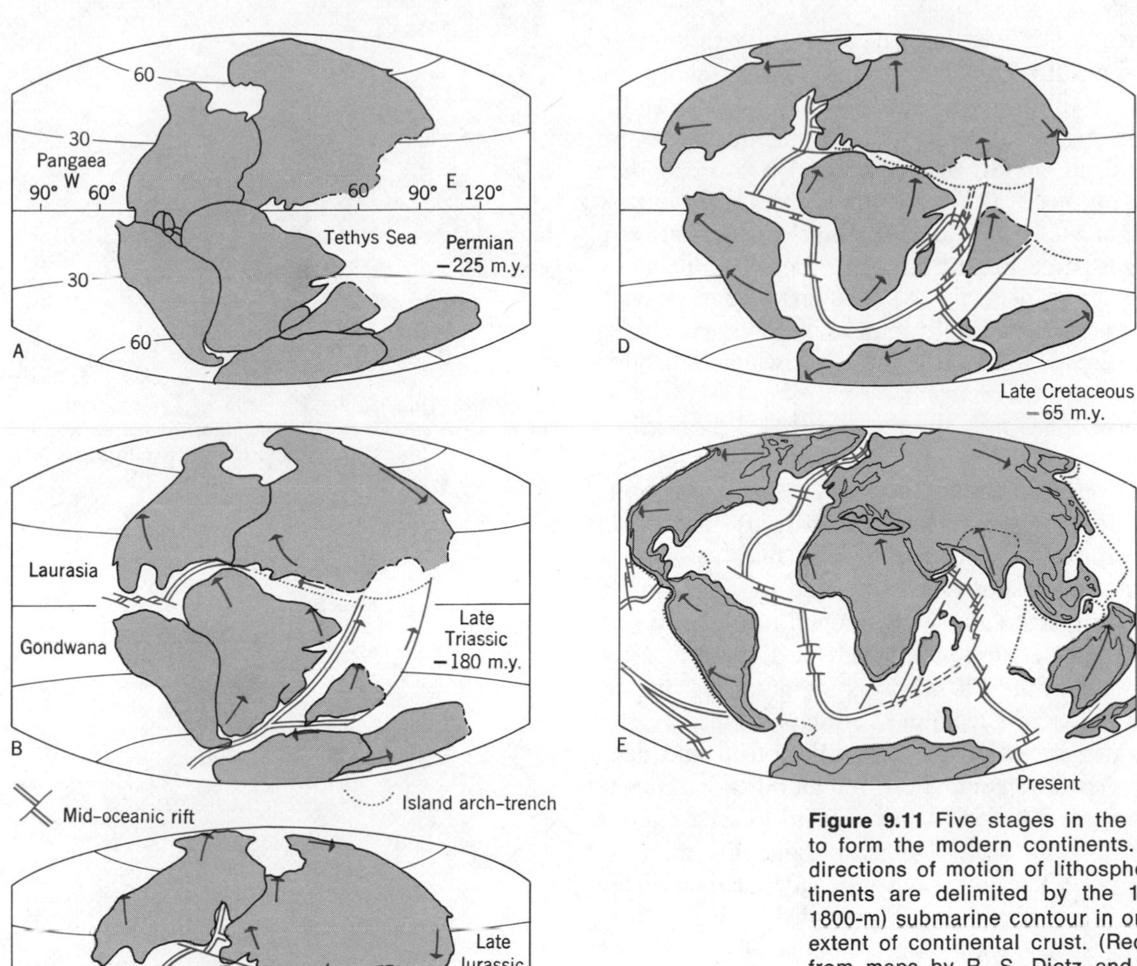

Figure 9.11 Five stages in the breakup of Pangaea to form the modern continents. Arrows indicate the directions of motion of lithospheric plates. The continents are delimited by the 1000-fathom (6000-ft; 1800-m) submarine contour in order to show the true extent of continental crust. (Redrawn and simplified from maps by R. S. Dietz and J. C. Holden, 1970, *Jour. Geophys. Research,* vol. 75, pp. 4943–4951, Figures 2–6.)

by the rise of mantle rock in the Mid-Oceanic Ridge axis. Thus the Atlantic Ocean crust has formed since about early Cretaceous time and cannot be much older than about −130 to −140 m.y. This inference as to the young age of the Atlantic Ocean floor is in line with radiometric ages of oceanic basalts and sediments, which have not been found much older than Cretaceous. Similarly, separation of Antarctica, Australia, and peninsular India from eastern Africa is depicted as having taken place to the accompaniment of crustal spreading along the Mid-Oceanic Ridge in the Indian Ocean.

Geologic evidence for the former unity of the continental shields takes a variety of forms. Matching of similar rock types and rock ages from the margin of one continent to another provides one line of evidence. The case of South America and Africa is particularly interesting. As Figure 9.10 shows, small fragments of continental nuclei in South America seem to fit with larger nuclei in Africa. Moreover, the trends of Precambrian tectonic structures in the area between these nuclei were then continuous from one continent to another. Today, in contrast, these structures project directly out toward the ocean basin and appear abruptly truncated. In the North Atlantic, tectonic structures of the Appalachian belt, passing through Nova Scotia and Newfoundland, appear to match up with corresponding structures of the same geologic age in the British Isles and Norway.

Matching of fragments of the Gondwana nuclei in a single continental mass has been based in part on similarities of sedimentary rocks and their contained plant fossils of Carboniferous age. According to those who first supported the hypothesis of continental drift, the simultaneous development of these plants on widely separated continents would have been an impossibility.

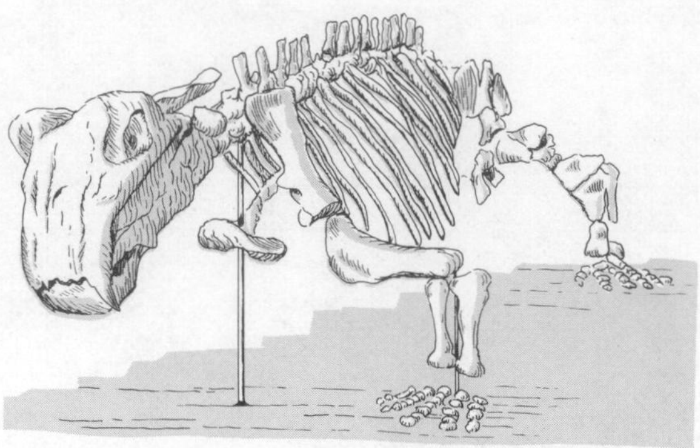

Figure 9.12 This partial skeleton of *Lystrosaurus*, a mammal-like reptile of Triassic age, measures about 3 ft (1 m) in length. (Sketched from a photograph.)

A similar argument for Gondwana has been based on distribution of animals supposed to be incapable of migrating from one continent to another over deep ocean water. Key evidence has come from a mammal-like reptile of the genus *Lystrosaurus*. This small animal somewhat resembled a hippotamus, with massive wide-set legs (Figure 9.12). Fossil remains of *Lystrosaurus* are abundant in Triassic strata of southern Africa and are also found in India, Russia, and China. The search for *Lystrosaurus* fossils in Triassic rocks of Antarctica was intensified in the 1960s because of increasing support for continental drift. It is considered almost an impossibility for this animal to have migrated to Antarctica across the broad and deep ocean basin that now separates Antarctica from all other continents. The search met with success in 1969 when remains of *Lystrosaurus* were found in the Transantarctic Mountains, about 400 mi (640 km) from the south pole. The fossil find was hailed as one of the most significant in modern times, for it presented strong paleontologic evidence of the existence of a unified landmass of Gondwana as late as the Triassic Period.

A word of caution: The term continental drift is archaic, and we should limit its use to the historical context of the Wegener theory and debate. Wegener's concept of granitic continental plates plowing through a stationary basaltic crust remains totally discredited. In modern plate tectonics, the continental crust is carried along with the basaltic crust; motion occurs in the soft layer of the mantle, much deeper than Wegener envisioned.

POLAR WANDERING AND A GREAT GLACIATION

Wegener and his supporters put forth a most interesting line of evidence that the continents were once united. This evidence consists of finding the markings and deposits of an ancient ice sheet on all parts of what are now the fragments of Gondwana. Rock surfaces show scratches and grooves unmistakably made by ice abrasion (Figure 9.13). Lithified glacial materials, known as **tillites**, are also found overlying the abraded rock surfaces.

Figure 9.13 This surface of dark igneous rock shows the scorings made by a glacier of Carboniferous time in South Africa. One can find similar features, produced by ice sheets of the Pleistocene Epoch, on many rock exposures in North America and Europe. The ice moved in a direction away from the observer up the sloping surface. The deep intersecting cracks are joint fractures enlarged by weathering. (Douglas Johnson.)

Geologists are agreed that this evidence points to a major glaciation. It occurred during the late Carboniferous Period, some 300 m.y. ago. The glacial period may also have extended into the Permian Period which followed. The question to be argued is this: Did the glaciation consist of individual ice sheets, each on its own isolated continent, or was there a single great ice sheet on Gondwana?

It is only reasonable to think that a great ice sheet would grow and spread only in a cold climate, and that such conditions would occur only in high latitudes, not too far from the earth's north pole or south pole. Evidence from paleomagnetism clearly shows that the position of the poles of rotation has migrated widely through geologic time, with respect to the continents. This phenomenon is called **polar wandering.** It requires that the entire global lithosphere slowly rotate over the soft asthenosphere beneath. As a result of polar wandering, parts of the continents that are now in warm, tropical locations may well have once been situated near one of the poles, in locations favorable to the growth of large ice sheets.

Figure 9.14 is a restoration of the nested continents of Gondwana, showing the places where evidence is found today of the Carboniferous-Permian glaciation. A dashed line shows the hypothetical limit of a single great ice sheet of that period. Also shown is the changing position of the earth's pole from the Devonian Period through the Permian Period. Notice that by upper Carboniferous time the pole lies near the center of the glaciated area.

Whatever may be the merits of geologists' arguments for and against a single ice sheet on Gondwana before the continents began to break up, the hypothesis is at least consistent with what is known about polar wandering in that span of geologic time.

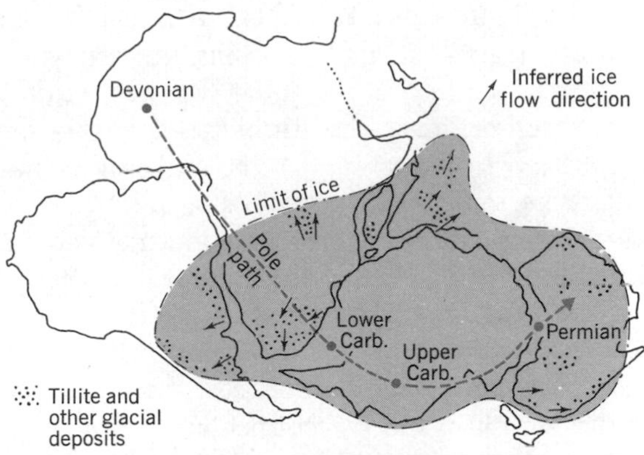

Figure 9.14 On this hypothetical restoration of the nested continents of Gondwana, dots show the locations of tillites and other glacial deposits of upper Carboniferous time; arrows show inferred directions of ice motion; a dashed line shows the limit of a single great ice sheet.

CONTINENTAL RUPTURING AND RIFT VALLEYS

The pulling apart of continents on new ruptures of the lithosphere initiated the splitting up of Pangaea into the modern continents. Some imagined details of a newly formed **continental rupture** are reconstructed in Figure 9.15. Diagram A shows the first phase of rupturing as involving an up-arching of the granitic crust, accompanied by **block faulting.** A central **rift valley** begins to form and is the site of accumulation of clastic sediment swept down from the adjacent block mountains. There are also eruptions of basaltic lava, occurring as flows interbedded with these sediments. This phase of rupture is illustrated today by the great African Rift Valley. It is over 1200 mi (2000 km) in length and runs north-to-south through Uganda, Tanzania, and Malawi. Lakes Nyassa and Tanganyika occupy parts of the rift valley. Highlands border the rift valley on both sides.

As the lithospheric plates continue to separate, a new belt of oceanic crust comes into existence, as shown in Diagram *B*. An example is the Red Sea trough, forming today as the African plate moves away from the Arabian Peninsula. In Diagram *C*, a broad belt of oceanic crust has formed; the characteristic Mid-Oceanic Ridge divides it into two parts.

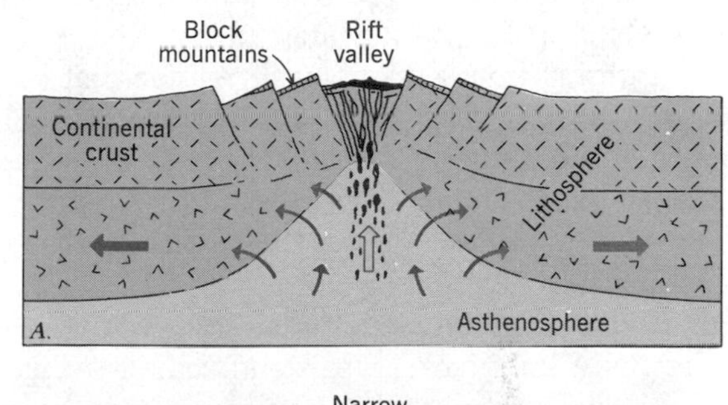

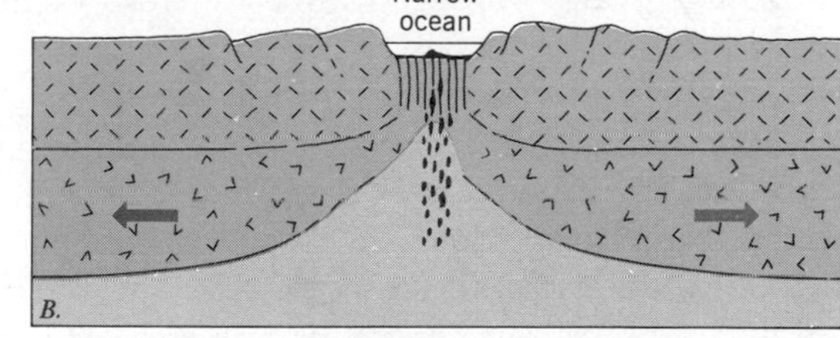

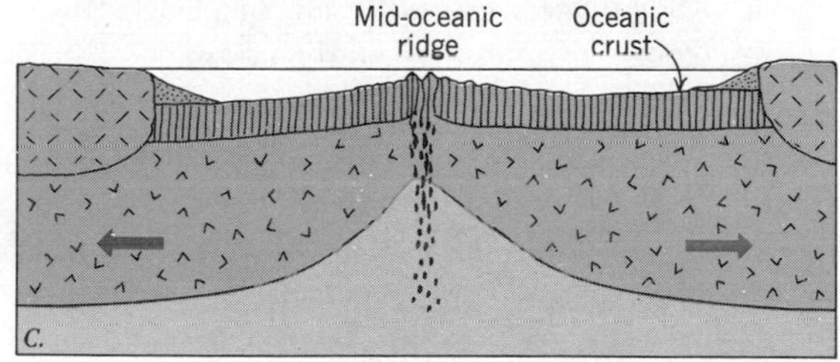

Figure 9.15 Stages in rupture of a continental plate.

COLLIDING CONTINENTS

We have seen that ocean-basin lithosphere can collide with continental lithosphere, producing an island arc-trench system in the subduction zone. But is it possible that two continental portions of lithospheric plates could meet? If this happened, what sort of tectonic feature would occur?

Geologists have drawn up a scenario for **continental collision.**

The details are shown in Figure 9.16. In Diagram A we see subduction of a plate in progress, causing the narrowing of an ocean basin. Along the continental margin at the right, sediments are accumulating as shelf deposits, while turbidites are being spread over the ocean basin floor. In Diagram B the ocean basin has narrowed greatly and undergone some up-arching. Folding and thrusting are occurring near the subduction zone, while more terrestrial sediments (delta deposits) are accumulating close to the continent at the right.

In Diagram C the ocean basin has completely disappeared. There is an up-arched alpine belt at the line of subduction, and only a very narrow segment of oceanic crust remains at depth. Finally, as shown in Diagram D, continental crust of both plates is in contact. The joining of two masses of continental crust has been quite appropriately named **suturing.** The alpine belt marking the line of junction is a **suture.**

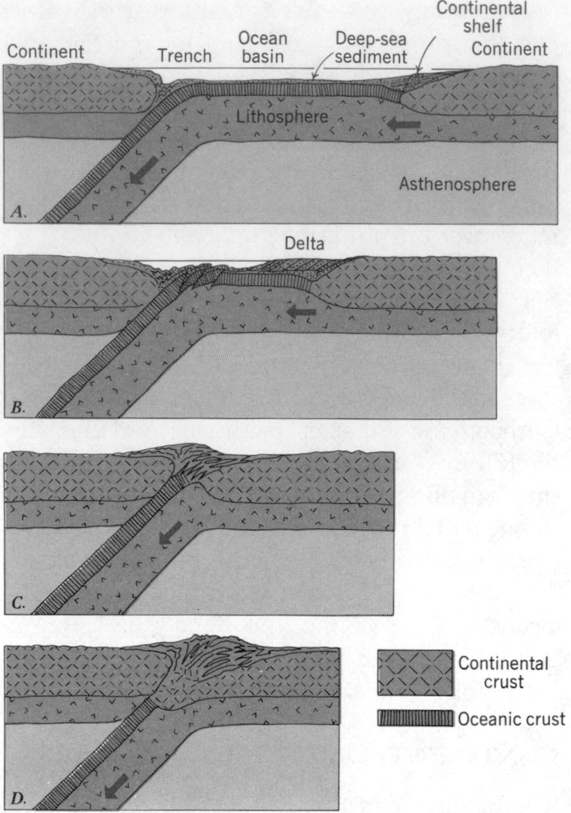

Figure 9.16 Stages in the collision of two continents.

Suturing may be in progress, or only recently completed, along the great Himalaya range of southern Asia. Here the continental crust of peninsular India is very close to the Asiatic continental crust. Just prior to the breakup of Pangaea, shown in stages in Figure 9.11, the Gondwana shield landmass was separated from the Asian continent by the Tethys Sea. In late Triassic time, the African plate began to move toward the Asiatic plate, narrowing the Tethys Sea. In Jurassic time the lithospheric plate beneath the Indian fragment of the Gondwana plate moved north, further narrowing the Tethys Sea. Finally, in the Cenozoic Era, the Indian continental crust was rammed into the thick sediments of the Tethys Sea, crumpling and thrusting them into the present-day Himalaya range.

The Appalachian mountain chain of North America may represent suturing from a continental collision that occurred at the close of the Paleozoic Era, long before the breakup of Pangaea. In the Appalachians, strata of Paleozoic age have been strongly folded as well as metamorphosed. Another possible example of an ancient suture is the Urals range cutting north-south across the Eurasian continental shield.

Evidence of ancient suturing leads us to suspect that the continents must have broken apart and come together repeatedly during the 3 billion years or so that continental crust was accumulating. The breakup of Pangaea, coming as it did late in the Mesozoic Era, may have been only the last of many such events.

Recent estimates suggest that the combined surface area of all continents is not measurably larger today than at the time Pangaea existed. If so, the continents on the average are no longer experiencing appreciable growth. To account for this steady-state condition, we must suppose that continents lose area at a rate balancing the rate of areal growth by production of new granitic igneous and metamorphic rock at the continental margins.

Just how such loss occurs is not known, but it seems possible that the continental crust could be carried down into the mantle in subduction zones, there to melt and become assimilated into the mantle. Subduction of a plate carrying continental crust is rarely shown in geologists' published diagrams, but might occur during suturing if the plates continued to move together after the continental crust had come into contact on both sides of the suture.

THE ROCK CYCLE IN A NEW LIGHT

In Chapter 5, we put together a cycle of rock transformation, taking into account the transport of mineral matter through both surface and deep environments (Figure 5.18). Plate tectonics theory offers the possibility of viewing the rock cycle in terms of plate motions. Figure 9.17 is a schematic diagram showing a single plate of the ocean-basin type, viewed as a conveyor belt. There is a complete cycle of magma rise and solidification in the spreading rift, transport as oceanic crust on the plate, sinking and remelting in a subduction zone, and eventual return by flowage deep within the asthenosphere. As this conveyor belt moves, it accumulates sediment and carries it along to the subduction zone. There, both oceanic and continental sediment is scraped off the descending plate and reformed into metamorphic and igneous rock in the orogenic belt, becoming a permanent addition to the continental crust.

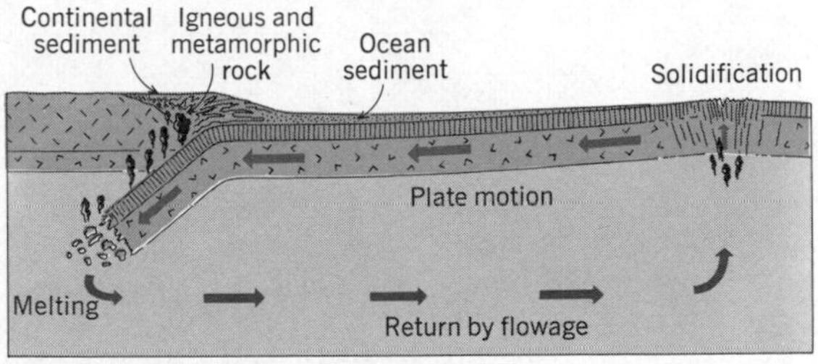

Figure 9.17 The rock cycle viewed in terms of the conveyor-belt action of a lithospheric plate.

YOUR GEOSCIENCE VOCABULARY

continental drift
Curie point
soft magnetization
hard magnetism
magnetometer
paleomagnetism
normal epoch
reversed epoch
magnetic event
magnetic anomaly
transform fault
tectonics
plate tectonics
convection currents
Pangaea
Laurasia
Gondwana
Lystrosaurus
tillite
polar wandering
continental rupture
block faulting
rift valley
continental collision
suturing
suture

SELF-TESTING QUESTIONS

1. Explain how permanent magnetism comes to be locked into lavas. What is the significance of polarity reversals determined from rock magnetism? Name the four most recent polarity epochs and give their approximate ages. How far back in time does the polarity reversal record extend?

2. Explain how magnetic anomalies provide evidence of sea-floor spreading. What rates of plate separation are indicated by this record?

3. Make a careful and precise distinction between a transcurrent fault and a transform fault as applied to transverse fractures of the mid-oceanic spreading zone. How did seismology play a key role in determining the existence of transform faults?

4. What are the three kinds of lithospheric plate boundaries? Name the principal lithospheric plates. For each plate tell which continents and ocean basins are included, as well as the nature of the plate boundaries.

5. Explain the role that convection currents in the mantle may play in plate tectonics. What driving mechanisms for plate motions have been proposed?

6. How were the continental nuclei grouped within Pangaea? Describe the stages in breakup of Pangaea and the opening up of new areas of ocean floor. Review the geological and paleontological evidence of a former union of the continents within Pangaea.

7. Describe the evidence for Carboniferous-Permian glaciation on a single continent of Gondwana. How is polar wandering involved as a cause of glaciation?

8. Describe the stages of tectonic activity during continental rupture. Where can examples of these stages be found today?

9. Describe the tectonic events involved in continental collision. Where can examples be found today? What is the larger significance of sutures dating back into the Paleozoic Era? Have the continents increased in surface extent since the breakup of Pangaea? How might continental crust be destroyed?

10. Synthesize the concept of the cycle of rock transformation (rock cycle) in light of plate tectonics.

(*Above*) The Ahnighito meteorite being excavated from its original position on Cape York, Greenland. After being raised on steel beams, the huge meteorite was chained to timbers and dragged to the water's edge.

(*Below*) The chained meteorite is loaded aboard the *Hope* for its journey to New York City. Both photographs were taken in 1897 by the polar explorer, Robert Peary. (American Museum of Natural History.)

10 ASTROGEOLOGY—THE GEOLOGY OF OUTER SPACE

GIFTS FROM HEAVEN

Debate, if you will, visits to our planet by living beings from outer space, but there's no denying that Earth receives a continued rain of nonliving visitors without return tickets. Most are silent nocturnal meteors, or shooting stars, that vaporize in the upper atmosphere. Perhaps 20 million particles of interplanetary dust, each no larger than a grain of sand, enter the atmosphere each day to become meteors.

On rare occasions, much larger bits of solid matter are trapped by the earth's gravity field. Up to many tons in weight, these pieces of space debris can make a spectacular fiery splash in our atmosphere, to the accompaniment of sounds like cannon fire or thunder. What is not consumed in the heat of atmospheric friction may reach the earth's surface intact, becoming a meteorite and a permanent addition to our lithosphere.

Meteorites have been a source of wonder and worship since the dawn of civilization. Chinese writers of the Han Dynasty, about the time of Christ, kept a careful log of meteorite falls, noting the time and place of each. Livy's history of Rome includes mention of a fall of several meteoritic stones upon Alban Hill, near Rome, in the sixth century B.C. A few decades later, meteorite fragments smashed some Roman chariots and killed 10 men.

It has been said that the Black Stone within the Kaaba at Mecca, the most sacred object of the Moslem world, is in fact a meteorite, and that it had been worshipped in Mecca long before the time of Mohammed. The Romans worshipped a meteorite which they brought from a temple at Pessinus in hopes that it would bring them victory over Hannibal's invading army. They were ultimately successful in turning back Hannibal, and the stone acquired a new aura of prestige. A fall of meteorite stones in Japan in the 1700s was believed to have dropped from the loom of a goddess residing in the Milky Way; she had used the stones to weight down her celestial loom.

The Eskimos of Greenland also revered meteorites as gifts from heaven. They named one great Iron meteorite Ahnighito and tried to pry loose slivers of the metal out of which to make knives, but with little success. In 1897 Admiral Peary brought the enormous meteorite back to the States, but only over the protests of the Eskimos, who thought grave misfortunes would ensue. Peary arranged for a return shipment of useful steel to compensate for the loss. Ahnighito is the second largest known meteorite. It now resides in the American Museum of Natural History in New York City. The 36½-ton monster is nearly 11 feet long, 6 feet high, and 5 feet thick, and is reasonably secure from shoplifters.

One topic we will take up in this chapter is the geologic interpretation of meteorites. Until the astronauts of *Apollo 11* brought back the first moon rocks, meteorites were our only samples of mineral matter from elsewhere in the solar system.

IS THERE A GEOLOGY OF OUTER SPACE?

It may seem absurd at first to extend the word geology to include a study of the Moon, planets, and other solid objects of the solar system. The prefix *geo*, after all, comes from the Greek word meaning "Earth."* Still, the Moon and closer planets are composed of minerals and rocks, and they can be expected to have some of the geological features of our Earth, such as volcanoes and lava flows, faults and earthquakes, mantles and cores, and perhaps even tectonic belts. There may even exist sediments and sedimentary strata of some sorts on these bodies. Perhaps, then, it is not at all unreasonable to extend the science of geology to outer space, applying it wherever we can.

Moon and Mercury, and to some extent Mars as well, have one feature in common that makes geological investigations especially fruitful: They have practically no atmosphere. None of the three has running streams or oceans of free water. On Earth, processes of rock weathering, erosion by streams, waves and wind, and sedimentation quickly remove or bury features produced by volcanism, tectonic activity, and the impact of large bodies from outer space. Where a planet has little or no atmosphere or hydrosphere, these processes do not act, or act very, very slowly. As a result, surface forms and structures produced by volcanic and tectonic activity, and by impact from outer space remain little changed through eons of geologic time on Moon, Mars, and Mercury. Study of these planetary features can shed much light on the early history of our own planet. For example, what was the composition of the earliest crust of the Earth? If none is preserved on our planet, perhaps it still remains on the Moon or on Mercury.

It is not surprising that earth scientists mounted a tremendous effort to explore the Moon's surface and to bring back samples for analysis. Lunar geology held great promise for new knowledge, not only of Moon itself, but of early Earth history as well. One leading scientist has likened the Moon to the Rosetta stone, which gave scholars the key to translation of Egyptian hieroglyphics, and so opened up the field of Egyptian history.

In this chapter we will apply geological principles to the understanding of the Moon, the inner planets, and to those fragments of a disrupted planet, the meteorites.

Astrogeology is the name established by the United States Geological Survey for the application of principles and methods of geology to all condensed matter and gases of the solar system outside the Earth.

THE EARTHLIKE PLANETS

Before embarking on a study of the Moon, it will be helpful to examine a few facts about the solar system. Our solar system consists of solid objects in orbital motion about the Sun. There are several varieties of orbiting objects; the largest are the major planets; of these there are nine. In order of distance from the Sun, these are

* In this chapter we capitalize the first letter of Earth, Moon, and Sun to make them proper names, consistent with the other planets.

Mercury, Venus, Earth, Mars, Jupiter, Saturn, Uranus, Neptune, and Pluto.

Our concern in this chapter is with the first four planets on this list; they are often called the inner planets, or the terrestrial planets. From the geological standpoint, the second label is the more interesting because it implies that Mercury, Venus, and Mars resemble the Earth in geological ways. You might say that the terrestrial planets are "rock" planets, for they consist largely of silicate minerals surrounding dense cores of iron. Average densities of the four planets in gm/cc are: Mercury, 5.0; Venus, 5.1; Earth, 5.5; and Mars, 3.9. Mars' lower density implies that it has only a small iron core, if any.

In comparing the sizes of the four terrestrial planets, we encounter major dissimilarities. Taking the Earth's mass as 100%, Venus is comparable in size, with 81% as great a mass as Earth. However, Mars is much smaller, 11%; Mercury is still smaller, 6%. Actually, Earth and Venus are closely matched in diameter, mass, and density.

The very small size of Mercury suggests that we might better compare it with our Moon than with the other three inner planets. Data are as follows:

	Diameter mi	(km)	Mass (Earth as 100%)	Average density gm/cc
Mercury	3000	(4900)	6.0%	5.0
Moon	2160	(3480)	1.2%	3.34

In terms of diameter, Mercury and Moon are fairly well matched. The major discrepancy comes in their masses; Mercury's is five times greater, because Mercury is the greater in volume and is denser. We find the comparison interesting because our Moon may well be classified as a small planet. Sometimes the Earth and Moon are linked together as a pair, called a binary planet. This linkage is a valid concept, because our Moon is exceptionally large in proportion to its master planet as compared with other planetary moons.

ORIGIN OF THE SOLAR SYSTEM

An overview of Earth history would not be complete without at least a quick sketch of Earth's origin as a member of the solar system. Modern thinking runs along the lines that our solar system is a byproduct of the formation of our Sun, a rather average-sized star with a history quite typical of many other stars of medium mass and temperature. A very diffuse body of hot interstellar dust and gas is taken as the starting point. This **solar nebula** began to contract under the mutual gravitational attraction of all its particles upon one another. Gradually the nebula began to cool and assumed a wheel-like shape—very thin, and somewhat larger in diameter than our present solar system (Figure 10.1). A dense, luminous mass of gas which was to become the Sun formed an enlarged central hub.

Figure 10.1 According to the modern condensation hypothesis, contraction of the solar nebula produced a thin disk of dust and gas that rotated about a central solar mass. The nebula is seen here in edge-on view.

Then, with further cooling of the nebular disk, condensation of gases began to occur rapidly. Minute particles of iron and silicates became solid first, then later the highly volatile compounds such as water, methane, and ammonia. As grains collided they began to stick together, a process called **accretion.** The resulting aggregations of matter are called **planetesimals.** They grew rapidly, but many were unstable, and disrupted into countless small, solid fragments; these orbited the Sun along with the remaining large planetesimals. By a series of collisions, the larger planetesimals swept up most of the smaller fragments and quickly grew to their final dimensions. The impacts were not so high in energy release as to heat the growing planets to the point of complete melting.

A possible tenth planet may have formed between Mars and Jupiter, then later fragmented into solid bodies known now as the **asteroids.** Most asteroids now orbit between the orbits of Mars and Jupiter. Possibly our Moon was yet another of the early planets, formed by accretion at the same time as Earth; at least, this is one hypothesis of origin of the Moon.

A timetable for this phase of Earth history has been suggested as follows: The solar nebula was in process of rapid contraction about 5 billion years ago (−5.0 b.y.); accretion of the planets was largely completed at about −4.7 b.y. to −4.5 b.y.

The Earth was probably not in a completely molten state at any time during its accretion. However, there must have been local melting where exceptionally large objects impacted the surface. The Earth at this stage was a mixture of silicates and iron, not as yet differentiated into a core and mantle. That change was to follow as radiogenic heat accumulated and caused major episodes of melting and overturn. In any event, the Earth had acquired its layered structure and was probably internally stable by about −3.6 b.y., when the first crustal rocks of which we have any record were formed.

This brief overview of planetary origin may raise more questions than it answers. What happened to the volatiles, such as methane and ammonia, abundantly present in early stages of accretion? These may have been driven off the growing planetesimals closest to the Sun through heating and the pressure of the solar wind. Some water, remained, however, locked up in the silicate minerals of the growing planet Earth. This water, with other gases, was later to emerge slowly by **outgassing** from volcanoes, ultimately to form the world ocean and atmosphere.

TESTIMONY OF THE METEORITES

Let's begin our geological investigation of outer space with the meteorites. These fragments of stone and metal have been studied for many decades by geologists. Consequently, much was known about meteorites long before manned space vehicles went to the Moon. We cannot obtain samples of the matter making up the Earth's mantle and core. However, meteorites have supplied interesting evidence, bearing on both the composition of the Earth's interior and the age of the members of the solar system.

Any fragment of solid matter entering the Earth's atmosphere from outer space is called a **meteoroid.** Most are extremely tiny particles and vaporize as they penetrate the atmosphere, leaving only a thin trail of light, called a **meteor.** Meteoroids may be the remaining planetesimals, or fragments of those bodies, left over from the accretion process. Occasionally, a very large meteoroid enters the Earth's atmosphere. Though partially vaporized, it may reach the Earth's surface. Such exotic rock objects are then called **meteorites.**

The fall of a meteorite is accompanied by a brilliant flash of light, called a **fireball.** There may also be loud sounds. Frictional resistance with the atmosphere causes the outer surface of the object to be intensely heated. However, this heat does not penetrate to the interior of a meteorite, which reaches the Earth with its original composition and structure unchanged. The single mass may explode before the impact, showering fragments over a wide area. The observed arrival of a meteorite and subsequent collection of the fragments is designated as a **fall.** Collection of a meteorite whose fall was not observed is designated as a **find.** Examples of very large meteorites are shown in Figure 10.2.

Meteorites have been intensively studied, not only as to chemical composition and structure, but also as to age. They fall into three classes. (1) The **irons** are composed almost entirely of a nickel-iron alloy, in which the nickel content ranges from 4% to 20%. (2) At the other end of the series are the **stones,** consisting largely of silicate minerals, mostly olivine and pyroxene, with only 20% or less nickel-iron. Plagioclase feldspar may also be present. (3) An intermediate class of meteorites consists of the **stony irons.** In these, silicate minerals and nickel-iron may form a continuous medium in which spherical bodies of silicate minerals are enclosed. Structures of the meteorites have aroused much interest in their similarities to and differences from terrestrial rocks. The nickel-iron of an iron meteorite typically shows crystalline structure. One group of stony meteorites possesses a coarse-grained structure that resembles the structure of plutonic igneous rocks. We can infer that meteorites with such structure solidified from a magma.

When the meteorites of observed falls are catalogued, their relative abundance turns out to be about as follows: stones, 94%; irons, 4.5%; stony irons, 1.5%. The stony meteorites are thus preponderant in bulk. The nickel-iron meteorites can be interpreted as cores of planetesimals, in which the process of differentiation had taken place. Stony meteorites having textures resembling terrestrial igneous rocks point to the possibility that melting and recrystallization had taken place to some degree in these original planetary bodies.

Age determinations of meteorites have been made, using the radiometric methods described in Chapter 5. There is a high degree of agreement in the results pointing to the time of formation of all types of meteorites as 4.5 b.y. before present. Note that this age is about 1 b.y. greater than that of the oldest known crustal rocks on earth. The composition and great age of meteorites has lent strong support to the accepted hypothesis that the Earth's core is composed of nickel-iron and the mantle of iron-magnesium silicates. This conclusion is greatly strengthened by independent evidence derived from the study of earthquake waves (Chapters 2 and 6).

Figure 10.2 Two kinds of meteorites. (*Upper*) This stony meteorite, weighing 745 lb (338 kg), is the largest single stony meteorite of which the fall has been observed. It struck the ground at Paragould, Arkansas, on February 17, 1930, forming a huge fireball visible over thousands of square miles. Height of the meteorite is about 2 ft (0.6 m). (Yerkes Observatory.) (*Lower*) The Williamette meteorite, an iron meteorite, weighs 15½ tons (14 metric tons) and is over 10 ft (3 m) long. The huge cavities were produced by rapid oxidation of the iron during fall through the atmosphere. (American Museum of Natural History—Hayden Planetarium.)

IMPACT FEATURES ON EARTH

Have meteoroids of great size struck the Earth's surface to produce recognizable impact features? The largest known meteorite is the Hoba iron, found in southwest Africa; it weighs about 66 tons (60 metric tons). The next five in size are irons weighing roughly half as much as the Hoba meteorite. However, no stony meteorite has been found whose weight is over one ton. The largest stony meteorite observed to fall, pictured in Figure 10.2, weighs one-third of a ton.

Rarely has an observed meteorite fall produced craters of measurable dimension. One of them was the Siberian Sikhote Alin fall of February 12, 1947, witnessed by many persons. The largest iron fragment recovered weighed 2 tons (1800 kg). Funnel-shaped craters as large as 92 ft (26 m) in diameter were produced by the larger iron fragments. Soviet scientists deduced from the observed trajectory of the meteoroid trail that it was a small asteroid traveling at a speed of 25 mi/sec (40 km/sec).

So we turn to prehistoric impacts of enormous meteorites capable of producing rimmed craters with diameters of 500 to 5000 ft (150 to 1500 m) and larger. Several fine examples of almost perfectly circular large craters with sharply defined rims have been found and examined over the continental surfaces of Earth.

We are, of course, excluding from this discussion all known volcanic craters constructed of lava and volcanic ash, as well as obvious calderas, the large craters produced by explosive demolition of a preexisting volcanic cone (Chapter 3).

Perhaps the finest example of a large circular-rimmed crater of almost certain meteoritic impact origin is the Barringer Crater (formerly known as Meteor Crater) in Arizona (Figure 10.3). The diameter of this crater is 4000 ft (1200 m), and its depth almost 600 ft (180 m). The rim rises about 150 ft (46 m) above the surrounding plateau surface, which consists of almost horizontal limestone strata. Rock fragments have been found scattered over a radius of 6 mi (10 km) from the crater center, while meteoritic iron fragments

Figure 10.3 Oblique air view of the Barrington Crater in northern Arizona. (Yerkes Observatory.)

numbering in the thousands have been collected from the immediate area. Carefully derived estimates of the impacting object specify that it was a 63,000-ton (56×10^6 kg) iron meteorite about 100 ft (30 m) across and that it was traveling at about 34,000 mph (15 km/sec) when impact occurred.

Figure 10.4 shows inferred steps in the formation of a simple meteorite crater of moderate size, such as the Barringer Crater. An enormous quantity of kinetic energy is almost instantly transferred to the ground by a shock wave, which intensely fractures and disintegrates the rock around the point of impact. The shock wave is reflected back to the meteorite body, which is fragmented into thousands of pieces and partly vaporized as well. Large amounts of fragmental debris are then thrown out, while the solid bedrock is forced upward and outward to create the crater rim. A great deal of rock material falls back into the crater, filling the bottom, where melted rock may be concealed.

The number of other craters of generally accepted meteoritic origin is not large—perhaps only a dozen or so. From these we can draw the conclusion that impacts by large meteorites have been extremely rare events in recent geologic time. Perhaps the number has been on the order of half a dozen occurrences per million years.

We may safely conclude that meteorites are not being produced in the modern solar system. If they represent space debris that continues to be swept up by planets and satellites, the frequency of impacts should be small in later geologic eras compared with a high frequency in the early stages of planetary formation.

So we must look to a celestial body on which impact features would have been preserved with little or no erosion for the entire 4.5 b.y. since Earth and other planets formed. Three such bodies are available for study. They are our Moon and the planets Mars and Mercury. Venus is so completely concealed by a dense cloudy atmosphere as to be an unlikely prospect.

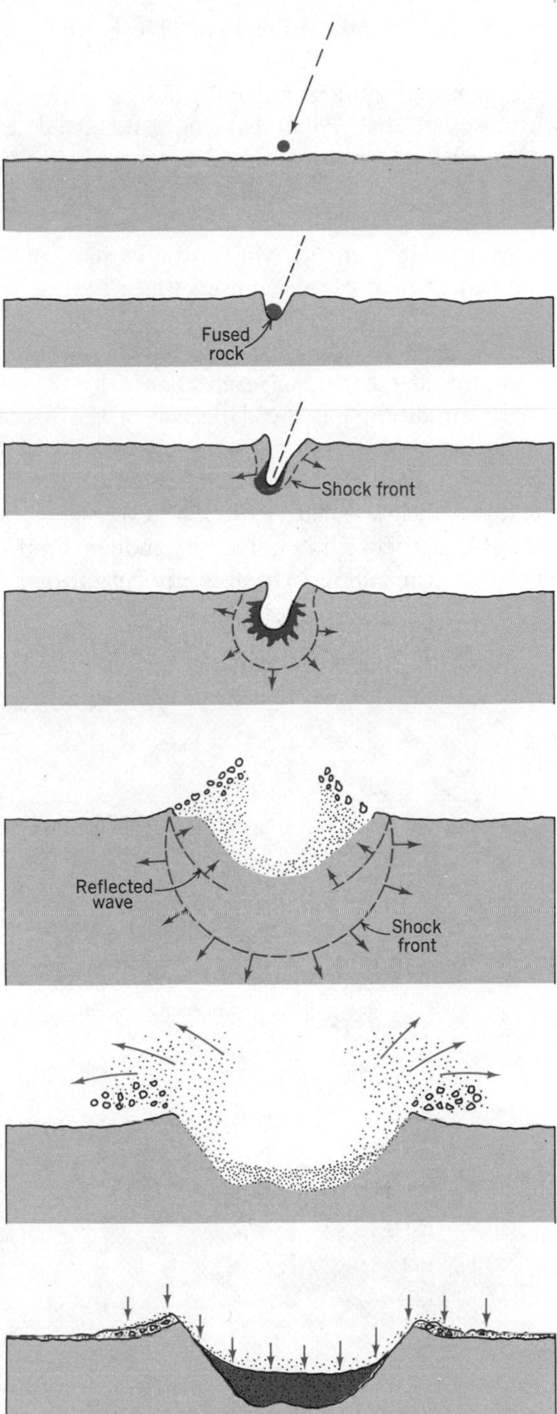

Figure 10.4 Hypothetical stages in the formation of a meteorite impact crater. (Data of E. M. Shoemaker and U.S. Geological Survey.)

THE LUNAR SURFACE ENVIRONMENT

To interpret correctly the physical features of the Moon's surface, we must first consider the surface environment of that satellite. Environmental factors include (1) the Moon's gravity field, (2) a lack of both atmosphere and hydrosphere, (3) intensity of incoming and outgoing solar radiation, and (4) surface temperatures. All of these factors show striking differences when compared with the environment of Earth.

Gravity is defined as the attraction which Earth or Moon exerts upon a very small mass located at its surface. Gravity on the Moon's surface is about one-sixth as great as on Earth. Therefore, an object that weighs 6 lb on Earth will weigh only 1 lb on the Moon. This relatively small gravity is of human interest when we watch films of our astronauts cavorting on the lunar surface. The small gravity is of great geologic importance in interpreting the Moon's surface and history. For example, rock of the same strength as rock on Earth could stand without collapse in much higher cliffs and peaks on the Moon than on Earth. Objects thrown upward at an angle from the Moon (as when the Moon is struck by a large meteoroid) will travel much higher and farther than objects under the same impetus on Earth.

Lack of sufficient gravity has cost the Moon the loss of any atmosphere that it may have once possessed, since gas molecules of any earlier atmosphere could have escaped readily into space.

Lacking an atmosphere, the Moon's surface intercepts the Sun's radiation on a perpendicular surface with the full value of the Sun's output. Moreover, the Moon absorbs most of this incoming energy. The effect is to cause intense surface heating during the long lunar day of about two weeks' duration. When the Sun's rays have been striking at a high angle for several days continuously, surface temperatures at lunar noon reach about 215° F (100° C). Correspondingly, conditions on the dark side of the Moon reach opposite extremes of cold during the long lunar night. With no atmosphere to block the escape of heat, surface temperatures drop to values estimated at below −280° F (−173° C).

Of particular interest is the sudden drop in lunar-surface temperature when sunlight is abruptly cut off by a lunar eclipse. In one

Figure 10.5 Sketch-map of the major relief features of the moon. Use this diagram as an aid in identifying areas and subjects in the lunar photographs of this chapter.

such instance, the surface temperature fell from 160° F (71° C) to −110° F (−79° C) in only 1 hr. This drop of 270 F° (150 C°) is vastly greater than any natural temperature drop on the Earth's surface in a comparable period of time.

From the geologist's standpoint, such vast temperature ranges are of interest because of the possible effect upon minerals exposed at the Moon's surface. The expansion and contraction that crystalline minerals undergo when heated and cooled can bring about the disintegration of solid rock into small particles. These volume changes can also cause loose particles to creep gradually to lower levels on a sloping ground surface. These effects may be unusually important on the Moon, because without an atmosphere and running water, ordinary terrestrial processes of weathering, erosion, and transportation cannot act upon the Moon. Without streams and standing water bodies, the Moon has no mechanisms or receiving areas for accumulation of water-laid sedimentary strata. Obviously, in comparison with Earth, surface changes must be extremely slow on the Moon.

Figure 10.6 The Moon as it would appear if it could be uniformly illuminated. This is a composite of many photographs. (NASA photograph.)

A DECADE OF LUNAR EXPLORATION

The final splashdown of *Apollo 17* in December 1972, brought to a close man's first exploration of the Moon's surface. The vast store of rock and soil samples, photographs, and other observational data accumulated during the Apollo program will take many years to analyze fully. Although certain basic geological facts about the Moon were established beyond a doubt, many more deep and complex scientific puzzles have emerged than could have been imagined.

The decade of firsthand lunar exploration began in 1963, when a United States Ranger spacecraft impacted the lunar surface, sending back thousands of pictures from a wide range of altitudes. There followed landings by Surveyor spacecraft, making direct physical and chemical tests of lunar surface materials as well as photographs of the ground immediately surrounding the vehicle. Then came a series of Lunar Orbiter vehicles on missions of photographing the lunar surface from distances as close as 35 mi (56 km). Vertical photographs at least 10 times sharper than the best taken by telescopes from Earth were obtained from both the nearside of the Moon, as well the previously unknown farside.

In 1969 and 1970, manned space vehicles of Apollo missions circled the Moon at low levels and descended to the lunar surface, permitting photographic negatives in color to be brought back.

THE LUNAR MARIA

As anyone can easily see, using a small telescope or binoculars, the first major subdivision of the Moon's surface is into light-colored areas and dark-colored areas (Figures 10.5 and 10.6). The former constitute the relatively higher surfaces, or **lunar highlands.** For many years, the light-colored areas were called collectively the *terrae*. This Latin word reflects the earliest interpretations of Galileo—that these areas were dry lands. The dark-colored areas, the low-lying, smooth lunar plains, are the **lunar maria,** plural of the Latin word *mare* for "sea"; Galileo applied this term to what he believed to be true seas of liquid. Of the nearside of the Moon, about 60% is terrae and about 40% maria.

Lunar highland areas exhibit a wide range of relief features. Most outstanding are the great mountain ranges, of which there are 20 major groups on the nearside. Perhaps the most spectacular of these are the Appenines rimming the Mare Imbrium on the southwest side (Figure 10.7). Several peaks within the Appenines rise to heights of 12,000 to 16,000 ft (4 to 5 km) above the nearby mare surface. The highest and most massive mountains are those of the Leibnitz range, near the lunar south pole, which have peaks rising to heights of 35,000 ft (11 km). Elsewhere, the highland terrain consists of gently rolling surfaces with low slopes, and of rough areas with steep slopes.

Lunar maria of the nearside are divided into 10 major named areas, in addition to a single vast area, Oceanus Procellarum (Figure 10.5). Although maria outlines are in places highly irregular, with many bays, a circular outline is persistent for several, and particularly striking for Mare Imbrium.

THE LUNAR CRATERS

Most spectacular of the Moon's surface features are the **craters.** Even under the low magnification of a small telescope the large craters form an awe-inspiring sight when seen in a partial phase of the Moon. Craters are abundant over both the highland and maria surfaces. Using telescopes alone, some 30,000 craters were counted on the nearside with diameters down to 2 mi (3 km). An estimated 200,000 have been identified with space-vehicle photography. Included are recognizable craters as small as 2 ft (0.6 m) across. On the lunar nearside, there are 150 craters of diameter larger than 50 mi (80 km). Largest of these is Clavius, 146 mi (235 km) in diameter and surrounded by a rim rising 20,000 ft (6 km) above its floor. (See south polar area in Figure 10.5.) Tycho and Copernicus are among the most striking of the large craters (Figure 10.8).

The forms of large craters fall into several types. In some, the floor is flat and smooth, and the rim is sharply defined and abrupt. In others, such as Copernicus, the floor is saucer-shaped, while the rim consists of multiple concentric ridges. Of particular interest are systems of **rays** of lighter-colored surface radiating from certain of the larger craters—for example, Copernicus (Figure 10.8). In several of the large craters there is a sharply defined central peak, which must be taken into account in interpreting the origin of craters. That of Eratosthenes shows up particularly well in Figure 10.8.

Figure 10.7 Mare Imbrium, with its bordering mountains, the Jura, Alps, Caucasus, Apennines, and Carpathians; and the great craters Plato, Aristillus, Archimedes, Eratosthenes. Note the spinelike peak, Piton. Identify these features with the aid of Figure 10.5. (The Hale Observatories.)

Many small lunar craters are almost perfectly circular and have a cup-shaped interior and a prominent rim (Figure 10.9). Others are less distinct and have low slopes or are mere shallow depressions. The impact origin for almost all of these smaller craters is generally accepted. A few are fresh in appearance, with a litter of boulders on the rim and within the crater itself.

The newer craters are typically lighter in color than the surrounding surface. The more subdued craters appear to be older and have lost their sharpness through slow processes of mass wasting. The oldest forms have lost their rims and show only a shallow depression. A few small craters are elliptical in outline and may represent the secondary fall of masses dislodged by much larger impacts.

It has been shown that the number of lunar craters per unit of area is inversely proportional to the square root of the crater diameter. Consequently, the number of large craters is small and that of very small craters is legion.

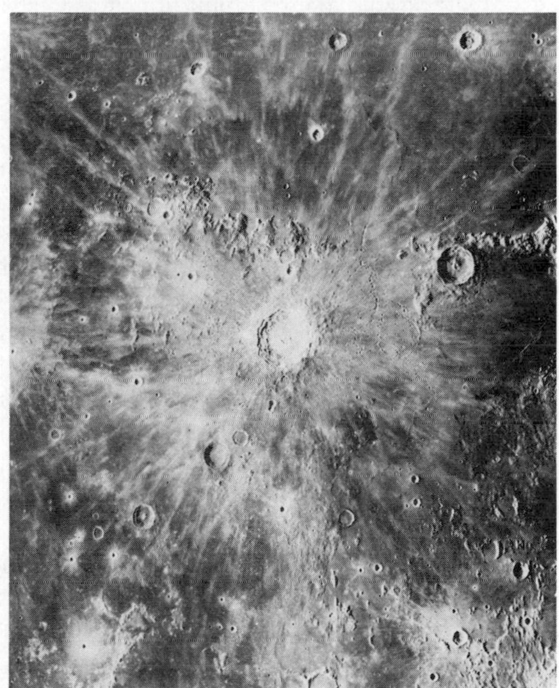

Figure 10.8 Copernicus, the great lunar crater lying south of Mare Imbrium (see Figure 10.5). Note the rays—radial streaks of lighter-colored material. The conspicuous crater lying east-northeast of Copernicus is Eratosthenes. (The Hale Observatories.)

RILLES AND WALLS

Yet another class of distinctive lunar-surface features is narrow, canyonlike features called **rilles.** Some are remarkably straight, taking the form of a trench up to 150 mi (240 km) long. Others are irregular in plan, and a few are sinuous, suggestive of terrestrial meandering rivers (Figure 10.10). Astronauts of Apollo 15 inspected the walls of Hadley's Rille at close range. Layers of rock are exposed, suggesting that an erosion agent carved the rille into older deposits. It has also been suggested that the sinuous rilles were formed over lines of vents from which volcanic gases were emitted under high pressure.

Most of the straight rilles are interpreted as fracture features in brittle rock of the lunar crust. Related features are the straight cliffs, or **walls,** which may be fault escarpments. A particularly striking example is the Straight Wall in Mare Nubium (Figure 10.5), which is 800 ft (240 m) high and may represent a fault that broke the mare surface.

Study of Lunar Orbiter photographs has revealed many classes of minor details that are as yet little understood. Particularly prevalent in areas of rough terrain are minute systems of parallel ridges and troughs. Many of the steep slopes exhibit steplike terraces near the base.

The farside of the Moon was photographed in detail by Lunar Orbiter spacecraft and by astronauts of Apollo spacecraft, the first human beings to see that side of the Moon. The lunar farside is heavily cratered and lacking in extensive maria (Figure 10.11).

Figure 10.9 Censorinus (*arrow*), one of the freshest craters on the Moon's nearside, shows a sharp rim and a surrounding zone of light-colored soil surface. Crater rim is about 4 mi (7 km) in diameter. (*Apollo 10* photograph by NASA.)

ORIGIN OF THE GREAT CRATERS AND MARIA

Origin of the great craters and the circular-rimmed maria brings into conflict two widely divergent hypotheses. Both have found supporters throughout the long history of lunar geology.

One extreme view held that the lunar craters and maria are largely of volcanic origin. This hypothesis implies a long history of intrusion and extrusion of molten rock from the Moon's interior. Circular depressions of the great craters were explained by the collapse of volcanoes as magma was removed from below. The maria were interpreted as vast lava fields produced by extrusion of basalt magma from deep sources.

The volcanic interpretation of craters encounters difficulty on a number of points. True volcanic cones, such as those found in abundance on Earth, are not present, and photographs rarely show features that can be interpreted as lava flows. Moreover, if the Moon's interior is sufficiently heated to produce plutonic rocks, there should also be some mountain belts of folding and faulting such as characterize the Earth. Except for fracture lines, which are numerous, mountain-making forms typical of zones of lithospheric plate spreading and subduction seem to be totally lacking. The high lunar mountain ranges seem, instead, to constitute rims of circular maria or of large craters. The volcanic hypothesis has few, if any, supporters today.

The second major hypothesis of the Moon's geology may be described as the **meteoritic hypothesis.** It includes the assumption that the Moon is a body without internal rock melting and volcanic action. One of the strong advocates of the meteoritic theory was the distinguished American geologist Grove Karl Gilbert, who published his explanation in 1893. Among the modern group of scientists who have contributed details to the impact hypothesis is Harold C. Urey, a Nobel prize winner. His modified explanation is often referred to as the Urey-Gilbert hypothesis.

Figure 10.10 Hadley's Rille, a sinuous, canyonlike feature, crosses a cratered plain and ends in a highland area. This *Lunar Orbiter V* photograph spans an area about 30 mi (50) km) wide. (NASA photograph.)

According to Gilbert, the great craters were produced by large meteorites, in the manner which we have previously discussed in connection with terrestrial meteorite craters. However, the lunar craters are much larger than those on Earth and the characteristic central peak requires explanation. It has been suggested that the central peak lay directly beneath the center of impact-explosion. Because the shock wave was directed downward, the underlying rock remained intact, while that surrounding it was blown outward (Figure 10.12). The rays which emanate from several large craters are explained as debris deposits thrown out over long distances from the explosion centers (Figure 10.8).

Under the Urey-Gilbert hypothesis, the rimmed maria, of which there are at least five on the lunar nearside, are the old impact scars of enormous masses, probably asteroids. The case of Mare Imbrium is particularly striking (Figure 10.9), as Gilbert pointed out. Urey also reconstructed the Imbrium collision, considering it as involving impact of an object about 125 mi (200 km) in diameter, approaching from a low angle. The impact raised an enormous wave of rock that spread outward, coming to rest in a great arc of mountain ridges. Gilbert had proposed that the heat of impact melted a vast quantity of lunar rock that subsequently solidified as basalt over the mare surface.

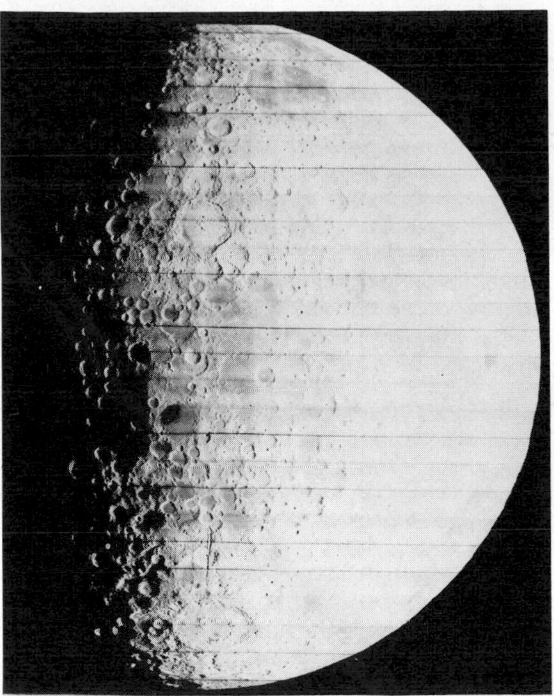

Figure 10.11 The moon photographed by *Lunar Orbiter IV* spacecraft from a distance of 1850 mi (3000 km). The heavily cratered lunar farside lies to the left. The lunar equator is represented by a horizontal line crossing the center of the photograph. (NASA photograph.)

LUNAR SURFACE MATERIALS

The first samples of lunar rock and soil were obtained in 1969 by astronauts of the *Apollo 11* mission. There, on Mare Tranquillitatis, as at subsequent Apollo landing sites, surface materials consist of unsorted fragmental debris. The fragments range in size from dust to blocks many feet in diameter. The layer of loose, dustlike mineral matter is called **lunar regolith.** The uppermost few inches are a brownish to grayish, cohesive, powdery substance; it consists of grains in the size range from silt to fine sand. The material is easily penetrated. Upon compaction it becomes stronger, easily supporting the weight of the astronauts and their equipment. Figure 10.13 shows the regolith compacted into a clear footprint.

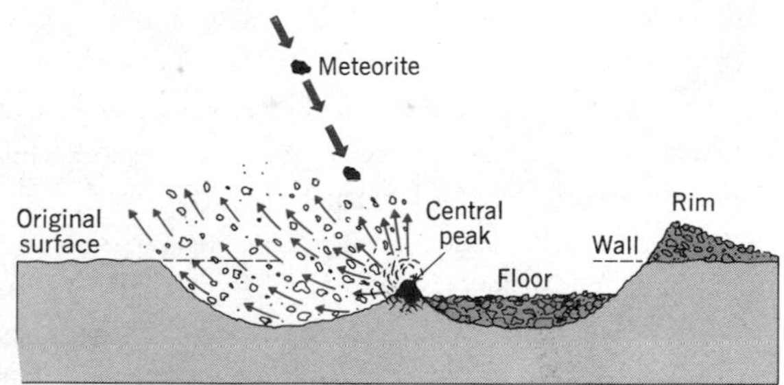

Figure 10.12 Origin of a large lunar crater, with its central peak, according to the meteorite-impact hypothesis of origin. The vertical scale is greatly exaggerated.

Figure 10.13 Astronaut's footprint in the lunar soil. Notice the many miniature craters in the surrounding surface. (*Apollo 11* photograph by NASA.)

All observed lunar material sufficiently hard to be called "rock" has proved to consist only of fragments, or clusters of fragments. Nowhere was there found a massive outcropping of a large body of solid rock in place. In other words, exposed bedrock as we know it on Earth does not exist on the Moon, so far as is known.

All of the regolith and rock fragments have proved to be of igneous origin, with pyroxene, plagioclase feldspar, and olivine being the most abundant minerals. There are many other accessory minerals; these are of secondary importance so far as abundance is concerned. Some examples familiar from your earlier study of igneous rocks are ilmenite, quartz, potash feldspar, and amphibole. Iron and titanium are constituent elements in quite a few of the accessory minerals. Some of the minerals contain uranium and thorium, which are radioactive.

Figure 10.14 A lunar rock sample collected by astronauts of *Apollo 11* mission at Tranquillity Base. It is an igneous rock of granular texture, showing glass-lined surface cavities. Specimen is about 8 in. (20 cm) high. (*Apollo 11* photograph by NASA.)

LUNAR ROCKS

One way in which to classify the lunar rocks is to divide them into two groups: First are fragments of igneous rocks, perhaps broken from some unknown parent mass of solidified magma. These come in both fine-grained and coarse-grained textures (Figure 10.14). Second are **lunar breccias.** A breccia is a rock put together of many small angular fragments. Individual pieces of rock comprising the breccia are themselves fragments of igneous rock. Consequently, a single chunk of lunar breccia may contain representatives of igneous rock from many different sources.

The great majority of lunar rocks collected by the astronauts are breccias. It is believed that the lunar breccias originated from intense shock, which bound together particles that had previously been fragmented by shock. This process has been called **impact metamorphism.** Meteoroid impacts are considered to be the shock mechanism, both for fragmenting rock into regolith and bringing it together into breccia.

Of the igneous rock fragments we will, for the sake of simplicity, identify only two major groups: basalt and anorthosite. Basalt is already familiar to you as a common terrestrial rock composed largely of calcic plagioclase feldspar, pyroxene, and commonly olivine as well. Lunar basalt, shown in Figure 10.15, is quite similar, except that the basalts underlying the maria are richer in iron and poorer in silica than Earth basalts. Maria basalt magma was highly fluid, more so than Earth basalt, and probably flowed very freely as it filled the maria basins. The second rock mentioned, **anorthosite,** is an igneous rock we did not include in the granite-gabbro series described in Chapter 1. Anorthosite occurs on Earth as a plutonic igneous rock. It consists largely of plagioclase feldspar. Consequently, anorthosite is a felsic rock, of lower density than basalt, which is classed as mafic.

Returning to the regolith, it proved to consist of tiny rock fragments, including both basalt and anorthosite, and numerous tiny glass beads, called spherules (Figure 10.16). The spherules are considered as "splash" phenomena, resulting from the sudden cooling of igneous rock melted by meteoroid impact and hurled from the impact crater as a spray of droplets.

Figure 10.15 This specimen of lunar basalt was named the Goodwill Rock by astronauts of *Apollo 17* mission. The cavities are original gas-bubble cavities (vesicles) in the basalt. The specimen is about 2 in. (10 cm) high. (*Apollo 17* photograph by NASA.)

A particularly interesting feature of the rock samples is the presence of surface pits lined with glass. These pits average less than 0.04 in. (1 mm) across. The pits, the rounding of gross shape of the fragments, and a lighter-colored surface than the inside rock suggest that some process of erosion has been in operation. Probably these effects resulted from impacts by small particles. The cavities are referred to as "zap pits."

If the Moon is smothered under a blanket of impact debris, how can we even guess at what kind of solid bedrock lies below, constituting the lunar crust? The answer lies in sampling the bouldery rim debris of craters (Figure 10.17). The idea is that a meteoroid, upon penetrating the regolith and exploding, tore loose and threw out boulders of the bedrock below. Acting on this supposition, geologists have inferred that the uppermost part of the lunar crust consists in some places of basalt and in other places of anorthosite.

As a broad generalization, we can say that basalt underlies the floors of the maria, while anorthosite underlies much of the lunar highlands. So, in a fashion somewhat like the Earth's twofold division into a felsic continental crust and a mafic oceanic crust, the Moon's crust has been differentiated into a denser basaltic crust and a less-dense anorthosite crust. At least, this arrangement applies for shallow depths, within the range of disturbance by meteoroid impacts.

Radiometric age determination has been a mainstay in attempts to work out a geological history of the Moon. One rather surprising finding from lunar samples is that, so far, no rock has turned up with an age of crystallization from its magma younger than −3 b.y. Impact fragmentation and formation of breccia has certainly been going on since that time, but it seems to have been only a mechanical reworking of the ancient igneous rocks. Practically all samples have shown an age of crystallization in the time span from −4.2 b.y. to −3.1 b.y. A single fragment, consisting of dunite, has yielded an age of −4.6 b.y.

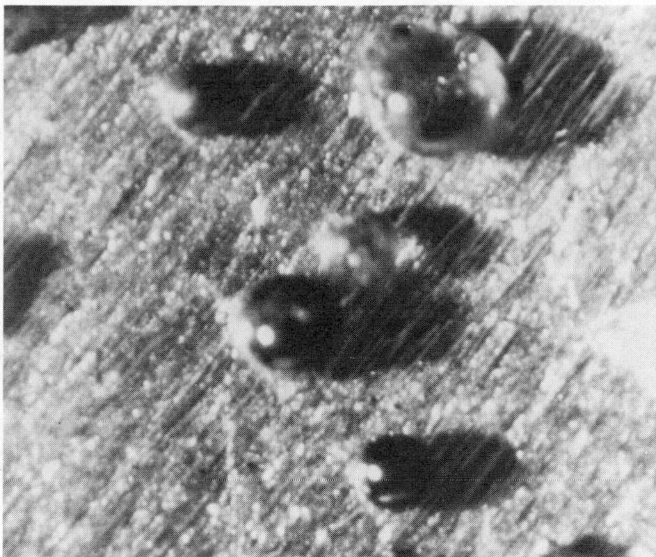

Figure 10.16 Glass spherules from the lunar regolith. Largest spherule is 0.016 in. (0.4 mm) in diameter. Spherules rest on an aluminum dish with a striated surface. (*Apollo 11* photograph by NASA.)

Figure 10.17 Large boulder and astronaut Harrison H. Schmitt, Station 6, Taurus Littrow. (*Apollo 17* photograph by NASA.)

ORIGIN OF THE MOON

One problem that seems to have remained entirely unsettled by the Apollo program is the Moon's origin. Suggested earlier in this chapter is the hypothesis that the Moon and Earth grew by accretion of nebular material as a binary planet, linked by gravitational attraction. However, the Moon's low density, and lack of iron and similar elements, are hard to explain if accretion occurred at the same point in time as Earth, under similar physical and chemical conditions.

A second possibility, widely debated, is known as the **capture hypothesis.** It is supposed that the Moon formed by accretion in another part of the solar nebula, where conditions were quite different from those prevailing where the inner planets were forming. Later the Earth captured the Moon, when by chance the Moon passed close to the Earth and was trapped by Earth's gravity field. It has been suggested that capture occurred about −3.7 to −3.6 b.y. and corresponded with the rise of the maria flood basalts. However, the mechanism of capture of so large a planetary body by Earth faces strong objections.

A third hypothesis requires that the Moon was formed of material broken away from a fast-spinning Earth. This **fission hypothesis** was put forward in the 1890s by Sir George Darwin, son of Charles Darwin and a noted authority on tides. He calculated that at some point in time the Earth was spinning much faster than today, and that the centrifugal force of rotation produced a large bulge in the Earth. The bulge then separated from the Earth and moved away to become the Moon. Darwin and others even went so far as to suggest that the Pacific Ocean basin was the source of the lunar material.

The fission hypothesis was revived in the early 1960s. Under this hypothesis, the Earth's iron core had already been formed, so that the material thrown off was part of the mantle. In this way, the Moon's low density and lack of a core are explained. Although the fission hypothesis seems satisfactory on a number of basic counts, there are strong objections. Certain minor elements show very different patterns of abundance when lunar and Earth rocks are compared from a chemical standpoint. Scientists argue that these differences make it most unlikely that the lunar material was ever part of Earth. And so the speculation reverts back to accretion of the Moon as a separate body, followed by capture. A satisfactory explanation seems almost as elusive as before the astronauts brought back their samples.

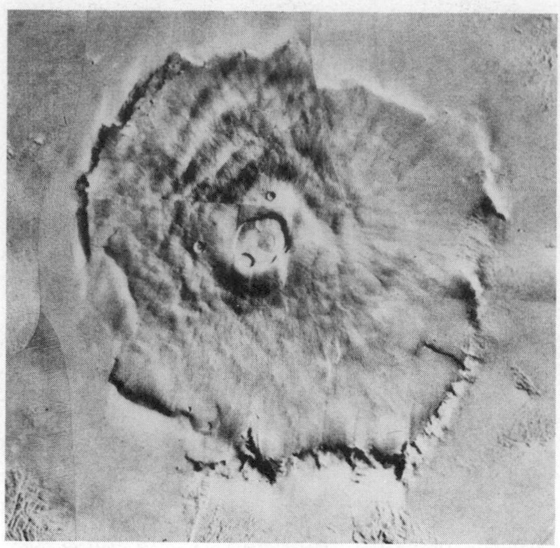

Figure 10.18 Nix Olympica, an enormous Martian shield volcano with a caldera at the summit. The basal diameter is about 300 mi (500 km). (NASA photograph.)

THE SURFACE OF MARS

Captured into orbit as a satellite of Mars, spacecraft *Mariner* 9 circled that planet from November 1971 to October 1972, sending back some 7000 photographs of the Martian surface. Many important Martian features are strikingly different from those found on Earth and Moon, although some others are recognizably similar.

Half of Mars, roughly equivalent to its southern hemisphere, is heavily cratered, much as is the Moon's surface, whereas the other hemisphere is only lightly cratered. Although no samples have been taken from Mars, it is thought that the history of cratering of Mars

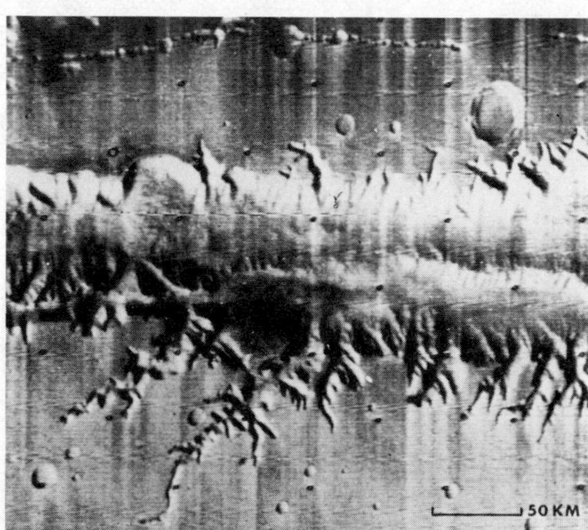

Figure 10.19 A deep, steep-sided trench near the equator of Mars. Width is about 60 mi (100 km); depths go to 4 mi (6 km). (NASA photograph.)

parallels that of the Moon in age and frequency of impacts. Just why much of the northern half of Mars consists of smooth, uncratered plains is a major puzzle.

Major geological features of Mars include volcanoes and troughlike rifts of enormous size by Earth standards. One huge shield volcano, Nix Olympica, is 300 mi (500 km) wide at the base and has a summit rising 15 mi (25 km) above the base (Figure 10.18). A caldera occupies the summit. Three other volcanoes are of comparable dimensions. These and lava flows indicate local accumulations of heat within the planet—possibly heat of radiogenic origin—leading to magma formation and extrusion recently in Mars' history.

Mars also has tectonic features in the form of numerous faults and a great rift valley system. The largest rift, Coprates, appears as a straight-walled trough 60 mi (100 km) wide and up to 4 mi (6 km) deep in places (Figure 10.19). The entire rift zone is some 3000 mi (5000 km) long. Although the Martian lithosphere seems to have been pulled apart along the rift zone, there is no sign of subduction zones and compressional mountain belts such as those on Earth.

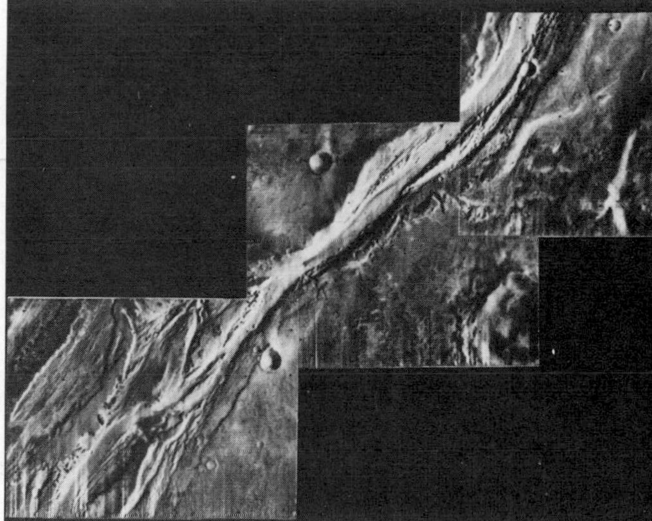

Figure 10.20 Amazonis Channel, a braided, channel-like feature possibly produced by a water stream. (NASA photograph.)

Among the most puzzling surface features of Mars are what appear to be channels formed by fluid flow. These are largely concentrated in one area, north of the rift zone. Some of the channels show a braided pattern in which flow lines repeatedly divide and join, like the shallow channels of Earth streams in dry regions (Figure 10.20). Mars is now devoid of surface water in liquid form, yet the channels suggest the action of great water floods, acting over short periods. One suggestion is that water ice, held frozen in the ground, underwent an episode of sudden melting, releasing great floods. These floods eroded the channels and left the channel floors strewn with rock debris.

Mars has an extremely thin atmosphere; the surface barometric pressure is only about 1% that on Earth. Nevertheless, winds on Mars are capable of raising great dust storms. One such storm obscured the Martian surface for weeks following the arrival of *Mariner* 9. Erosion of rock by impact of windblown sand and silt is considered a possibly important agent in shaping Martian surface forms. Dune areas, somewhat like those on the Earth's deserts, testify to the importance of wind as a geologic agent on Mars.

Perhaps the strangest of the many strange Martian features are the polar caps (Figure 10.21). The white polar caps consist of a thin surface coating of carbon-dioxide ice. Each cap undergoes a large seasonal change in size, expanding greatly during the winter of the respective pole and shrinking to a small area during the summer. The small residual polar cap, left at the end of summer, is thought to consist of an accumulation of water ice. The residual cap areas have strange, platelike terrain features, looking something like a collapsed stack of poker chips arranged in a spiral form. The plates consist of light and dark laminations, which may be alternating layers of ice and dust.

To summarize, volcanic, tectonic, and erosional activities seem to have been active on Mars, but to a much lesser degree than on Earth. Thus, in some respects, Mars is geologically intermediate in activity between our Earth and an inactive Moon.

Figure 10.21 The Martian polar cap, with what appear to be overlapping terrain plates surrounding the pole. (NASA photograph.)

THE SURFACE OF MERCURY

Mariner 10 passed close to the planet Mercury on March 1974, sending back the first photographs ever to reveal the planetary surface. Strikingly like the Moon, the heavily cratered surface of Mercury shows countless lunar-type craters of many sizes and ages (Figure 10.22). Some craters are comparatively recent in age—light in color and having well-defined ray systems. Others are obviously very old, large craters with flat floors. Also seen are dark plainlike areas suggestive of the lunar maria. Infrared radiation analyzed by instruments on the space vehicle suggest that Mercury has a regolith of low density, perhaps much like that on the Moon. Many long, straight scarps were identified, along with long, narrow ridges suggestive of crustal compression (Figure 10.23). However, no sinuous rilles like those on the Moon, were observed.

As expected in the absence of water and all but an extremely tenuous atmosphere, the surface of Mercury revealed no signs of the eroding action of fluids, such as the braided channels of Mars. The magnetic field of Mercury is extremely weak. Although Mercury may have an iron core, the planet's extremely slow rotation would probably not give rise to a dynamo effect such as that generated in the Earth's liquid core.

Figure 10.22 The surface of Mercury, compiled from *Mariner 10* photographs. Notice the striking similarity between this surface and that of the farside of the moon, shown in Figure 10.11. Rays emanate from some of the younger craters at the right side of the photograph. (*Mariner 10* photograph by NASA.)

YES, THERE IS A GEOLOGY OF OUTER SPACE

Within only a decade of space exploration the extension of geology as a science to the Moon and inner planets has passed from infancy to vigorous youth. Perhaps the greatest gain in our knowledge of the Earth's history comes from the very ancient lunar rocks, almost all of which crystallized in a billion-year period predating any known terrestrial rock. Earth's record of the first billion years of geologic time is totally missing, because Earth has had an active lithosphere and vigorous atmospheric processes.

While Earth rocks were continually remelted and recycled, destroying all of the ancient crust, the primitive crustal features of the Moon, Mars, and Mercury went into cold storage, so to speak. Only meteoroid impacts have disturbed the surface of the Moon and Mercury since about −3.0 b.y., and even on Mars, half the planet retains the early impacted surface. We thus have not one, but three Rosetta stones from which to read the early history of our planet.

Of the inner planets, only Venus remains completely obscured from geological examination by satellite cameras. Beneath the dense, clouded atmosphere of Venus may lie a varied surface showing intense tectonic and volcanic activity, for Venus is a close match to Earth in size and density and may well prove to have a "plate tectonics" of its own.

Figure 10.23 Between craters, the surface of Mercury shows irregular fractures and ridges forming a blocklike pattern. The width of the area shown is about 300 mi (500 km). (*Mariner 10* photograph by NASA.)

YOUR GEOSCIENCE VOCABULARY

astrogeology
solar nebula
accretion
planetesimals
asteroids
outgassing
meteoroid
meteor
meteorite
fireball
fall
find
irons
stones
stony irons
gravity
lunar highlands (terrae)
lunar maria
lunar craters
rays
rilles
walls
meteoritic hypothesis
lunar regolith
lunar breccia
impact metamorphism
anorthosite
capture hypothesis
fission hypothesis

SELF-TESTING QUESTIONS

1. Name the nine major planets in order of distance from the Sun. Of these, which are the inner planets? Compare the inner planets in terms of their diameters, masses, and densities. What are the major similarities and differences between Moon and Mercury? Why are Earth and Moon referred to as a binary planet?

2. Sketch the important stages in evolution of the solar system from a nebula. How did the planets form? How do the Moon and asteroids fit into this history?

3. Describe the principal types of meteorites. What is the composition of each type? How does the age of meteorites fit into the timetable of origin of the planets? How do meteorites give evidence as to the composition of the Earth's interior?

4. Describe the Barringer Crater and explain how it originated. About how frequently are major impact craters formed on Earth?

5. Describe the lunar surface environment, including lunar gravity, surface temperatures, and temperature contrasts. At what time of the lunar day did the Apollo astronauts explore the Moon's surface? Give a brief review of the first decade of lunar exploration by space vehicles.

6. Describe the major terrain subdivisions of the Moon. What relief features are found on the highlands? Describe the floors and outlines of the maria.

7. Describe the lunar craters in terms of size, internal form, and surface appearance. What origin has been suggested for rilles? What lunar features give evidence of crustal fracturing?

8. What origin has been suggested for the lunar maria? With what material have the maria basins been filled?

9. Describe the materials of the lunar surface. How did the lunar regolith originate? Interpret the lunar breccias in terms of lunar history. What are the principal types of lunar igneous rock, classified by composition? What is the significance of spherules? What problems do geologists face in trying to determine the nature of lunar bedrock?

10. What hypotheses have been advanced for the origin of the Moon as a satellite of Earth? Evaluate these hypotheses in terms of geological evidence. How can you prove that the Moon did not originate from the Earth's mantle, torn from what is now the Pacific Ocean basin?

11. Describe the major features of the surface of Mars. Which of these features are like those found on the Earth? on the Moon? Which features are unique to Mars? What geologic history can be inferred from the surface features of Mars?

12. What features have been photographed on the surface of Mercury? Compare Mercury with Earth, Mars, and Moon in terms of tectonic and volcanic activity.

(*Upper*) Surface complex at the Western Deep Levels gold mine, southwest of Johannesburg, South Africa. Dumps of rock waste rise as prominent landmarks. Shaft headgears, housed in square towers, show where the mine shafts are located. (*Lower left*) A seven-story building would fit into this excavation 10,000 ft (3000 m) below the surface. It will house the hoisting gear for a new shaft to reach a rich ore layer 12,000 ft (3700m) below the surface. This is the world's deepest mine. (*Lower right*) Drillers at work on the ore face. Free State Gehuld gold mine, Orange Free State. (Photographs by courtesy of the Anglo American Corporation of South Africa, Limited.)

11 RESOURCES OF THE SOLID EARTH

THE RIDGE OF WHITE WATERS

In spite of an old saying, gold is not where you find it—at least, not much of the world's gold is waiting to be picked up. Instead, you have to go to the Witwatersrand, or "Ridge of White Waters," in the Transvaal of South Africa. Here, and in the adjoining Orange Free State, lies the modern source of about two-thirds of the world's gold production. Leave behind any romantic ideas about nuggets and flakes of pure gold glistening in the bottom of a prospector's pan. Gold of the Witwatersrand Basin—or Rand—is highly dispersed into minute particles embedded in an extremely hard conglomerate. This rock is one of the hardest found in the earth's crust.

Gold mines of the Rand are among the deepest openings man has yet driven into the crust. Already, shafts have penetrated to a depth of 10,000 feet below the surface and are being carried even deeper. Rock temperatures at this depth often exceed 120° F, and men can work there only with the aid of powerful air-conditioning systems. Air is cooled by refrigeration in rock chambers deep beneath the surface, then pumped down further to the working levels. It is said that each year, 10 billion tons of air are pumped through the gold mines of South Africa.

No single elevator can efficiently be run from bottom to top of the deep gold mines, for the weight of the cables alone would be too great. Instead, three separate lift systems are used, one beneath the next. Rising at speeds of up to 35 miles per hour, one of these elevators can lift 100 men or 22 tons of ore at a time through a vertical distance as great as 4500 feet. In one mine a new shaft is being driven to reach the Carbon Leader Reef, at depths to 12,000 feet, and eventually it may go to 14,000 feet. The enormous cavity created to house the hoisting gear for this new shaft could hold a seven-story building.

Gold of the Witwatersrand Basin was deposited some 2½ billion years ago by water currents spreading sand and gravel over an ancient surface, perhaps a floodplain or delta, or even a shallow sea. In many ways, these accumulations may have then resembled the gold-bearing gravels of the Sacramento River, where Sutter first discovered California gold. The strata were then deeply buried and solidified, and finally warped into a basin-shaped structure. The gold-bearing pebble beds, called reefs, are only a few feet thick, but they extend for miles. The ore runs in long, thin "pay streaks" within the reefs. The Main Reef has been mined continuously for more than 20 miles. Gold makes up only about 10 parts per million of the reef ore, so that enormous quantities of rock have to be brought to the surface. Some 100 million tons of rock are excavated annually, and it yields only about 10 grams of gold per ton. Altogether, the South African mines have yielded over a billion ounces of gold. At today's price of about $150 per ounce, the entire haul comes to $150 billion worth of the precious metal.

Extraction of Rand gold so finely divided and so thinly diffused through the reef rock would not have been possible without a major advance in technology. In the 1890s, a process was perfected in which cyanide solutions dissolve the gold. Up to that time, the standard method of using mercury to form an amalgam with the gold had come to failure, because of the huge quantities of mercury that were required.

This little essay about South African gold carries two messages we will repeat in this chapter on earth resources. First, the gold of the Witwatersrand Basin took millions of years to accumulate in the reefs; no more will be produced there by nature when the mining is over. Like practically all of our metallic ores, and all of our mineral fuels, the resource is nonrenewable, but we are taking it out of the ground as if there is no tomorrow. Second, we must in future decades turn to ores of such low grade that they can be mined only by processing enormous quantities of rock and by application of increasingly expensive, energy-consuming technologies.

THE DEMAND FOR EARTH RESOURCES

Man is rapidly consuming earth resources that required geologic spans of time to accumulate. Natural rates of replenishment by geological processes are infinitesimally small in comparison with the present rates of consumption. Geological resources are therefore finite. Once we know approximately the world extent of a particular resource, we can predict its expiration according to any number of use schedules.

A major factor to consider is that world demand for geologic resources is increasing rapidly. What will happen if the consumption of mineral resources by the United States increases by 5% per year—a rate considered desirable by some economists for the annual increase of our total national output of goods and services? At this rate, our consumption of mineral resources will double each 14 years. But this is not the whole picture. The large developing nations of the world now operate primarily on an agricultural economy. Should these nations set as their goal the achievement of industrial production at the same level as the highly developed nations, demands upon the earth's mineral resources will be further increased many times over.

A distinguished geologist, Preston Cloud, points out that merely to raise the economic level of all the world's 3.6 billion persons to the same level enjoyed by the 200 million persons now living in the United States would require from 100 to 200 times the present annual production of metals such as iron, lead, zinc, and tin.* He goes on to note that world energy demands would be increased in a similar ratio. Most of our energy comes, of course, from crustal materials requiring millions of years to accumulate. Fuels such as petroleum, natural gas, coal, and uranium are nonrenewable earth resources, along with the ores of iron, lead, zinc, and tin.

Does serious trouble lie ahead for the human race? This question is currently being debated. While some scientists believe we are on a disaster course, others argue that the discovery of new mineral resources, combined with the advances of technology, will take care of our needs far into the foreseeable future. Entering the debate is not the purpose of this chapter. Instead, we shall try to present basic facts needed to evaluate the pros and cons of the argument.

DISPERSING AND RECYCLING OF GEOLOGIC RESOURCES

An important concept in the study of geologic resources is the change from concentrated states to dispersed states. This change is exactly the opposite of geological processes that produced the initial concentrated states. For example, coal represents an extremely dense concentration of hydrocarbons in large quantities. Combustion of coal disperses this matter into the atmosphere, and the stored energy is transformed and dissipated as heat into outer space. We have no way to reverse this process without expending an equivalent amount of energy.

* *Environment: Resources, Pollution and Society*, W. W. Murdoch, ed., Sinnauer Associates, Inc., Stamford, Conn., p. 83.

For example geologic processes have concentrated one comparatively rare metal—lead—into rich ores. We have been dispersing much of it into the atmosphere through combustion of leaded gasoline. On the other hand, a great deal of used lead is recovered and used again. Aluminum provides an example of a metal of which only a small proportion (about 3%) is reused; the remainder is widely dispersed, as witness the ubiquitous beer can.

Actually, man himself usually must carry out the final stages of mineral concentration, as in the case of nuclear fuels, copper ores, or most iron ores. In terms of percentages by weight in the earth's crust, these deposits as mined represent an extraordinary degree of concentration, but it is still not enough.

NONRENEWABLE EARTH RESOURCES

Nonrenewable earth resources covered in this chapter can be grouped about as follows:

Metalliferous deposits (examples: ores of iron, copper, tin)
Nonmetallic deposits, including
  Structural materials (examples: building stone, gravel, and sand)
  Materials used chemically (examples: sulfur, salts)
Fossil fuels (coal, petroleum, and natural gas)
Nuclear fuels (uranium, thorium)

Notice that the last two groups represent sources of energy, whereas the first two groups are sources of materials.

All of these nonrenewable resources come from the earth's crust. Their scientific study is a major branch of geology, often called **economic geology.** This is the practical and applied side of geology, without which there would be no industrial civilization such as we have today.

METALS IN THE EARTH'S CRUST

Metals occur in useful concentrations as **ores.** An ore is a mineral accumulation that can be extracted at a profit for refinement and industrial use. A number of important metallic elements are listed in Table 11.1 with their abundances, as percentage by weight, in the average crustal rock. Magmas are the primary source of many metals. Our concern here is with the natural geological processes of concentration of metallic elements and compounds into ores of various kinds. Whereas aluminum and iron are relatively abundant, most of the essential metals of our industrial civilization are present in extremely small proportions—witness mercury and silver, with abundances of only 0.000008% and 0.000007%, respectively.

In a classification of metals by uses, iron stands by itself in terms of total tonnage used in the production of steel. Related to iron is a group of **ferro-alloy metals,** which are used principally as alloys with iron to create steels with special properties. The ferro-alloys include titanium, manganese, vanadium, chromium, nickel, cobalt, molybdenum, and tungsten, listed in order of appearance in Table 11.1. Other important metals (nonferrous metals), standing apart

Table 11.1 **Average metallic abundances in crustal rock**

Symbol	Element name	Abundance (percentage by weight)
Al	Aluminum	8.1
Fe	Iron	5.0
Mg	Magnesium	2.1
Ti	Titanium	0.44
Mn	Manganese	0.10
V	Vanadium	0.014
Cr	Chromium	0.010
Ni	Nickel	0.0075
Zn	Zinc	0.0070
Cu	Copper	0.0055
Co	Cobalt	0.0025
Pb	Lead	0.0013
Sn	Tin	0.00020
U	Uranium	0.00018
Mo	Molybdenum	0.00015
W	Tungsten	0.00015
Sb	Antimony	0.00002
Hg	Mercury	0.000008
Ag	Silver	0.000007
Pt	Platinum	0.000001
Au	Gold	0.0000004

* Source: Data from B. Mason, *Principles of Geochemistry,* 3rd ed. (New York: Wiley, 1966), pp. 45–46, Table 3.3.

individually with respect to industrial uses, are aluminum, magnesium, zinc, copper, lead, and tin. A minor group listed in Table 11.1 includes antimony, silver, platinum, and gold. Finally, there are metals which are radioactive, including uranium, thorium, and radium.

CONCENTRATION OF METALS IN ORES

While a few metals, among them gold, silver, platinum, and copper, occur as elements, that is, as **native metals,** most occur as compounds. Oxides and sulfides are the most common forms, but more complex forms are present in many ores.

The crustal abundance of a metal is an abstraction of no practical value to the economic geologist. He is interested in its abundance in the form of the ore of its usual occurrence, either as an element or a compound.

Obviously, most metals must occur greatly concentrated in ores to be extractable at a profit, as compared with the extremely small average crustal abundances shown in Table 11.1. For example, chromium has an average crustal abundance of only 0.01%; it must be concentrated by a factor of about 1500 times to become an ore sufficiently rich to be extracted. For lead, the crustal abundance must be concentrated by a factor of 2500 times to become an ore. Ores of sufficient richness to be extracted have depended upon very special geological processes to come into existence.

METALLIFEROUS ORES

One major class of ore deposits is formed within magmas by **magmatic segregation,** in which mineral grains of greater density sink through the fluid magma while crystallization is still in progress. Masses or layers of a single mineral accumulate in this way. One example is chromite, the principal ore of chromium, with a density of 4.4 gm/cc. Bands of chromite ore are sometimes found near the base of an igneous body.

Another example is seen in the nickel ores of Sudbury, Ontario. These sulfides of nickel apparently became segregated from a saucer-shaped magma body and were concentrated in a basal layer (Figure 11.1). Magnetite is another ore mineral that has been segregated from a magma to result in an ore body of major importance.

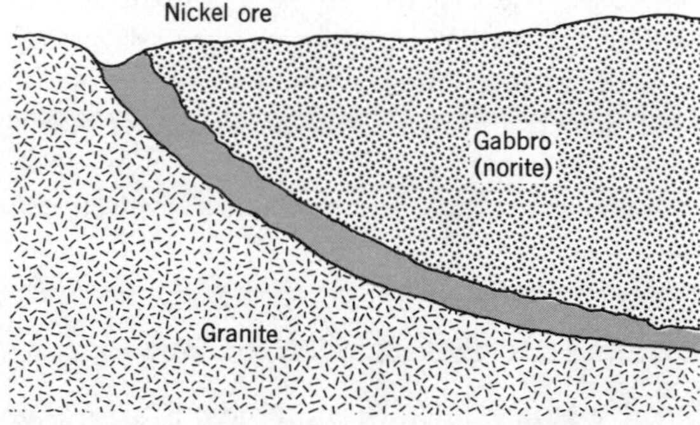

Figure 11.1 Cross section of a sulfide nickel ore deposit at Sudbury, Ontario. The ore layer lies at the base of a body of gabbro, overlying an older granite basement.

The process of contact metamorphism, referred to in Chapter 5, is a second source of important ore deposits. High-temperature fluids from within the magma soak into the surrounding country rock, introducing ore minerals in exchange for components of the rock. For example, a limestone layer may have been replaced by iron ore consisting of hematite and magnetite (Figure 11.2). Ores of copper, zinc, and lead have also been produced in this manner. Valuable deposits of nonmetallic minerals have also resulted from contact metamorphism.

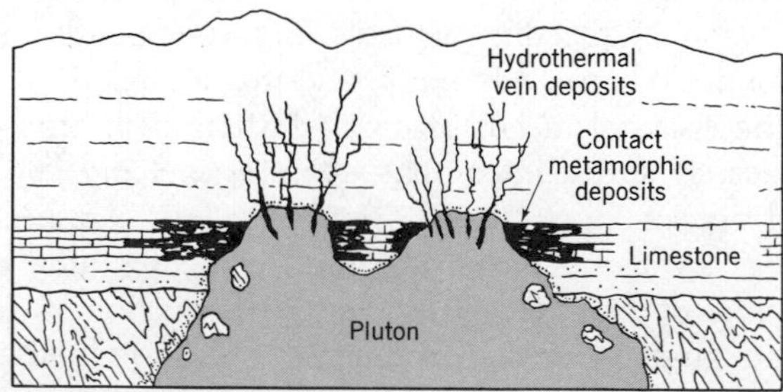

Figure 11.2 Schematic cross section of vein deposits and contact metamorphic deposits adjacent to an intrusive igneous body.

A third type of ore deposit is produced by the effects of high-temperature solutions, known as **hydrothermal solutions.** These are watery solutions that leave a magma during the final stages of its crystallization and are deposited in fractures to produce mineral **veins** (Figure 11.2). Some veins are sharply defined and evidently represent the filling of open cracks with layers of minerals. Other veins seem to be the result of replacement of the country rock by the hydrothermal solutions. Where veins occur in exceptional thicknesses and numbers, they may constitute a **lode.**

Hydrothermal solutions produce yet another important type of ore accumulation, the **disseminated deposit,** in which the ore is distributed in minute masses throughout a very large mass of rock. Certain of the great copper deposits of the disseminated type are referred to as **porphyry copper** deposits. Here the ore has entered a large body of igneous rock of a class known as a porphyry. The rock mass had in some manner been shattered into small blocks that permitted entry of the solutions. One of the most celebrated of these is at Bingham Canyon, Utah (Figure 11.3).

Hydrothermal solutions may also rise toward the surface, making vein deposits in a shallow zone and even emerging as hot springs. Many valuable ores of gold and silver are deposits of the shallow type. Particularly interesting is the occurrence of mercury ore, in the form of the mineral cinnabar, as a shallow hydrothermal deposit. Most renowned are the deposits of the Almaden district in Spain, where mercury has been mined for centuries and has provided most of the world's supply of that metal.

A fourth category of ore deposits results from the effects of solutions moving downward from the ground surface. Mineral deposits enriched in this manner to produce ores are described as **secondary ores.** Rainwater entering the soil moves downward through the bedrock, dissolving various minerals and transporting them to lower

Figure 11.3 Open-pit mine at Bingham Canyon, Utah. (Kennecott Copper Corporation.)

Figure 11.4 A gold dredge scoops up gravel from the river flood plain, washes the gravel to secure gold particles, then dumps the waste to form curious piles like fallen stacks of poker chips. Yukon River, Alaska. (Bradford Washburn.)

levels. In some places these minerals have accumulated in sufficient concentrations to be mined as ores. Important ores of this type are oxides of zinc, copper, iron, and lead. At greater depth, concentrations of sulfides of iron, copper, lead, and zinc have accumulated as valuable ores. Oxides remaining behind near the ground surface have also accumulated as ores. A good example is bauxite, the principal ore of aluminum. It occurs in the warm, wet tropics in rocklike crusts just below the soil surface. This type of ore deposit is known as a **laterite.**

A fifth category of ore deposit is that in which concentration has occurred through fluid agents of transportation: streams and waves. Certain of the insoluble heavy minerals derived from weathering of rock are swept as small fragments into stream channels and carried down-valley with the sand and gravel. Because of their greater density, these minerals become concentrated in layers of gravel to become **placer deposits** (Figure 11.4). Native gold is one of the minerals extensively extracted from placer deposits; platinum is another (Figure 11.5). A third is an oxide of tin, the mineral cassiterite. Diamonds, too, are concentrated in placer deposits, as are other gem stones. Transported by streams to the ocean, gravels bearing the heavy minerals are spread along the coast in beaches, forming a second type of placer deposit, the **marine placer.**

Finally, we can recognize a sixth group of ore deposits in the nonclastic category of sediments, explained in Chapter 4. Sediment deposition is the principal source of nonmetallic mineral deposits, considered below, but some important metalliferous deposits are also of this origin. Iron, particularly, occurs as **sedimentary ore** in enormous quantities. Sedimentary iron ores are oxides of iron—usually hematite.

Figure 11.5 This great gold nugget was found near Greenville, California, during the gold rush of 1848. It is about 8 in. (20 cm) long and weighs 82 oz (2.3 kg). (Smithsonian Institution Photo No. 2254.)

U.S. METAL DEMANDS AND THE WORLD SUPPLY

In the last quarter-century of the 1900s, demands by the United States for essential industrial metals will increase greatly. A doubling of need will be felt for many metals, among them iron, manganese, zinc, cobalt, lead, antimony, mercury, and silver. For others, the factor of increased demand will be much greater. For example, our need for aluminum will increase about sixfold, for titanium twelvefold, and uranium twentyfold. Can these increased demands be met from reserves within the United States? The answer is an emphatic NO!

Even at the present moment, the United States is dependent upon foreign imports for a large proportion of nearly all industrial metals consumed. For example, of the aluminum we consume, we produce within the United States only about 4%; of manganese, 5%; of zinc, mercury, and silver, about 50%; of iron and lead, about 75%. We produce practically none of the platinum, chromium, and strontium we use. We do well on uranium, mining substantially more than we need at the moment. Molybdenum is about the only essential metal we produce with an important exportable surplus—about as much extra as we consume.

Figure 11.6 shows the apportionment of known recoverable reserves of ores of five metals between the Sino-Soviet bloc and the United States in comparison with the total world resources. In the case of the essential metal mercury, for which there is no known substitute in many industrial uses, most of the world reserves lie in Spain, Italy, and the Sino-Soviet bloc. The dependence of the United States upon foreign sources is heavy, indeed, for a long list of metals. International tensions are bound to arise where mineral resources lie in developing nations and in nations with politically unstable or unfriendly governments.

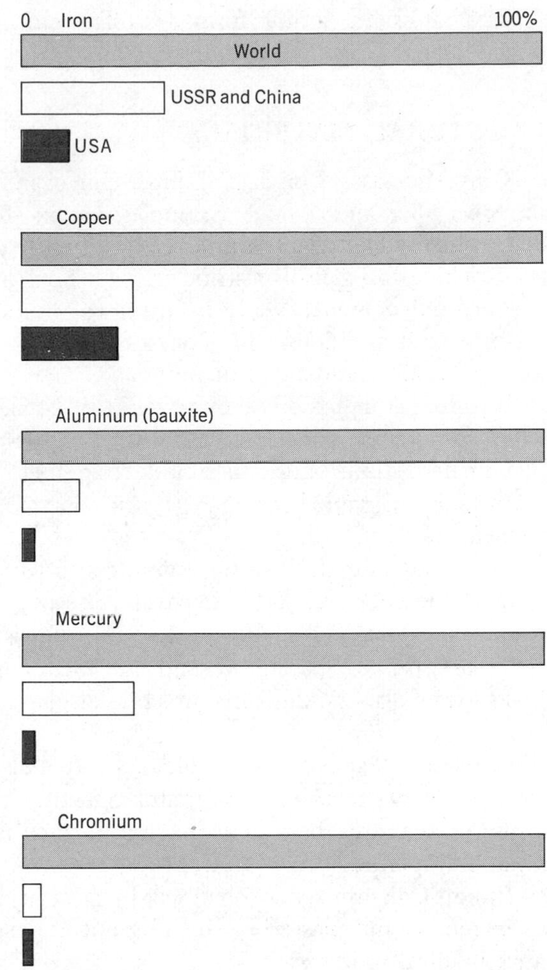

Figure 11.6 Apportionment of world recoverable reserves of ores of iron, copper, aluminum, mercury, and chromium between the Sino-Soviet nations and the United States.

RECYCLING OF METALS

Of increasing importance in manufacturing today is the secondary production of metals through reprocessing of durable metal goods manufactured in the past 10 to 100 years. We are not here referring to **new scrap** metal, derived as cuttings during initial manufacture, but to **old scrap,** those salvaged materials from discarded products, such as automobiles and refrigerators.

Metals can be reclaimed from old scrap by processes of distillation, electrometallurgy, mechanical separation, and chemical processes. As the total output of manufactured goods increases through time, the input of metals from secondary sources will also rise in volume.

Recycling of metals is rising in importance as national mineral resources are becoming depleted at increasing rates and as the grade of ores being mined is declining. **Metals recycling** is defined as the ratio of old scrap metal used to the total use of primary metal plus new scrap. Currently, metals recycling in the United States is highest for silver, about 65%. Lead is next with about 35%; most of this is lead recovered from plates of discarded batteries. Copper recycling is about 25%; that of iron, tin, and mercury between 15% and 20%. For aluminum, recycling is very low, about 3%.

NONMETALLIC MINERAL DEPOSITS

Nonmetallic mineral deposits (not including fossil and nuclear fuels) include a large and diverse assemblage of substances and cover a wide range of uses. It would be impossible to do the subject justice in a few paragraphs. In outline form, we offer some examples of these mineral deposits classified by use categories:

STRUCTURAL MATERIALS

Clay: For use in brick, tile, pipe, chinaware, stoneware, porcelain, paper filler, and cement. Examples: kaolin (for china manufacture) from residual deposits produced by weathering of felsic rock; shales, marine and glaciolacustrine clays for brick and tile.

Portland cement: Made by fusion of limestone with clay or blast-furnace slag. Suitable limestone formations and clay sources are widely distributed and are of many geologic ages.

Building stone: Many rock varieties are used, including granite, marble, limestone, sandstone. Slate is used as a roofing material.

Crushed stone: Limestone and "trap rock" (gabbro, basalt) are crushed and graded for aggregate in concrete and in macadam pavements.

Sand and gravel: Used in building and paving materials such as mortar and concrete, asphaltic pavements, and base courses under pavements. Sources lie in fluvial and glaciofluvial deposits and in beaches and dunes. Specialized sand uses include molding sands for metal casting, glass sand for manufacture of glass, and filter sand for filtering water supplies.

Gypsum: Major use is in calcined form for wallboard and as plaster, and as a retarder in Portland cement. Source is largely in gypsum or anhydrite beds in sedimentary strata associated with red beds and evaporites.

Lime: Calcium oxide obtained by heating of limestone, has uses in mortar and plaster, in smelting operations, in paper, and in many chemical processes.

Pigments: Compounds of lead, zinc, barium, titanium, and carbon, both manufactured and of natural mineral origin, are widely used in paints.

Asphalt: Asphalt occurs naturally, but most is derived from refining of petroleum. It is used in paving, and in roofing materials.

Asbestos: Fibrous forms of four silicate minerals, used in various fireproofing materials.

MINERAL DEPOSITS USED CHEMICALLY AND IN OTHER INDUSTRIAL USES

Sulfur: Principal source is free sulfur occurring as beds in sedimentary strata in association with evaporites. Chief use is for manufacture of sulfuric acid.

Salt: Naturally occurring rock salt, or halite, is largely sodium chloride, but includes small amounts of calcium, magnesium, and sulfate. It occurs in salt beds in sedimentary strata and in salt domes. Major uses include manufacture of sodium salts, chlorine, and hydrochloric acid.

Fertilizers: Some natural mineral fertilizers are phosphate rock, of sedimentary origin; potash derived from rock salt deposits and by treatment of brines; and nitrates, occurring as sodium nitrate in deserts (Atacama Desert of Chile).

Sodium salts: Found in dry lake beds (playas) of the western United States are various salts of sodium, such as borax (sodium borate). These have a wide range of chemical uses. Also important are sodium carbonate and sodium sulfate, found in other dry lake accumulations.

Fluorite: The mineral fluorite is calcium fluoride. It is found in veins of both sedimentary and igneous rocks. Uses are metallurgical and chemical, e.g., to make hydrofluoric acid.

Barite: Barite is barium sulfate and occurs as a mineral in sedimentary and other rocks. It is used as a filler in many manufactured substances and as a source of barium salts required in chemical manufacture.

Abrasives: A wide variety of minerals and rocks have been used as abrasives and polishing agents. Examples are seen in garnet, used in abrasive paper or cloth; and diamond, for facing many kinds of drilling, cutting, and grinding tools.

The above list is by no means complete. It can serve only to give you an appreciation of the strong dependence of industry and agriculture upon mineral deposits and the products manufactured from them.

MINERAL RESOURCES FROM THE SEA BED

If the prospect of eventually running out of various mineral resources from the lands seems all too real, we may want to consider possible substitutions of mineral resources from the sea. Sea water has always been available as a resource, and it has long provided the bulk of the world's supply of magnesium and bromium, as well as much of the sodium chloride.

The list of elements present in sea water includes most of the known elements and despite their small concentrations, these are potential supplies for future development. It is thought that sodium, sulfur, potassium, and iodine lie in the category of recoverable elements. But it seems beyond reason to hope for extraction of ferrous metals (principally iron) and the ferro-alloy metals in significant quantities to provide substitutes for ore deposits of the continents.

The continental margins, with their shallow continental shelves and shallow inland seas, are already being exploited for mineral production. An example is the working of placer deposits of platinum, gold, and tin in shallow waters. The petroleum resources of the North American continental shelf are already under development along the Gulf coast; zones of potential development are believed to exist on the shelf off the Atlantic coast as well. The possibility exists of finding and using mineral deposits of continental crystalline rocks submerged to shallow depths, although this has not yet happened.

Exploration of the deep ocean floor as a source of minerals is still in an early stage. Already, the layer of manganese nodules found in parts of all of the oceans is regarded by some as a major future source of manganese, and of a number of other metals in lesser quantities (Figure 11.7). Presence of substantial amounts of silica with the manganese oxide may render the nodules unfit for exploitation of manganese by present extraction methods, but this does not rule out the possibility of future use.

Figure 11.7 This manganese nodule, about 4 in. (11 cm) high, was dredged from the floor of the Atlantic Ocean. Its cross section reveals a nucleus of volcanic rock (*light color*) surrounded by a layer of manganese and iron oxides. (From Karl K. Turekian, *Oceans*, © 1968. By permission of the author and Prentice-Hall, Inc.)

In reviewing the overall prospects of mineral resources from the oceans and ocean basins we are only being realistic in concluding that contributions from sea water itself are limited only to a few substances. Most of the contributions of the sea floor will be from shallow continental shelves, where petroleum and natural gas are the major resources. Prospects of substantial mineral contributions from the deep ocean floor are rather poor at this time. In the light of these conclusions, the need for conservation and careful planning for the use of mineral resources of the lands becomes all the more evident.

SOURCES OF ENERGY

Before looking into the sources of energy that are derived from the solid earth, let us review the full picture of world energy resources to gain a better perspective. Sources of energy are found in both sustained-yield and exhaustible categories. A **sustained-yield energy source** is one that undergoes no appreciable diminution of energy supply during the period of projected use. Consider first the sustained-yield sources.

Solar energy is a sustained-yield source of extreme constancy and reliability. Stated in terms of power, solar radiation intercepted by one hemisphere is calculated to be about 100,000 times as great as the total existing electric power-generating capacity. The problem is, of course, that solar radiation derived from a large receiving area must be concentrated into a very small distribution center. To produce power equivalent to that of a large generating plant (about 1000 megawatts capacity) would require at an average location a collecting surface of 4 sq mi (10.4 sq km). While there seems to be no technological barrier to building such a plant, the cost at present is far too high to make this energy source a practical one.

Water power under gravity flow, or **hydropower,** is a second source of sustained energy and has been developed to a point just over one-quarter of its estimated ultimate maximum capacity in the United States. Presently water power supplies about 4% of the total energy production of the United States (see Figure 11.14). Since we have a good knowledge of stream flow, the estimate of maximum capacity is probably not much in error. For the world as a whole, present development is estimated to be about 5% of the ultimate maximum capacity. Potential hydropower is particularly great in South America and Africa, where coal is in very short supply.

A serious defect of hydropower estimates is that the capacity of artificial reservoirs declines through sedimentation. Most large reservoirs behind big dams have an estimated useful life of a century or two at most. Perhaps, after all, water power should not be categorized as a "sustained" source of energy.

Tidal power is a third sustained-yield energy source. To utilize this power, a bay is chosen along a coast subject to a large range of tide. Narrowing of the connection between bay and open ocean intensifies the differences of water level that are developed during the rise and fall of tide. A strong hydraulic current is produced and alternates in direction of flow. The flow is used to drive turbines and electrical generators, with a maximum efficiency of about 20% to

25%. Assessment of the world total of annual energy potentially available by exploitation of all suitable sites comes to only 1% of the energy potentially available through hydropower development.

GEOTHERMAL ENERGY

Yet another form of sustained-yield energy is **geothermal energy** drawn from heat within the earth at point of locally high concentration. Usually the surface manifestations of such heat are hot springs, fumaroles (holes from which steam issues), and active volcanoes. Wells drilled at such places yield superheated steam, which can be used to power turbines. Electric power is presently being generated from a number of geothermal fields. The most important fields at present are in Italy, New Zealand, and the United States. New developments now under construction or planned will about double the present total. One authority estimates that the ultimate development of known sources would give a world geothermal energy output on the order of 10 times that in 1969. This total is about the same as for the ultimate yield of all potential tidal power projects. Both energy sources represent but a small fraction of existing energy requirements.

COAL AND LIGNITE

We turn next to our major source of energy for the past century, the **fossil fuels:** coal and lignite, petroleum, and natural gas. The composition, geological occurrence, and origin of these hydrocarbon substances were discussed in Chapter 4.

The oldest important occurrences of coal are in strata of Carboniferous age (refer to Table 7.1 for geologic periods and their ages). During this period and in the following Permian Period, coals of great thickness and extent were formed in many parts of Pangaea. These now have widespread distribution on the fragmented continents. Coal is also found in Triassic and Jurassic strata, but in limited distribution. Second in importance to the Carboniferous coals are those deposited in the Cretaceous Period, at which time continental separation was in progress. Strata of Cenozoic age contain most of the world's lignite, but some coals of high quality were also produced in that era.

World coal reserves are most unevenly distributed among the continents. Figure 11.8 shows a breakdown into eight world regions; figures are in billions of metric tons. This estimate shows the USSR and the United States to be in very strong positions, while Canada and Western Europe are also very favorably endowed with coal. In contrast, Africa, Australia, Japan, and the Latin American countries are poorly endowed.

Note that these figures include coal lying at depths as great as 4000 ft (1200 m) and occurring in seams as thin as 14 in. (36 cm). With present mining technologies, only about one-tenth of this amount of coal can be mined on an economically successful basis. So the figure of 1500 billion tons for the United States reduces to 150 billion tons, in terms of existing mining methods. Converted into crude oil energy equivalent, 150 billion tons represents about

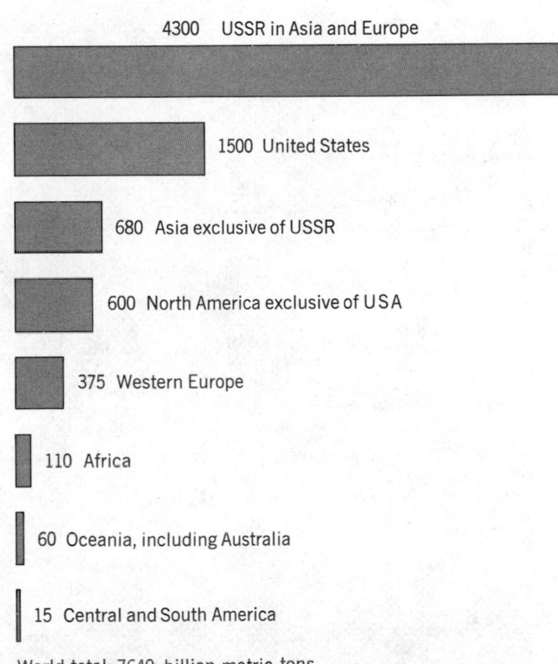

Figure 11.8 Estimated world coal reserves, in billions of metric tons, based on data of the United States Geological Survey.

640 billion barrels. This quantity is roughly 18 times larger than the proven crude oil reserves of the United States.

COAL MINING

To extract coal from flat-lying strata lying deep beneath the surface, vertical shafts are first driven down to reach the coal. Horizontal **drifts** are then opened out along the coal seams. Where terrain is mountainous, coal seams outcrop along the contour of the mountainsides; here the horizontal drifts are driven directly into the exposed coal seam.

Where coal seams lie close to the surface, **strip-mining** methods are used. Earth-moving equipment removes the overlying soil and rock, called **overburden,** to bare the coal seam. Power shovels then remove the coal. Two types of strip mining are used, depending upon terrain conditions. Where the land is level and strata are flat-lying, overburden is removed from a long, straight trench and piled in a ridge to one side (Figure 11.9). After the coal has been removed from the floor of the trench, another belt of adjacent overburden is removed and piled in the first trench. In this system, known as area strip mining, the overburden accumulates in a series of parallel spoil ridges, as shown in Figure 11.10. In mountainous terrain, coal is removed by contour strip mining, also shown in Figure 11.9. The overburden is excavated as far back into the mountainside as eco-

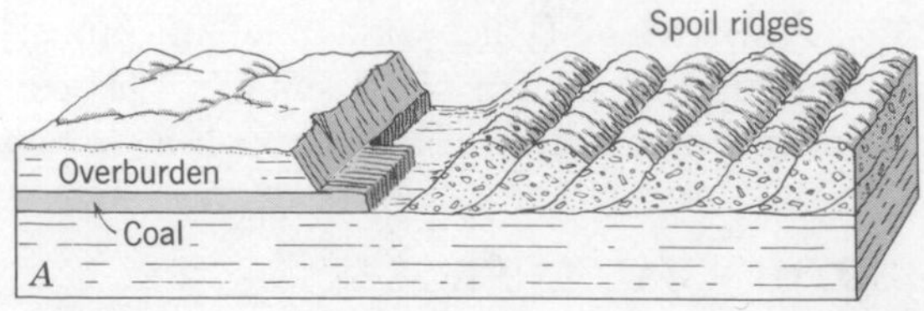

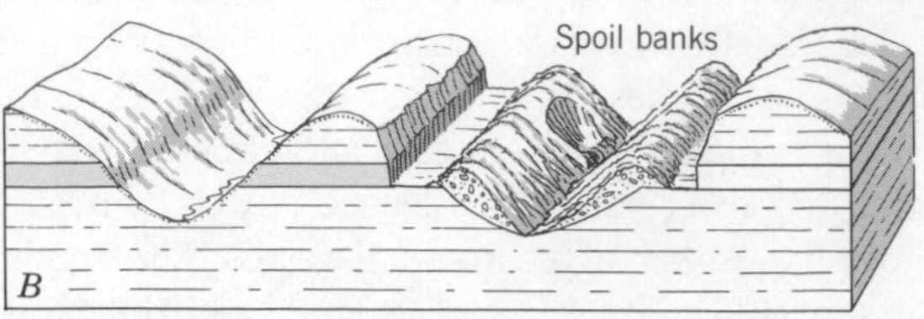

Figure 11.9 (*A*) Area strip mining. (*B*) Contour strip mining. (From A. N. and A. H. Strahler, 1973, *Environmental Geoscience,* Santa Barbara, California, Hamilton Publ. Co., Figure 10.11, p. 250.)

Figure 11.10 Area strip mining in Ohio County, Kentucky. Dragline in background is removing overburden and piling it at the right. Loader in foreground is removing the exposed coal. (TVA)

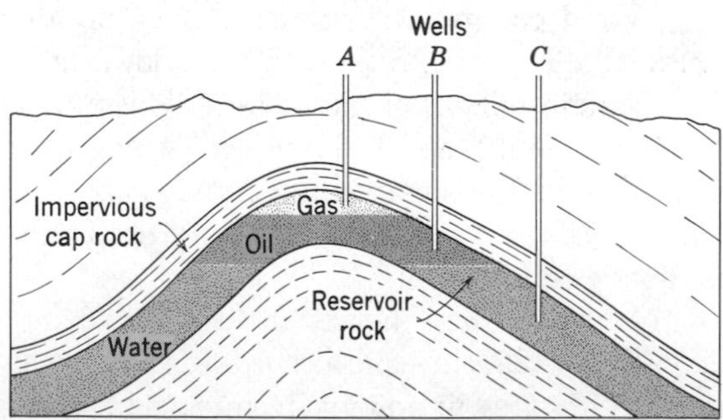

Figure 11.11 Idealized cross section of an oil pool in an anticline or a dome structure in sedimentary strata. Well *A* will draw gas; well *B* will draw oil; and well *C* will draw water. The cap rock is shale; the reservoir rock is sandstone.

nomically feasible, and the exposed coal removed. Additional coal is secured by means of large augers, driven as far as possible into the moutainside. The spoil from contour strip mining is deposited in a bank on the downslope side of the excavation, where it may slide far down into the valley below. Between the spoil bank and the clifflike rock face, a trench remains, and this may follow the contour of the mountainside for miles, winding in and out of one valley-head after another.

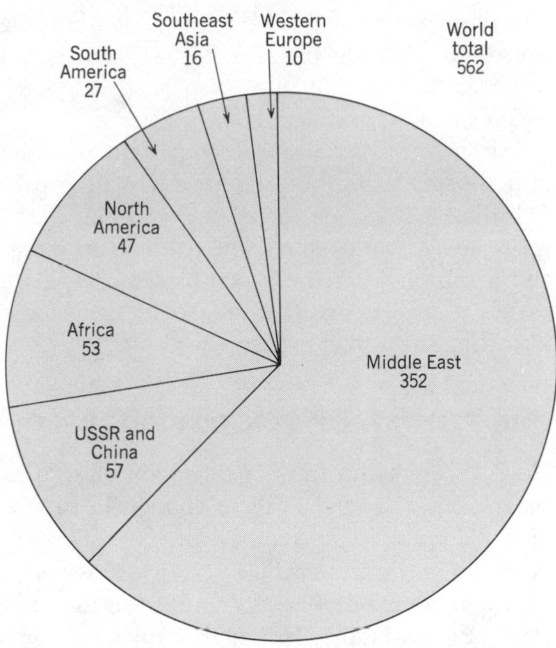

Figure 11.12 The world petroleum pie. These figures by the National Petroleum Council give proven reserves of crude oil in billions of barrels.

THE OCCURRENCE OF PETROLEUM AND NATURAL GAS

Accumulations of petroleum that can be tapped by drilling of wells are known as **oil pools.** The crude oil fills the interconnected pore spaces of a **reservoir rock,** which is usually a sand or a formation of sandstone, and sometimes limestone or dolomite. Natural gas accumulates above the oil, while water saturates the zone beneath the oil. It is essential that the reservoir rock be overlain by an impervious **cap rock;** typically this is a shale formation. The cap rock prevents the upward movement of the petroleum.

One of the simplest arrangements of sedimentary rocks favorable to oil accumulation is an up-arching of strata in either a dome or an anticline (Figure 11.11). Various other favorable arrangements of rock units exist and are referred to as **petroleum traps.**

WORLD PETROLEUM RESERVES

Proven petroleum reserves are mostly concentrated in a few world regions. As Figure 11.12 shows, the Middle East holds more than half the known world reserves of crude oil. Of this enormous accumulation—some 350 billion barrels—Saudi Arabia has about 40% (140 billion barrels), which is almost four times as much as United States reserves (36 billion barrels). Kuwait, Iran, and Iraq hold most of the remainder of the Middle East oil. Another important center of oil accumulation is in lands surrounding the Gulf of Mexico and Caribbean Sea, with major reserves in the U.S. Gulf Coast region and Venezuela.

OIL SHALE

Almost everyone who reads the papers has heard of **oil shale** and of the tremendous reserve of hydrocarbon fuel it holds. The truth is that this sedimentary rock in the Rocky Mountain region is not really shale at all, and the hydrocarbon it holds is not really petroleum. Strata of the Rocky Mountains called oil shales are of calcium-carbonate and magnesium-carbonate composition and were formed as lake deposits of lime mud (marl) in a Cenozoic lake. These soft, laminated deposits belong to the Green River Formation; it is of Eocene age (Figure 11.13).

The hydrocarbon material of the Green River Formation occurs in a particular bed, the Mahogany Zone, about 70 ft (20 m) thick. It is a waxy substance, called **kerogen,** which adheres to the tiny grains of carbonate material. When oil shale is crushed and heated

Figure 11.13 These cliffs of oil shale, near Rifle, Colorado, are the site of test mining operations. (U.S. Bureau of Mines.)

to a temperature of 900° F (480° C), the kerogen is altered to petroleum and driven off as a liquid. The shale may be mined and processed in surface plants, or burned in underground mines, from which the oil is pumped to the surface.

It is estimated that the equivalent of some 120 billion barrels of petroleum lie in the prime beds—those capable of yielding 30 gal (115 liters) of oil per ton of shale. When we compare these figures with proven world petroleum reserves of about 600 billion barrels and a proven United States petroleum reserve of about 35 billion barrels, we realize that the prime oil shale deposits of the Green River formation are a major energy resource.

THE FUTURE OF FOSSIL FUEL RESOURCES

How well will the world reserves of fossil fuels serve mankind in generations to come? Is there enough to supply an annual increase in world energy consumption of, say, 5% per year—doubling our consumption in the next 14 years? Let us first look at graphs showing the production of energy from fossil fuels plus water power since 1900 (Figure 11.14). Notice that the upper graph is scaled in units of heat energy. Water power has, over the past 70 years, amounted to about 3% to 4% of the yearly production of energy, so that only a small allowance needs to be made for its inclusion.

Both actual contributions and relative contributions of the sev-

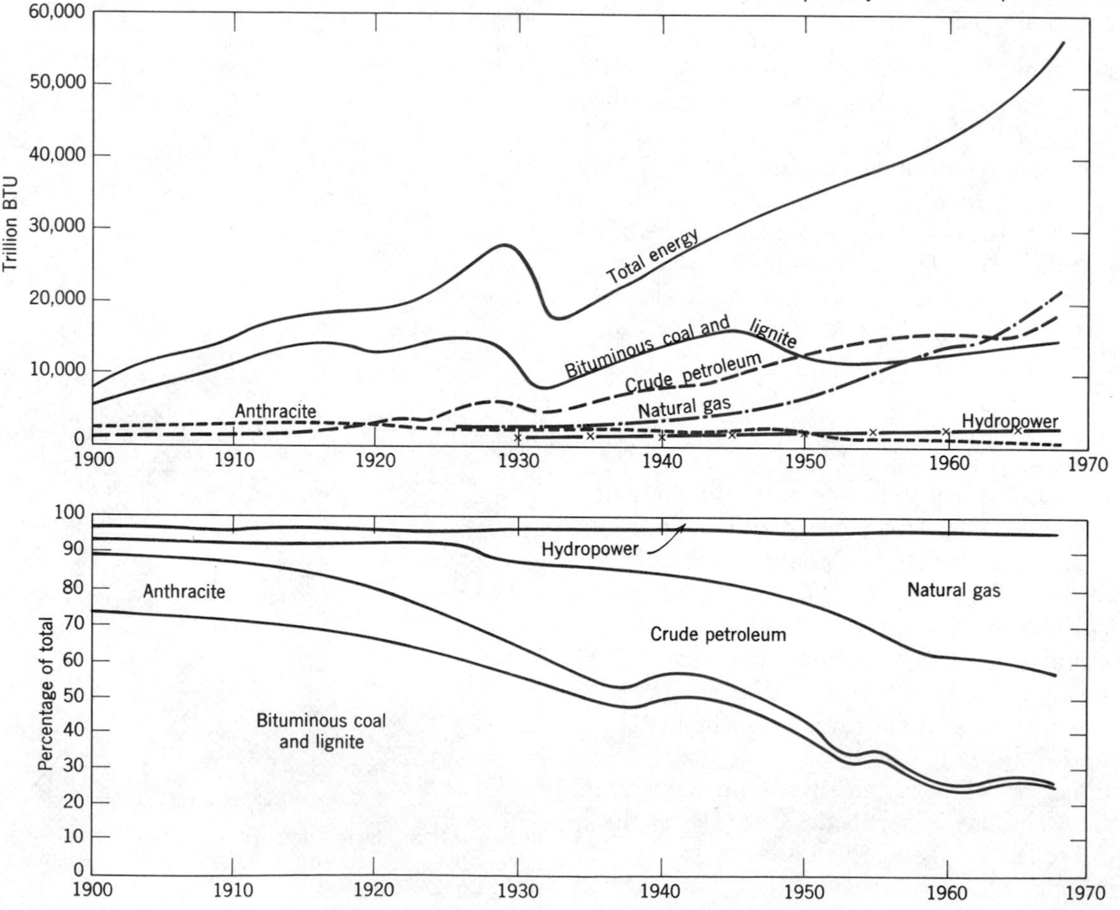

Figure 11.14 Production of mineral energy resources and of electricity from hydropower since 1900. The quantity of nuclear power is too small to be shown.

eral sources of energy since 1900 are shown in Figure 11.14. The two graphs tell us that while the total production of energy has increased over fivefold since 1900, the contributions of the several sources have changed markedly in ratio. Coal and lignite together have been reduced to less than half of the starting percentage, while anthracite has declined to almost nothing. Both petroleum and natural gas have increased in proportion in the same period, but of the two, natural gas has greatly expanded its ratio and is now about equal with petroleum.

Because the quantity of stored hydrocarbons in the earth's crust is finite, and the rate of geologic production and accumulation of new hydrocarbons is immeasurably small in comparison with the rate of their consumption, the ultimate exhaustion of this energy source is inescapable.

When exhaustion will happen is a difficult thing to predict, since we have to project into the future two independent curves. First is the rate of production, which at present is increasing by about 6% per year for petroleum (crude oil). Second is the rate of world discoveries of new oil and gas reserves, which so far have more than kept pace with production, so that there has been a moderate increase of known reserves. But in due time the rate of increase in new discoveries must slow down and then begin to decline. The reserves will then dwindle and eventually be entirely used, after which point production itself must begin a decline and will ultimately approach zero.

Are there other petroleum sources that can be developed? One possibility is the use of **heavy oil** enclosed in sands and not as yet exploited. Estimated reserves of heavy-oil accumulations show that they are an important fraction as compared with petroleum reserves.

As to oil shales as a source of energy, we have noted that the known resources of oil shales in prime beds of the western United States alone amounts to about one-fifth of the proven crude oil reserves of the world. For the entire world, the oil shale resource is is considerably greater. The exploitation of all prime oil shale would substantially extend the petroleum resources, but the general picture of depletion would not be greatly changed.

What of the future of our coal resources? The picture is quite different from that of petroleum and natural gas. Estimates of coal resources are regarded as quite realistic, since the existence and thickness of coal seams can be diretly sampled from borings. World coal reserves are equivalent in energy to from 16 to 25 times the world crude oil reserves (depending upon the oil reserve estimate chosen). For the United States alone, the energy equivalent of coal minable under present technology is roughly 18 times greater than the proven crude oil reserves. The message is clear: Coal is a vastly greater available source of energy than petroleum.

Projected curves of coal production indicate a peak around the year 2100 to 2200, which is a century or two beyond the peak for petroleum. Furthermore, the present rate of production (about equal to that of petroleum) is only a small fraction of the peak value. Coal will become our main hydrocarbon energy source by the year 2050 or thereabouts and will continue in that role thereafter until it, too, is exhausted.

NUCLEAR ENERGY AS A RESOURCE

The controlled release of energy from concentrated radioactive isotopes can be achieved through one of two processes: fission and fusion. **Nuclear fission** makes use of uranium-235. The fission of 1 gm of this substance yields an amount of heat equivalent to the combustion of about 3 metric tons of coal or about 14 barrels of crude oil. Uranium-235 is a rare isotope of a very rare element (see Table 11.1).

The United States reserve of uranium ore profitably recoverable with existing technology is estimated to have the equivalent energy of 85 billion barrels of crude oil. Compare this figure with a United States proven crude oil reserve of 35 billion barrels. Compare this figure with our estimate of recoverable United States coal reserves, equivalent to about 640 billion barrels of oil. Coal appears to be much the larger U.S. energy resource.

The world supply of uranium ore would be rapidly exhausted if nothing else were used. It is, however, possible to induce fission in other isotopes, notably other isotopes of uranium and isotopes of plutonium and thorium. This induced fission, known as **breeding,** can greatly reduce the expenditure of uranium-235. Geologists have stressed the necessity to develop breeder reactors to conserve uranium. If such development is successful, low-grade deposits of uranium can be exploited, making available a source of energy judged to range from hundreds to thousands of times greater than all reserves of fossil fuels.

Energy from **nuclear fusion** depends upon fusing isotopes of hydrogen into helium, with a consequent large release of energy. As everyone knows, the explosive release of enormous quantities of energy has been achieved through the hydrogen (thermonuclear) bomb. As yet, controlled release of energy through hydrogen fusion has not been achieved, although research is in progress. In theory, the quantities of energy available through fusion could exceed that of all fossil fuels by a factor ranging into the hundreds of thousands.

THE ENVIRONMENTAL IMPACT OF MINERAL EXTRACTION

Extraction of mineral resources from the earth has serious environmental impact. We shall mention only a few of its effects here.

Deep scarring of the land and the accumulation of great heaps of waste rock (spoil) go hand in hand with open-pit mining of ores, quarrying of structural materials, and the strip mining of coal, clay, and phosphate rock. The strip mining of coal, in particular, poses one of the most formidable environmental problems. It seems that much of our energy supply in the next few decades will come from coal. While the greatest part of this resource lies too deep for strip mining, there is strong pressure to strip mine all shallow coal as soon as possible. Besides deeply scarring the natural land surface and burying it under great masses of rock spoil, strip mining adversely affects the quality of water of streams and lakes deriving flow from strip mining areas. Deep mining for coal also produces large spoil heaps, and there may be subsidence of the land above mine workings.

The extraction of petroleum may seem to have little impact on the land surface above an oil pool, but there are important undesirable side effects in some instances. As petroleum is withdrawn, compaction of the strata may set in, causing the land surface to be lowered. **Land subsidence,** as this phenomenon is known, amounted to more than 25 ft (8 m) in the harbor area of Long Beach, California, as a result of oil production over a quarter of a century. In a low-lying coastal area such as this, subsidence of the land may lead to flooding by ocean water. Expensive dikes and levees must be constructed to keep the salt water out.

Another side effect is the **oil spill,** occurring from offshore wells drilled in search of petroleum beneath continental shelves. Oil under pressure may force its way through faults in weak rock surrounding the well, to emerge uncontrolled from the ocean floor. This type of break is often called a **blowout.** Figure 11.15 is a schematic drawing of such oil seepage. Most persons are familiar with the Santa Barbara, California, accident of 1969, in which a blowout occurred during drilling operations from an offshore platform. A large quantity of crude oil was released, and severely polluted the beaches and harbor of that city. Renewal of drilling operations from offshore platforms threatens to bring more oil spills to this coastline.

Subsurface mining for any mineral resource, whether it be metallic ores or coal, has always been a hazardous business. Serious health hazards to miners exist through the inhalation of dusts consisting of pulverized coal or silicate minerals. In the case of uranium mining, there is the added hazard of exposure to ionizing radiation from the radioactive radon gas released in the natural decay of uranium. Another dangerous daughter product of radioactive decay is radium; it may be leached from the rock waste of uranium mining and processing operations to reach streams and render the water toxic. Added to these mining hazards are the possibilities of pollution of streams and lakes by leakage of radioactive isotopes generated in the processing of nuclear fuels and in the operation of nuclear power plants.

Smelting of ores, particularly the sulfide ores, has been an important source of air pollution in areas surrounding smelters. Destruction of vegetation can be severe from such sources. The fallout can contaminate surface water. The combustion of fossil fuels, as we all know, is our major source of air pollution generally. Besides the emission of pollutant substances such as hydrocarbon compounds and oxides of nitrogen and sulfur, fuel combustion releases large amounts of heat into the atmosphere. Urban climates are known to be substantially changed as a result, and there may be important global effects upon climate in the future, as the quantities of fuel burned annually increase greatly.

Yet another environmental impact of use of fossil fuels and nuclear fuels arises from the disposal of heat from large power-generating plants. Where water is used in large quantities as a coolant, it is released into streams, lakes, or estuaries at a much higher temperature than when it is withdrawn. The release of heated water gives rise to **thermal pollution,** often with serious consequences to aquatic life.

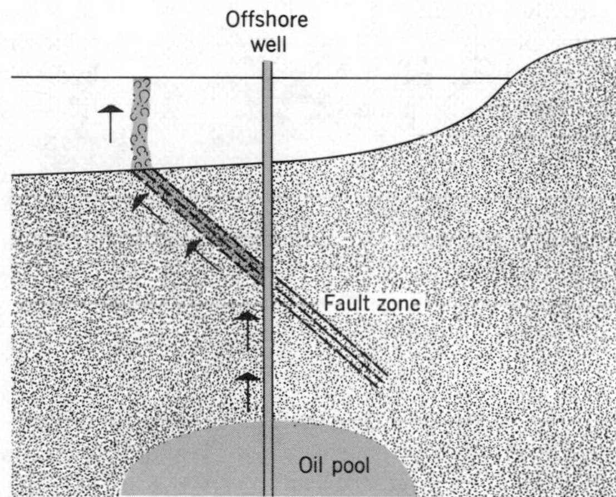

Figure 11.15 Schematic diagram of petroleum seepage along a fault zone penetrated by an offshore well. (From A. N. and A. H. Strahler, 1973, *Environmental Geoscience,* Santa Barbara, California, Hamilton Publ. Co., Figure 10.21, p. 255.)

MINERAL RESOURCES IN REVIEW

The salient message of this chapter deserves repeating. Industrial nations are ever more rapidly consuming natural resources that required millions of years of geologic time to bring into existence. As demands for materials and energy rise steeply in decades to come, rates of extraction of ore deposits and mineral fuels will be intensified. The richer, more easily extractable deposits will become exhausted, sending us in search of new deposits. Ores of poorer grade, and those occurring at greater depths, will be put into use, but with higher production costs and greater environmental damage. New oil reserves will be sought by geologists in localities much more difficult to develop, as in the case of offshore drilling in deeper waters, and in the case of the North Slope of North America, under a forbidding arctic climate.

Will technology solve all the problems that will arise? Will substitutes be found for the scarcer minerals? Will recycling greatly reduce the demand upon primary mineral sources? Opinions differ on the ability of technology to solve our problems in the future. Opinions also differ as to the quantities of mineral resources as yet undiscovered. In following the debate, your knowledge of the nature and occurrence of mineral resources should give you increased ability to evaluate accurately the arguments. You will also be more keenly aware of some of the environmental impacts to follow from increased intensity of mineral extraction.

YOUR GEOSCIENCE VOCABULARY

nonrenewable earth resources
metalliferous deposits
nonmetallic deposits
fossil fuels
nuclear fuels
economic geology
ore
ferro-alloy metals
native metals
magmatic segregation
hydrothermal solutions
veins
lode
disseminated deposit
porphyry copper
secondary ores
laterite
placer deposit
marine placer
sedimentary ore
new scrap
old scrap
metals recycling
sustained-yield energy source
solar energy
hydropower
tidal power
geothermal energy
fossil fuels
drifts (in mines)
strip mining
overburden
oil pool
reservoir rock
cap rock
petroleum trap
oil shale
kerogen
heavy oil
nuclear fission
nuclear fusion
breeding
land subsidence
oil spill
blowout
thermal pollution

SELF-TESTING QUESTIONS

1. Compare consumption of mineral resources today with that we may expect by the year 2000. What will be the impact of industrialization of the developing nations upon the world's finite mineral resources? In what way is the dispersion of concentrated resources an irreversible change?

2. List the principal groups of nonrenewable earth resources. What branch of geology deals with the finding of mineral resources?

3. Approximately how abundant are the following metals in the earth's crust: aluminum, iron, manganese, chromium, lead, uranium, mercury, platinum? (Give as a percentage by weight to the nearest power of 10.) Name several ferro-alloy metals and several nonferrous metals. Name three metals which occur as native metals.

4. What is the primary source of most ore minerals? Give an example of the degree to which an uncommon metal, such as chromium or lead, must be concentrated by ore-forming processes in order to make it commercially profitable to extract.

5. Explain how magmatic segregation by density can produce one type of ore deposit. What kinds of ores are produced by contact metamorphism? How are veins and lodes of ore minerals produced by hydrothermal solutions? Give an example of a disseminated ore deposit. Describe the processes of formation of secondary ores.

6. Describe the occurrence of placer deposits. Name some of the common metals and metallic ores found in placer accumulations. Give an example of an important sedimentary ore of iron.

7. Describe the position of the United States with respect to self-sufficiency in the primary production of metals from ore deposits within the nation. Give some specific examples of metals. Compare our position with respect to that of the Sino-Soviet bloc for two or three important metals. How can metals recycling relieve the problem of primary metals shortages? How effective is recycling today?

8. Give several examples of nonmetallic mineral deposits in the structural-materials category. Give examples of some nonmetallic minerals of importance in the chemical industry.

9. Evaluate the sea bed as a potential source of important mineral resources. Can you give a reason why the bedrock of the ocean basins is not abundantly endowed with ore deposits such as are found on the continental shields?

10. Evaluate the following sources of energy in terms of our present and future needs: solar energy, hydropower, tidal power, geothermal power.

11. In strata of what geologic periods is most of our coal and lignite found? Give a broad picture of the distribution of world coal reserves in terms of continents and subcontinents. Which nations are richest in coal? Which are poorest?

12. Compare the known U.S. reserves of coal presently minable on an economically profitable basis with the proven U.S. petroleum reserves. (Compare in terms of crude oil equivalent of coal.)

13. Describe coal mining methods. Under what circumstances is strip mining feasible? Describe some of the undesirable kinds of environmental impact associated with coal mining.

14. Describe the occurrence of petroleum in a geological trap. What kinds of strata serve as reservoir rocks? as cap rocks? What undesirable form of environmental impact can result from petroleum withdrawal?

15. Give a rough rundown of the world distribution of proven petroleum reserves. Where does the United States stand in this list?

16. Describe the occurrence of hydrocarbon compounds in the oil shale of the Rocky Mountain region. How can these strata yield petroleum? Compare the estimated reserves of petroleum in prime oil shale beds of the western United States with the proven petroleum reserves of the United States and of the world.

17. Compare both natural gas and coal resources with petroleum on the same terms. How would the inclusion of oil shale and heavy oil affect this picture?

18. Evaluate nuclear energy as an energy resource in comparison with coal and crude oil. What impact can the breeder reactor have on future nuclear energy supplies? What are the prospects of nuclear fusion becoming a major source in the near future?

19. What forms of air and water pollution arise form the combustion of fossil fuels? What are some of the hazards to health associated with the mining, processing, and use of nuclear fuels?

(*Above*) Seen from the air, the Madison Slide forms a great dam of rubble across the Madison River Canyon. (U.S. Geological Survey.)

(*Left*) A reporter examines a camper's automobile that was first crushed by the landslide, then swept along the riverbed by the flood wave. Of the family who owned this vehicle, both parents and three of their four children perished in the slide. The debris which buried the campground forms a small mountain in the background. (UPI Telephoto.)

12 WEATHERING AND MASS WASTING

DISASTER IN THE DARKNESS

For some two hundred vacationers camping near the Madison River in a deep canyon not far west of Yellowstone Park the night of August 17, 1959, began quietly, with everyone safe in their tents or house trailers. But at precisely 11:37 PM, Mountain Standard Time, not one, but four, terrifying forms of disaster were set loose upon the sleeping vacationers—earthquake, landslide, hurricane-force wind, and raging flood. The earthquake was the primary cause of it all; it measured 7.1 on the Richter scale, which is about as big as they come. The first shock, lasting several minutes, rocked the campers violently in their trailers and tents. Those who struggled to go outside could scarcely stand, let alone run for safety.

Then came the landslide. A dentist and his wife watched through the window of their trailer as a mountain seemed to move across the canyon in front of them, trees flying from its surface like toothpicks in a gale. Then, as rocks began to bang into the roof and sides of the trailer, they got out and raced for safer ground. Later they found that the slide had stopped only 75 feet from the trailer. Pushed by the moving mountain came a vicious blast of wind. It swept upriver, tumbling trailers end over end.

Then came the flood. Two women school teachers, sleeping in their car only 15 feet from the river bank, awoke to the violent shaking of the earthquake. Like other campers, they first thought they had a marauding bear on their hands. They turned on the headlights, but there was no bear. They headed the car for higher ground, and as they did so were greeted by a great roar coming from the mountainside above them. An instant later the car was completely engulfed by a wall of water that surged up the river bank, then quickly drained back. With the screams of drowning campers in their ears, the two women managed to drive the car to safe ground, high above the river. After the first surge of water, generated as the landslide mass hit the river, the river began a rapid rise. This rise was aided by great surges of water topping the Hebgen Dam, located upstream, as earthquake aftershocks rocked the water of Hebgen Lake back and forth along its 30-mile length. But of course, in the darkness of night, the terrified victims of the flood had no idea what was happening. In the panic that ensued a 71-year-old man performed an almost unbelievable act of heroism to save his wife and himself from drowning. As the water rose inside their house trailer, he forced open the door, pulled his wife with him to the trailer roof, then carried her up into the branches of a nearby pine tree. Here they were finally able to reach safety. The water had risen 30 feet above ground level in just minutes.

The Madison Slide, as the huge earth movement was later named, had a bulk of 37 million cubic yards of rock. It consisted of a chunk of the south wall of the canyon, measuring over 2000 feet in length and 1000 feet in height. The mass descended over a third of a mile to the Madison River, its speed estimated at 100 miles per hour. Pulverized into bouldery debris, the slide crossed the canyon floor, the momentum carrying it over 400 feet in vertical distance up the opposite canyon wall. At least 26 persons died beneath the slide and their bodies have never been recovered. Acting as a huge dam, the slide caused the Madison River to back up, forming a new lake. The photograph on the opposite page shows the rising lake shortly after the slide occurred. In three weeks' time the lake was nearly 200 feet deep. Today it is a permanent feature, named (you guessed it!) Earthquake Lake.

Landslides are one topic of this chapter. Rare as they are, and usually occurring in rather sparsely settled mountains, great landslides have caused some notable disasters. As in the case of the Madison Slide, earthquakes have been the triggering mechanism in many cases. Similarly, the falls of huge rock masses into deep mountain lakes and coastal fiords have generated enormous water waves, overwhelming entire towns and drowning their inhabitants within minutes.

LANDFORMS

With this chapter we enter a new area of geology to study **landforms,** the varied relief features of the land surface we see about us at all times. Examples of landforms are hills and valleys, plains, and mountains. Volcanoes and fault scarps, which we studied in earlier chapters, are also landforms.

Volcanoes and fault scarps are created by **internal earth processes:** volcanism and tectonic activity. In contrast, the landforms under examination in this and later chapters are shaped by **external earth processes.** External processes act through the atmosphere and oceans, where air and water come in contact with the lithosphere. Solar energy drives these external processes, whereas volcanic and tectonic activity are driven by energy sources deep within the earth.

You've already been introduced to rock weathering as one phase of the interaction between the atmosphere and lithosphere. Weathering in the form of alteration of silicate minerals produces the clay minerals (Chapter 4). The production of sediments and sedimentary rocks depends upon mineral alteration, which is part of the overall process of rock weathering.

Weathering is a passive process, in the sense that the products of rock decay and decomposition tend to remain in place where formed, although these same products, and in some cases bedrock as well, often yield to the force of gravity and move downslope by rolling, sliding, or flowage to lower levels. These spontaneous movements under gravity are referred to collectively as **mass wasting;** it is also a passive process.

The land surfaces are also subjected to a group of active processes by which weathered rock is transported, often for long distances, and eventually deposited to form new sedimentary accumulations. The active processes are carried out by several agents: running water, glacial ice, waves and their associated currents, and wind. These **active agents** are all fluids, strictly speaking. In moving over the land surface they perform **erosion,** which is simply the forceful removal of mineral matter from the parent mass of soil or rock. Erosion is always accompanied by transportation and must always eventually end in deposition.

DENUDATION

With a few exceptions, erosion, transportation, and deposition by the active agents persistently carry mineral matter from higher places on the continents to lower places of accumulation. The overall process of lowering of the lands is called **denudation,** a very convenient term to cover at once all of the kinds of work done by both active and passive agents.

Allowed to go on for millions of years, denudation would eventually lower the surface of a continent almost to sea level. Waves and currents of the oceans—also agents of denudation—might be expected to complete the job and the continent would end up as a shallow submarine platform. This has been the end result over some large continental areas at certain points in geologic time. However, in most places, tectonic and volcanic activity have periodically elevated the crust and thwarted the progress of denudation.

Man and all other forms of terrestrial life evolved upon continental surfaces of denudation. These surfaces are extremely varied from place to place in terms of physical environmental qualities. In combination with climatic factors of available heat and water, the nature of a given type of land surface determines the capacity of that surface to support life. Since denudation never ceases on the lithospheric surface, there has been a continual change in life environments through geologic time. These long-term environmental changes have determined the course of organic evolution.

The study of the origin and evolution of landforms is a branch of geology known as **geomorphology.** In the study of landforms, the various landscape features are sorted out according to processes of origin.

INITIAL AND SEQUENTIAL LANDFORMS

Using the concepts offered in previous paragraphs, we find that all landforms fall into two great groups. Those formed directly by volcanic and tectonic processes belong to the group known as **initial landforms.** Those landforms shaped by the agents of denudation are **sequential landforms.** Sequential means "following after" and suggests that these landforms must be derived from the initial landforms. Figure 12.1 illustrates this concept. The upper diagram shows an uplifted fault block produced by internal earth forces. This enormous mass of bedrock has been elevated by forces active in moving and dislocating lithospheric plates; it is an initial landform. Once raised, this initial mass is carved into sequential landforms by the active agents which operate through solar energy.

The sequential landforms can in turn be subdivided into two varieties: erosional and depositional. **Erosional landforms** are those resulting from the progressive removal of earth materials; **depositional landforms** are those resulting from the accumulation of the products of erosion. In the example shown in Figure 12.1, erosion by water flowing on slopes and in streams has carved a host of erosional landforms, consisting of canyons and the intervening divides and peaks. Deposition of sediment by streams has at the same time been forming a type of depositional landform known as a fan. Each agent of erosion produces a characteristic assemblage of erosional and depositional landforms.

All landscapes of the continents reflect an unending conflict between internal and external processes. Where internal processes have been active recently, through mountain making, there exist rugged alpine mountain chains and high plateaus. Where external processes have been given an opportunity to operate with little disturbance for vast spans of time, the land surfaces have been reduced to low plains.

In many periods of the geologic past, continental surfaces stood low. These surfaces were the peneplains described in Chapter 7 and illustrated by the great unconformities of the inner Grand Canyon. At such times the subsidence of continental crust allowed shallow seas to spread far inland. The continents today are, in contrast, generally high in elevation and rugged in relief. The internal processes have temporarily gained the upper hand on a worldwide scale.

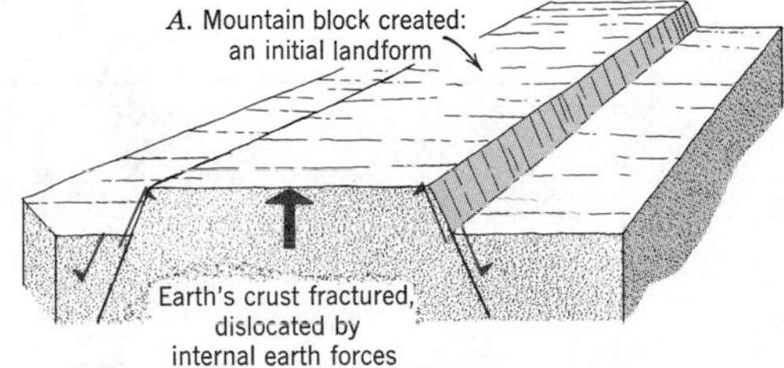

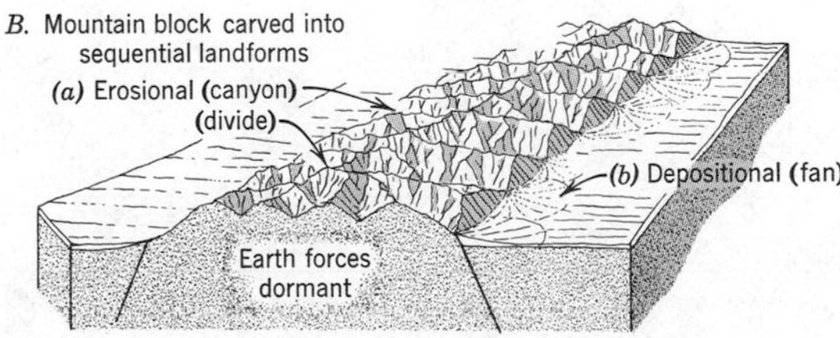

Figure 12.1 Two great classes of landforms. (*A*) Initial. (*B*) Sequential. (© 1973, John Wiley & Sons, New York.)

BEDROCK AND REGOLITH

When you inspect a freshly cut cliff, such as that in a new highway excavation or quarry wall, you may find several kinds of earth materials (Figure 12.2). Solid hard rock that is still in place and relatively unchanged is called **bedrock.** It grades upward into a zone where the rock is partly decayed and has disintegrated into clay and sand particles. This material is called the **regolith.** The prefix *rego* comes from the Greek word for "blanket"; literally, then, the regolith is a "blanketing layer over the bedrock."

At the very top is a layer of true **soil,** often called topsoil by farmers and gardeners. Over the soil may be a protective layer of grass, trees, or other vegetation.

In some places the soil and regolith have been stripped off down to the bedrock, which then appears at the surface as an **outcrop** (Figure 12.3). In other places, after cultivation or forest fires, only the true soil is stripped off, exposing the regolith.

Thicknesses of both soil and regolith are variable. The true soil is rarely more than a few feet thick. In contrast, **residual regolith,** formed in place by rock decay, may extend down tens or even hundreds of feet. Formation of regolith is greatly aided by the presence of innumerable cracks, called **joints,** in the bedrock (Figure 12.2). Joints are usually a result of tectonic stresses and few masses of bedrock are free of joints. Sets of intersecting joints break the bedrock into blocks.

The term regolith is broad in scope; it may refer to any sort of relatively loose or soft mineral particles lying on the bedrock. Gravels, sands, or floodplain silts laid down by streams, or rubble left by a disappearing glacier are all forms of regolith, but they are unique in having been transported by such agents as streams, ice, wind, or waves. We call such material **transported regolith** to distinguish it from residual regolith. Figure 12.3 shows a deposit of transported regolith of a type known as alluvium, covering the bottom of a valley, where it has been left by a stream. Many kinds of

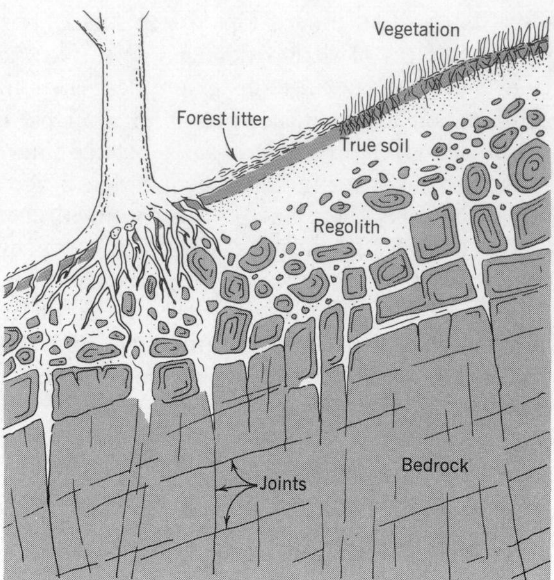

Figure 12.2 Soil and regolith overlying bedrock. (© 1973, John Wiley & Sons, New York.)

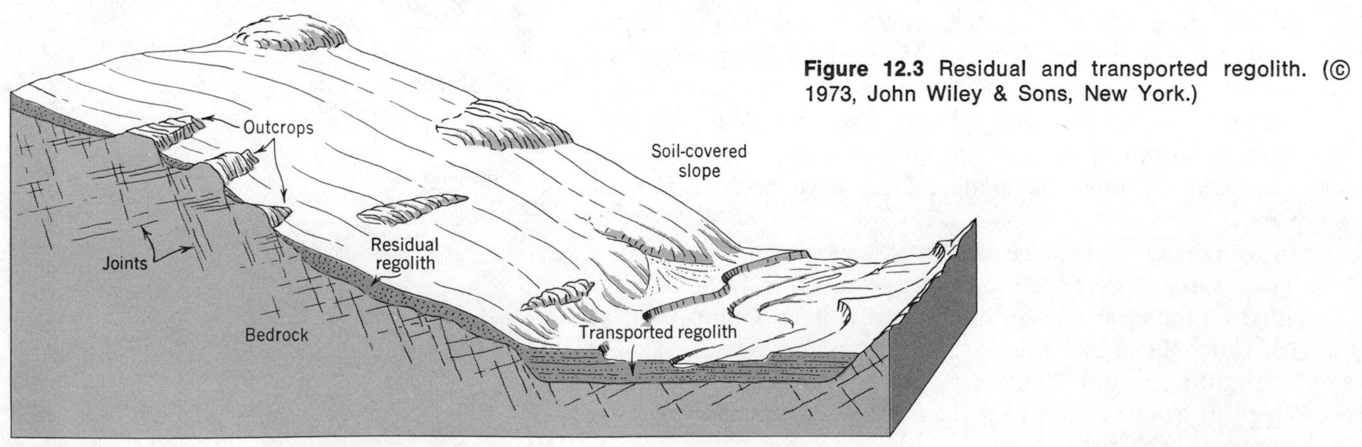

Figure 12.3 Residual and transported regolith. (© 1973, John Wiley & Sons, New York.)

transported regolith can be identified, and we shall learn more about them when studying streams, glaciers, waves, and winds.

PHYSICAL WEATHERING

We will treat first the processes of **physical weathering**, in which mechanical stresses act upon rock, causing disintegration. These processes perform the primary breakdown of bedrock into fragments, whose surfaces are in turn exposed to chemical weathering.

In climates with cold winters, and at high altitudes, alternate freezing and melting of water in the soil and rock provides a powerful mechanism of rock breakup. Water that has penetrated the joint planes, bedding planes, foliation planes, and other openings of the rock is transformed into ice crystals of needlelike form. Growing masses of such crystals exert great pressures upon the confining rock walls, causing joint blocks and bedding layers to be heaved up and pried free of the parent mass.

Disintegration of bedrock by freezing water is a process commonly referred to simply as **frost action.** Its effects are most conspicuous above the timberline of high mountains and, at lower levels, in arctic latitudes. The ground may be covered with large angular blocks of fresh rock. Figure 12.21 shows such an environment, dominated by frost shattering.

Formation of ice bands and layers in the soil is a widespread occurrence in winter in cold climates. Where the soil is rich in fine-grade silt and clay, soil water tends to freeze in the form of horizontal ice layers. As an ice layer thickens, a strong upward pressure is exerted upon the overlying soil layer, lifting or heaving the soil. Because of the irregularity of growth of the ice layers, **frost heaving** is uneven and produces mounds of soil.

An interesting effect of ice-crystal growth in soils is the moving of larger rock fragments, in the size range of pebbles, cobbles, and boulders, upward toward the surface. In arctic environments, particularly in the tundra climates, the larger fragments tend to be moved sidewise as well as upward by such frost action and to be sorted out into narrow bands. These bands intersect to form a net-like pattern consisting of **stone rings** (Figure 12.4).

Throughout vast arctic regions of North America and Siberia the average annual temperature is below freezing. This region comprises the treeless arctic tundra. Except for a shallow surface layer in which soil moisture thaws during the short summer, all water below the surface is continually frozen. This condition of perennially frozen ground is called **permafrost.** Ice bodies are found in various shapes within alluvial deposits. Particularly interesting are the **ice wedges,** perpendicular in orientation, embedded in silty materials (Figure 12.5). Ice wedges are linked into polygonal systems, much like stone polygons in their pattern. Permafrost also extends southward into forested regions of subarctic latitudes, but here it occurs in discontinuous patches.

The arctic tundra is extremely vulnerable to man-made disturbances, because these can easily lead to deep melting of the permafrost. Stripping away of tundra vegetation and forest exposes the mineral soil to summer thawing. Ground ice layers and wedges melt,

Figure 12.4 Stone rings near Thule, Greenland. (A. E. Corte, Geology Department, Universidad Nacional del Sur, Bahia Blanca, Argentina.)

Figure 12.5 A V-shaped ice wedge surrounded by alluvial silt is seen here exposed in the banks of a stream near Livengood, Alaska. (T. L. Péwé, U.S. Geological Survey.)

producing a soft, water-saturated mud. Highways and buildings can sink deep into the mud. A major concern of environmentalists is that construction and operation of the Trans-Alaska Pipeline will cause serious disturbance to the permafrost areas which it crosses. Some fear that the hot oil within the pipeline will melt the surrounding permafrost, leading to breakages in the line and serious oil spills.

In dry climates an important agent of rock disintegration is **salt-crystal growth,** a process quite similar physically to ice-crystal growth. Such climates have long drought periods in which evaporation can occur. Water films are drawn surfaceward and moisture is steadily evaporated, permitting dissolved salts to be deposited in openings in the rock and soil. The growing salt crystals are capable of exerting powerful stresses. Even the hardest rocks (also concrete, mortar, and brick) can be reduced to a sand by continued action of the process. The grain-by-grain disintegration of rocks in dry climates is often a conspicuous process. Sandstones are particularly affected. Water may emerge gradually near the base of a sandstone cliff, supplying moisture with dissolved salts for continual evaporation (Figure 12.6). As the rock disintegrates and the sand particles are blown away or washed out in rainstorms, the rock wall recedes to produce a niche, or in some cases a shallow cave. Such well-protected recesses were used by Indians of the southwestern United States as sites for cliff dwellings.

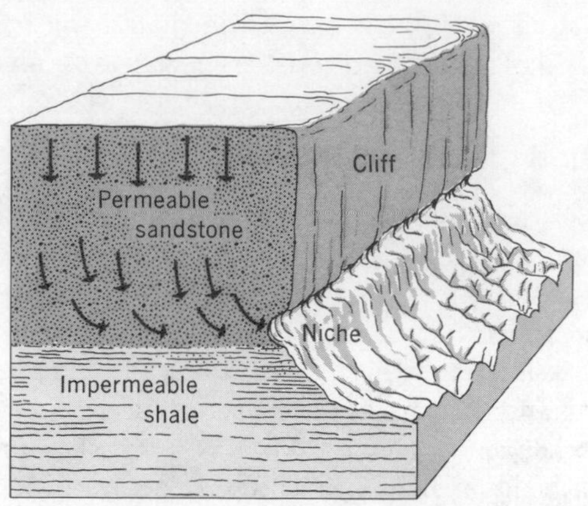

Figure 12.6 Sandstone has disintegrated at the base of a high cliff, resulting in a niche and cliff overhang. (© 1973, John Wiley & Sons, New York.)

The simple process of wetting and drying of soil and rock can result in forces capable of agitating soil and disintegrating rock. Clay consisting of colloids has a strong affinity for water and will swell greatly when permitted to absorb water. Certain varieties of shales and those siltstones and sandstones containing clay particles tend to disintegrate by moisture absorption on exposed surfaces. Clay-rich soils and sediments—swelling when wet, contracting when dry—are continually affected by changes in moisture content. Shrinkage produces soil cracks and **mudcracks** (Figure 12.7).

Rock disintegration is sometimes attributed to temperature changes alone. It is well known that most crystalline solids expand when heated and contract when cooled. The theory is that because the heating of rock will cause expansion of the minerals, the rock may be broken. Sudden and intense heating by forest and brush fires causes severe flaking and scaling of exposed rocks. Also, we know that primitive mining methods included the building of fires upon a quarry floor to cause slabs to break free. But geologists are uncertain that the daily temperature cycle under solar heating and nightly cooling produces sufficiently great stresses to cause fresh hard rock to break apart. Laboratory tests have shown that rocks can stand the equivalents of centuries of daily heating and cooling without showing signs of disintegration. We may be safe in supposing, however, that expansion and contraction through daily temperature changes may assist in breaking up rocks already weakened by other stresses and by chemical decay.

Figure 12.7 Mudcracks on an exposed riverbed. Individual soil blocks are about 1 ft (0.3 m) across. (G. K. Gilbert, U.S. Geological Survey.)

Closely related to physical weathering, and commonly included as one of the processes, is the rupturing of otherwise solid bedrock as a result of **spontaneous expansion.** The rock undergoes increase in volume when it is relieved of the confining pressure of overlying and surrounding rock. In quarries of such massive rocks as marble and granite, a well-known phenomenon is the rifting loose of rock in great slabs or sheets, sometimes with explosive violence. When a slab is cut into a block, the block expands measurably.

This evidence shows that most massive rocks, such as the igneous and metamorphic types, are under a state of slight compression when deeply buried. As the denudation of the landmass proceeds, these rocks are gradually brought nearer the surface. Free of load or of confining rock on the sides, the mass expands. Usually the expansion results in the splitting off of shells or sheets of rock up to several feet thick. This process is called **exfoliation.**

Where great bodies of granitic rock are subject to exfoliation, there remain domelike mountain summits, known as **exfoliation domes.** Fine examples are seen in Yosemite National Park, where individual rock shells are as much as 50 ft (15 m) thick (Figure 12.8).

Still another physical weathering process is that of the action of growing plant roots, exerting pressure upon the confining walls of soil or rock. This process is important in the breakup of rock already weakened or shattered by other physical and chemical processes.

Figure 12.8 North Dome (*left of center*) and Basket Dome (*right of center*), two great exfoliation domes of Yosemite National Park, California. (Douglas Johnson.)

FORMS PRODUCED BY CHEMICAL WEATHERING

Chemical weathering processes were explained in Chapter 4 under the subject of mineral alteration. Recall that chemical weathering consists mainly of oxidation, hydrolysis (combination with water), and reaction with carbonic and other acids. The silicate minerals of igneous and metamorphic rocks decay largely through oxidation and hydrolysis, forming clay minerals. Sedimentary rocks are in many cases almost immune to chemical weathering because they are formed of clastic materials already fully altered by exposure to the atmosphere and water. An important exception is the carbonate group of sedimentary and metamorphic rocks—limestone, dolomite, and marble. These rocks are made up of calcium and magnesium carbonate and are readily acted upon by weak solutions of carbonic acid found in all rainwater, soil water, and stream water.

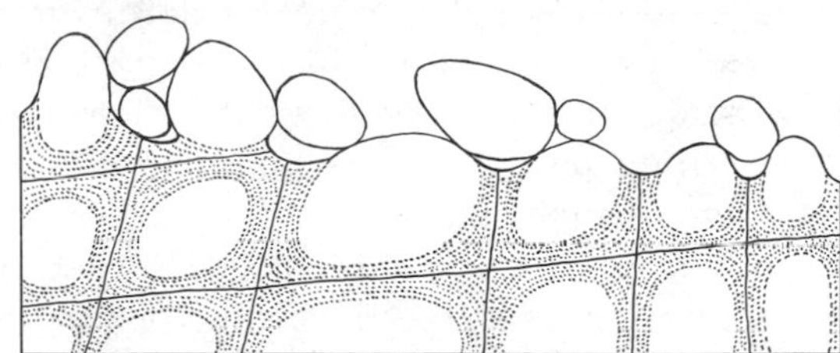

Figure 12.9 Rounded boulders produced by granular disintegration of rectangular joint blocks of granite. (Based on a drawing by W. M. Davis; © 1973, John Wiley & Sons, New York.)

Chemical decay of joint blocks of igneous rock takes two forms. **Granular disintegration** (grain-by-grain disintegration) commonly affects the coarse-grained igneous rocks such as granite or gabbro. Falling away of grains at the edges and corners of the blocks produces rounded, egg-shaped boulders (Figure 12.9). The products of disintegration, in the form of a coarse sand or gravel of individual mineral crystals, are swept away by rainwater, to become the sediment load of streams. Rounded boulders still in place, surrounded by coarse, granular decayed rock, are shown in Figure 4.4.

The finer-grained igneous rocks are commonly affected by **spheroidal weathering,** a chemically caused form of exfoliation in which the joint blocks are modified into spherical cores surrounded by shells of decayed rock (Figure 12.10).

Figure 12.10 Spheroidal weathering, shown here, has produced many thin concentric shells in a basaltic igneous rock. (U.S. Geological Survey.)

In the warm, humid climates, chemical decay of igneous and metamorphic rocks extends to depths as great as 300 ft (90 m). The residual regolith is a thick layer of soft, clay-rich material known as **saprolite.** Examples are common throughout the Piedmont and Appalachian regions of the southeastern United States. Saprolite is easily removed by power shovels, with little or no blasting required. Presence of abundant clay minerals with plastic behavior greatly reduces the ability of the saprolite to support heavy structures.

Carbonic-acid action, called **carbonation** for brevity, plays an important role in the decomposition of many mineral and rock varieties. Its effects are most striking in the weathering of the carbonate rocks—limestone, dolomite, and marble. Carbonic acid combines readily with calcium carbonate to produce a highly soluble salt, calcium-bicarbonate, which is carried away in streams. Limestone surfaces commonly show elaborate pits, grooves, and cup-shaped hollows on exposed surfaces (Figure 12.11). These are effects of etching by carbonic acid in rainwater and soil water. Deep below the surface, carbonic acid acts upon limestone strata to produce cavern systems (Chapter 13).

WEATHERING AND SOILS

All plant and animal life of the lands depends upon the soil layer; it is rarely more than a few feet thick and is entirely missing from vast desert areas, where much sterile bedrock or dune sand lies at the surface. True soil capable of supporting the larger forms of plant life—forests, grasses, crops—typically shows distinctive layers, called soil horizons. Horizons are a result of many centuries (or even many thousands of years) of the combined activity of chemical and physical weathering, plant growth and decay, and slow removal of mineral elements in rainwater that moves down through the soil.

A mature soil with well-developed horizons is an extremely complex natural system and is easily damaged or destroyed. You might say that the fertile soil layer is a nonrenewable mineral resource, like the metallic ores or coal. Careless use of the land, resulting in excessively rapid soil erosion, can cause a serious depletion or loss of this resource. As world population grows by leaps and bounds, demanding more and more food, prime agricultural soils are being depleted by poor farming practices, and totally destroyed by the growth of cities and highways, and by strip mining for coal.

Figure 12.11 Deeply pitted surface of limestone, near Fremantle, Western Australia. (Douglas Johnson.)

MASS WASTING

The force of gravity acts constantly upon all soil, regolith, and bedrock. In most places the internal strength of these materials is sufficient to keep them in place. Consequently, we rarely see soil or rock moving spontaneously except when carried by an active agent of erosion. Wherever the ground surface is sloping, a proportion of the force of gravity is directed downslope parallel with the surface. Every particle has at least some tendency to roll or slide downhill and will do so whenever the downslope force exceeds the resisting forces of friction and cohesion that tend to bind the particle to the rest of the mass.

The forms of mass wasting range from the catastrophic slides in alpine mountains, involving millions of cubic yards of rock and capable of wiping out a whole town, down to the small flows of water-saturated soil seen commonly along the highways in early spring. But extremely slow movement of soil, imperceptible from one year to the next, also acts on almost every hillside.

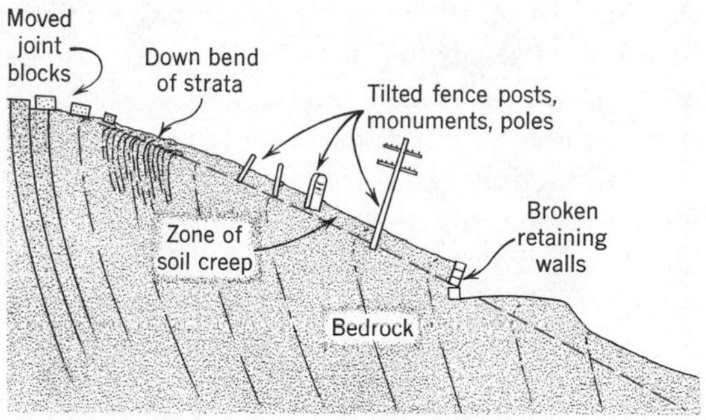

Figure 12.12 Commonplace evidences of the almost imperceptible downslope creep of soil and weathered rock. (© 1973, John Wiley & Sons, New York.)

SOIL CREEP

Careful inspection of a hillside often discloses evidence that the soil has been very slowly moving downslope rather steadily over a long period of time. This phenomenon is termed **soil creep** (Figure 12.12). Where a distinctive type of rock outcrops high up on a hillside you may find that the larger joint blocks have moved away from their original locations and that smaller fragments of the rock have been carried far down the slope in the soil mass. Yet it is unlikely that these particles have at any time slid or rolled rapidly.

Where steeply dipping layered rocks such as slates or shales underlie a hillside, the upper edges of the layers are commonly turned downhill as if bent. This phenomenon is the result of slow creep distributed on countless joint fractures and bedding or cleavage surfaces in the rock (Figure 12.13). Trees, posts, poles, and monuments may be found tilted downhill, suggesting rotation as the soil has crept downslope, the surface layers moving more rapidly than those at depth. Masonry retaining walls paralleling the slope are often found to be tipped over and broken, yielding to the pressure of soil creep.

Figure 12.13 Downturning by creep of the upper edges of near-vertical shale strata, Maryland. (U.S. Geological Survey.)

The mechanism of soil creep is a combination of the various weathering processes that agitate the soil, acting in concert with the force of gravity. Whatever mechanism disturbs the soil induces downslope movement of the particles, because gravity exerts an influence on the motions and its influence is in the downslope direction.

EARTHFLOWS

In hilly and mountainous regions of humid climate, yielding of water-saturated soil and regolith rich in clay minerals takes the form of **earthflows.** These are tonguelike masses that have flowed a limited distance down the hillside, perhaps coming to rest before reaching the base, or in some cases turning and flowing down-valley for a short distance. At its upper end the earthflow leaves a depression bounded on the uphill side by a curved scarp (Figure 12.14). At its

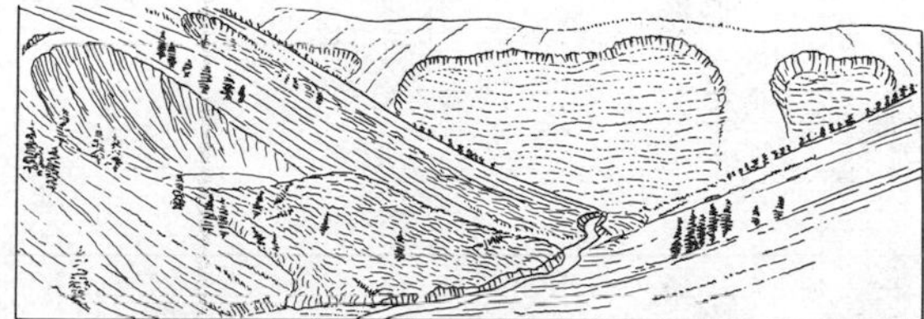

Figure 12.14 Earthflows on steep mountain slopes in a humid region. (© 1973, John Wiley & Sons, New York.)

lower end the earthflow bulges convexly downslope in a toe. Where the hillside flattens to a broad valley floor, the toe spreads out into a broad, rounded mass resembling a pancake. Where the valley is narrow, the toe forms a dam, sometimes creating a lake.

Solifluction is an arctic variety of earth flowage important in the treeless tundra. In these regions the permanently frozen subsoil acts as a barrier to downward percolation of water released in the spring by the melting of the snow cover and ice in the surface layer of the soil. Unable to escape by drainage, moisture builds up until the thawed soil is saturated, resulting in slow flowage of a shallow layer. The result is a succession of **solifluction lobes** (Figure 12.15). Motion is on the order of a few yards per year, with a speed of a few inches per day at most.

Figure 12.15 Solifluction on this Alaskan tundra slope has produced lobelike masses of soil, locally called "earth runs." (U.S. Geological Survey.)

MUDFLOWS

Where the proportion of water to mineral matter is large, mass wasting takes the form of a rather fluid mixture capable of traveling rapidly in streamlike masses down the channels of streams. This wasting is termed **mudflow.** In nature one finds all gradations between earthflow and mudflow, and it is not practical to try to draw a precise line of distinction.

One type of mudflow, common in arid regions, originates in the watersheds of streams high in a mountain range. Here, torrential thunderstorm rains sometimes wash large quantities of loose soil and regolith down steep mountain slopes into the adjacent canyons. As the heavily laden streams progress down-valley, the loss of water by seepage and the increasing proportion of solid matter picked up from channel floors and banks cause the stream to thicken into a mudflow. The mudflow may attain the consistency of ready-mix concrete and will continue down-valley, where it commonly spreads out upon the plain at the foot of the mountain range (Figure 12.16). Eventually, thickening of the mud by loss of water causes flowage to cease.

Mudflows of the mountainous deserts are a serious threat to life and property because they may emerge from canyons with little warning and spread over populated lands. The **debris flood,** a disastrous flood occurring in such localities as Los Angeles and Salt Lake City, is a form of flowage intermediate between the turbid flood of a mountain stream and a true mudflow (Figure 12.17).

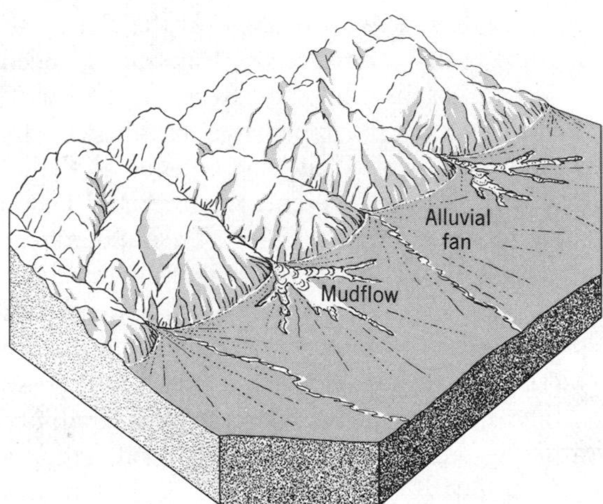

Figure 12.16 Mudflows issuing from the mouths of canyons in a semiarid region. (© 1973, John Wiley & Sons, New York.)

Figure 12.17 Debris deposits at the mouth of Benedict Canyon, Los Angeles County, California, following a January flood. (U.S. Army Corps of Engineers.)

Mudflows are also produced on the slopes of active volcanoes where torrential rains saturate freshly fallen volcanic ash. Herculaneum, a Roman village at the western base of Mount Vesuvius, was buried by mudflows in the eruption of A.D. 79.

Both earthflows and mudflows are common on the flanks of spoil heaps built from coal mines, including strip mines. In recent times, a major disaster occurred at Aberfan, Wales, when a great waste heap, 600 ft (180 m) high, built during coal-mining operations, spontaneously lost strength and moved quickly downslope. The tongue of debris overwhelmed buildings in the town below, causing the loss of over 150 lives (Figure 12.18).

Figure 12.18 A destructive debris flow at Aberfan, Wales. (Sketched from a photograph.)

LANDSLIDES

A **landslide** is the rapid downslope sliding of a large mass of bedrock. In the early stages of the movement, the mass travels as a unit, but typically breaks up into rock rubble before coming to rest. A landslide moves with one of two basic types of motion. In a **rockslide** a single block slides on its lower surface on a bedding plane, joint plane, or fault plane. In a **slump** a block slips on a curved fracture plane, rotating backward upon a horizontal axis as it sinks (Figure 12.19).

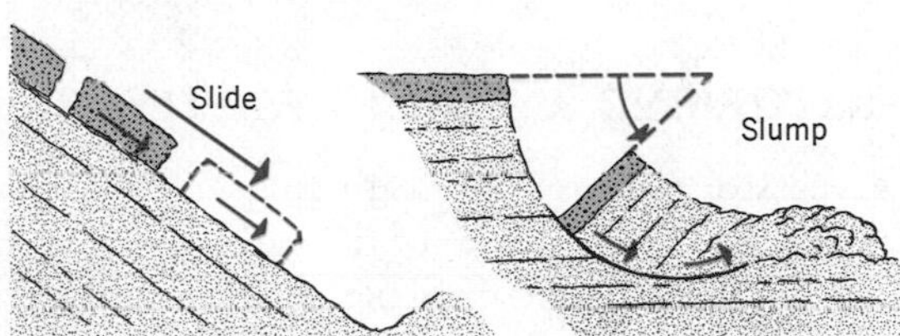

Figure 12.19 Slide and slump are two basic forms of landsliding. (© 1973, John Wiley & Sons, New York.)

Rockslides are found in high mountain ranges of the alpine type where steep rock walls have previously been formed by glacial erosion. Many rockslides have been reported in the Alps, Rocky Mountains, and mountains of Norway. One slide may involve many millions of cubic yards of rock and can travel with the speed of a freely falling object. Towns, highways, and railroad lines in the path of the slide are obliterated. Immediate causes of rockslides are not often evident, and the time of their occurrence cannot be predicted. An earthquake shock may set off a slide, an example being the Madison Landslide caused by the Hebgen Lake earthquake of 1959 in Montana.

Slumping of bedrock masses on a vast scale is found along cliffs of sedimentary strata or lavas (Figure 12.20). Rotated blocks as long as a mile or more and with a thickness of a thousand feet are known. Steepening of the cliff by running-water erosion of shale formations in the cliff base precedes the slump. Hundreds of such large slump blocks are known in the Colorado Plateau and Columbia Plateau.

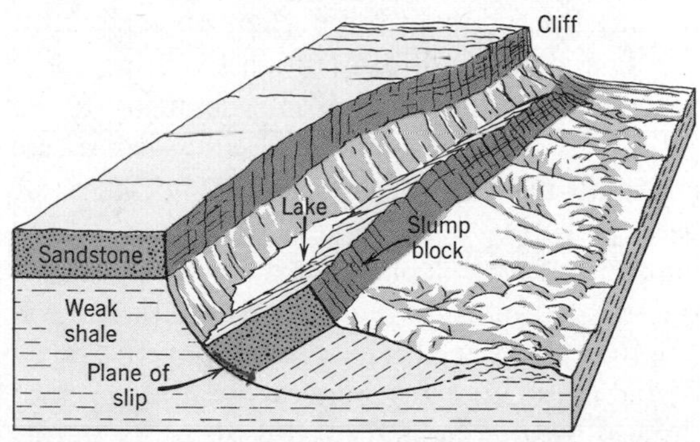

Figure 12.20 A slump block descended from a cliff of horizontal strata has rotated backwards on an axis parallel with the cliff. (© 1973, John Wiley & Sons, New York.)

ROCKFALL AND TALUS

On any near-vertical rock cliff you will find that fragments are continually being pried free by the processes of physical weathering already described. The fall of any such fragment, whether it be merely a sand grain or a huge mass weighing hundreds of tons, is described simply as **rockfall.** Although the fragment may bounce or roll down the cliff face, its descent is approximately at the speed of a freely falling body. Most large falling rock masses shatter into many smaller fragments, strewing the base of the slope.

Rockfall continued from a cliff face over many centuries eventually builds a slope of loose rock fragments at the cliff base (Figure

Figure 12.21 Talus cones of quartzite fragments, Snowy Range, Wyoming. (A. N. Strahler.)

12.21). The accumulation is called a **talus.** Normally, the cliff has recesses and ravines tending to funnel the fragments into chutelike exits, causing the piles of fragments to take the shape of **talus cones.** The surface of a talus slope is remarkably constant at an angle of close to 35° with the horizontal. Active talus cones will be seen forming today above timberline in mountain ranges that were deeply carved by glaciers during the last ice age. Here frost action is intense on steep rock walls.

WEATHERING AND MASS WASTING IN REVIEW

The passive processes of weathering and mass wasting act continuously on the land surfaces of the earth. Usually inconspicuous, these contributors to denudation perhaps rank first in importance to life on earth, in comparison with all of the agents involved in denudation. Without weathering there would be no soil, without soil no plant life, and without plant life no terrestrial animal life. Even the life of the oceans owes its existence to weathering, because the essential nutrients of plant life found in sea water are ions derived from rock weathering on the lands. Soil creep assists in continual renewal of the soil profile, allowing new parent matter to become exposed. In this way the supply of soil nutrients is replenished.

The more rapid forms of mass wasting are often destructive and cataclysmic in action. Not only do many forms of mass wasting—earthflows, mudflows, and rockslides—constitute an environmental hazard to man and his structures, but they are often induced by man's activities. Both the removal of support by excavation and the piling up of rock debris contribute to instability of slopes and lead to unwanted mass movements of regolith and soil.

Repeatedly in this chapter we have referred to the role of water. Water is essential to chemical weathering and to many forms of physical weathering and mass wasting. Water is, of course, essential to formation of mature soils and their horizons. In the next chapter we will investigate the geologic role of water in the soil, regolith, and bedrock beneath the land surface.

YOUR GEOSCIENCE VOCABULARY

landforms
internal earth processes
external earth processes
mass wasting
active agents
erosion
denudation
geomorphology
initial landforms
sequential landforms
erosional landforms
depositional landforms
bedrock
regolith
soil
outcrop
residual regolith
joints
transported regolith
physical weathering
frost action
frost heaving
stone rings
permafrost
ice wedge
salt-crystal growth
mudcracks
spontaneous expansion
exfoliation
exfoliation dome
granular disintegration
spheroidal weathering
saprolite
carbonation
soil creep
earthflow
solifluction
solifluction lobes
mudflow
debris flood
landslide
rockslide
slump
rockfall
talus
talus cone

SELF-TESTING QUESTIONS

1. In what way do landforms represent the interaction of internal and external earth processes? Which of the external earth processes are active, which are passive? What would the ultimate result of denudation be? Why is this result never reached?

2. Give examples of initial and sequential landforms. What are the two classes of sequential landforms?

3. Describe a typical exposure of earth materials in a quarry wall or highway cut, starting at the top and working down. How does transported regolith differ from residual regolith?

4. Name and describe the action of the important physical weathering processes. What surface forms or landforms does each produce? Explain how climate determines which of these processes dominates in a given region. Under what conditions of climate and in what parts of the world does permafrost occur?

5. Which kinds of rocks are susceptible to chemical weathering generally? Which are particularly susceptible to carbonation? Describe some forms in regolith and bedrock caused by chemical weathering.

6. What causes soil creep? What are some observable effects of soil creep? Compare the external form of an earthflow with that of a mudflow. Where do mudflows originate? What is solifluction?

7. What are the two basic types of motion in landslides? Where do large landslides most frequently occur? In what kinds of rocks is slump a common form of mass wasting?

8. How is a talus cone constructed? Where can talus cones in process of growth be found in abundance today?

The Geysers power plant in Sonoma County, California. The two generating units seen here have a capacity of 24,000 kilowatts. (Pacific Gas and Electric Company.)

13 GROUND WATER AND ITS GEOLOGIC ACTIVITY

THE GEYSERS

Imagine a beautiful California valley, its verdant slopes clothed in oak trees, chaparral, and pasture. Before your eyes, great jets of steam begin to arise, one by one, from scattered points on the valley bottom. The pure white plumes reach upward many tens of feet. What is going on here? Has Nature gone mad?

What you are witnessing is a geothermal energy field. The steam jets issue from wells driven deep into the rock beneath. Normally, the steam is taken by pipes to an electricity generating plant. Periodically, though, the steam is released directly from the well heads to remove accumulated debris. The geothermal field is called The Geysers; it is located about 90 miles north of San Francisco. The steam is a natural phenomenon. Evidently there is some very hot rock not far below the surface. Water saturating the rock is raised to the boiling point. Deeper down, there is undoubtedly a magma body, slowly cooling to become igneous rock.

In this day of energy shortages we are looking to geothermal energy to contribute to the total supply. At The Geysers, total electrical production from a steam-powered generating plant is about 600 megawatts. This is only a little less than the output of a typical large generating plant using coal, oil, or natural gas. The Geysers has one major advantage over the plant burning fossil fuels—there is no air pollution to contend with. Moreover, the energy source will continue its supply for many decades to come.

Opportunities for further development of geothermal power are considered very good for certain parts of the West. Under the Imperial Valley, within a 2000-square-mile area, is an underground reservoir of water heated to over 500° F. It is estimated that this field alone could produce enough electric power to meet the needs of all of southern California for several decades. Water from the condensing steam, after being used in power generating plants, would be a major source of irrigation for crops in the area.

Our subject for this chapter is water beneath the earth's continental surfaces. Only in a few places in the world is this water hot, or even warm. In many regions, fresh water beneath the lands is drawn upon for the community water supply. Underground water is a vital natural resource; but it is more than just that. The slow movement of water beneath the surface performs important geological functions and creates many interesting geological features.

THE HYDROLOGIC CYCLE

The total plan of movement, exchange, and storage of the earth's water is called the **hydrologic cycle.** Water moves from the world ocean to the lands and back, following various paths (Figure 13.1). This water can move and be stored in all three of its natural states: liquid, water vapor, and ice. Most of the world's water, 97%, is held in the oceans. Ice sheets and mountain glaciers hold about 2%; water held in streams and lakes on the continents amounts to only about 0.6%. The amount of water present as vapor in the atmosphere is an extremely tiny fraction of the total, or about 0.001%.

Some idea of the quantities of water passing through the hydrologic cycle each year can be had from Figure 13.2. As the principal reservoir of the earth's water, the oceans form a convenient point at which to start. An estimated 109,000 cu mi of water evaporates annually from the ocean surface, and about 15,000 cu mi evaporates from the lands, including lakes and marshes. Thus a total of 124,000 cu mi of water evaporates, and an equal amount must be returned to the earth's surface annually by precipitation. Of this, about 26,000 cu mi falls as rain or snow upon the land surfaces. These figures show that the quantity precipitated upon the lands is some 73% greater than the amount of water returned to the atmosphere by evaporation from the lands. We conclude that the remaining precipitation on the lands, 11,000 cu mi, or about 46%, is returned annually to the oceans by liquid or glacial flow over and beneath the ground. This flow to oceans is collectively called **runoff.** The most obvious part of this return flow is, of course, by streams emptying into the oceans, but some water seeps into the ground and travels beneath the lands into the coastal ocean waters.

The study of the earth's water and its motions through the hydrologic cycle makes up the science of **hydrology,** which, like geology, is one of the geosciences. Geologists work closely with hydrologists in studying water on and beneath the lands. In fact, the United States Geological Survey is responsible for measuring the flow of streams and investigating the movement and storage of all water lying beneath the ground surface. Water on and beneath the lands is a geologic agent, performing geologic work. The hydrologist is interested in "where water goes"; the geologist in "what water does."

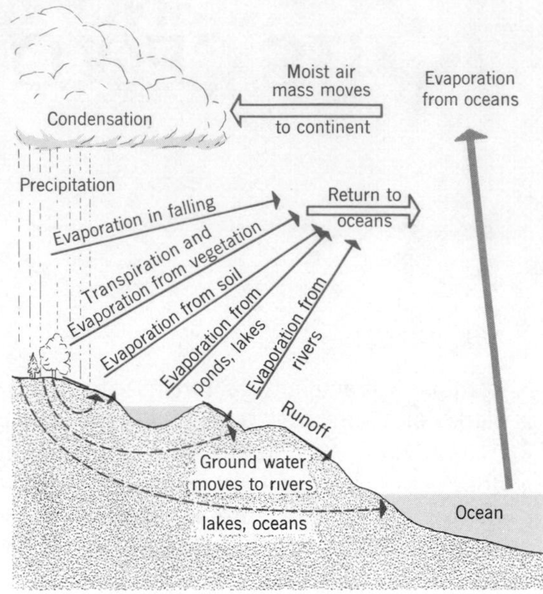

Figure 13.1 The hydrologic cycle. (© 1973, John Wiley & Sons, New York.)

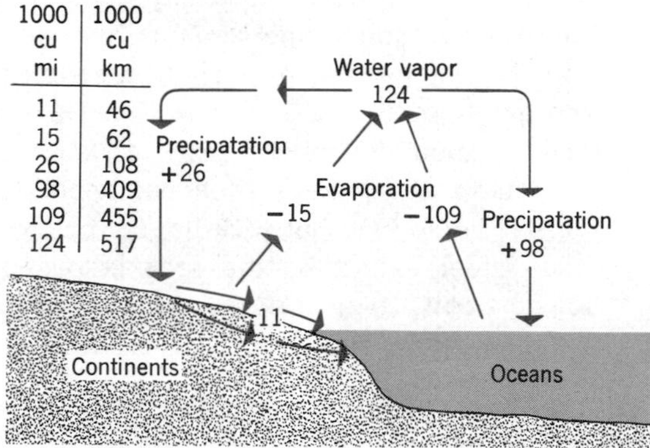

Figure 13.2 Schematic diagram of the world's water balance. (© 1973, John Wiley & Sons, New York.)

SURFACE WATER AND SUBSURFACE WATER

Water that flows on the land surface or lies impounded in lakes and marshes is **surface water.** Water that lies beneath the land surface—enclosed in pores of the soil, regolith, or bedrock—is **subsurface water.**

Most soil surfaces can absorb the water from light to moderate rains and transmit it downward by a process called **infiltration.** Natural passageways are available between individual soil grains and between the larger aggregates of soil. These passages may be soil cracks caused by previous drying; borings made by worms and burrowing animals; openings left by the decay of plant roots or created by the alternate growth and melting of ice crystals. Such openings tend to be kept clear by the protective mat of decaying leaves and plant stems, which also acts to break the force of falling raindrops. When rain falls too rapidly to escape downward through the soil passages, the excess quantity escapes as a surface layer of water following the slope of the ground. This escaping water is known as **overland flow.**

We have a choice now as to whether to trace first the paths of surface water in overland flow and in streams, or the underground paths. Because water emerges from the ground in many places to feed into streams, it will be best to begin with subsurface water.

Water is held in the soil and underlying rock in different ways. For simplicity, let us refer to an idealized body of mineral matter beneath the ground surface. A good model will be a densely packed mass of pure sand (such as beach sand or dune sand) extending downward indefinitely. In such sand about 35% of the bulk volume is open space; the voids are fully interconnected and are large enough to permit movement of water through the mass.

Under typical conditions of a humid climate, there exists an upper zone of soil and rock in which water is held in the form of small films or droplets clinging to mineral surfaces with a force stronger than the force of gravity. This water is described as **capillary water,** because the adhesive force is capillary force. Air occupies the remaining volume of the pore spaces, either as a connected network of air spaces or as separate air bubbles. This subsurface region of capillary water and air is known as the **unsaturated zone** (Figure 13.3). It is geologically significant as a zone in which oxidation of mineral matter can take place. Hydrologists also recognize a soil water belt, a shallow layer holding capillary water within reach of plant roots.

Below the unsaturated zone lies the **saturated zone,** in which all pore space is occupied by water. The upper surface of this zone is the **water table.** Water in the saturated zone is **ground water** and moves slowly in response to gravity.

The capillary state of water above the water table is temporarily destroyed whenever heavy rain or rapidly melting snow allows large amounts of water to infiltrate. At such times the soil openings are fully saturated, and the water moves down under the influence of gravity to reach the water table. In this way the ground-water body is replenished, a process called **recharge.** Most of the excess water drains out of the unsaturated zone soil in a period of days, and the capillary state takes over again.

GROUND-WATER MOVEMENT AND RECHARGE

The position of the water table can be determined by noting the level at which the water surface stands in a well penetrating the saturated zone (Figure 13.3). Where many wells are closely spaced over an area, the configuration of the water table can be shown by connecting the levels of standing water in the wells (Figure 13.4). The water table will be found to be highest in elevation under the hill summits. From these high points, it slopes toward the nearest valleys, intersecting the surface in the channels of streams or at the shores of lakes and marshes.

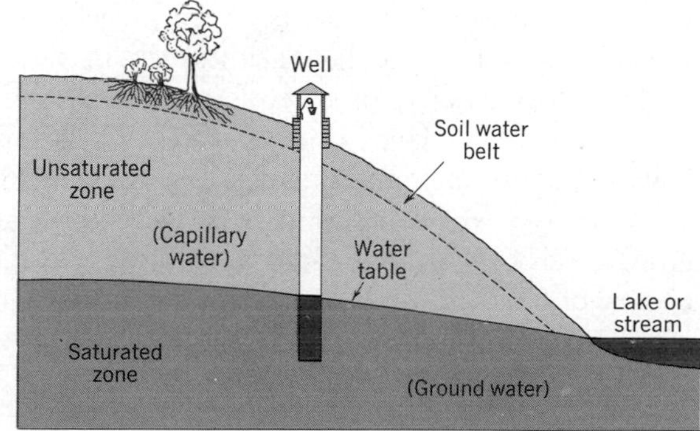

Figure 13.3 Zones of subsurface water.

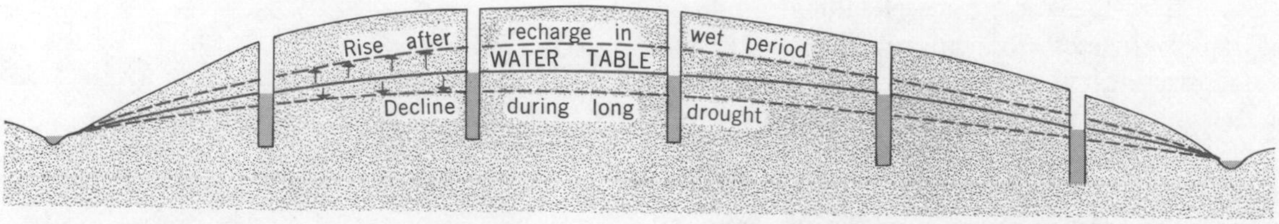

Figure 13.4 Relation of the water table to the ground surface. (© 1973, John Wiley & Sons, New York.)

Water in an open body, such as a lake, assumes a horizontal surface because there is little resistance to flowage. In the saturated zone, however, gravity movement of ground water is through the very tiny spaces between mineral grains, and along thin cracks in bedrock. These narrow spaces impede flowage. Consequently, percolating water reaching the water table under the hill summits cannot escape readily. The ground water tends to accumulate, raising the water table to higher levels than at the streams, where water is escaping. The difference in level between the water table under a summit and at the low point of a valley sets up a pressure called the **hydraulic head.** Under this pressure the water flows very slowly within the ground-water body, following curved paths as shown in Figure 13.5. A particular molecule of water, if it could be traced, might follow a curving path that carries it deep into the rock and returns it by upward flow to the line of a stream channel. Here the water escapes into the stream as surface runoff. Flow velocity is fastest near the line of escape, as suggested by the close crowding of arrows in Figure 13.5.

With a generally constant climate the water table will become approximately fixed in position. The rate of recharge will, on the average, balance the rate at which water is returned to surface flow by seepage in streams, lakes, and marshes. But when there is a period of unusually dry years, the water table slowly falls. During a period of unusually wet years, it will gradually rise (Figure 13.4).

Smaller seasonal fluctuations in the level of the water table are normal in certain climates. A seasonal decline in the water table results from the cutting off of recharge when a dry season occurs or when soil moisture is solidly frozen. A seasonal rise results from recharge by the percolation of excess rainfall or snowmelt through the unsaturated zone.

POROSITY AND PERMEABILITY

Ground water can saturate any type of geologic material, whether it be bedrock or regolith. The bedrock may be of any variety of igneous, sedimentary, or metamorphic rock, and the regolith may range from dense clays to coarse gravels. Consequently, the speed of flow of ground water and the quantity which can be held in the rock are subject to wide variations. It is such variations that concern the ground-water geologist, or **hydrogeologist.** He must apply the laws of flow of fluids to varied and complex geologic conditions.

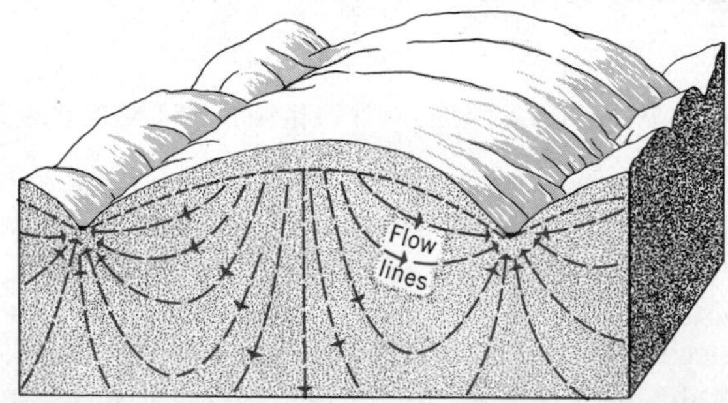

Figure 13.5 Ground water follows strongly curved paths of travel where the subsurface material is uniform throughout. (© 1973, John Wiley & Sons, New York.)

The total volume of pore space within a given volume of rock is termed the **porosity.** It gives an indication of the ability of a rock to hold a fluid in storage. Clastic sedimentary rocks, such as sandstone and conglomerate, can have high porosity because of the relatively large openings possible between the well-sorted, rounded grains (Figure 13.6A). Transported regolith of stream-laid gravels and sands, or beach deposits of hard, well-rounded, coarse mineral grains have a high porosity. Where sands are mixed with silts, the porosity is reduced because the fine particles fit into the openings between the large grains (Figure 13.6*B*).

Soft clays and muds have a high original porosity, but when compacted into shale they have extremely low porosity. Some rock masses have very large openings. Limestone, for example, may have cavernous openings resulting from solution. Scoriaceous lavas have numerous bubblelike cavities formed by expanding gases.

Certain of the very dense rocks, such as igneous and metamorphic varieties, have negligible pore space in the fresh, unweathered state because the mineral crystals are tightly intergrown.

Even though a rock has a high porosity, water cannot move freely through the rock mass unless the openings are interconnected and of sufficient diameter to permit flow. The property of **permeability** is the relative ease with which water will move through the rock under unequal pressure. Permeability is of primary importance in determining the rate of ground-water movement and the amount of water that can be withdrawn by pumping from wells.

Permeability of unconsolidated sands and gravels is extremely high, while that of dense clays and shales is extremely low. In fact, clay and shale can be described as **impermeable rocks** for all practical purposes. Permeability of sandstone formations is commonly high, but may be greatly reduced if the pores are closed by cementation of mineral matter. Igneous and metamorphic rocks, where they are broken by numerous joint fractures and faults, may have high permeability.

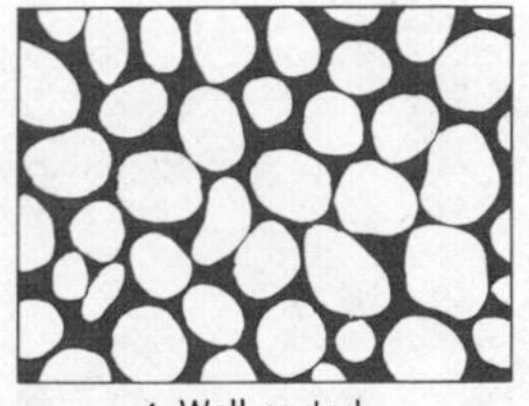

A. Well sorted

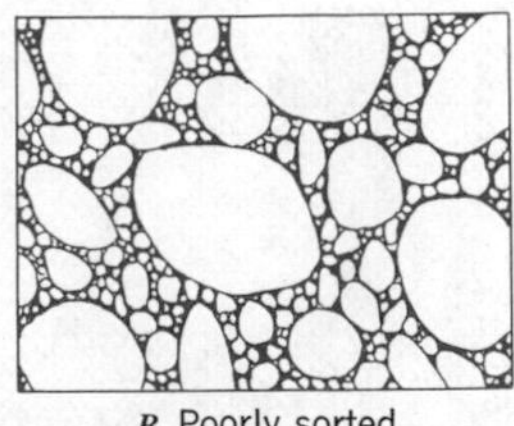

B. Poorly sorted

Figure 13.6 Pore space is relatively greater in a well-sorted sediment than in poorly sorted sediment.

AQUIFERS AND AQUICLUDES

Layered rocks can offer strongly contrasting zones of permeability, particularly where sedimentary strata of varying types are interlayered. In this way the geology of an area exerts a strong control on the movement of ground water. Figure 13.7 shows thick layers of sandstone with a shale bed between.

A sandstone layer, being typically high in porosity and permeability, can hold and transmit large quantities of water; it is designated as an **aquifer.** A shale layer may be almost impermeable, preventing or greatly retarding flow of ground water through itself. Such a layer is called an **aquiclude.**

The shale layer in Figure 13.7 is blocking the downward percolation of water to the main water table below, causing a **perched water table** to be formed in the overlying aquifer. Water from this table emerges in the valley side along a horizontal line in the form of slow seepages and trickles; these are **springs.** The main water table, intersecting the stream channel, is being recharged at more distant places where the overlying aquiclude is absent or broken.

Springs are usually insignificant features, going unnoticed because of a concealing cover of vegetation. Many small springs cease to flow in the summer season. Some large springs yield copious, sus-

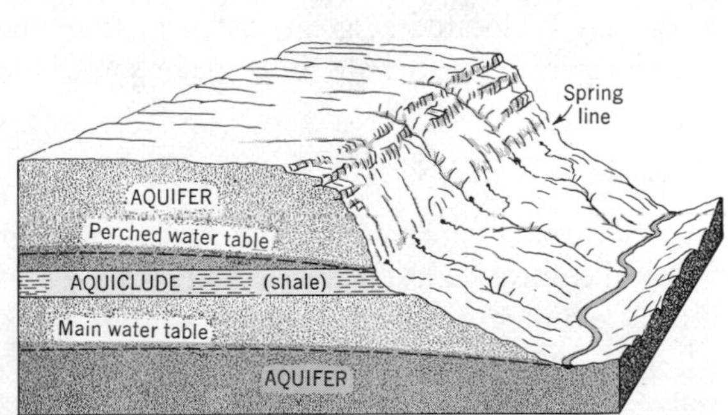

Figure 13.7 A perched water table and spring line. (© 1969, John Wiley & Sons, New York.)

Figure 13.8 Thousand Springs, Idaho. Located on the north side of the Snake River Canyon, nearly opposite the mouth of the Salmon River, these great springs extend for a half mile (0.8 km) along the edge of a layer of scoriaceous basalt. The copious discharge is nearly constant. (U.S. Geological Survey.)

tained flows of water. An impressive example is the Thousand Springs of the Snake River Canyon in Idaho, where a large volume of flow emerges from highly permeable layers of scoriaceous basalt (Figure 13.8).

ARTESIAN FLOW

An interesting type of ground water flow is found in the **artesian well**, in which water is forced upward under natural pressure to rise above the surface of the ground. Geologic conditions necessary to artesian flow are illustrated in Figure 13.9. The aquifer consists of an inclined sandstone layer; it receives water along its exposed outcrop at a relatively high position on a ridge or mountain. Ground water moves through the aquifer and is confined between aquicludes of shale. As a result, ground water beneath the valley is under hydraulic pressure greater than normal for this depth. When a well is drilled into the aquifer under the valley floor, the water is forced to the surface and emerges as a copious flow, continuously maintained by the pressure difference, or head, between the well and the intake area. Natural artesian springs can occur where faulting has produced a natural passageway for water to make its way up through the aquiclude.

Where the sedimentary strata have a very gentle dip, artesian wells may be located at great distances from the intake region. In many regions artesian wells are a major source of water supply.

GRAVITY WELLS

Most water wells are simply tubes or shafts drilled or dug to a depth below the water table. These wells are supplied by gravity flow from the surrounding ground-water zone and are called **gravity wells.** If left undisturbed, the water level comes to rest at the level of the water table. Water must be lifted to the surface. For effective and prolonged use, a well must be protected from collapse by means of a solid casing.

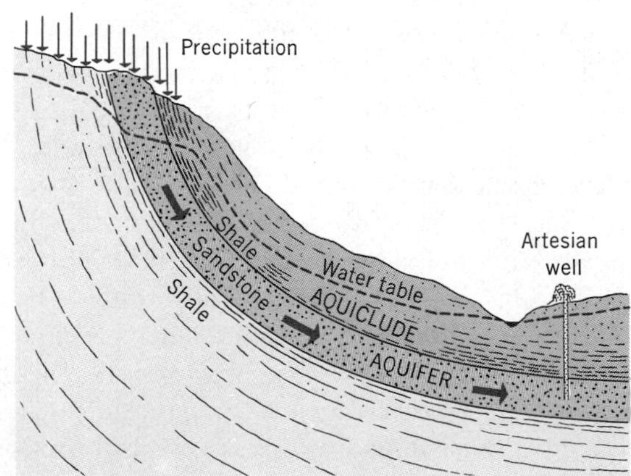

Figure 13.9 Artesian flow requires an inclined aquifer overlain by an aquiclude. (© 1969, John Wiley & Sons, New York.)

Dug wells are usually less than 50 ft (15 m) deep and are effective where the water table lies close to the surface. Most wells for domestic use are drilled and may yield as much as 500 gallons (2000 liters) per minute in a highly permeable deposit. Large drilled wells, which may have diameters up to 18 in. (45 cm), can be driven to depths of over 1000 ft (300 m) and, with powerful pumps, will furnish as much as several million gallons per day if located in highly permeable material.

Pumping of water from a well commonly exceeds the rate at which water can enter the well, so that the water level drops progressively lower (Figure 13.10). As this is done, the water table is drawn down into a conical form, or **cone of depression,** surrounding the well. The **drawdown** is the difference in height between water table and water level in the well. Formation of the cone of depression actually increases the rate at which water flows into the well, and so increases the yield of the well, but this effect is limited. Beyond a critical limit of drawdown the yield no longer increases.

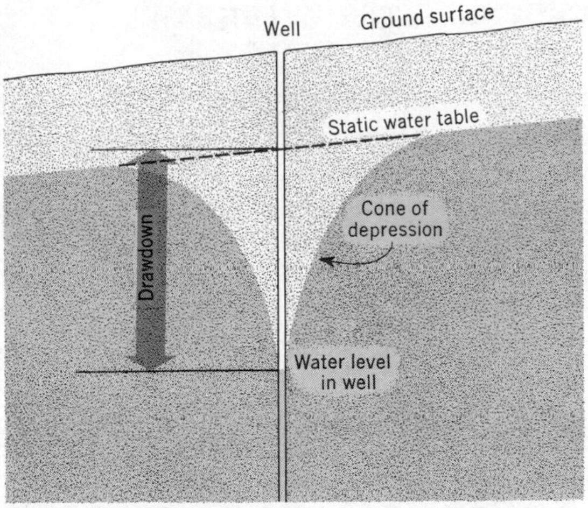

Figure 13.10 Rapid pumping of a well causes drawdown and a cone of depression. (© 1973, John Wiley & Sons, New York.)

The cone of depression may extend several miles from a large well, and the combined effect of closely spaced wells is to depress the water table generally. Continued heavy pumping can eventually lead to a serious decline in water yield. In arid regions, particularly, the ground water stored in valley deposits can be withdrawn much faster than it is restored by seepage from the beds of surface streams. In humid regions, careful regulation of pumping and the artificial recharge of the ground-water table by waste water, pumped down into recharge wells, can bring about a balance in the rates of withdrawal and recharge.

Ground-water supplies are subject to pollution by various kinds of man-made wastes that infiltrate the ground and become a part of ground-water recharge. Discharge of sewage and liquid industrial wastes into infiltrating basins forms one potential pollution source, while landfill waste disposal sites (dumps) are arother. In both cases, because of the abnormally high rate of infiltration, the water table is built up into a mound. From this high point the infiltrating fluid moves outward to surrounding parts of the ground-water body where it may emerge in streams or lakes, or may be drawn into nearby wells (Figure 13.11).

Another common ground-water pollutant is road salt, which is used on highways and city streets in winter, and can pollute shallow wells close to the roadways. Spills of petroleum fuels and chemicals from highway accidents and from leaking storage tanks are yet another source of ground-water pollution.

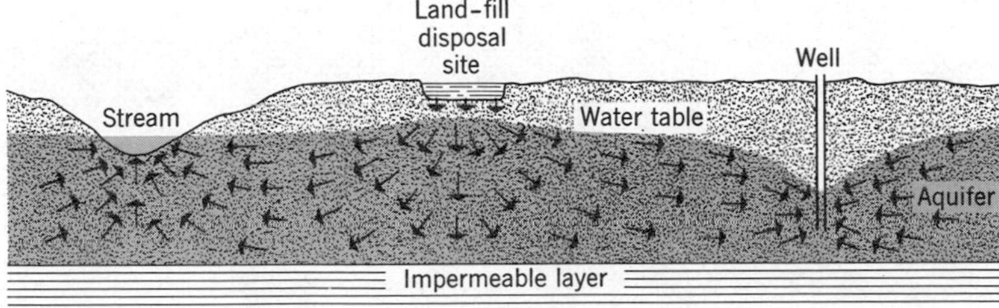

Figure 13.11 Pollution from a solid-waste disposal site can move within the ground water beneath to reach a well or a stream.

GROUND-WATER WITHDRAWAL AND LAND SUBSIDENCE

One serious environmental effect of excessive ground water withdrawal is land subsidence. Recall that this effect has been induced in oil fields by withdrawal of petroleum (Chapter 11). Land subsidence by ground-water withdrawal is particularly severe where water is pumped from thick beds of unconsolidated sediment. These are beds of gravel, sand, and silt lying beneath broad valley floors. As the water table declines, the silty sediments which come into the unsaturated zone become compacted under the load of overlying layers.

A now-classic example of subsidence is that of the San Joaquin Valley in California. Over a period of many years, the water table beneath the valley has been drawn down over 100 ft (30 m). Land subsidence has been as much as 10 ft (3 m) in a 35-year period and has resulted in damage to wells in the area. Water withdrawal beneath the city of Houston, Texas, has resulted in subsidence of from 1 to 3 ft (0.3 to 1 m) over an area 30 mi (50 km) in diameter, with serious damage to pavements, building foundations, airport runways, and flood-control works. Minor faults have been activated by the subsidence and can be traced through the city. Mexico City has suffered even greater subsidence—from 13 to 23 ft (4 to 7 m) in the period between 1891 and 1959—as a result of ground-water withdrawals from aquifers beneath the city.

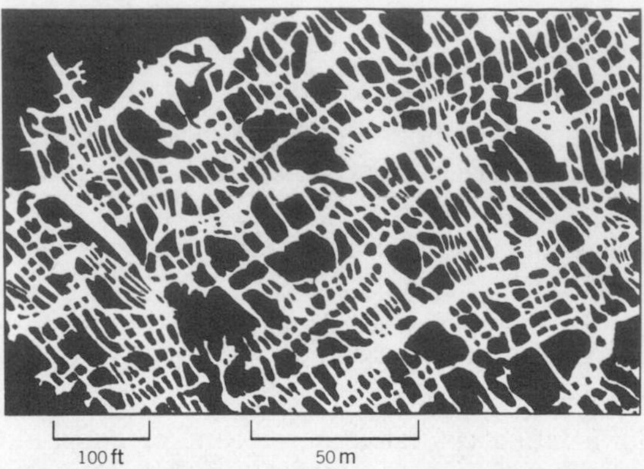

Figure 13.12 In this map of a portion of Anvil Cave, Alabama, solid rock is shown in color. (Courtesy of W. W. Varnedoe and the Huntsville Grotto of the National Speleological Society.)

LIMESTONE CAVERNS

Limestones, composed of carbonate of calcium and magnesium, are highly susceptible to the action of carbonic acid found in rainwater and ground water. As we noted in Chapter 12, this acid action is called carbonation. Its effects are obvious in the etched surfaces of exposed limestone masses. Lowering of the land surface by carbonation is a comparatively rapid process in areas underlain by limestone or marble where rainfall is abundant. As a result, carbonate rocks typically form valleys and lowlands in humid climates.

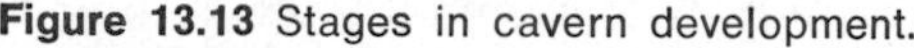

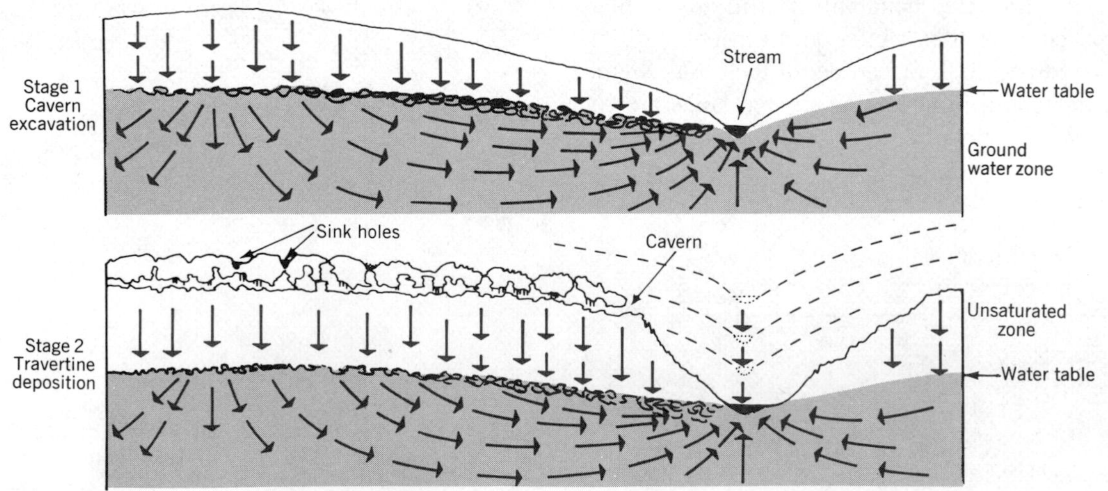

Figure 13.13 Stages in cavern development.

Figure 13.14 Stalactites and stalagmites. Indian Echo Cave, near Hummelstown, Pennsylvania. (D. L. Babenroth.)

Limestone caverns consist of subterranean passageways and rooms forming complex interconnected systems. Patterns of joints in the limestone exert a strong control over the network of passages (Figure 13.12). Variations in composition among individual limestone beds also help to control size and spacing of passages.

Most cavern systems are excavated in the ground-water zone. As shown in Figure 13.13, water percolating downward in the unsaturated zone reaches the water table. This water moves as ground water with extreme slowness along curved paths indicated by the arrows. Carbonic acid reaction with limestones is concentrated in the uppermost part of the ground-water zone; here passageways are enlarged. The dissolved carbonate matter is carried into streams and out of the region.

In a second stage, shown in Figure 13.13, the stream has deepened its valley and the water table has been lowered accordingly. New caverns are being formed at the lower level, while those previously formed are now in the unsaturated zone. Here percolating water is exposed to the air on ceilings and walls of the caverns. The carbonic acid reaction is reversible, so that under favorable conditions calcium carbonate is precipitated as a mineral deposit.

Calcium carbonate accumulated on the inner surfaces of caverns occurs as **travertine,** a banded form of calcite. Encrustations forming where water drips from cave ceilings are called **dripstone,** while encrustations made in moving water of pools and streams on cave floors are **flowstone.** Dripstone and flowstone accumulate in elaborate shapes known popularly as "formations," that give many caves their great beauty.

Some scenic features of caverns are illustrated in Figure 13.14. From ceiling points where the slow drip of water takes place, spikelike forms called stalactites are built downward. From points on the cavern floor receiving a steady drip of water, postlike columns called stalagmites are built upward. Stalactites and stalagmites often join into columns, and the columns forming under a single joint crack may fuse into solid walls. Growth downward below a joint crack may produce a drip curtain. Blocks of limestone fallen from the ceiling impound the runoff along the cavern floor, making pools in which travertine terraces are formed.

After cavern development has been in progress for a long period of time the land surface above is deeply pocked with depressions. These are parts of the old cavern system lying close to the surface. Such depressions are called **sinkholes** (Figure 13.15). They may be partly filled with clay soil and can hold ponds or marshes, or may be cultivated if the soil is well drained. Some sinkholes are gaping holes leading down to open caverns beneath.

Figure 13.15 A sinkhole about 60 ft (18 m) deep on the Kaibab Plateau, Arizona, elevation 8500 ft (2500 m). Horizontal limestone strata outcrop on the far wall of the depression. (A. N. Strahler.)

GROUND WATER AS A GEOLOGIC AGENT

Cavern systems are spectacular examples of the action of groundwater, which usually performs its important geologic work with more ordinary results. This work consists of removal and transport of mineral matter in solution throughout large rock masses. Cementation of coarse clastic sediments into solid rock depends upon ground water movement, which brings in silica or calcium carbonate, depositing it in pore spaces.

A common phenomenon of ground-water action is **mineral replacement,** in which molecules of one compound are removed, to be replaced by molecules of another compound. Fossils often show the effects of mineral replacement. A good example is petrified wood. Here the original wood fibers were gradually replaced by silica, retaining all details of the cell structure of the tree trunk.

Another commonplace example of the geologic action of ground water is the **geode,** a rock mass with a spherical cavity lined with mineral crystals (Figure 13.16). Geodes commonly occur in limestones and shales. They can be explained by a two-stage development. First the cavity is excavated by removal of rock in solution in ground water. Then the cavity is filled by mineral matter carried in by ground water. These minerals—commonly quartz, calcite, or fluorite—accumulate very slowly and form perfect crystals, pointing inward toward the center of the geode.

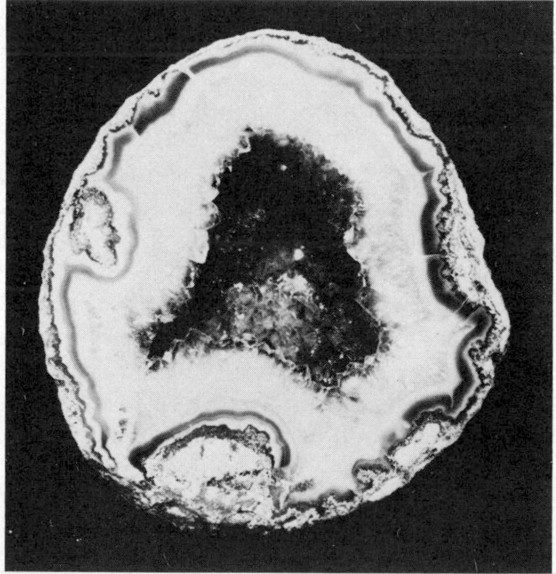

Figure 13.16 This geode has a rim of agate and is lined with quartz crystals. (Ward's Natural Science Establishment, Inc., Rochester, N.Y.)

HOT SPRINGS, GEYSERS, AND FUMAROLES

At a number of widely separated localities, ground water emerges in **hot springs,** at temperatures not far below the boiling point of water, which is 212° F (100° C) at sea-level pressure. In certain of these localities, periodic jetlike emissions of steam and hot water occur from small vents and are known as **geysers** (Figure 13.17). The heated water emitted by hot springs and geysers is largely ground water that has been heated by contact with rock. The rock itself is heated by rising hot gases released from magmas at depth.

Vents which emit volcanic gases are known as **fumaroles.** Gas temperatures at fumaroles are often far above the boiling point of water and have been measured as high as 650° F (320° C). Hot

Figure 13.17 The Waikite Geyser, Rotorua, North Island, New Zealand. (New Zealand Tourist Bureau.)

springs, fumaroles, and geysers were the source of the earth's atmosphere and hydrosphere by outgassing during early geologic time (Chapter 10). Most of the emission from these sources (over 99%) is water. The remaining gases are largely carbon dioxide, sulfur, nitrogen, chlorine, fluorine, and hydrogen.

As we found in Chapter 11, the superheated steam of fumaroles is a valuable source of energy—geothermal energy. Most of this steam and its associated gases are believed to be volcanic in origin, in contrast with hot springs, which consist largely of recirculated ground water. Probably only 1% or less of the water in hot springs and geysers is water of magmatic origin.

Because heated ground water is highly active chemically, it dissolves unusually large amounts of mineral matter, most of which is silica or calcium carbonate, depending upon the composition of the bedrock through which the solutions move. Upon reaching the surface, the hot water is rapidly cooled and must precipitate much of the dissolved mineral matter. In this way encrustations are built up close to hot springs and gradually spread laterally in flat-topped terraces, which may be stepped, one above the other. Terraces formed of silica, known as **siliceous sinter,** are typical of hot springs in which water rises through igneous rocks. Cones of siliceous sinter are built around the orifices of geysers.

Where limestone bedrock furnishes calcium carbonate, the deposits of hot springs are of travertine. An example is the Mammoth Hot Springs, pictured in Figure 13.18. Certain algae thrive in the pools of hot water, and these may also precipitate calcium carbonate.

GROUND WATER AS BOTH A RESOURCE AND A GEOLOGIC AGENT

Subsurface water plays divergent roles in geoscience, as we have found in this chapter. As a reservoir of fresh water, ground water is a resource of enormous value to man. Unfortunately, the supply of

Figure 13.18 This photograph of Mammoth Hot Springs, Yellowstone Park, Wyoming, was taken in 1870 by the pioneer photographer, W. H. Jackson. (U.S. Geological Survey.)

ground water can too easily be withdrawn more rapidly than it can be recharged.

In arid lands, particularly, ground water is best described as a nonrenewable mineral resource being "mined" with reckless abandon for irrigation and urban use. Rapidly falling water tables in these areas of overdraft are a recorded fact, which goes largely unheeded because people cannot see the water table with their own eyes. Ultimately these depleted ground water supplies will fail.

In humid regions of abundant rainfall, ground water is copiously recharged by nature, and a substantial proportion can be withdrawn on a year-to-year basis with relatively little environmental damage. In these regions, ground water is a renewable resource, when its use is carefully regulated and its quality protected.

The geologic role of ground water as a cementer of rock and a former of ore deposits is an inconspicuous activity, carried out with incredible slowness over long periods of geologic time. As a subterranean sculptor of grotesque and beautiful cavern forms, subsurface water plays out a still different role. Finally, emerging waters of hot springs and geysers deposit their mineral load to construct surface forms unlike anything else seen in nature.

In the next chapter we go back to surface water. We will be particularly interested in the geologic role of streams, as agents of erosion, transportation, and deposition.

YOUR GEOSCIENCE VOCABULARY

hydrologic cycle
runoff
hydrology
surface water
subsurface water
infiltration
overland flow
capillary water
unsaturated zone

saturated zone
water table
ground water
recharge
hydraulic head
hydrogeologist
porosity
permeability
impermeable rock
aquifer
aquiclude
perched water table
spring
artesian well
gravity well
cone of depression
drawdown
limestone caverns
travertine
dripstone
flowstone
sinkhole
mineral replacement
geode
hot springs
geysers
fumaroles
siliceous sinter

SELF-TESTING QUESTIONS

1. Describe the workings of the hydrologic cycle. In what forms is global water held in storage, and in what relative proportions? Compare the annual total precipitation over the continents with the total evaporation from the same areas. What happens to the excess?

2. Differentiate between surface water and subsurface water. Under what conditions does overland flow occur?

3. Describe the zones of subsurface water and explain how water moves in each zone.

4. What causes ground water to move through rock? Describe the usual configuration of the water table beneath hills and valleys in a humid region. How is the water table influenced by drought? by seasonal cycles of precipitation?

5. Distinguish between porosity and permeability, illustrating each concept with specific types of rock and regolith. What kinds of geologic materials are impermeable, or nearly so?

6. How does the layered arrangement of rocks affect the movement and accumulation of ground water? How do springs originate? Explain how artesian flow occurs.

7. Describe the extraction of water from gravity wells. What effects are produced in the water table by rapid pumping? Describe the forms and processes of ground-water pollution. How can land subsidence result from withdrawal of ground water? Give two examples.

8. How are limestone caverns formed? What is their relationship to ground water and the water table? Describe some of the forms of travertine deposition in caverns.

9. What geologic work does moving ground water perform? Give examples.

10. What is the source of heat for hot springs and geysers? What is the typical composition of steam from a fumarole? What is the source of water for these phenomena?

In this 1910 photograph, steam locomotives labor to haul a freight train up the steep grade of the Royal Gorge of the Arkansas River. (A. K. Lobeck.)

14 THE GEOLOGIC WORK OF RUNNING WATER

THE GREAT RAILROAD WAR

The winning of the West involved a lot of shooting. One notable shooting incident took place over a deep canyon, and the rights of a railroad to use it. This was the Great Railroad War waged between 1876 and 1880 for possession of the Royal Gorge of the Arkansas River in central Colorado. The gorge is only a few miles long, but it is over 1000 feet deep, with near-perpendicular walls of granite. The river at the bottom runs in rapids over a bouldery bed, with scarcely room for a man to stand beside it, let alone for a railroad line.

The contestants in the Great Railroad War were the Denver & Rio Grande and the Atchison, Topeka & Santa Fe, the latter an outfit of unlimited resources, financed by dudes from Boston. The Rio Grande was a local outfit, the brainchild of Gen. William G. Palmer. He planned to open up the Rocky Mountain region with a new system of railroad lines.

The Royal Gorge, or Big Canyon, as it was known locally, was to be used by General Palmer's line as a means of crossing the Colorado Front Range, a formidable barrier in the path of east-west travel. Meantime, the Santa Fe people had designs on the Royal Gorge for their own railroad. They planned a mass attack upon the gorge so as to occupy it before the Rio Grande could move and defend its property. Through the telegraph, which the Rio Grande luckily controlled, the attack plans leaked out. To forestall its attackers, the Rio Grande made plans to start grading the line through the gorge on April 20, 1878. But when the general manager of the Santa Fe got wind of this strategy, he sent his civil engineer to occupy the gorge immediately with a survey and grading party. Because the Rio Grande wouldn't let the engineer ride on their trains, he had to ride horseback to the mouth of the gorge at Canyon City. His exhausted horse fell dead within three miles of Canyon City, and he ran the rest of the way. Once there, he recruited a force of several hundred armed men, led them to the mouth of the gorge, and fortified the position. He had popular support because the local citizenry hated the Rio Grande folks and were glad of a chance to get even.

A long period of sporadic armed conflict set in, with shootings and bloodshed. The record is not specific about engagements and casualties, so we can't say much more. It's reported that Rio Grande engineers and graders descended into the canyon by ropes, to stand waist-deep in ice-cold water as they surveyed the grade and began blasting the hard granite. Even mules and carts were lowered by rope, so that construction camps could be set up in the bottom of the gorge. With both companies trying to work at the same time in these narrow confines, it is little wonder that many hand-to-hand fights broke out and many shots were exchanged.

The Great Railroad War finally came to an end in 1880, when the Federal Supreme Court ruled in favor of the Rio Grande. We are told that the legal battles involved some of the finest lawyers of the country and that "the encounters in the field were marked by deeds of heroism and bloodshed that were worthy of a better cause."* In this chapter we will learn how rivers carve great gorges like the Royal Gorge and the Grand Canyon.

* M. R. Campbell, 1922, *Guidebook of the Western United States,* Part E, Bulletin 707, U.S. Geological Survey, U.S. Govt. Printing Office, Washington, D.C. This publication is the source of the account given here.

FLUVIAL DENUDATION

To the geologist, streams are much more than just mechanisms for the disposal of runoff from the lands; they are major agents of land sculpture. Streams create a vast array of erosional and depositional landforms. Streams also transport sediment and deposit it in basins and shallow seas, where it can be transformed into sedimentary rock. Stream action, in combination with weathering, mass wasting, and overland flow, is responsible for a total process we call **fluvial denudation.** It is the process responsible for creating most landscapes seen on the surfaces of the continents. True, glacial ice is a dominant agent in high mountains, wind creates conspicuous forms in a few desert and coastal localities, and wave action shapes the shorelines. But from the standpoint of total surface area affected, fluvial denudation is the predominant landscape former.

OVERLAND FLOW AND CHANNEL FLOW

Running water as a geologic agent acts in two basic forms. First is **overland flow,** the movement of runoff downhill on the ground surface in a more or less broadly distributed sheet or film. Second is **channel flow,** or **stream flow,** in which water moves to lower levels in a long, narrow, troughlike feature called a **stream channel.** This channel is bounded on both sides by rising slopes, called **banks;** these contain the flow. A **stream** is a body of water flowing within a channel, and includes all sediment carried by the water. Overland flow starts near the hill summits and converges upon the stream channels, supplying them with both water and sediment (Figure 14.1).

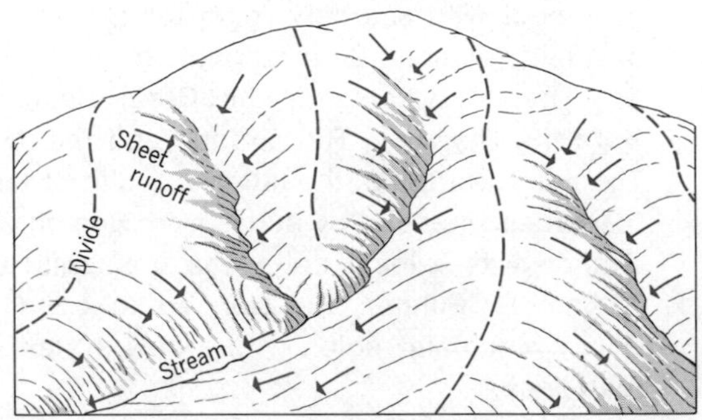

Figure 14.1 Stream channels are fed by overland flow. (© 1973, John Wiley & Sons, New York.)

EROSION BY OVERLAND FLOW

When overland flow is taking place during a heavy rainstorm, or when snow is melting rapidly, the sheet of water moving downhill exerts a drag force over the ground surface. Progressive removal of mineral grains by this force, together with downhill transport, is called **soil erosion.**

A good plant cover, particularly a grass sod, breaks the force of falling raindrops and absorbs the energy of the overland flow, reducing the rate of soil erosion. Even under heavy and prolonged rains, a thickly vegetated hillside yields very small quantities of mineral solids. In contrast, the barren slopes of a desert landscape or the unprotected surface of a cultivated field will produce large quantities of sediment with each rainstorm.

Steepness of the ground surface also strongly influences the rate of soil removal by overland flow. On steep hillsides the flow moves swiftly, and its power to erode and transport is much greater than on a gentle slope.

NORMAL AND ACCELERATED EROSION

Erosion of the soil is a process of nature by which many landscape features of the earth's surface are slowly carved. For a geologist, soil erosion is a normal geologic process associated with the hydrologic cycle. In humid climates, where vegetation is dense, soil erosion is normally very slow, and the natural soil profiles are maintained as erosion proceeds. In arid climates, the normal rate of erosion of soft rocks, such as shale, is rapid and furnishes large quantities of sediment to streams. We can conclude that there is a **geologic norm** of soil erosion appropriate to the particular conditions of climate and bedrock prevailing in an area.

Where man has cut the forests and converted the land to agricultural uses, or where a fire has destroyed the vegetation, there may be a sudden large increase in the erosion rate, producing a condition known as **accelerated erosion.** Soil is usually removed in thin, uniform layers by a process termed **sheet erosion.** Soil horizons are removed at a much faster rate than they can be formed, and there is a rapid decline in fertility of the soil. Eventually only the infertile regolith or bedrock remains. Streams become burdened with quantities of sediment far in excess of the normal amounts to which their courses have been adjusted.

Soil removed in sheet erosion is carried to the base of the slope. Here, much of it accumulates in thin layers to form a deposit known as **colluvium.** Some particles will, of course, be carried into stream channels, to be deposited as alluvium on the valley floor. **Alluvium** is a broad term for any stream-laid sediment deposit. **Valley sedimentation** is the term applied in agricultural engineering studies to the accumulation of both colluvial and alluvial deposits as a result of accelerated soil erosion. Sedimentation can bury fertile agricultural land under a coarse, permeable layer unfit for cultivation.

Figure 14.2 Shoestring rills on a steep, barren slope, Ventura County, California. At the left, weeds are taking hold to form a protective cover. (Soil Conservation Service, U.S. Department of Agriculture.)

On steep slopes laid bare of vegetation, intense runoff forms long narrow channels called **shoestring rills** (Figure 14.2). These rills merely score the surface. They are typically seasonal features, obliterated by freeze and thaw in winter or by plowing of the land. Shoestring rills can deepen to form **gullies** of awesome proportions (Figure 14.3). Gully development is particularly striking in regions underlaid by a thick layer of saprolite, wind-transported silt (loess), or soft shale bedrock.

We should not lose sight of the fact that our natural soil is a nonrenewable earth resource. A fertile soil takes many thousands of years to form, but only a few years to destroy by accelerated erosion.

STREAM CHANNELS AND STREAM FLOW

Stream channels range in size from insignificant brooks one can step across in a single stride, to the trenches of great rivers hundreds of feet wide. Many channels of desert streams are dry most of the time, yet they bear the unmistakable markings of rapidly flowing water and are occupied by raging torrents on the rare occasions of flood. Even in a humid region, small channels normally become dry in the summer season, when the water table falls so low that no ground water can seep into the stream.

An essential feature of every stream is that it descends to lower elevations along its length. The rate of this descent is the **stream gradient.** Of course, when you examine a stream bed in detail it will usually prove to have deeper places, called pools, separated by shallower places, called riffles. Because of these irregularities, the bed itself can locally rise in level in the downstream direction. How-

Figure 14.3 A great gully system in Stewart County, Georgia, in 1936. Such severe gullying has now been largely controlled. (Soil Conservation Service, U.S. Department of Agriculture.)

ever, the water surface must have a downstream gradient at all points, or flow would not be possible.

A stream moves downgrade in its channel because a fraction of the force of gravity acts in a downstream direction, parallel with the bed. The water responds freely to this force, each water layer moving over the layer beneath it. **Stream velocity** is the speed of downstream water flow, as measured at any selected point above the bed. Because of friction with the bed and banks, stream velocity varies from zero in contact with the bed to a maximum in midstream, some distance above the bed.

Arrows in the lower part of Figure 14.4 show the tracks that would be taken by water particles starting out together on a vertical line. We see that velocity increases very rapidly from the bed upward, then decreases, but less rapidly, so that the maximum velocity is found at a point about a third of the distance from the surface. The upper part of the figure shows that on the stream surface, velocity increases from zero at the banks to a maximum near the center line.

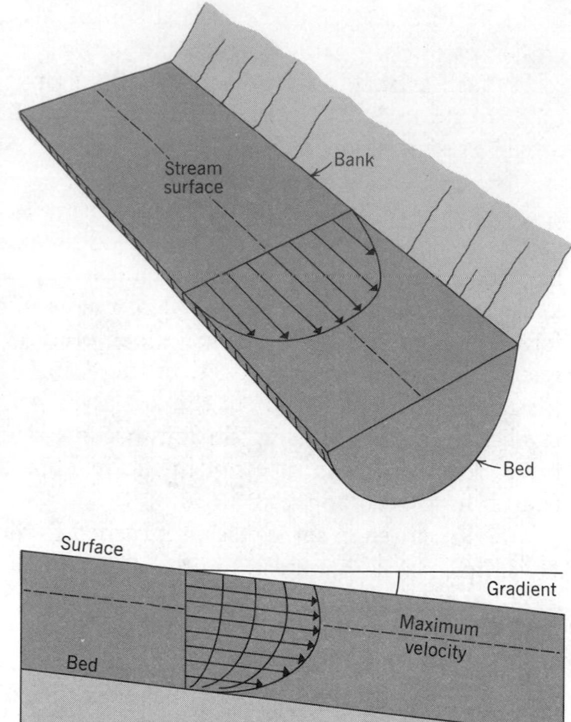

Figure 14.4 Speed of water flow in a stream channel is fastest near the center. Velocity is proportional to the length of the arrow.

TURBULENCE IN STREAMS

Statements we made in the previous paragraph imply that each particle of water moves downstream in a direct simple path. That would be the case in true **streamline flow,** which occurs in fluids when their motion is very slow. In most forms of runoff, including most overland flow and nearly all stream-channel flow, the water particles follow highly irregular paths of travel. Their twisted corkscrew motion includes sideways and vertical movements. Such motion is called **turbulent flow.** It consists of innumerable eddies of various sizes and intensities continually forming and dissolving. Velocity, as we referred to it earlier, and the simple paths of flow shown by the parallel arrows in Figure 14.4, are merely the average velocities and average paths of the particles at given levels in the stream.

Turbulent flow in fluids is of vital importance in the process of stream erosion, because the transportation of fine particles held suspended in the fluid depends upon the upward movement of currents in turbulence to support the particles. Without turbulence, particles could only be rolled or dragged upon the bed, or lifted a short distance above it.

The quantity of water that flows through a stream channel in a given period of time is the **discharge.** Discharge is defined as the volume of water passing through the stream cross section each second, and is commonly measured in cubic feet per second.

STREAM EROSION

Streams perform three closely interrelated forms of geologic work: erosion, transportation, and deposition. **Stream erosion** is the progressive removal of mineral matter from the surfaces of a stream channel. The material of the channel may consist of bedrock, residual or transported regolith, or soil. **Stream transportation** is the movement of eroded particles in chemical solution, in turbulent suspension, or by rolling and dragging along the bed. **Stream deposition** consists of the accumulation of any transported particles on the stream bed, on the adjoining floodplain, or on the floor of a body of standing water into which the stream empties. These phases of geologic work cannot be separated from each other, because where erosion occurs, there must be some transportation, and eventually the transported particles must come to rest.

The nature of stream erosion depends upon the materials of which the channel is composed and the means of erosion available to the stream. One simple means of erosion is **hydraulic action,** the pressure and drag of flowing water exerted upon grains projecting from the bed and banks. Weak bedrock and various forms of regolith are easily carved out by hydraulic action alone, but the process has little effect on strongly bonded bedrock.

Hydraulic action is the dominant process of stream erosion in weak alluvial deposits. In flood stage the swift, highly turbulent flow undermines the channel wall, causing masses of sand, gravel, silt, or clay to slump and slide into the channel. This activity is **bank caving.** Huge volumes of sediment are incorporated into the stream flow in times of high stage, and the channel can shift laterally by many yards in a single flood (Figure 14.5).

Figure 14.5 Hydraulic action by a stream in flood removed glacial sands and gravels from an area 1 mi (1.6 km) wide and 3 mi (5 km) long, destroying eight large farms and leaving the trench shown here. Cavendish, Vermont, November 1927. (Wide World Photos.)

Mechanical wear, termed **abrasion,** occurs when rock particles carried in the current strike against the exposed bedrock of the channel surfaces. Small particles are further reduced by crushing and grinding when caught between larger cobbles and boulders. Chemical reactions between ions carried in solution in stream water and exposed mineral surfaces result in a form of erosion called **corrosion.** It is essentially the same as chemical rock weathering.

Abrasion of hard-rock channels yields a variety of minor erosional forms such as chutes, plunge pools, and a type of cylindrical pit known as a **pothole** (Figure 14.6). The pothole is deepened by a spherical or discus-shaped stone, a grinder, rotated by the force of corkscrew water currents in the cylinder.

STREAM TRANSPORTATION

We can distinguish three forms of stream transportation of mineral matter. First, corrosion yields dissolved solids. These can travel downstream indefinitely and may reach the ocean. They do not affect the mechanical behavior of the stream.

Second, particles of clay, silt, and sometimes fine sand are carried in **suspension.** In this form of transport the upward currents in eddies of turbulent flow are capable of holding the particles indefinitely in the body of the stream. Material carried in suspension is referred to as the **suspended load.** It constitutes a large share of the total load of most streams. Clay particles, once lifted into suspension, are so readily carried that they travel long distances. Silts settle rapidly when turbulence subsides. Coarse sands are rarely transported in suspension except in floods. As a result, suspension provides a means of separating solid particles of various sizes and carrying each size category to a different location, a process known as **sorting.**

Figure 14.6 These potholes were carved into granite in the channel of the James River, Henrico County, Virginia. (U.S. Geological Survey.)

Third of the modes of transportation is that of rolling or sliding of grains along the stream bed. These motions can be conveniently included in the term **traction.** Fragments moved in traction are

Figure 14.7 This bouldery debris represents the bed load of a desert stream, swept down-valley on the rare occasions when the narrow channel is in flood. Gonzales Pass Canyon, Arizona. (Mark A. Melton.)

referred to collectively as the **bed load** of the stream (Figure 14.7). Traction results both from the direct pressure of the water flow against the upstream face of a grain and from the dragging action of the water as it flows over the grain surface. In bed-load movement, individual particles roll, slide, or take low leaps downstream, then come to rest among other grains.

LOAD OF STREAMS

We measure a stream's load in terms of the weight of sediment moved past a fixed cross section in a unit of time; for example, tons of sediment per day. Geologists refer to the maximum load a stream can carry as the **stream capacity.** The increase in load with increase in discharge is very striking. In a typical case, a 10-fold increase in discharge brings a 100-fold increase in load. Increasing discharge increases the stream velocity; an increase in velocity increases the capacity. Obviously, the great bulk of stream transportation occurs at high stages, including floods, while little is moved at low stages. Many rivers which are very turbid (murky) at high stage are quite clear at low stage, with only a small amount of sand being dragged along the bed.

As you would expect, streams differ greatly in the typical loads they carry. Given two rivers of the same discharge, the river that drains a forested region will have much less solid load than one draining a semiarid region with extensive areas of bare soil. Accelerated soil erosion brought about by poor crop cultivation practices causes an enormous increase in sediment load.

As the load of a stream first increases and then decreases, accompanying a cycle of stream rise and fall, the stream channel also undergoes important changes. During a rising stage, caused by an increase in discharge, the channel is scoured and deepened. Because the water is very turbid, we can't see this deepening taking place. As the discharge decreases and the water surface becomes lower, deposition occurs on the channel floor, building up the level of the bed. Sediment alternately removed and redeposited on the channel floor is one of the common forms of alluvium. This alluvium is said to be **reworked** by successive rising and falling of stream level. Of course, this material is also moved down-valley as it is reworked.

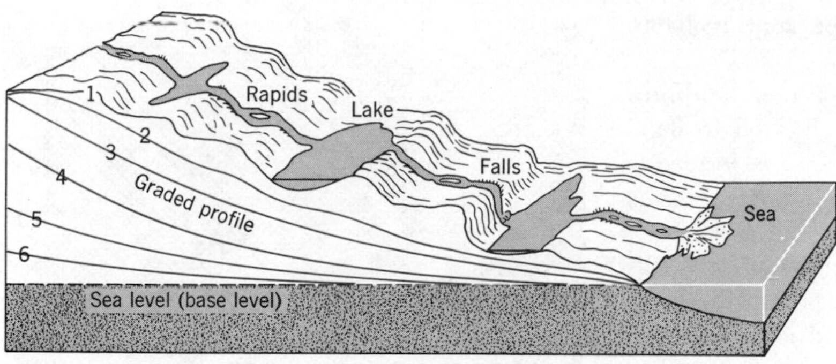

Figure 14.8 The irregular profile of a stream becomes graded into a smoothly upconcave form after the original falls and rapids have been removed. With the passage of time, the profile is lowered and gradually approaches baselevel.

STAGES OF STREAM GRADATION

The way in which a stream shapes its landforms is nicely illustrated by a model life history of stream development. We can imagine that this history starts with a newly formed block of land, which we will call a **landmass.** For example, a section of the smooth continental shelf, rapidly upraised from beneath the sea and faulted into blocks, makes a good starting landmass for our purpose. As shown in the upper surface of Figure 14.8, this landmass consists of closed depressions and steplike alternations of steep and gentle slopes.

Surface runoff immediately organizes itself into a crude and inefficient drainage system. The flow path consists of a series of lakes connected by narrow streams passing over the steep fault steps in a succession of falls and rapids.

Intense channel abrasion is concentrated at those points where the stream passes over steep **waterfalls.** Here, narrow **gorges,** or **canyons,** are quickly formed. As the rock barriers are reduced, lakes are lowered and finally drained. Although the stream now has a continuous, narrow channel, it has many rapids along its course. The transporting capacity of the stream exceeds the sediment load available, and so the stream channel consists largely of exposed bedrock. Abrasion continues rapidly and the channel is lowered, causing the steep-walled rock gorge to be deepened. The result may be a canyon of spectacular proportions (Figure 14.9). Weathering, sheet erosion, and mass wasting act upon the rock walls to widen the canyon. Rock fragments that roll, slide, or are washed down to the stream are swept away as suspended and bed load.

As time passes, many new stream branches are formed and feed into the main stream. As these branches cut their valleys, new sources of sediment are created and the total load of the main stream steadily increases. At the same time the gradient of the main stream is becoming less steep, so that the stream's capacity to transport load is becoming less. Obviously it is only a matter of time before the increasing supply of coarse sediment being fed into the stream matches the stream's capacity for bed load transportation.

This point in the life history of the stream is a most important one, for when the stream is receiving and transporting sediment to the limit of its capacity, the period of rapid channel downcutting comes to an end and the channel is said to be **graded.** The stream has now become a **graded stream.** (Level 3 of Figure 14.8.) Rapids will have been removed by abrasion and the channel will have formed a smooth gradient throughout its length. A layer of alluvium

Figure 14.9 The Grand Canyon of the Yellowstone River, viewed from over Inspiration Point. The canyon is carved into the surface of a lava plateau. (U.S. Army Air Service.)

will normally cover the channel floor and will be continually reworked as the stream stage rises and falls.

Once graded, the stream begins to produce a new landform. This feature is a **floodplain,** a flat strip of land subjected to flooding about once a year (Figure 14.10). The floodplain is formed as the stream cuts horizontally on the outsides of the stream bends. This activity, termed **lateral planation,** resembles the action of a saw turned on its side. Planation takes place in times of large discharges, when the stream has great energy and scours both the banks and the bed. On the insides of the bends, bed load is deposited in the form of sand and gravel bars, creating a widening belt of low, nearly flat ground. When extreme floods occur, this low ground is inundated and is the site of deposition of silt that settles out from the turbulent water. The silt layers thus accumulate to produce a flat fertile floodplain.

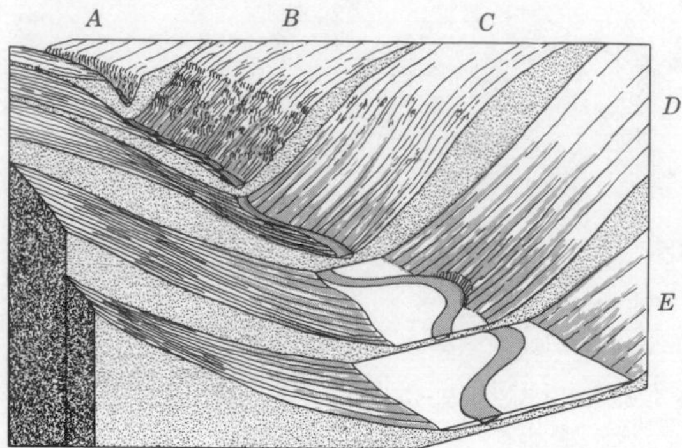

Figure 14.10 During stages *A, B,* and *C,* the stream deepens its valley and its profile is graded. In stages *D* and *E* the graded stream carves a floodplain and begins to develop meanders. (© 1973, John Wiley & Sons, New York.)

In stage *D* of Figure 14.10 the main stream is shown to be graded and to be producing a floodplain. As the floodplain is widened by further lateral planation the valley walls are less rapidly undermined and decline in steepness through the action of weathering and mass wasting.

In stage *E* of Figure 14.10 the floodplain has widened to the degree that wavelike channel bends are able to form. Widening of the valley occurs only where the outside of a bend impinges upon the valley wall.

THE EQUILIBRIUM PROFILE; BASELEVEL

The downstream changes in gradient of a stream may be studied by means of the **longitudinal profile,** a display in which stream elevation is plotted on the vertical axis and the horizontal distance on the horizontal axis, as shown in Figure 14.8. The longitudinal profile of a graded stream is known as the **graded profile.** It is upwardly concave and shows a decreasing gradient from head to mouth.

The stream mouth enters a body of standing water at a very low gradient. The level of the body of standing water effectively limits the downcutting of the stream and constitutes the **baselevel** of the stream. Sea level, projected inland beneath the stream system, forms the baselevel for fluvial denudation of the lands.

Successive profiles numbered in Figure 14.8 show how the graded profile is gradually lowered and flattened. The surrounding land surfaces are also being reduced in height and steepness, supplying less and less load to the stream. The whole process can be expected to take from one to three million years, or more.

ALLUVIAL RIVERS

Many of the world's graded rivers occupy broad floodplain belts. (A river is simply a large stream. We use the word river here because it agrees with popular usage.) Designated **alluvial rivers** by hydraulic engineers, these streams flow with very low gradients on thick deposits of alluvium; they have extremely sinuous bends known as **alluvial meanders** (Figures 14.11 and 14.12).

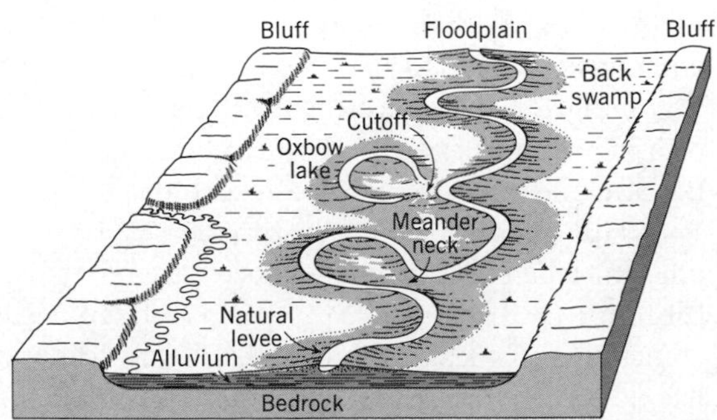

Figure 14.11 A meandering alluvial river produces many interesting landforms on its floodplain.

Meanders originate from the enlargement of bends in the path of flow of the stream. Once a bend is produced, centrifugal force continues to thrust the flow toward the outside of the bend. Undermining continues the enlargement of the bend until a meander loop is formed.

A meander may become constricted, creating a narrow **meander neck.** As shown in Figure 14.11, the neck may be cut through by bank caving or by overflow in the time of flood, permitting the stream to bypass the bend and to produce a **cutoff.** The cutoff meander bend is quickly sealed off from the main stream by silt deposits and becomes an **oxbow lake.** Gradual filling of the lake results in an **oxbow swamp** (Figure 14.12).

Most alluvial rivers have a yearly flood of such proportions that the water can no longer be contained within the channel and spreads out upon the floodplain (Figure 14.13). Such **overbank flooding** permits fine-grained sediment (silts and clays) to be deposited from suspension in the relatively slow-moving water covering the floodplain. The sediment is laid down in layers, which are called **overbank deposits.**

Adjacent to the main channel, in which flow is relatively swift because of greater depth, the coarsest sediment—sand and coarse silt—is deposited in two bordering belts. Repeated floods build up lateral zones of somewhat higher ground, called **natural levees** (Figure 14.11). The highest points on the levees lie close to the river bank. There is a gentle slope away from the river down to the low-lying marshy areas of floodplain some distance from the river. In a big flood, such as that pictured in Figure 14.13, the natural levees can be identified by two tree belts, one on each side of the submerged channel.

FLOOD HAZARDS

Alluvial rivers have long posed an environmental problem because of the overbank flooding that is normal on an annual basis and the much higher floods that occur a number of years apart. A conflict of interest arises over the fact that overbank flooding adds a layer

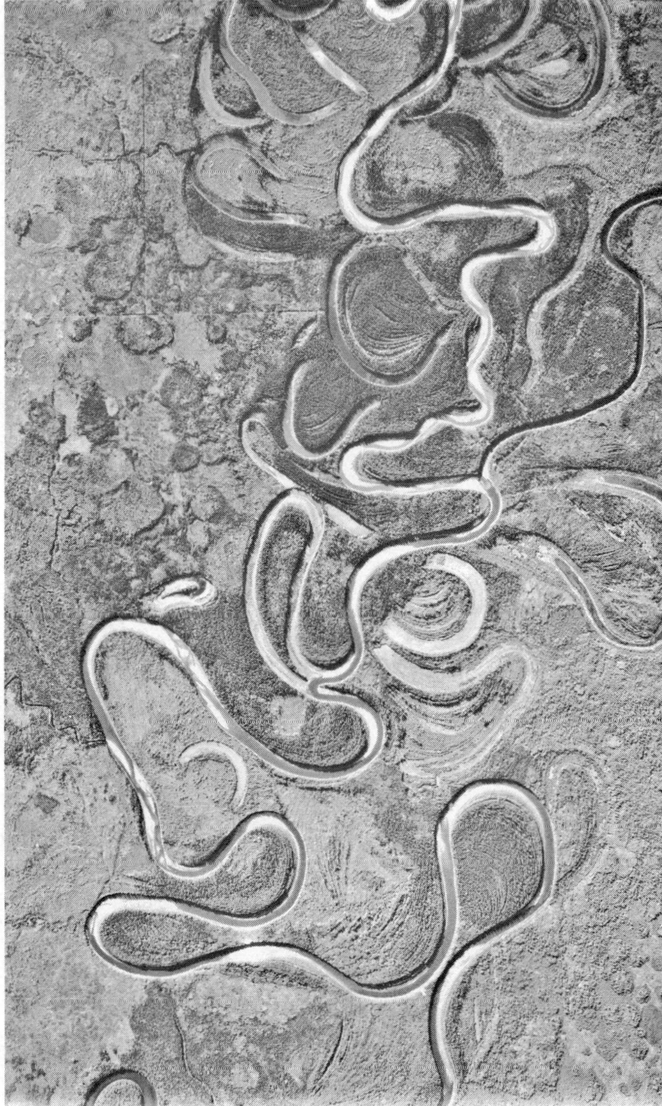

Figure 14.12 Seen from an altitude of about 20,000 ft (6100 m), the Hay River in Alberta is replete with meanders, cutoffs, oxbow lakes, and oxbow marshes. (National Air Photo Library, Surveys and Mapping Branch, Canada Department of Energy, Mines, and Resources.)

Figure 14.13 The floodplain of the Washita River, near Davis, Oklahoma, was almost completely inundated in this flood of 1950. A meander bend of the river channel is marked by a double line of trees growing on the natural levees. The channel carries most of the discharge because of its much greater water depth and velocity of flow, whereas water spread over the floodplain is moving very slowly. (Soil Conservation Service, U.S. Department of Agriculture.)

of silt rich in plant nutrients. These maintain fertility of the floodplain soils, particularly in regions having heavy rainfall that tends to leach out these nutrients. Counterbalancing this beneficial effect is the destruction of life and property caused by floodplain inundation. Flood-control engineering has been practiced for decades in hopes of reducing to the minimum the hazards of overbank flooding.

AGGRADING STREAMS; ALLUVIAL FANS

Coarse rock waste is sometimes supplied to a stream by its tributaries and by runoff from adjacent hillsides in greater quantity than the stream is capable of transporting. When this happens the excess load is spread along the channel floor, raising the level of the entire channel, a process termed **aggradation.**

A stream channel in which aggradation is in progress is typically broad and shallow. Aggradation takes the form of deposition of long, narrow bars of sand and gravel, which tend to divide the flow into two or more lines. The stream subdivides and rejoins in a manner suggesting braided cords and is called a **braided stream** (Figure 14.14). Where the braided stream flows between narrow confining walls, aggradation steadily raises the level of the floodplain.

Perhaps the most common cause of aggradation in stream channels is the combination of arid climate and mountainous relief, as we find in the southwestern United States. Barren, steep mountain

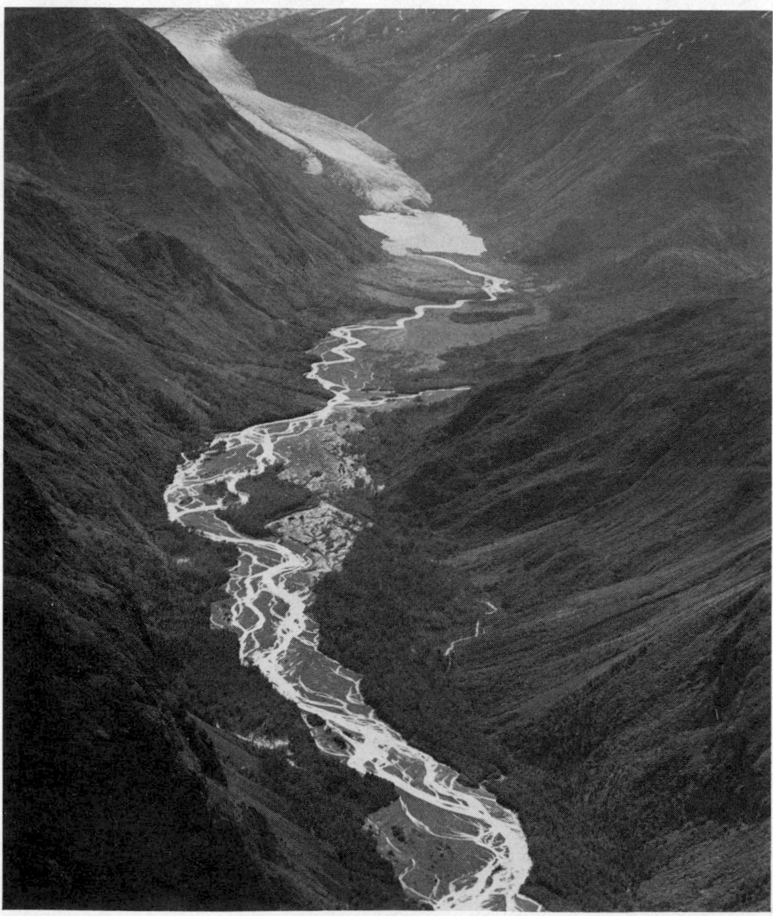

Figure 14.14 This braided stream derives its coarse bed load from a melting glacier far up the valley. Peters Creek, Chugach Mountains, Alaska. (Alaska Pictorial Service, Steve & Dolores McCutcheon.)

slopes shed large quantities of coarse debris when eroded by runoff of torrential rains. Floods in the mountain valleys are characterized by a large proportion of coarse bed load; this is carried down-valley on steep gradients (see Figure 14.7). Where a canyon emerges upon a valley floor of gentle slope, aggradation occurs because the stream is not able to transport its load on a sharply reduced gradient. Shifting from side to side as aggradation occurs, the stream spreads its excess load in the form of an **alluvial fan** (Figure 14.15).

The alluvial fan takes the form of a low, upwardly concave sector of a cone, steepening in gradient toward an apex situated at the canyon mouth. At its outer edge the fan slope grades imperceptibly into the flatter plain. As one might suspect, the size of rock particles making up the fan is greatest near the apex, where much bouldery material accumulates. The alluvium grades to progressively finer particles toward the base. Large fans of mountainous deserts may be several miles in radius from apex to outer edge (Figure 14.16).

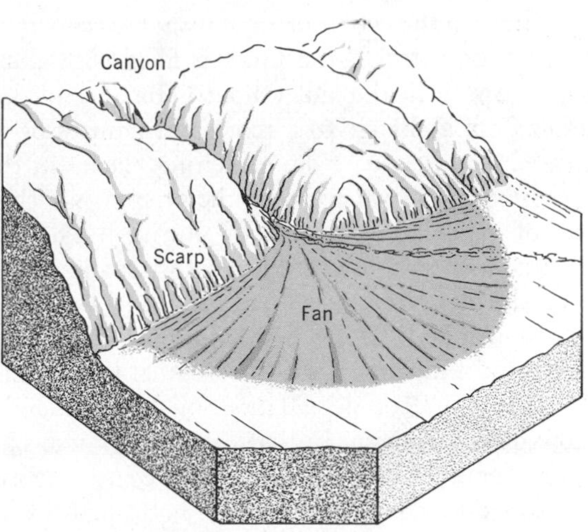

Figure 14.15 Idealized diagram of a simple alluvial fan. (© 1973, John Wiley & Sons, New York.)

ALLUVIAL TERRACES

A **terrace** is a steplike landform, bounded by steeply rising slopes on one side and by descending slopes on the other. There are many kinds of terraces. The **alluvial terrace** is one important type, carved out of alluvial deposits previously deposited in a valley.

Figure 14.16 A great alluvial fan in Death Valley, built of debris swept out of a large canyon. Notice the many braided stream channels. (Spence Air Photos.)

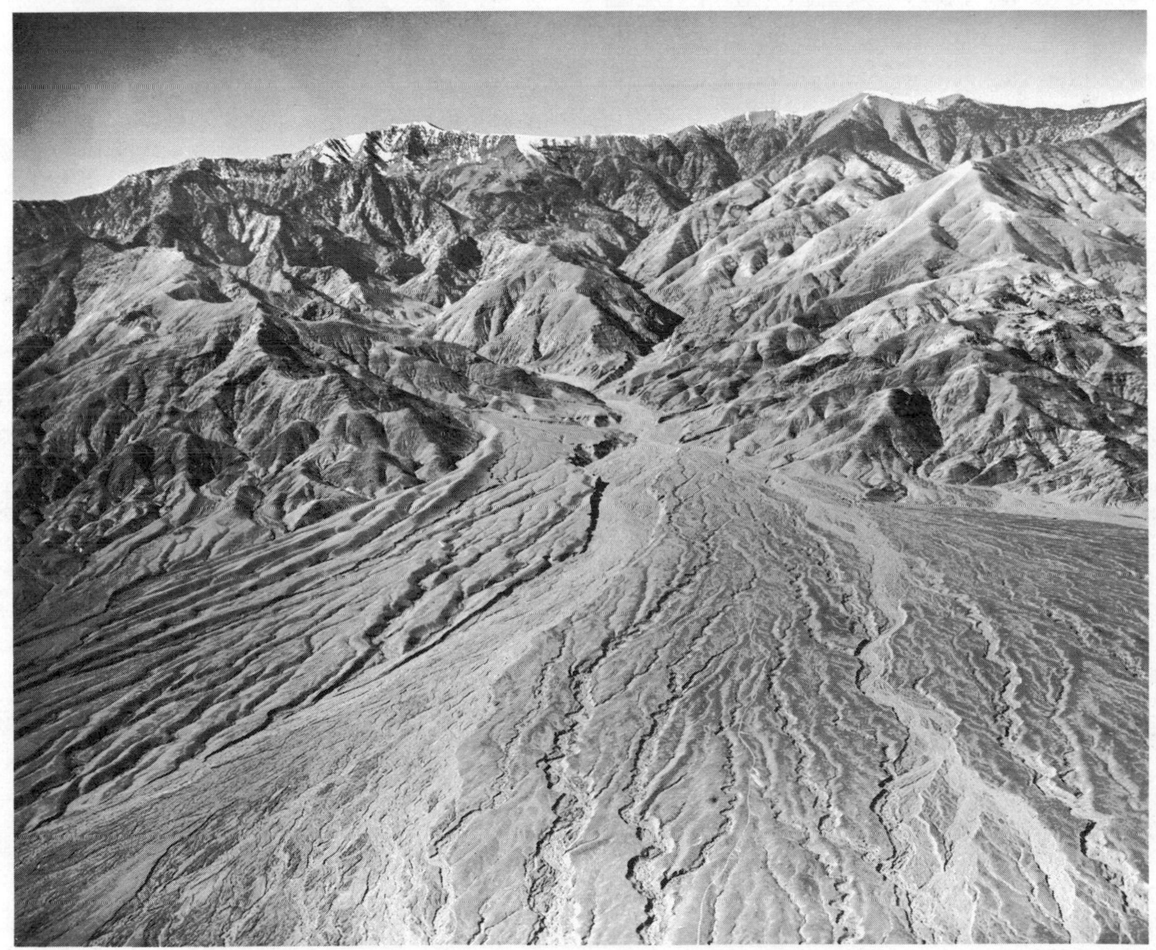

Steps in the evolution of alluvial terraces are illustrated in Figure 14.17. First (block A), a valley is filled with alluvium by an aggrading stream until the valley floor is broad and flat. Aggradation can result from a change to a more arid climate or from the greatly increased load derived from a melting glacier in the upstream region.

In block *B* we see that the stream has changed its role from one of aggradation to one of downcutting, or **degradation.** The stream has begun to cut down into the alluvial deposit, at the same time shifting laterally and carving out its floodplain in the easily eroded material. There now appears a single terrace level. Its surface is the remnant edge of the original floodplain.

The cause of degradation might be found in a change to a more humid climate. In that case, the growth of a denser vegetative cover would tend to hold back the coarser debris from the streams and so reduce the stream loads in both quantity and size of particles. A similar effect follows the disappearance of glaciers.

In block *C* of Figure 14.17 degradation has proceeded further. Now we have a series of alluvial terraces, resembling a flight of broad stairs on the valley wall (Figure 14.18). Once formed, a terrace may be protected from later undermining by the presence of an outcrop of resistant rock at the base of the terrace (letter R in block C).

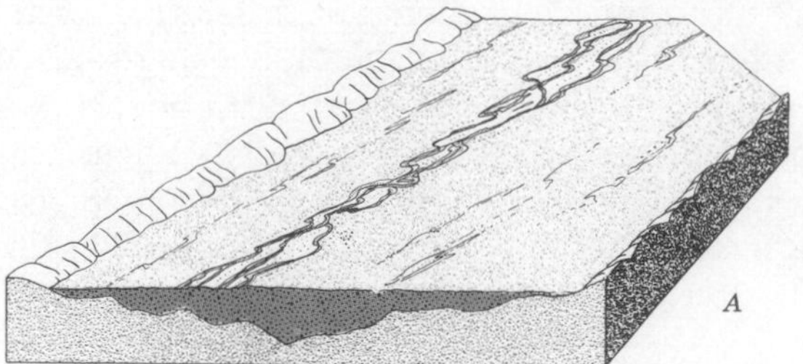

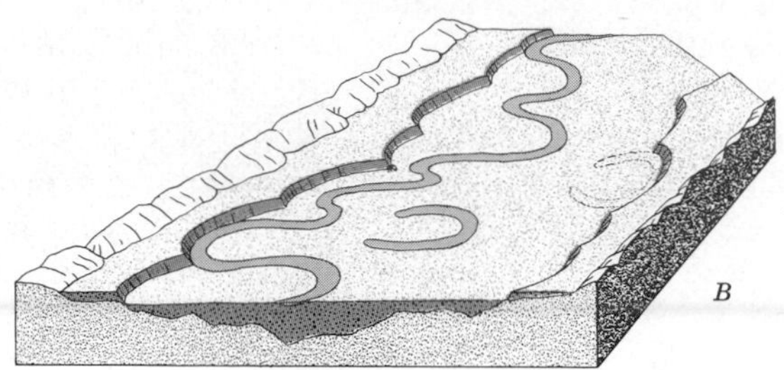

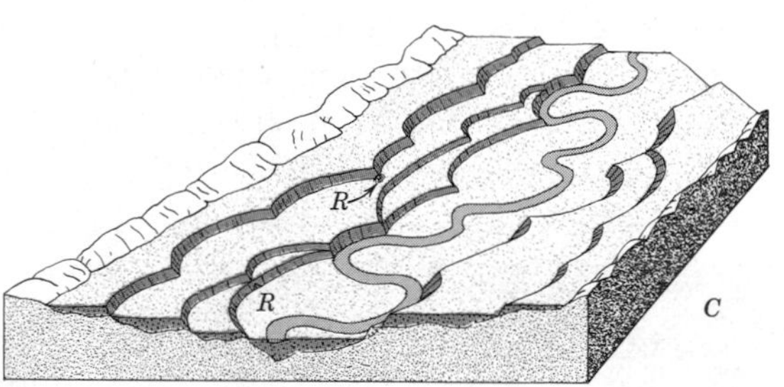

Figure 14.17 Formation of alluvial terraces. The letter *R* in block *C* refers to a point where a terrace is defended by a rock outcrop. (© 1973, John Wiley & Sons, New York.)

STREAM TRENCHING AND ENTRENCHED MEANDERS

At any time in the life of a graded stream, sea level may drop, or the earth's crust may rise. Either change causes a drop in baselevel. The establishment of a lower baselevel has a profound effect upon the graded stream; it scours the bed deeply, making a trench.

Channel trenching forms a series of rapids. Degradation extends rapidly upstream, and a narrow, steep-walled gorge or canyon is produced. On either side is a **rock terrace,** representing the former broad-floored valley (Figure 14.19).

Figure 14.18 Terraces cut by the Shoshone River, west of Cody, Wyoming, rise like broad flights of steps. (Frank J. Wright.)

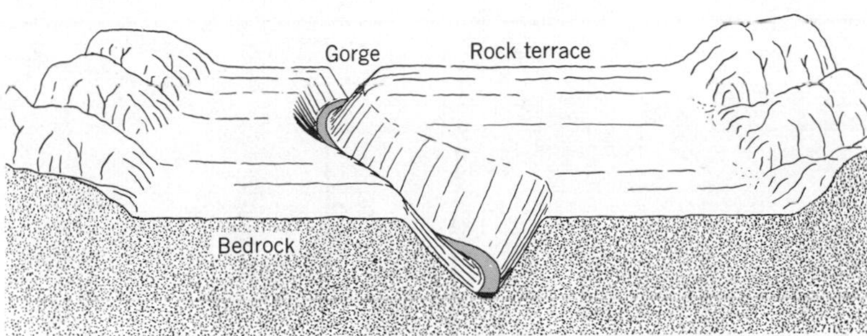

Figure 14.19 A rock terrace, formerly a floodplain belt, borders a steep-walled inner gorge resulting from stream trenching.

In an extreme case of trenching, the floodplain meanders of an alluvial river become deeply carved into the underlying bedrock, producing **entrenched meanders** (Figure 14.20). These form a winding river gorge. Cutoff sometimes occurs at the narrow meander neck, leaving a rock arch, or natural bridge.

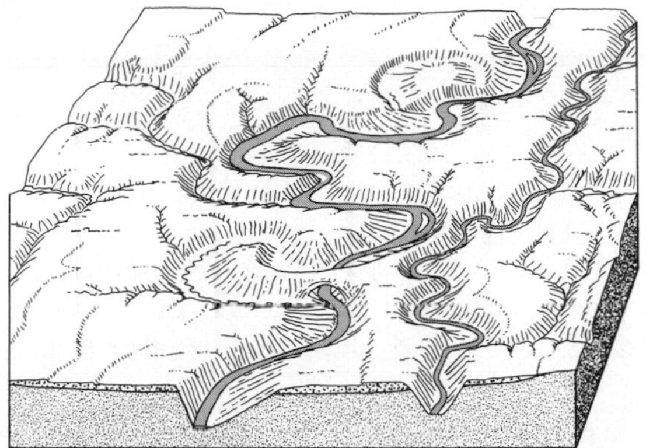

Figure 14.20 Entrenched meanders with cutoffs and a natural bridge. (© 1973, John Wiley & Sons, New York.)

DELTAS

A stream reaching a body of standing water builds a sediment deposit, the **delta,** composed of the stream's load. The growth of a simple delta can be followed in stages, shown in Figure 14.21. For simplicity, imagine that the water body is a small lake, lacking in tides and not greatly affected by waves. The stream enters the standing water body as a jet, but its velocity is rapidly checked (Figure 14.22).

Sediment is deposited in lateral embankments in zones of less turbulence on either side of the jet. In this way the channel is extended into the open water. The stream repeatedly breaks through the embankments to occupy different radial positions. In time it produces a deposit of semicircular form, much like the alluvial fan (which is, in a sense, a terrestrial delta).

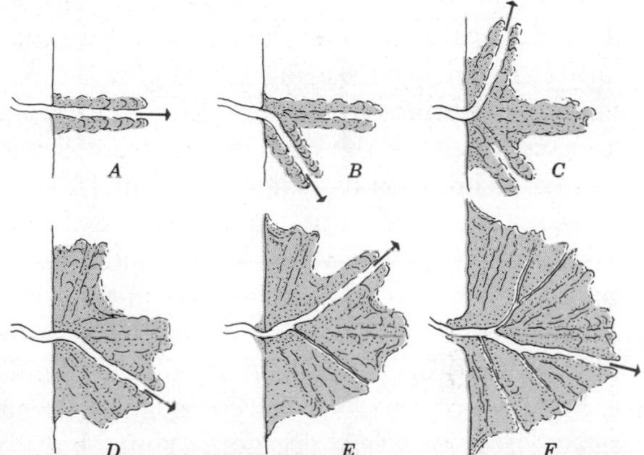

Figure 14.21 Stages in the development of a simple delta built into a lake in which wave action is slight. (© 1973, John Wiley & Sons, New York.)

Figure 14.22 In this air view of the Kander River delta, Switzerland, we see a jet of sediment-laden water being projected into the lake. (Swissair photograph.)

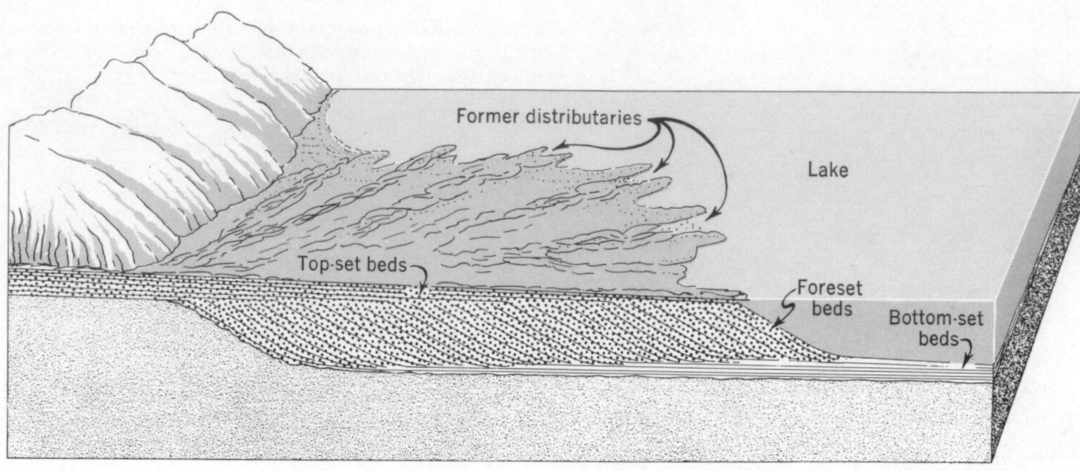

Figure 14.23 Internal structure of a simple delta built into a lake. (© 1973, John Wiley & Sons, New York.)

In cross section the simple delta consists largely of steeply sloping layers of sands, called foreset beds (Figure 14.23). These sand and gravel layers grade outward into thin layers of silt and clay, the bottomset beds. As the delta grows, the stream aggrades slightly and spreads new layers of alluvium, the topset beds.

SURFACE WATER AS A GEOLOGIC AGENT AND A NATURAL RESOURCE

In this chapter we have investigated the geologic work of running water on the lands. Overland flow on slopes and stream flow in channels work together to lower the lands. But we should not forget that weathering and mass wasting are part of the process of fluvial denudation, preparing the rock for transportation and helping to bring it to lower levels. In the next chapter we will put together a model for fluvial denudation of a large continental landmass.

More than three times as much water from streams and lakes is consumed as ground water in the United States, for public water supplies, industry, and irrigation. Only in the arid southwestern states is ground water used in greater amount than surface water. Streams and lakes are easily polluted, as we all know, and the record to date has been a sorry one. We have enough fresh surface water in the United States for several decades to come, but it must be protected from pollution, and it may need to be diverted long distances to supply the communities of greatest need. The successful use of this renewable natural resource will require careful planning.

YOUR GEOSCIENCE VOCABULARY

fluvial denudation
overland flow
channel flow, stream flow
stream channel
banks
stream
soil erosion
geologic norm
accelerated erosion
sheet erosion
colluvium
alluvium
valley sedimentation
shoestring rills
gullies
stream gradient
stream velocity
streamline flow
turbulent flow
discharge
stream erosion
stream transportation
stream deposition
hydraulic action
bank caving
abrasion
corrosion

pothole
suspension
suspended load
sorting
traction
bed load
stream capacity
reworked alluvium
landmass
waterfall
gorge, canyon
graded channel, graded stream
floodplain
lateral planation
longitudinal profile
graded profile
baselevel
alluvial river
alluvial meanders
meander neck
cutoff
oxbow lake, oxbow swamp
overbank flooding
overbank deposits
natural levees
aggradation
braided stream
alluvial fan
terrace
alluvial terrace
degradation
rock terrace
entrenched meanders
delta

SELF-TESTING QUESTIONS

1. How important is fluvial denudation, as compared to denudation by other erosional agents? What are the two basic forms of flow taken by running water on the lands?

2. Describe the processes of soil erosion. What factors influence the rate of soil erosion? Distinguish between the geologic norm of soil erosion and accelerated soil erosion. Describe the forms of accelerated soil erosion.

3. Describe the variations in size and form of natural stream channels. Where is stream velocity highest? Why? Distinguish between streamline flow and turbulent flow. What is the geologic importance of turbulence in streams?

4. Name and describe in detail the three forms of geologic work performed by streams. What are some obvious features created by abrasion? How does stream flow act as a sorting mechanism for sediment?

5. Explain the relationship between a stream's discharge and the load it transports. How do stream channels change with rising and falling stages? What happens to channel alluvium during these changes?

6. Follow an ideal stream through the normal series of changes in activity and valley forms during its life history. What is the importance of the graded condition, so far as stream activity is concerned? How is baselevel related to the condition of grade?

7. Name and describe the important landforms of an alluvial river and its floodplain. Describe the kinds of deposits comprising a floodplain. In what way is overbank flooding beneficial to agricultural lands on a floodplain?

8. What landforms are produced by aggrading streams? Under what climate conditions is aggradation most likely to be found? Explain how alluvial fans are built. How are alluvial terraces formed?

9. What effect has a drop of sea level or a rise of the crust upon the activities of a graded stream? What landforms result?

10. Explain how a simple delta is built where a stream enters a lake. What kinds of beds comprise such a delta?

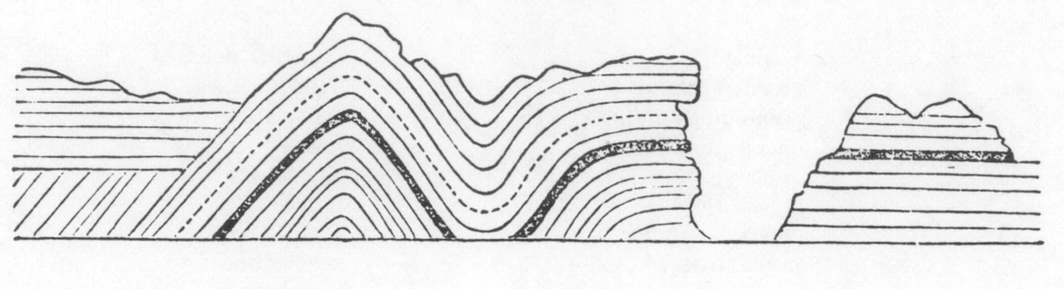

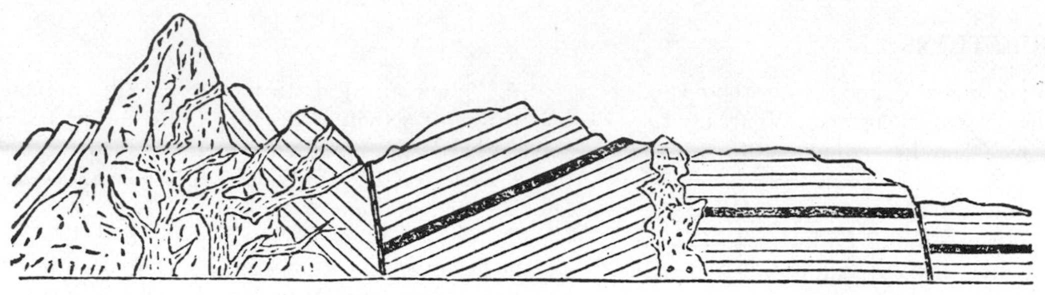

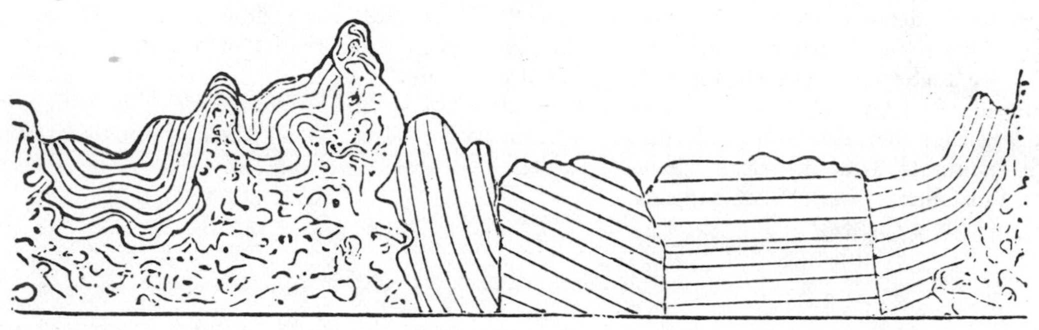

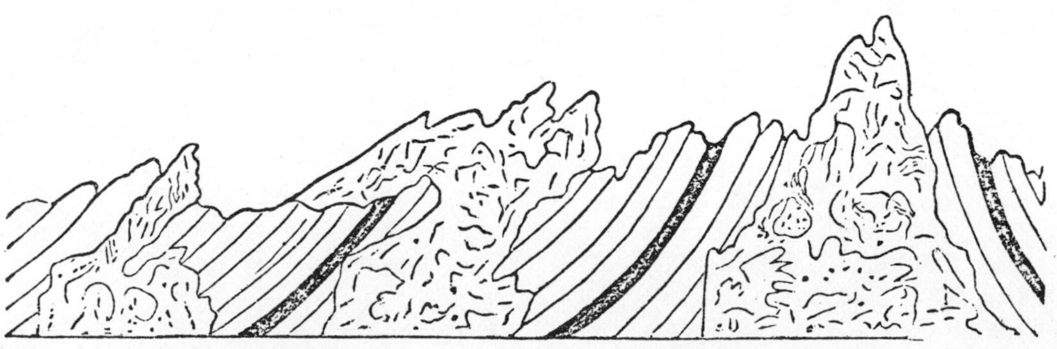

These figures from an early geology textbook illustrate faulting, folding, and intrusion, and their expression as landforms. (From *Elements of Geology,* 1860, by David Page, New York, A. S. Barnes & Burr.)

15 DENUDATION AND ROCK STRUCTURE

INSTANT SCENERY

Until the middle 1700s, interpretation of scenery was a very simple matter. In Christian lands, the Church had it all explained. Hills, valleys, and mountains were "instant scenery," created by a great cataclysm accompanying the formation of the earth. It was all part of the six-day Creation scenario, occurring in the year 4004 B.C. Great earthquakes wrenched the crust apart to make spectacular gorges; alpine mountains were thrust up thousands of feet in only hours. The Noachian Flood added the finishing touches, soaking things down and leaving layers of sediment with fossils in unlikely positions far above today's sea level.

Adherents to the theory of instant scenery—or catastrophism—were not entirely within the Church. A strong advocate was Baron Cuvier, a French zoologist and paleontologist who made major advances in the classification of animals. As permanent secretary to the French Academy of Sciences, his scientific influence was powerful, and his views were dominant in a period spanning the late 1700s and early 1800s.

Pitted against Baron Cuvier were a handful of no-nonsense Scots. One of them was James Hall of Edinburgh, whom we mentioned in the opening paragraphs of Chapter 5. His geological colleague and close friend was James Hutton. Together, these geologists had traced the folded strata of the Scottish coast to discover a major unconformity. Hutton's field observations convinced him of the impossibility of explaining sediment accumulations and unconformities by a single catastrophic event. Vast spans of time were needed for such features to be formed; the processes involved were, he reasoned, no different from those seen in action today in river valleys, along coastlines, and in shallow seas.

Hutton countered catastrophism with the doctrine of uniformitarianism. Applied to landforms, this doctrine holds that the slow processes of denudation seen in action today can account for all landscape features, except of course those obviously produced by recent faulting and volcanism. Hutton published his observations in 1795 under the title *Theory of the Earth.* Hutton's writing style was ponderous and obscure, so a younger friend, John Playfair, undertook to popularize Hutton's principles. He succeeded with the lucid and charmingly written *Illustrations of the Huttonian Theory of the Earth,* published in 1802.

Playfair is best remembered for his explanation of the origin of stream valleys. The following sentence from his book has become known as Playfair's law of streams:

> Every river appears to consist of a main trunk, fed from a variety of branches, each running in a valley proportioned to its size, and all of them together forming a system of valleys, communicating with one another, and having such a nice adjustment of their declivities (gradients) that none of them join the principal valley either on too high or too low a level; a circumstance which would be infinitely improbable if each of these valleys were not the work of the stream which flows in it.

The efforts of James Hall and James Hutton made slow headway against both the catastrophism of Cuvier and the plutonism of Abraham Werner (see Chapter 4), but ultimately they succeeded. Their cause was championed by a leading geologist, also of Scottish birth, Sir Charles Lyell. A successful textbook writer, whose three-volume *Principles of Geology* went through 12 editions, Lyell came to dominate English geology in the mid-1800s. Thanks to Lyell's efforts, stratigraphy became established on a sound basis, to serve as the foundation for Charles Darwin's theory of organic evolution. In this chapter we apply uniformitarianism to models of landmass denudation.

LANDMASS DENUDATION

The last three chapters have given us details of the fluvial denudation process, consisting of the combined action of weathering, mass wasting, and running water. In this chapter we will develop the total picture of fluvial denudation, using idealized models of landform evolution. We will take a large landmass, uplifted above sea level, and follow its changes as the surface is lowered toward baselevel.

A DENUDATION MODEL

We choose a region of humid climate, in which there is a large surplus of annual precipitation over annual evaporation. Under these conditions, runoff flows by streams to the nearest ocean, exporting sediment to the continental shelves. For these streams the baselevel is sea level. The landmass we choose for our model is a massive block about 100 mi (160 km) wide; it has steep sides and a broadly arched summit rising to 20,000 ft (6 km) above sea level, as shown at the top of Figure 15.1. We use these dimensions because they are about the size of a large mountain range, such as the Sierra Nevada, Rocky Mountains, Appalachians, or Cascades of North America. Other modern ranges of the alpine type have comparable dimensions. Length of the block is many times more than its width.

Next, we postulate that the entire block was rapidly elevated to its final position, taking 5 million years (m.y.) altogether, but with most of the uplift occurring in a span of 2 m.y. There is good geologic evidence to support these rough figures. Denudation occurs over any land surface exposed to the atmosphere. Consequently, we must allow for denudation taking place during uplift of the block. The elevated mass will have been carved into a complex maze of steep-walled gorges and narrow ridges by the time uplift is completed. The original upper surface of the block will have been largely destroyed, and many of the mountain summits considerably lowered. In stage A of Figure 15.1, a greatly exaggerated profile of the landmass, we see conditions at the cessation of uplift. The average elevation of the block is now approximately 15,000 ft (4.6 km). Stream profiles are upwardly concave, and have steep gradients. The mountainsides are also very steep. The total erosion rate is very rapid in this stage.

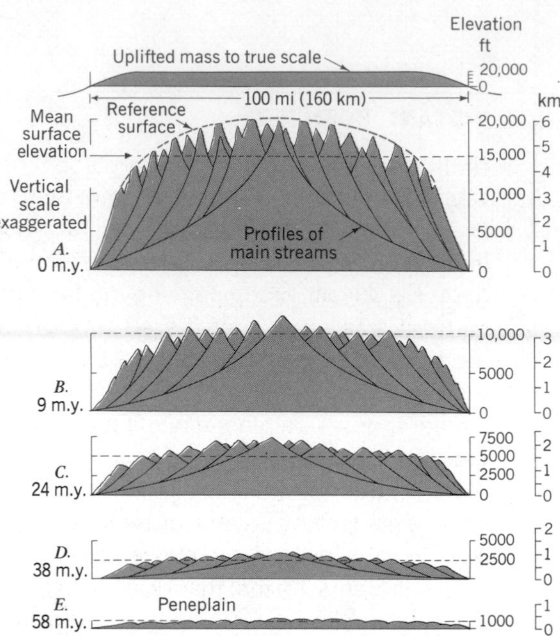

Figure 15.1 Schematic diagram of denudation of a landmass following crustal uplift.

Figure 15.2 The Idaho batholith of Cretaceous age, exposed in the Sawtooth Mountains of south central Idaho. (From *Geology Illustrated* by John S. Shelton. W. H. Freeman and Company. Copyright © 1966.)

Figure 15.3 The Blue Ridge upland in North Carolina. (Frank J. Wright.)

As time passes and denudation continues, summit elevations and stream profiles are lowered within the landmass, as shown in stages *B*, *C*, and *D* of Figure 15.1. Erosion slows down, because stream profiles are decreasing in gradient. Using a figure for the original erosion rate which is arbitrary but reasonable in light of geologic evidence, we have been able to assign hypothetical time spans to given amounts of lowering. Profiles *B*, *C*, and *D* represent conditions at 9, 24, and 38 m.y. after profile A. After 58 m.y., we predict an average elevation of only 1000 ft (0.3 km). The surface at this point in time is a peneplain, a gently undulating surface on which stream gradients are very low. Recall from Chapter 7 that peneplains are preserved as unconformities. Radiometric dating of rocks above and below an unconformity can give a rough estimate of the time required for a peneplain to form.

SLOPES AND DENUDATION

Figure 15.2 shows a landmass in a fairly early stage of fluvial denudation. The mountains are high and rugged; they have steep sides and narrow divides. The valleys are deeply cut and their gradients are steep. Soils are thin and stony, while bedrock outcrops are numerous. Rugged mountains like this are typical of the continental margins close to active subduction zones, such as those surrounding the Pacific Ocean.

Figure 15.3 shows a landmass well advanced in the fluvial denudation schedule. These features would be called hills in comparison with the mountains shown in Figure 15.2. The hillsides have moderate slopes and can be used for pastures, and even for farming. Soils are well developed and fertile, the regolith beneath is thick. There is danger of severe soil erosion if conservation practices are not strictly adhered to. A landmass in this stage is typical of the stable continental margins that are far removed from active lithospheric plate margins. The eastern margins of North and South America are such regions.

Figure 15.4 shows a peneplain, produced in the near-final stages of the denudation schedule. Because peneplains lying close to sea level are featureless and not photogenic, we have shown instead a peneplain that has been uplifted and trenched by a major stream.

Figure 15.4 The St. John peneplain in Puerto Rico is represented by a rolling upland surface at an elevation of about 2000 ft (600 m). The peneplain, of Miocene age, is now deeply trenched by the Rio Usabón flowing in the steep-walled Canyon de San Cristobal. (U.S. Geological Survey.)

The rolling upland in the distance is the former peneplain surface. Its gentle slopes contrast with the steep walls of the river gorge in the center. On a peneplain, soils are thick and often fertile. They are not easily damaged by soil erosion. Before uplift, a peneplain is characterized by streams of very low gradient, with many broad alluvial floodplains. Large parts of the Amazon-Orinoco lowland of South America illustrate a peneplain lying close to sea level. So faint are the stream gradients in this area that rivers as far distant as 1900 mi (3000 km) from the mouth of the Amazon are less than 650 ft (200 m) above sea level.

Figure 15.5 Dendritic drainage pattern on igneous rock of the Idaho batholith. (© 1973, John Wiley & Sons, New York.)

LANDFORMS AND ROCK STRUCTURE

Our model of fluvial denudation has been based upon the simplest possible landmass—one in which the rock is uniform throughout in its physical and chemical properties. In this respect the imaginary rock in our model resembles a sugar cube—you can turn any side up and see no difference in the structure or texture. A rock mass answering this description is said to have **homogeneous structure.**

Truly homogeneous rock masses are a rarity in nature. Some large plutonic masses may qualify, and some thick shale formations behave as if homogeneous. However, most continental crustal rock comes in layers, which may be inclined or folded, or in blocklike masses of unlike rocks faulted one against the other. Intrusion of magma in the form of stocks, dikes, and sills creates still another set of arrangements of rock masses. Some rock bodies are shattered by countless closely set joints; others are almost free of joints.

Denudation is a selective process, rapidly removing weak rock to form low places and leaving harder rock masses to stand high as hills, ridges, or mountains. Streams seek out the weaker belts of rock, occupying them in preference to the harder rock masses. In this way the landscape features come to reflect the rock inequalities, or rock structure of the landmass.

In the remainder of this chapter, we will investigate some distinctive landforms controlled by rock structure. These landforms lend variety and beauty to the scenery of the continents, they make one region distinctively different from another.

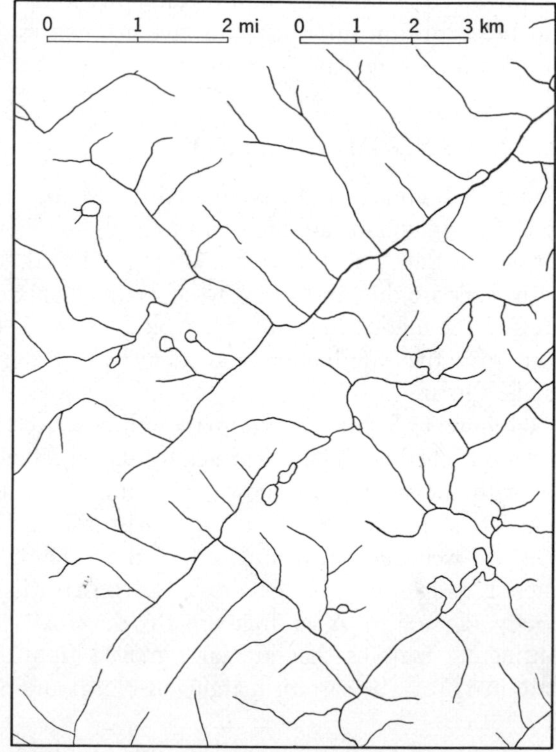

Figure 15.6 A rectangular drainage pattern consisting of subsequent streams located on fault zones. Adirondack Mountains, New York. (© 1973, John Wiley & Sons, New York.)

DRAINAGE PATTERNS

Particularly sensitive to rock structures are **drainage patterns,** the distinctive forms of complete stream channel networks viewed on a flat map. The total pattern consists not only of the main stream, but also all its branches, traced to their very tips. The drainage pattern associated with homogeneous structure is described as **dendritic.** Figure 15.5 is a stream network map illustrating the dendritic pattern. The branching is often described as treelike. The dendritic pattern shows no grain; the individual segments of channel between junctions do not favor any single trend.

The dendritic pattern is inherited from the early stages of denudation, in which streams competed for space on a new landmass. In homogeneous structure, no single direction was favored over another, so that the smaller channels came to be oriented randomly.

Figure 15.5 is the stream network of the same granite batholith pictured in Figure 15.2. The granite is nearly homogeneous, and the pattern is dendritic. Although this rock is broken by countless minor joints, they do not noticeably affect the pattern.

Some areas of plutonic rock are broken by major faults. The rock is crushed along the fault plane and is easily eroded. Streams seek out these fault lines. Figure 15.6 is the drainage pattern of a faulted plutonic mass in which the faults intersect at about right angles. A stream following a fault runs straight for a long distance, then makes a sharp right-angle bend to follow another fault. Any stream controlled by a zone or belt of weak rock is classed as a **subsequent stream.** Streams following major faults are one variety of subsequent stream. The resulting network pattern is called a **rectangular pattern.**

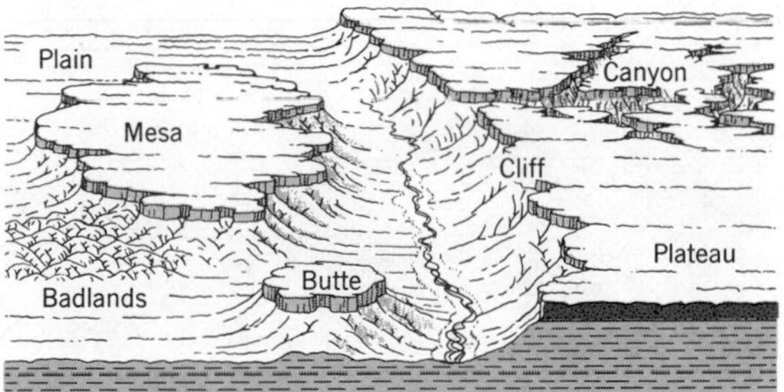

Figure 15.7 Landforms of horizontal strata evolving in an arid climate. (© 1973, John Wiley & Sons, New York.)

EROSION FORMS OF HORIZONTAL LAYERS

Vast areas of the continental shields bear covers of marine sedimentary strata. These beds were laid down upon a basement of ancient igneous and metamorphic rocks that had been reduced to a peneplain and submerged to become the floor of a shallow sea. The thickness of these strata does not exceed several thousand feet, which is relatively thin in comparison with the strata accumulated in subsiding geosynclinal troughs.

Over large parts of the continental shields, strata of Paleozoic, Mesozoic, and Cenozoic ages have been raised in broad crustal uplifts, involving little or no tilting, folding, or faulting. A particular sandstone layer, for example, may have been raised to an elevation of a few thousand feet above sea level with so little disturbance that it is inclined only a fraction of a degree from the horizontal; the landmass consists of **horizontal strata.**

In the United States much of the region lying between the Rockies and the Appalachians is underlain by nearly horizontal strata. Another large region is that of the Colorado Plateau of Arizona, Utah, New Mexico, and Colorado. It is from these areas that we draw examples of the landforms developed in horizontal strata.

The plateau basalts, described in Chapter 3, also make up large expanses of near-horizontal layered rocks. An example is the Columbia Plateau region of Washington, Oregon, and Idaho (see Figures 3.6 and 3.7).

Details of landscape development in horizontal strata are illustrated in Figure 15.7. A cap rock of resistant sandstone maintains a nearly flat surface called a **plateau.** The exposed edges of this hard layer stand as nearly vertical **cliffs,** kept in sharp definition by constant undermining of weak shale beds beneath. Continued erosion causes a portion of the plateau to become detached. The resulting flat-topped mountain, bounded on all sides by steep cliffs, is a **mesa.** As a mesa shrinks by wastage of its surrounding cliffs, it assumes the form of a small, flat-topped hill, a **butte** (Figure 15.8).

Drainage patterns developed upon horizontally layered rocks are dendritic. A given layer is homogeneous in all horizontal directions, so that it does not appreciably control the direction taken by the growing fingertip streams in the early stages of denudation.

Figure 15.8 This early photograph (circa 1900) shows a butte of horizontal red sandstones capped by a gypsum layer, near Cambria, Wyoming. (U.S. Geological Survey.)

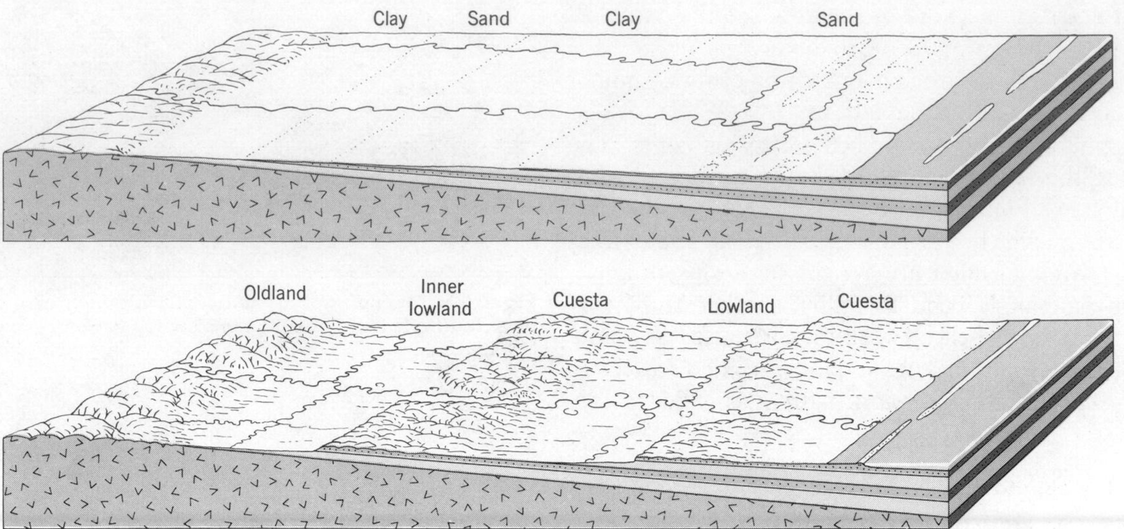

Figure 15.9 Erosional development of a coastal plain. The upper block shows the plain recently emerged. Consequent streams flow directly to the new shoreline. The lower block shows a later stage, in which erosion has produced lowlands and cuestas. Subsequent streams occupy the lowland. (© 1973, John Wiley & Sons, New York.)

GENTLY INCLINED STRATA—COASTAL PLAINS

Marine sedimentary strata of the United States Atlantic and Gulf coastal region were deposited upon a sloping continental shelf during the Mesozoic and Cenozoic Eras. They became exposed by crustal uplift in late Cenozoic time. The strata have a persistent seaward dip of perhaps 1° or 2° at most. Such strata form a **coastal plain** (Figure 15.9). Because the deposition of the strata took place to the accompaniment of a series of marine invasions and retreats, individual formations tend to thin to a featherlike edge toward the land and to thicken seaward.

When brought above sea level for the last time, the coastal plain presents a very smooth, sloping plain. Across the plain flow streams bringing runoff from the former mainland. Any stream that comes into existence upon a newly formed land surface by following the

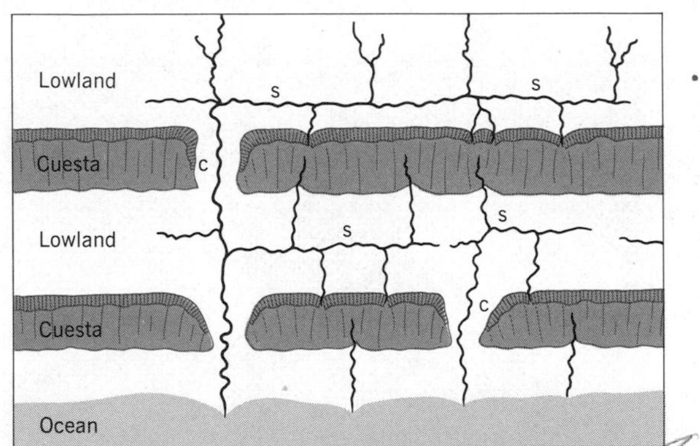

Figure 15.10 The trellis drainage pattern of a coastal plain. The major streams are consequent (*C*) and subsequent (*S*). Shorter tributaries flow from the cuestas to join the subsequent streams.

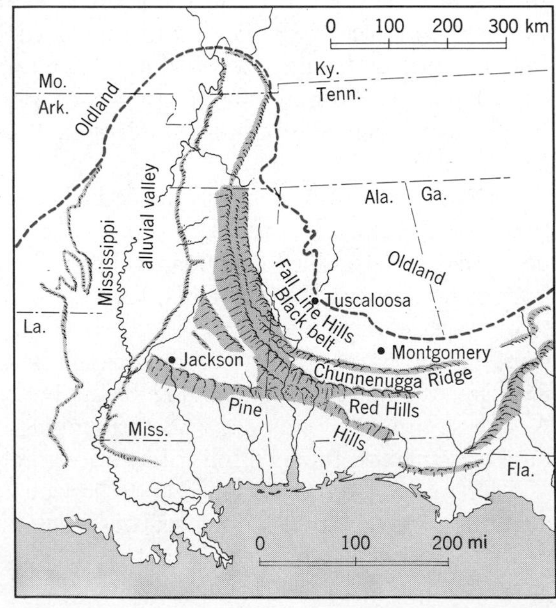

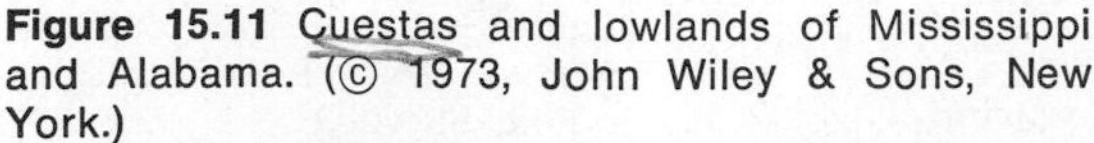

Figure 15.11 Cuestas and lowlands of Mississippi and Alabama. (© 1973, John Wiley & Sons, New York.)

direction of the initial slope is a **consequent stream.** Figure 15.9 shows two consequent streams flowing to the new shoreline.

After fluvial denudation has begun to act upon the poorly consolidated clays and sands of the coastal plain, there emerge belts of low hills. These belts are **cuestas;** they represent exposed sand layers. Although the sand is not lithified, it resists erosion because of its high capacity for infiltration, for even heavy rains produce little overland flow on sand. Separating the hill belts are broad, low valleys where the surface is underlain by layers of soft clays; these are **lowlands.**

As the lowlands are excavated, a new set of streams develops, following the lowlands and joining the consequent streams at right angles. These are subsequent streams, formed by extending themselves headward along a belt of weaker rock.

The drainage pattern of a dissected coastal plain is a **trellis pattern,** in which the most important elements are the consequent and subsequent streams (Figure 15.10). The trellis form is completed by short tributaries; these drain the cuestas and join the subsequent streams. A good example of a dissected coastal plain with multiple cuestas is found in Mississippi and Alabama (Figure 15.11).

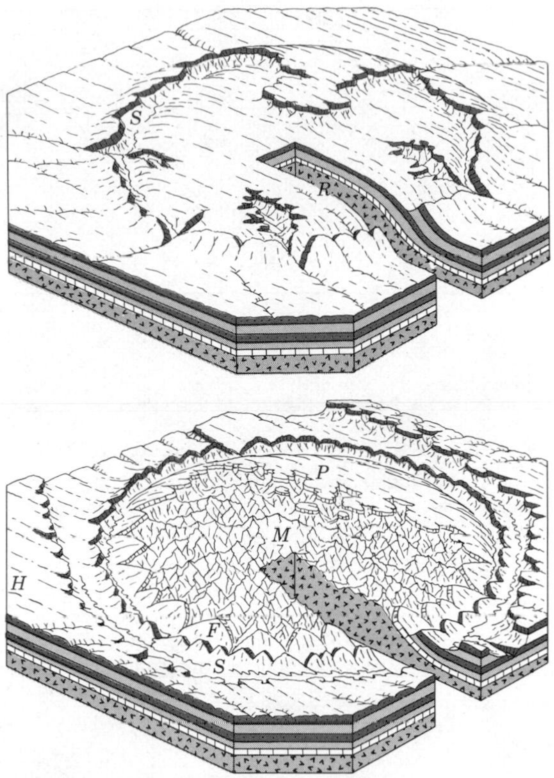

Figure 15.12 Landforms of a mountainous dome. (*Upper*) Early stage in removal of sedimentary cover. (*Lower*) Fully dissected dome with exposed core. *S:* subsequent stream. *R:* stripped sandstone formation. *P:* stripped limestone formation. *H:* horizontal strata. *F:* flatiron. *M:* crystalline *rocks.* (© 1973, John Wiley & Sons, New York.)

DOMES IN LAYERED ROCKS

An interesting tectonic structure of shield areas bearing a sedimentary cover is sharp upward doming. The result is a structure called a **sedimentary dome.** Typically, these domes are nearly circular (Figure 15.12). The sedimentary strata are sharply flexed upward around the base, and the central part is elevated to heights of several hundred or even a few thousand feet. On the dome flanks, the strata dip at angles from 30° to 60°, or even more. Sedimentary domes may be from 5 to 100 mi (8 to 160 km) across the base.

As denudation progresses, strata are removed from the center of the dome and are eroded outward to become sharp-crested ridges called **hogbacks** (Figure 15.13).

Figure 15.13 This hogback of steeply dipping sandstone was photographed in the early 1900s by a geologist, whose horse and buggy are on the roadside. Today, Interstate Highway 40 occupies the same gap, a few miles east of Gallup, New Mexico. (U.S. Geological Survey.)

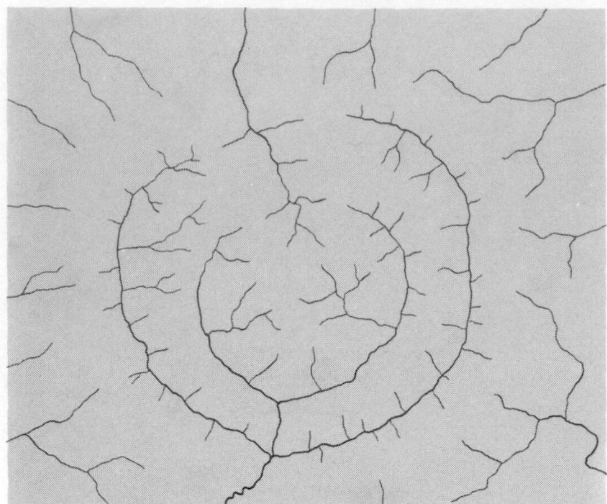

Figure 15.14 Radial and annular drainage on a dissected dome. (© 1973, John Wiley & Sons, New York.)

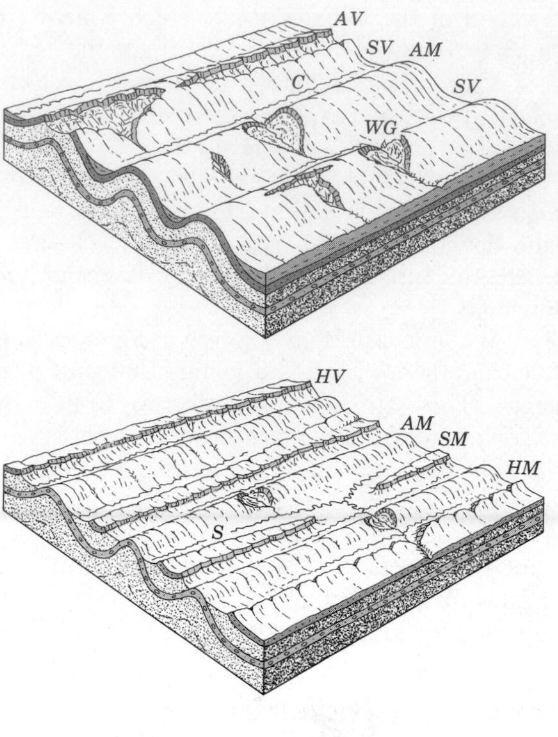

Figure 15.15 Landforms on open, parallel folds. *AV:* anticlinal valley. *SV:* synclinal valley. *AM:* anticlinal mountain. *WG:* watergap. *C:* consequent stream. *HV:* homoclinal valley. *SM:* synclinal mountain. *HM:* homoclinal mountain. *S:* subsequent stream. (© 1973, John Wiley & Sons, New York.)

When a mountainous dome is deeply dissected, the central region may have been completed stripped of sedimentary strata, leaving exposed a mass of older igneous or metamorphic rock. This central mass is the core of the dome (Figure 15.12). A sandstone layer resting directly upon the older rock now develops a series of triangular-shaped plates known as **flatirons.** Beyond these are hogback ridges separated by narrow circular valleys occupied by subsequent streams.

The drainage pattern of a mountainous dome consists of **radial streams** and **annular streams**—the former are consequent streams, the latter are subsequent streams. The total effect is that of a trellis pattern bent into a circular form (Figure 15.14).

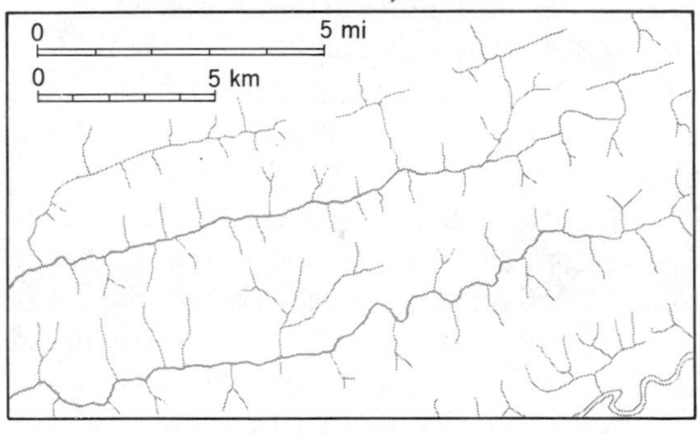

Figure 15.16 Trellis drainage pattern on dissected folds. (© 1973, John Wiley & Sons, New York.)

EROSION OF FOLDED STRATA

Erosional landforms developed on a series of simple open folds are illustrated in Figure 15.15. As shown in the upper block, a very massive, resistant formation of sandstone has been almost stripped clean of overlying weak formations of shale. Bold **anticlinal mountains** are formed by the anticlinal arches, and deep **synclinal valleys** coincide with the synclinal troughs. Major streams that originally crossed the anticlines may maintain their transverse courses by cutting deep steep-walled **water gaps** across the anticlinal mountains.

Next, the crests of the anticlines become breached by narrow valleys that grow along the mountain crests. These **anticlinal valleys** are occupied by subsequent streams and continue to lengthen and deepen until the anticline is completely opened out along its length. On the flanks of the anticline, the sandstone formation now presents its upturned edges to form **homoclinal mountains,** which are essentially the same features as the hogback ridges previously described.

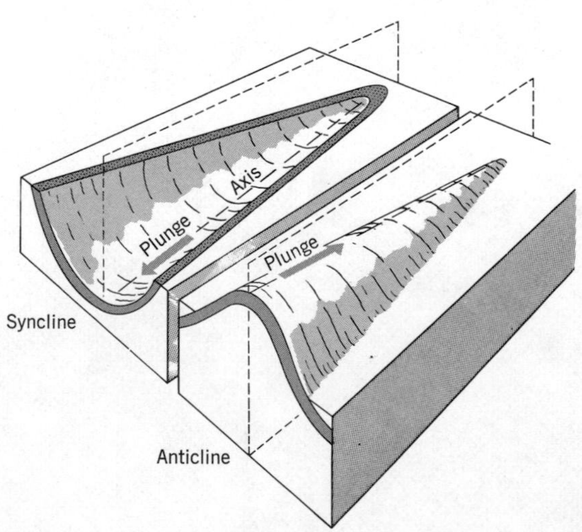

Figure 15.17 Plunging folds.

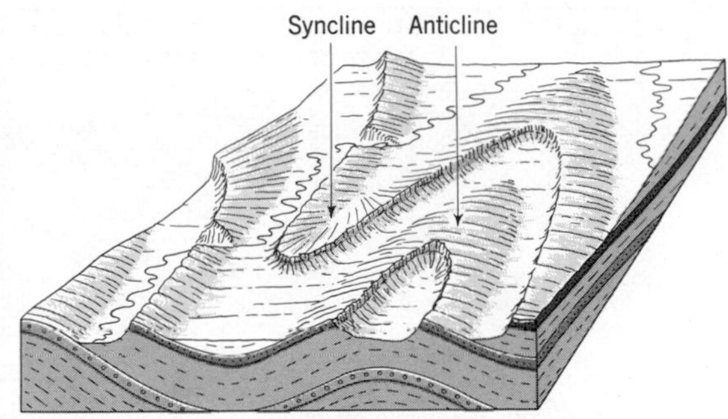

Figure 15.18 Zigzag ridges developed on plunging folds. (© 1973, John Wiley & Sons, New York.)

Figure 15.19 (*Left*) Zigzag ridge in plunging folds, south of Hollidaysburg, Pennsylvania. The nearer bend is a plunging syncline. The more distant bend (*at right*) is a plunging anticline enclosing a cove. The resistant bed forming the ridge crest is the Tuscarora formation, a quartzite of Silurian age. (John S. Shelton.)

The more deeply eroded folds are shown in the lower block of Figure 15.15. The sandstone formation referred to above is entirely eroded from above the anticlines, but persists along the synclinal axes, where it forms long, narrow, flat-topped **synclinal mountains.** These features are like long mesas, but with shallow summit valleys extending along the center line of the mountain. A trellis drainage pattern is associated with dissected folds (Figure 15.16).

The folds in Figure 15.15 idealize the crests of the anticlines and troughs of the synclines as maintaining horizontality. Most fold crests and troughs, when followed for long distances, alternately descend and then rise again in a series of undulations. Descent of the fold axis is referred to as **plunge** (Figure 15.17).

Where plunging folds are eroded, the homoclinal mountains form **zigzag ridges** (Figure 15.18). Where the ridge crosses a plunging anticline, a steep-walled cove is formed. Where a syncline is crossed we find a long, spadelike mountain with steep walls around the end and sides. Good examples of zigzag ridges are seen in the Appalachians of central Pennsylvania (Figure 15.19).

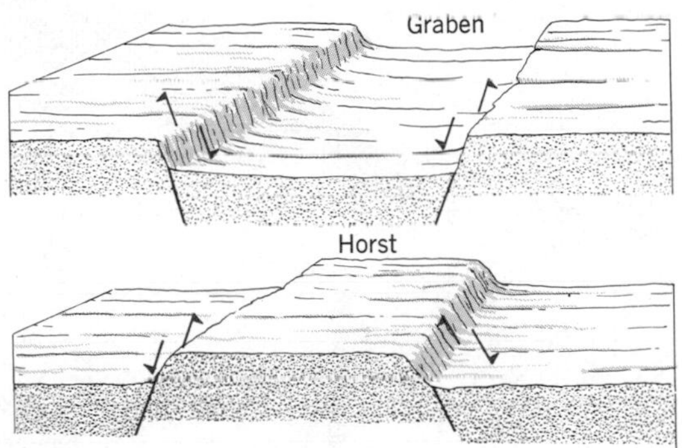

Figure 15.20 Graben and horst. (© 1973, John Wiley & Sons, New York.)

LANDFORMS DEVELOPED BY BLOCK FAULTING

Normal faults are commonly arranged in pairs, running parallel with one another, but with opposite sides downthrown or upthrown. The result is a downdropped fault block, or **graben,** lying between the faults; or an uplifted block, a **horst,** between them (Figure 15.20). Grabens can form broad flat-floored valleys, many miles wide and tens of miles long, bounded by bold, straight scarps.

Normal faulting affecting large masses of the crust has involved such large vertical displacements that **fault block mountains** have resulted (Figure 15.21). Individual blocks, commonly tilted during uplift, may be several miles wide and several tens of miles long. Between the upfaulted blocks are downfaulted blocks which form

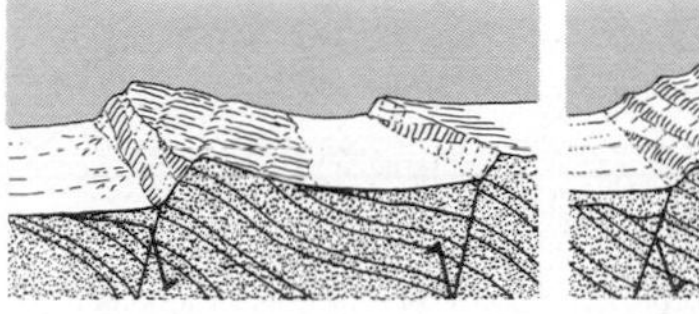

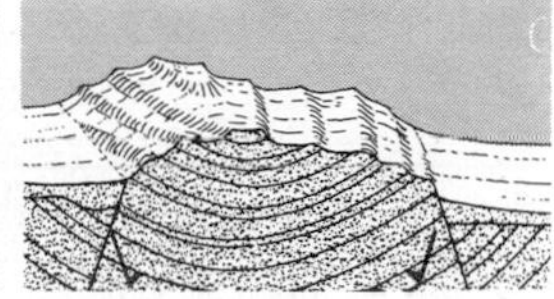

Figure 15.21 Tilted and lifted mountain blocks. (© 1960, John Wiley & Sons, New York.)

Figure 15.22 A tilted fault block is deeply eroded, leaving only triangular facets to represent the fault scarp.

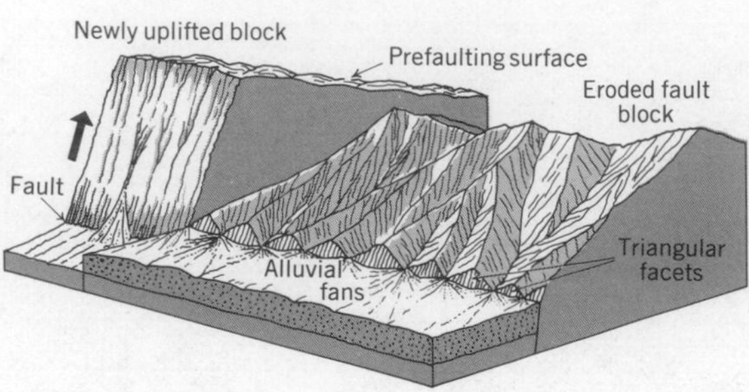

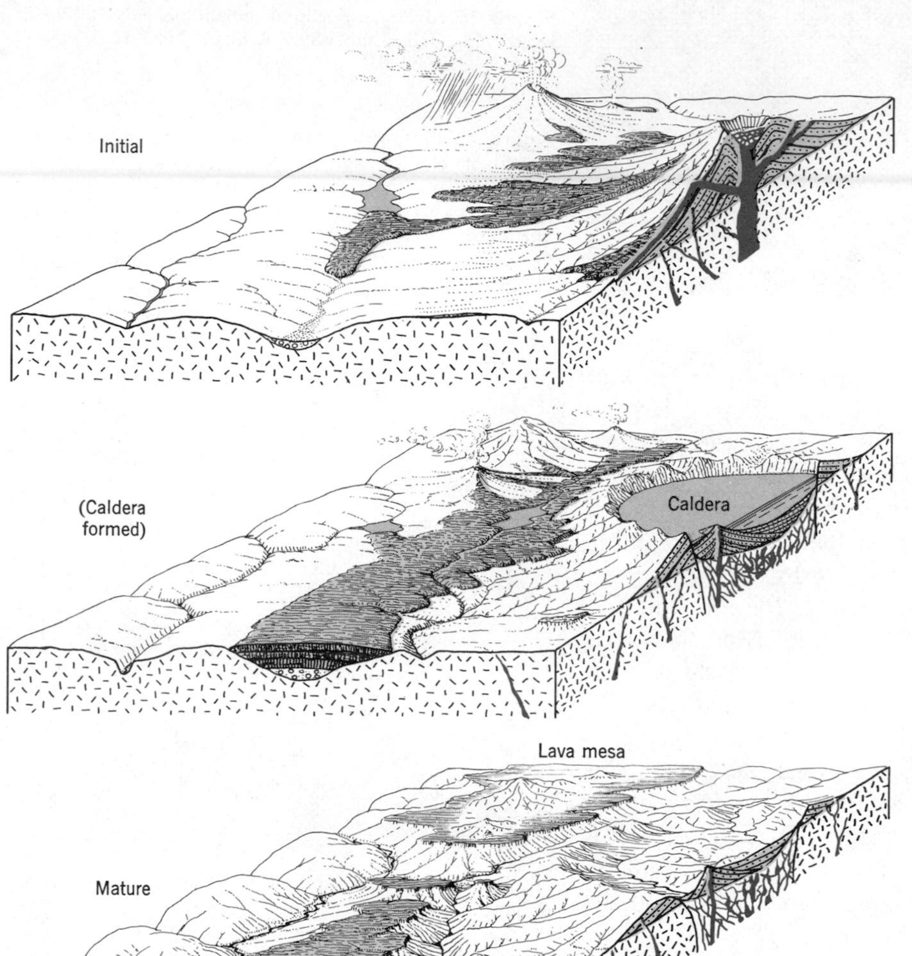

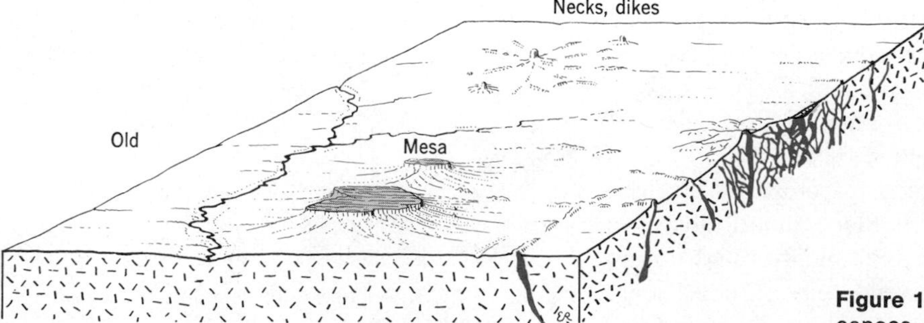

Figure 15.23 Stages of erosional development of volcanoes. (© 1960, John Wiley & Sons, New York.)

Figure 15.24 Shiprock, New Mexico, is a volcanic neck. The radiating narrow ridges are dikes. (Spence Air Photos.)

topographic basins. Figure 15.22 shows erosional development of a large fault block. After the block has been deeply dissected, all that is left of the fault scarp is a row of **triangular facets** along the mountain base.

One of the most extensive regions of fault block mountains and fault basins is the basin-and-range region of Nevada, Arizona, western Utah, southeastern California, southern Oregon, and parts of New Mexico.

EROSIONAL DEVELOPMENT OF VOLCANOES

Landforms resulting from the erosion of an extinct composite volcano and a caldera are illustrated in a series of block diagrams in Figure 15.23. The top block shows newly formed volcanoes and associated lava flows. The second block shows one of the cones to have been destroyed by explosion, leaving a caldera. More lava flows have been extruded, filling a stream valley.

In a later stage of dissection (third block), the cones and caldera have been cut up into systems of radial valleys and ridges. The crater rim is eventually breached by streams, so as to conduct the runoff out of the central depression by surface flow and to drain a lake that had occupied the crater.

Long after the volcanic cone has been entirely removed by denudation, there remain projecting above the land surface erosion remnants of the magma that solidified within the central **volcanic neck** and in near-vertical radial **volcanic dikes** (bottom block). Where the neck and dikes are surrounded by weak shale they rise as a conspicuous steep-sided peak with walls radiating like spokes from the hub of a wagon wheel (Figure 15.24).

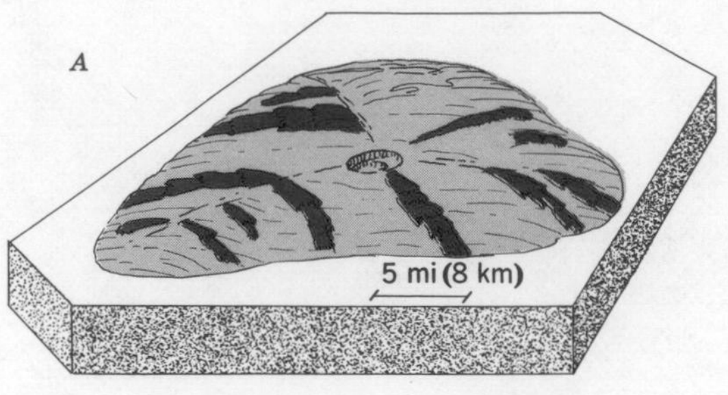

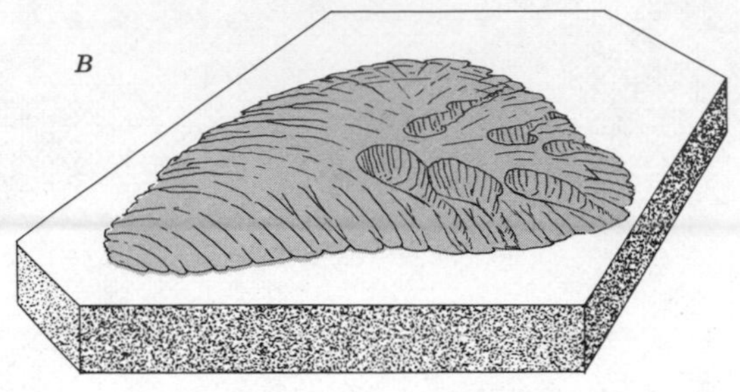

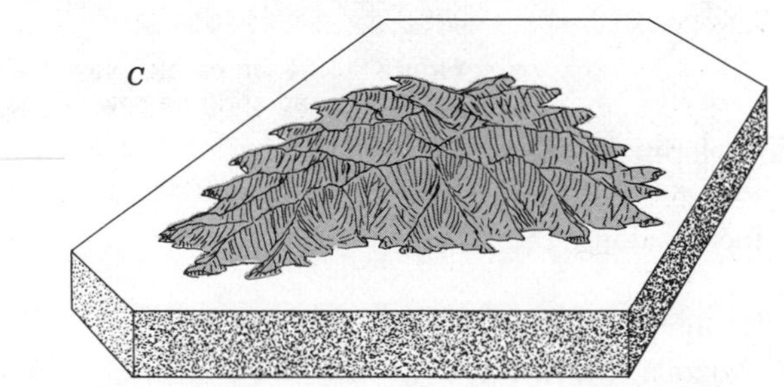

Figure 15.25 Stages in the erosion of Hawaiian shield volcanoes. (*A*) Initial stage, lava dome with central depression and fresh flows. (*B*) Valley heads deeply eroded. (*C*) Advanced stage of erosion. (© 1960, John Wiley & Sons, New York.)

Stages of erosion of Hawaiian shield volcanoes are illustrated in Figure 15.25. After volcanic action has ceased, streams begin to carve deep canyons into the flanks. Ultimately these canyons occupy the entire island, making a mountainous landscape with steep mountainsides and numerous sharp-crested ridges.

FLUVIAL DENUDATION IN REVIEW

In this chapter we first followed the idealized stages in denudation of a model landmass. While tectonic activity lifts a landmass rapidly, the process of denudation is very slow. The final stage, development of the peneplain, drags on for a very long time.

Rock structure exerts strong control over landforms shaped by fluvial denudation. Horizontal strata, coastal plains, domes, and folds are dissected into a remarkable variety of landforms as the weaker strata are etched from between resistant strata. Faulting in block structures gives another class of structurally controlled landforms; the erosion of volcanic structures provides still another set of landforms. Landforms controlled by structure are among the finest scenic features of our land; many are set aside as national parks and national monuments.

YOUR GEOSCIENCE VOCABULARY

homogeneous structure
drainage pattern
dendritic pattern
subsequent stream
rectangular pattern
horizontal strata
plateau
cliff
mesa
butte
coastal plain
consequent stream
cuesta
lowland
trellis drainage pattern
sedimentary dome
hogback
flatiron
radial streams
annular streams
anticlinal, synclinal, homoclinal mountains
anticlinal, synclinal valleys
water gap
plunge of fold
zigzag ridges
graben
horst
fault block mountain
triangular facets
volcanic neck
volcanic dike

SELF-TESTING QUESTIONS

1. Describe the stages in denudation of a landmass under a humid climate. What time spans are involved in these stages? Where can we see a peneplain today?

2. What kind of drainage pattern forms on a landmass of homogeneous rock? Describe the pattern and explain how it is developed. What effect does the presence of major faults have upon a stream pattern? What class of streams is developed on fault zones?

3. Describe the characteristic landforms developed on horizontal strata. Where can these be seen best developed? What drainage pattern develops on horizontal strata?

4. What landforms are distinctive of coastal plains? Where can a coastal plain be found today? What drainage pattern develops on a dissected coastal plain?

5. Describe the succession of landforms produced during the erosion of a sedimentary dome. What drainage pattern is developed? Of what classes of streams does the pattern consist?

6. Describe the evolution of mountains and valleys during erosion of a series of open folds. What effect does plunge of folds have upon landforms?

7. What landforms result from the pairing of normal faults? Describe a deeply eroded fault block. What evidence of faulting remains?

8. Describe the erosional features of composite volcanoes. How does erosion alter the form of a shield volcano?

THE MER DE GLACE.

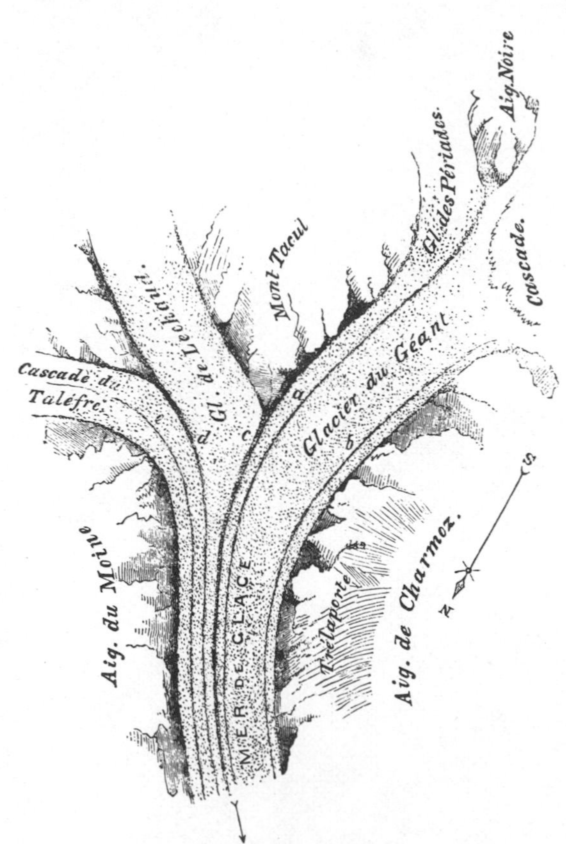

MORAINES OF THE MER DE GLACE.

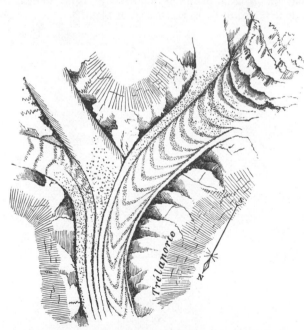

DIRT-BANDS OF THE MER DE GLACE, AS SEEN FROM THE CLEFT STATION, TRÉLAPORTE.

Illustrations from a book by John Tyndall, titled *Glaciers of the Alps* and published in 1861. Tyndall was a physicist interested in mountain climbing and glaciology. He made observations on the flow of Swiss glaciers during the 1850s, verifying the earlier results of studies by Louis Agassiz.

16 LANDFORMS SHAPED BY GLACIAL ICE

THE MAN WHO CHANGED HIS MIND

Conversions are often spectacular, the convert becoming a zealous apostle of his new viewpoint. One such convert was Jean Louis Rodolphe Agassiz, a leading nineteenth-century zoologist. Born in Switzerland in 1807, Agassiz' early preoccupation was with research on fossil fishes. While amassing enough information on the fishes to fill a five-volume treatise, Agassiz' interest became diverted to the glaciers of the Alps.

To understand what was going on, we must look back some decades and pick up the threads of a controversy. The objects of speculation in this controversy were boulders out of place, or erratics. Over many parts of the British Isles and northern Europe, boulders were observed to consist of rock different from that comprising the bedrock on which they rest. In many cases the boulder could be identified as having a source many miles away. Early in the 1800s, the view was widely held by geologists that erratics were carried by icebergs drifting in a sea that once covered the lands. Upon melting, the bergs dropped their boulders at odd places. Erratic boulders and layers of finer sediment associated with them came to be known as drift, an expression quite appropriate if they were truly of iceberg origin. To this day we refer to such deposits as drift. Charles Darwin was among the adherents to the iceberg hypothesis in his younger years. He had good company in such noted geologists as Lyell and von Buch, whom we have mentioned in an earlier chapter.

Getting back to the Alps, there are countless erratics over the Swiss upland meadows and in the deeply carved valleys. Many of these bear scratches and grooves. Upon seeing these, James Hutton (our hero in the preceding chapter) concluded that the erratics had been carried by glaciers, which he said were formerly much longer and thicker than we find them today, and that the marks were produced during transport.

Thus the germ of a new theory of erratics had its beginning. Others took up the idea and during the 1820s and 1830s the glacial theory, as it was to become known, was debated in sessions of the Helvetic Society, a prestigious science forum. It was here, in 1834, that Louis Agassiz entered the picture. He had attended a meeting of that society at which a Swiss colleague, Jean de Charpentier, read a paper strongly supporting the glacial theory, even going so far as to argue in favor of a former great polar ice cap as the bearer of erratics now found spread over the British Isles and northern Europe. Agassiz felt sure that de Charpentier was wrong, and in the summer of 1836 he examined for himself some glaciers and the valley deposits associated with them, hoping to find evidence against the glacial theory. Instead, after weeks of study, Agassiz became a convert. Waxing enthusiastic, he addressed the Helvetic Society the following year, arguing for de Charpentier's view.

Three years later, Agassiz published his views in a book entitled *Étude sur les glaciers,* and it received widespread publicity. Unfortunately, Agassiz did not give much credit to de Charpentier, who had priority in the matter, and the latter felt miffed. However, the iceberg theory of erratics lingered on well after Agassiz' exposition, and it was not until the 1860s that it was put to rest by leading geologists.

Agassiz came to America in 1846 and became a professor of zoology and geology at Harvard University. He was well received, particularly by Edward Hitchcock, state geologist of Massachusetts, who had already espoused the glacial theory. Even so, as in Europe, the glacial theory was not fully accepted in the United States until the 1860s. Exploration of the Greenland Ice Sheet in the 1850s helped to sell the glacial theory, since it proved that a vast ice sheet can exist and that glacial drift is being formed today.

In this chapter we don't even bother to argue the case for the glacial theory of drift. Over a century has passed since the matter was settled, once and for all.

GLACIAL ICE

Because most of us have dealt with ice only in small pieces or thin layers, we think of it as a brittle solid. Ice fractures easily and often shows crystalline structure, for it is a true mineral. But where an ice layer has accumulated on land to a depth of about 300 ft (100 m) and rests on a sloping surface, the basal part of the ice becomes plastic, yielding by slow internal flowage. The entire ice mass is carried downslope and becomes a **glacier.**

For a glacier to form, the mass of snowfall added annually must exceed the mass of ice lost annually by melting and evaporation. Geologists use the term **ablation** for the combined ice losses through both melting and evaporation.

Freshly fallen snow on a glacier surface rapidly becomes more compact. Under the load of new snow, older layers reach a dense granular state called **firn,** with about half the density of pure ice. Deeper burial causes further compaction, expelling most of the air and closing the voids between grains. The resulting glacial ice undergoes recrystallization and comes to resemble a coarse-textured plutonic rock.

Today, glaciers are sustained on high mountains and plateaus, where air temperatures are low and there is abundant snowfall. The west coasts of continents in high latitudes are favored by abundant snowfall, and here glaciers flourish. Even in the equatorial zone of South America, we find glaciers thriving on the Andes, because of their very high altitude. Another tropical belt of large glaciers is within the Himalayas of southern Asia, the loftiest of the earth's mountain chains.

Glaciers formed in high, steep-walled mountain ranges are shaped into long, narrow ice streams. These glaciers occupy previously carved stream valleys. They have many of the basic elements of a fluvial drainage system, such as tributary channels leading downgrade to a trunk stream, and are classified as **alpine glaciers** (Figure 16.1). Ice masses of continental proportions, such as those of Greenland and Antarctica, are called **ice sheets.** A great ice sheet may reach a thickness of several thousand feet and may cover an area of several million square miles.

ALPINE GLACIERS

The form of a simple alpine glacier is illustrated in Figure 16.2 by cross sections. Snow accumulates at the upper end of the glacier in a bowl-shaped depression, the **cirque.** The cirque holds a broad expanse of smooth glacier surface, the **firn field.** By slow flowage at depth, the glacial ice moves toward the exit of the cirque and may pass over a steepened gradient in the rock floor of the valley, resulting in an **ice fall.** Here the rigid surface ice is deeply broken by gaping fractures, called **crevasses.** A large ice fall can be seen on the Swiss glacier in Figure 16.3.

From the standpoint of relative gain and loss of mass, the glacier may be divided into two parts. The high-altitude part lies in the **zone of accumulation.** Here the glacier receives more snow than it loses by ablation, so it has a smooth firn surface. The lower part

Figure 16.1 This branching alpine glacier flows westward along the northern edge of the Juneau Ice Field in Alaska. The dark bands are medial moraines. (U.S. Army Air Force.)

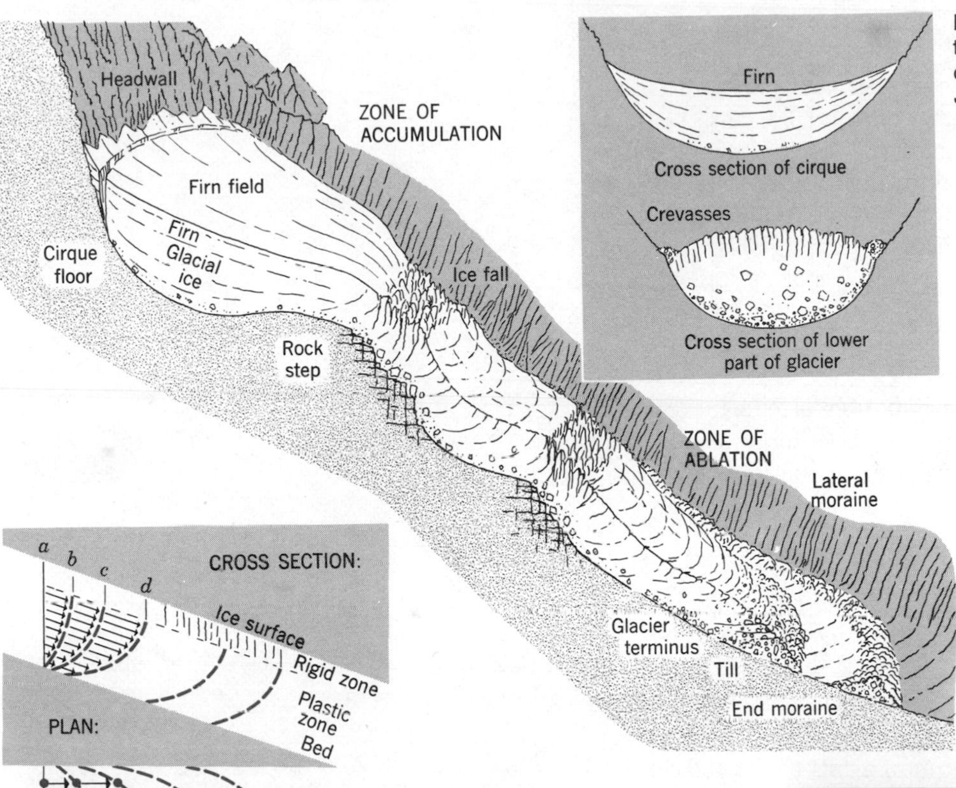

Figure 16.2 Idealized longitudinal and transverse sections show the anatomy of a simple alpine glacier. (© 1973, John Wiley & Sons, New York.)

lies in the **zone of ablation.** Here the glacier loses more of its surface to ablation than it gains yearly by snowfall. This zone is characteristically rough and pitted. Typically the cross profile of the glacier is concave upward in the zone of accumulation and convex upward in the zone of ablation (Figure 16.2).

As the glacier extends farther and lower down-valley, it encounters progressively warmer climatic zones and suffers progressively greater losses from ablation. At the **glacier terminus** the rigid ice is thrust forward over its own deposits of debris. We may think of a valley glacier as a moving stream, nourished at the upper part, but disappearing by surface loss in the lower part. When rates of gain and loss are balanced, the glacier surface remains approximately constant in form and position, despite the fact that ice is moving from one end to the other. This condition is called **glacier equilibrium.** As long as glacier equilibrium is maintained, the terminus of the glacier holds to a fixed position.

Glacier equilibrium is easily upset. Suppose, for example, that the rate of nourishment by snowfall is sharply increased. The ice of the upper end of the glacier thickens and consequently flows more rapidly downgrade. This action moves the glacier terminus farther down-valley, in what is described as a **glacier advance.** But the advance brings the glacier terminus to lower levels, where temperatures are higher and ablation more rapid. Soon the rate of ablation

Figure 16.3 Glacier d'Argentière in the Swiss Alps has a great ice fall and deeply crevassed surface in the lower section. The higher part in the background has a smooth firn surface, extending back into large cirque at the upper lett corner of the photograph. (Swissair Photo.)

balances the rate of ice advance, and the terminus again becomes stabilized.

Suppose, instead, that the rate of glacier nourishment decreases, as it might if summers become warmer or snowfall is less. The thinning glacier will flow less rapidly and will not be brought to the terminus as rapidly. Now the rate of ablation will be excessive, and the glacier terminus and surface will melt away rapidly, causing the terminus to retreat up-valley. This change is called **glacier recession.** But as the glacier recedes up-valley less and less of its surface lies in warmer climatic zones; as a result, ablation is reduced. Again equilibrium will be established, with the terminus stabilized at a higher level.

Glaciers are extremely sensitive indicators of climatic changes. Scientists are documenting the changes in glacier form in an attempt to interpret climatic changes that have been responsible.

GLACIER FLOW

The rate of glacier flow has long been a subject of interest to **glaciologists,** those scientists specializing in the study of glaciers. Pioneers in this field—Louis Agassiz, the Swiss-born naturalist, and J. D. Forbes, an English physicist—made observations on glaciers of the Swiss Alps from 1840 to 1842. They planted rows of stakes across the glacier surface and surveyed the movements of the stakes. Both men came independently to the conclusion that surface motion is most rapid near the center, decreasing toward the sides. If the line of stakes were to be mapped at equal intervals of time, it would be found to be bent convexly down-valley into a series of parabolic curves, as suggested by lines in Figure 16.2.

Slow surface motion of this type normally goes on at the rate of a few inches to 3 ft (1 m) per day in a valley glacier. Agassiz found that his fastest moving stakes had traveled over 250 ft (75 m) downstream in one year. A large boulder resting on the ice moved almost 500 ft (150 m) in two years. It should be remembered, of course, that the upper zone of the glacier is composed of brittle ice in the rigid zone. The relative movements of surface points, although resembling flowage of a fluid on a large scale, actually occur by slippage between small blocks formed by the fracturing of the brittle ice. Crevasses are produced in this way. Below the rigid zone, the ice within the glacier moves less rapidly with increasing depth, as indicated in Figure 16.2.

At certain times and for very short periods the entire glacier is thrust forward as a block and apparently is undergoing a slipping movement over its bed. This phenomenon is termed **basal slip.** The sudden movement is caused by strong push from upstream sections of the ice.

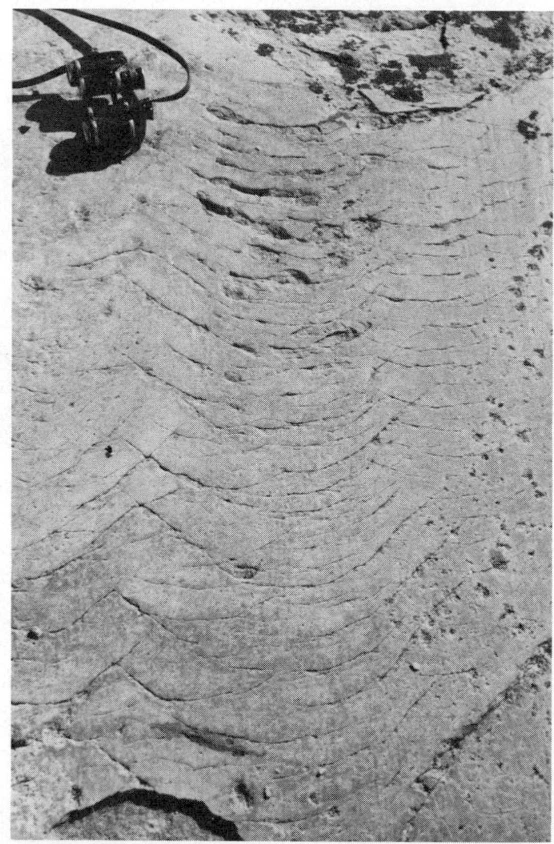

Figure 16.4 These sets of chatter marks indicate that ice movement was away from the observer. Elmer Creek, Sierra Nevada, California. (D. L. Babenroth.)

GLACIER EROSION AND TRANSPORTATION

Like streams of water, glaciers represent to the geologist much more than mere systems of water disposal in the hydrologic cycle. Glaciers perform erosion, transportation, and deposition of mineral matter.

Rock fragments become enclosed in glacial ice from the rock floor. Close to the headwall of the cirque, meltwater pouring down from snowbanks above the glacier enters the joint fractures in the headwall rock. Here it freezes into seams of ice, breaking loose joint blocks. These blocks become incorporated into the upper end of the glacier. Beneath the glacier, ice flows plastically around joint blocks, then drags them loose when a sudden basal slip occurs. This activity is **glacial plucking.** Blocks of rock being carried within the glacial ice are scraped and dragged along the rock floor, gouging and grooving the bedrock and chipping out fragments of rock. This process of abrasion is **grinding.**

The valley floor formerly beneath a glacier shows a number of interesting erosional features resulting from grinding and plucking. Rock surfaces are generally of hard fresh rock and bear numerous fine scratches called **glacial striations.** Striations mark the lines where sharp corners of large fragments have scraped the surface. Where pressure was strongly applied, the impinging boulders created curved fractures in the bedrock. One fracture type, the **chatter mark,** is bent concavely toward the downstream direction of ice flow (Figure 16.4).

In particularly susceptible types of bedrock, such as some limestones, glacial abrasion produces long, deep **glacial grooves.** These troughlike features parallel the direction of ice flow and are scored by numerous parallel striations.

A common small landform produced beneath a glacier is the **glaciated rock knob.** It is a hill of bedrock strongly shaped by abrasion and plucking (Figure 16.5). The side of the knob facing upstream with respect to ice flow is smoothly rounded by ice abrasion; the downstream side is strongly plucked.

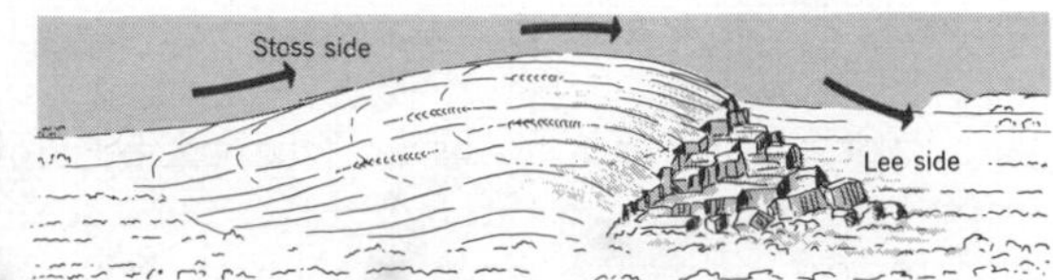

Figure 16.5 Sketch of a glacially rounded and plucked rock knob. Arrows indicate the direction of ice flow.

Still another source of glacier load is the rolling and sliding of rock fragments down the steep sides of the cirque and the valley walls adjacent to the ice stream. Such slide rock takes the form of talus cones and sheets (Figure 16.6). Rock fragments reaching the glacier margin are dragged along by the moving ice. These marginal embankments of debris are called **lateral moraines.** After the glacier has disappeared, these embankments form ridges parallel with the valley walls (Figure 16.6).

Where two ice streams join, the debris of the inner lateral moraines is dragged out into the middle of the combined ice streams to form a long narrow line of debris, a **medial moraine.** Several of these features are visible in Figure 16.1, appearing as parallel lines dividing the glacier into narrow bands. Debris supplied from marginal slopes remains largely on the glacier surface.

Near the glacier terminus the proportion of solid load to ice increases greatly, until at the very end there is more solid debris than ice. This residual mass of rock debris constitutes the **terminal moraine** of the glacier (Figure 16.6).

Seeing the huge boulders composing glacial moraines, you may not realize that much rock is also ground by the glacier into extremely fine particles. This fine silt and clay is called **rock flour.** It gives to meltwater streams issuing from a glacier a characteristic milky appearance. Settling out in lakes beyond the glacier limit, the rock flour forms layers of silt and clay.

Figure 16.6 The shrunken remnant of the Black Glacier is almost buried in its own morainal debris. Talus cones have been built from the valley walls. The terminal moraine at the lower left extends up-valley in ridgelike lateral moraines. Bishop Range, Selkirk Mountains, British Columbia. (H. Palmer, Geological Survey of Canada.)

LANDFORMS CARVED BY ALPINE GLACIATION

Many spectacular landforms are shaped by alpine glacier erosion. These can be explained through the use of a series of block diagrams (Figure 16.7). Block A shows a mountain region shaped by processes of fluvial denudation in a period of milder climate preceding glaciation. Thick accumulations of soil and regolith mantle the mountain slopes; divides are broad and somewhat subdued in appearance.

Block *B* shows the same mountain mass occupied by valley glaciers that have been in action for many thousands of years. The higher central summit area has been carved into steep-walled cirques. These meet in sharp, knife-edge crests, culminating in toothlike peaks, called **horns.** Where two cirques are arranged back to back on opposite sides of a divide, the intervening rock wall may be cut through to form a deep pass. Notice that the somewhat lower mountain summit at the far right of block *B* is only partly consumed by cirque development and that the preglacial mountain surface remains intact over the summit.

Tributaries enter the main glacier with smooth ice-surface junctions. Abrasion of the valley walls has planed away projecting spurs of the preglacial stream valleys. Where the main glacier blocks a stream valley at low altitude, discharge is impounded in temporary lakes.

Block *C* shows this same region after the glaciers have entirely disappeared because of a general climatic warming. This condition is found today in many high mountain ranges of the middle latitudes. The trunk glacier eroded a deep U-shaped **glacial trough.** Its smaller tributaries also carved troughs, but because these ice streams were of smaller cross section, their floors were not so deeply cut. These tributary troughs now enter at levels high above the main trough floor; they are **hanging troughs.** In the upper reaches of the troughs are many irregularities of gradient, giving a succession of **rock steps** and **rock basins;** the latter often hold lakes. Depressions in the floors of cirques hold lakes, called **tarns.**

Grandest of all glacially carved landforms are the glacial troughs; they may be thousands of feet deep and tens of miles long. Troughs cut below sea level in a coastal mountain range are now invaded by the sea to become **fiords,** characterized by the steepness of their walls and the great depth of their floors (Figure 16.8). The water is often several hundred feet deep. Fiords are widely found along mountainous coasts of arctic and subarctic latitudes. Important fiord coasts occur in Alaska and British Columbia, southern Chile, Scotland, and Norway.

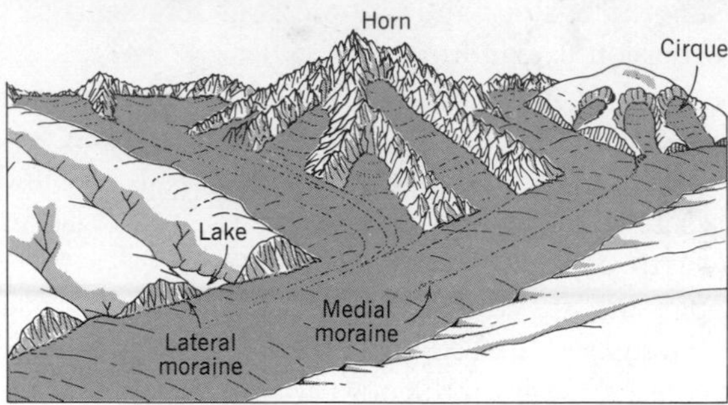

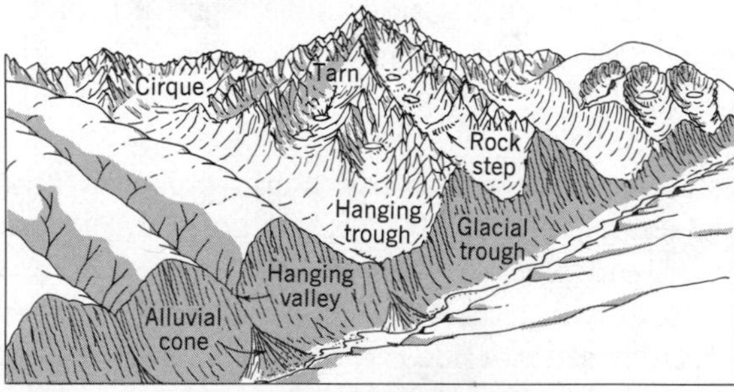

Figure 16.7 Landscape evolution under alpine glaciation. (*Top*) Preglacial topography formed by fluvial denudation. (*Center*) Stage of maximum alpine glaciation. (*Bottom*) Postglacial landscape, following disappearance of all glacial ice. (© 1973, John Wiley & Sons, New York.)

GREENLAND AND ANTARCTIC ICE SHEETS

From the small, streamlike alpine glaciers we turn our attention to the two enormous ice masses of subcontinental size that exist today. Greenland and Antarctica bear these great ice sheets.

The Greenland Ice Sheet occupies some 670,000 sq mi (1,740,000 sq km), which is 80% of the entire area of the island of Greenland.

Figure 16.8 This Norwegian fiord has the steep rock walls of a deep glacial trough. (Mittet and Co.)

Ice covers all but narrow land fringes (Figure 16.9). The ice sheet contains almost 700,000 cu mi (3,000,000 cu km) of ice. In a general way the ice forms a single broadly arched, doubly convex ice lens. It is smooth surfaced on the upper side and considerably rougher and less strongly curved on the underside. The mountainous terrain of the coast passes inland beneath the ice, but gives way to a central lowland area close to sea level in elevation. The ice surface is characterized by wind-eroded and drifted features (Figure 16.10).

The ice thickness measures close to 10,000 ft (3 km) at its greatest. It is not surprising that the center of Greenland is actually depressed under such a load, in conformity with the principle of isostasy (see Chapter 2), since 10,000 ft (3 km) of glacial ice is roughly equivalent to a rock layer at least 3000 ft (0.9 km) thick.

Because the Greenland ice surface slopes seaward, the ice creeps slowly downward and outward toward the margins, where it discharges by glacial tongues. These are **outlet glaciers**—closely resembling alpine glaciers, but fed from a vast ice sheet rather than from a cirque. Outlet glaciers reach the sea in fiords and are the source of North Atlantic icebergs.

Like Greenland, the Antarctic continent is almost entirely buried beneath glacial ice. This is an ice area of just over 5 million sq mi (13 million sq km), or about 1½ times the total area of the contiguous 48 states of the United States (Figure 16.11). Ice volume is about 6 million cu mi (25 million cu km), which is over 90% of the total volume of the earth's glacial ice. (In comparison, the Greenland Ice Sheet has about 8%.)

The Antarctic Ice Sheet reaches its highest elevations, almost 13,000 ft (4 km), in a broadly rounded summit. Surface slope is gradual to within 200 mi (320 km) of the edge of the continent, where a marked steepening occurs. The greatest ice thickness is over 10,000 ft (3000 m). Although, in general, the ice is thickest where surface elevation is highest, there are important exceptions to this statement. There is a great subglacial channel, or valley, named the Byrd Basin (see profile in Figure 16.11). Here an ice thickness of nearly 13,000 ft (4000 m) has been measured, and the rock floor lies some 6500 ft (2000 m) below sea level.

The subglacial topography of Antarctica is partly mountainous. Mountain peaks and ranges rise above the ice in several belts, mostly close to the continental margins. Here the ice moves in outlet glaciers from the interior polar plateau to reach the coast (Figure 16.12).

A characteristic feature of the Antarctic coast is the presence of numerous **ice shelves,** which are great plates of floating ice attached to the land (Figure 16.11). Largest is the Ross Ice Shelf, about 200,000 sq mi (520,000 sq km) in area, with its surface at an average elevation of about 225 ft (68 m), (Figure 16.13). Almost as large is the Filchner Ice Shelf. Smaller ice shelves occupy most of the bays of the Antarctic coast and in places form a continuous but narrow ice fringe. The ice shelves represent those parts of the ice sheet that have been pushed seaward into water of sufficient depth that the ice is floated off the bottom. The ice shelves are largely maintained by snow accumulation on their surfaces. These ice shelves are the source of enormous tablelike icebergs of the Antarctic Ocean.

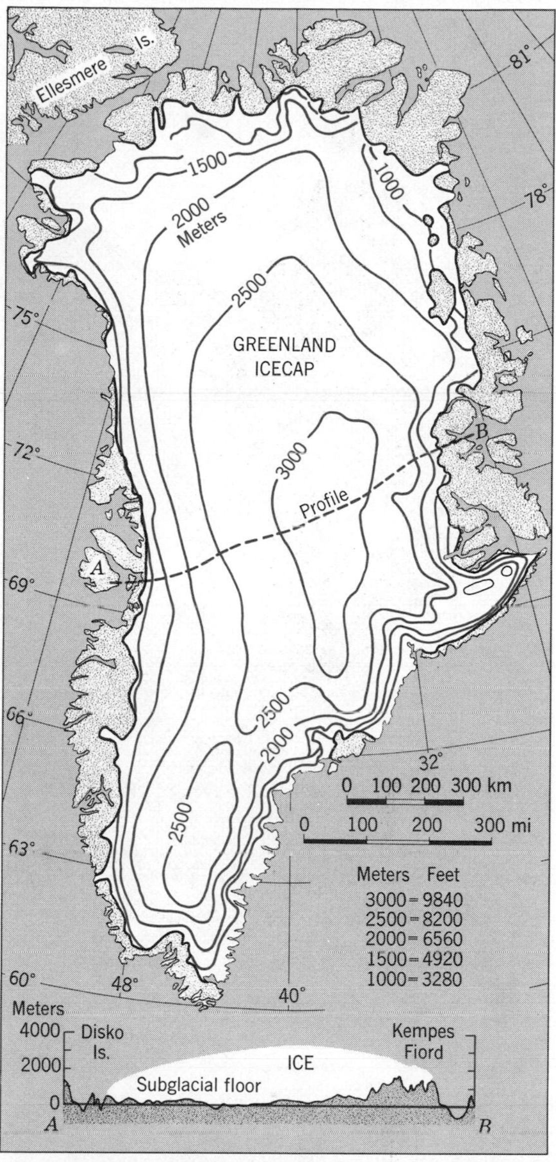

Figure 16.9 Greenland and its ice sheet. (© 1973, John Wiley & Sons, New York.)

Figure 16.10 Blizzard winds have eroded the packed snow surface of the Greenland Ice Sheet. (L. H. Nobles.)

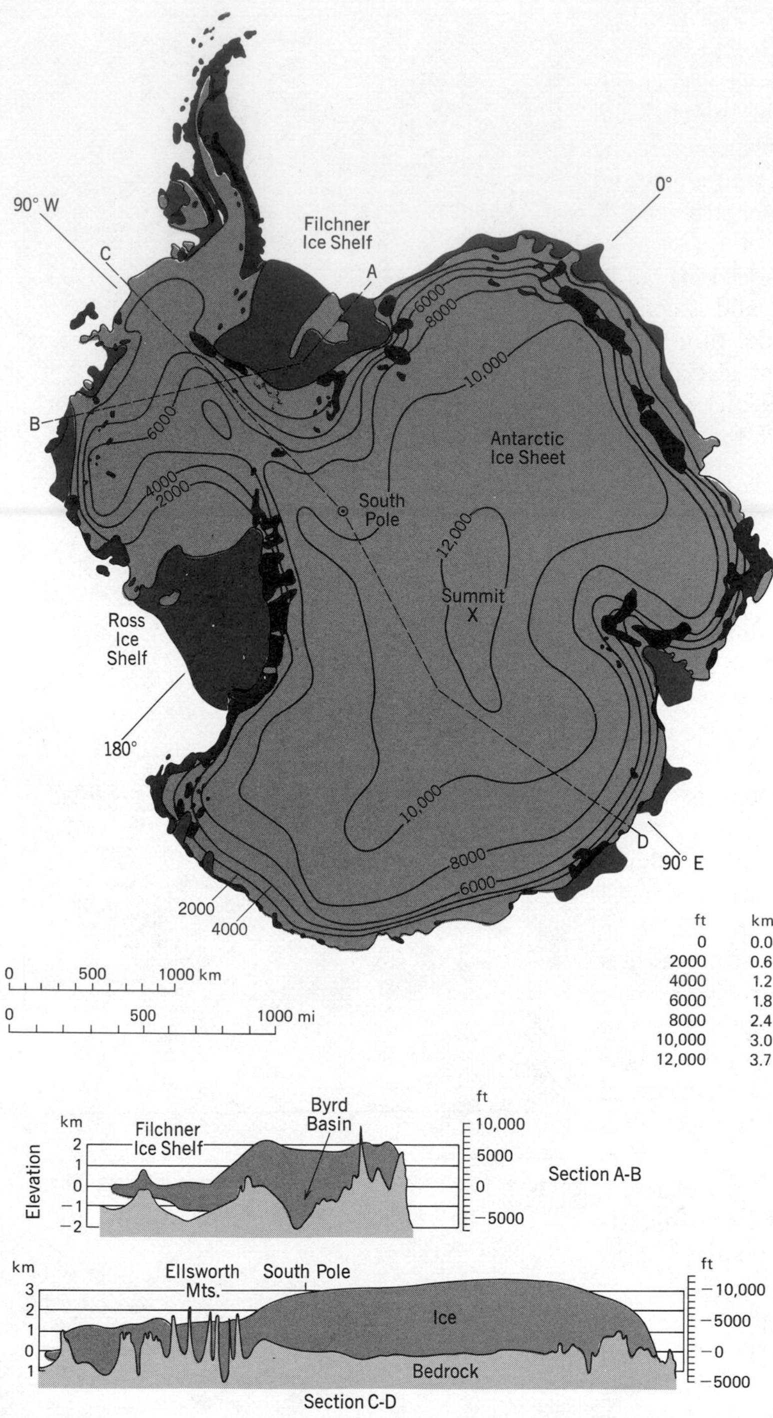

Figure 16.11 Map of Antarctica showing ice surface elevations by contours. Areas of exposed bedrock shown in black; ice shelves by darker color.

Table 16.1 **North American glaciations**

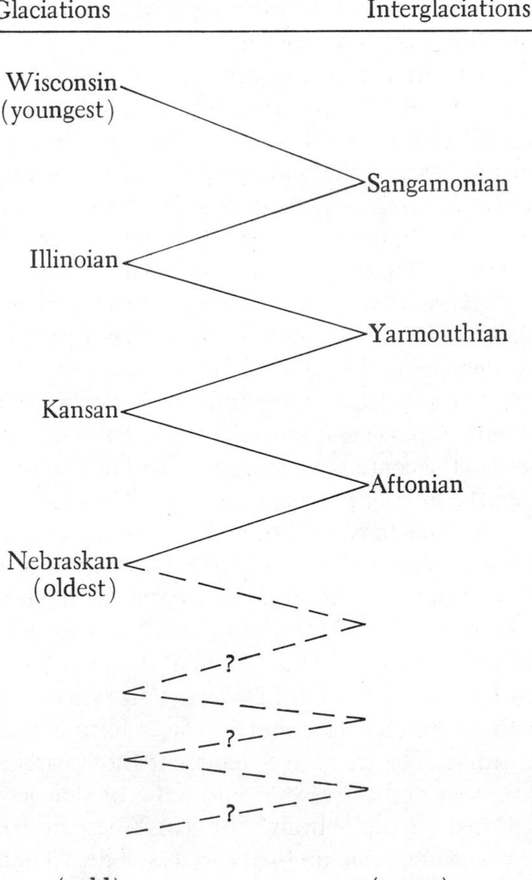

PLEISTOCENE ICE SHEETS

We now know beyond doubt that large parts of North America, Eurasia, and South America were covered by great ice sheets in the Pleistocene Epoch. This unit of geologic time spanned approximately the last 2 million years. Only within the last 10,000 to 15,000 years did these ice sheets disappear from the now heavily populated lands of North America and Europe. The landforms resulting from glacial erosion and deposition are in many places extremely fresh in appearance, so that the former ice limits can be mapped in great detail.

The maximum extent of ice sheets of the Pleistocene Epoch in North America and Europe is shown in Figures 16.14 and 16.15. In North America, all of Canada and the mountainous areas of Alaska were covered. Over the Cordilleran Ranges alpine glaciers coalesced into a single ice body which spread westward to the Pacific shores and eastward down to the foothills of the mountains. Much larger was the great Laurentide ice sheet, which was centered over Hudson Bay and spread radially. The Laurentide ice sheet inundated the Great Lakes area and spread south into the United States about as far as the line of the Missouri and Ohio Rivers.

In Europe, the Scandinavian ice sheet was centered over the Baltic Sea. It covered all of the Scandinavian peninsula and reached southward and eastward into the Low Countries, Germany, Poland, and Russia. This ice mass also spread westward across the North Sea, where it joined with an ice sheet that covered much of the British Isles. The Alps and Pyrenees ranges bore small caps of ice formed by the coalescence of many individual valley glaciers. As you would expect, glacial activity in all the world's high mountain ranges was greatly intensified. Valley glaciers were then much larger and extended into lower altitudes than they do today.

The only large ice sheet of the Southern Hemisphere (exclusive of Antarctica) was in South America. It was formed over the Andean range of Chile and Argentina, largely south of the 40th parallel, by coalescence of valley glaciers.

Figure 16.12 In this air view of the head of Shackleton Glacier, Antarctica, the polar ice plateau is seen in the distance. (U.S. Geological Survey.)

GLACIATIONS AND INTERGLACIATIONS

The growth and spread of ice sheets over the continents is called a **glaciation.** During mild climatic periods between successive glaciations the ice sheets largely disappeared. These episodes are called **interglaciations.**

For several decades four glaciations have been recognized in North America, while a similar and possibly equivalent four-glaciation history has been established for Europe on the basis of studies in the Alps. The total number of glaciations of the Pleistocene Epoch remains to be established and may even exceed six. Names of the four established glaciations and interglaciations of North America are given in Table 16.1.

Maximum southern extent of ice in each glaciation in the north central United States is shown in Figure 16.16. The ice fronts advanced south in great lobes. Notice in Figure 16.16 that an area in southwestern Wisconsin escaped inundation by Pleistocene ice sheets. Known as the Driftless Area, it was apparently bypassed by glacial lobes moving on either side.

Figure 16.13 The Ross Ice Shelf, Antarctica. The steep ice cliff, from 50 to 150 ft (15 to 46 m) high, presents a formidable barrier. (Official U.S. Coast Guard photograph).

CAUSES OF CONTINENTAL GLACIATION

The cause of glaciations remains uncertain despite all efforts of modern science to find a satisfactory explanation. We can only touch the topic lightly.

We know that glaciation has occurred a number of times in the geologic past. One of the best documented is the Carboniferous-Permian glaciation, from which there remains lithified glacial debris, including striated and faceted stones and a striated surface of older rock upon which the indurated glacial clay rests. The hypothesis that this glaciation occurred while Pangaea remained intact and was centered over the south pole was discussed in Chapter 9. Good geologic evidence is also available of glaciation in earliest Cambrian time. Any theory of glaciations must therefore account for occasional and seemingly sporadic repetitions of glaciation throughout all recorded geologic time.

A general requirement of glaciation is a lowering of the earth's average atmospheric temperature, along with sustained or increased levels of precipitation. It is well established by worldwide evidence that during the Pleistocene Epoch, the **snow line** (elevation above which snowbanks remain throughout the year) was lowered in elevation by about 2000 ft (600 m) in equatorial latitudes and by 3000 to 4000 ft (900 to 1200 m) in middle and high latitudes. This worldwide phenomenon clearly indicates a generally colder climate for the earth as a whole at times of glaciation. This conclusion is strongly reinforced by the data of deep-sea cores. A reduced average temperature would, in general, reduce rates of ablation at those places where snow could accumulate in large quantities. Reduced ablation would aid the growth of glacial ice bodies.

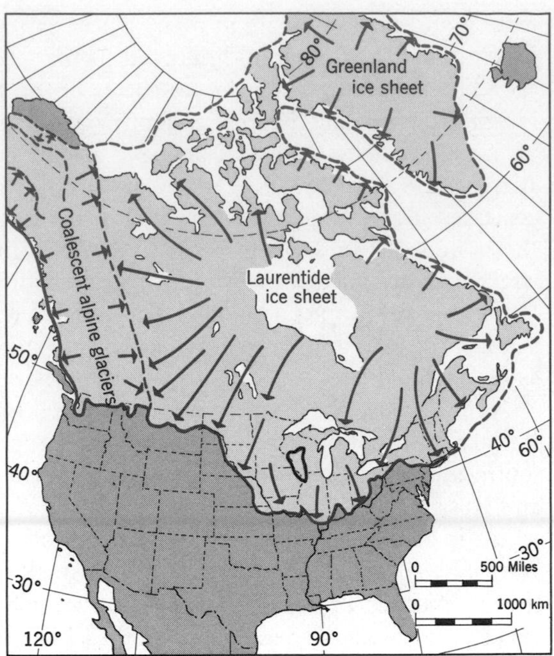

Figure 16.14 Maximum extent of Pleistocene ice sheets of North America. (© 1973, John Wiley & Sons, New York.)

Another requirement of ice sheet growth is that there be present an elevated landmass—a plateau or mountain range—favorably situated to receive snowfall. Low-lying continental plains would not be likely to have enough snowfall to initiate ice sheet growth, even if the climate were sufficiently cold. For this reason some large arctic areas of Siberia were never glaciated.

Favorable topographic conditions are found in Greenland and Antarctica, where ice sheets exist today, and in the Labrador Highlands, the northern Cordilleran ranges, Scandinavia, and the southern Andes. These were the centers from which Pleistocene ice sheets grew. Once formed, a small body of ice might be expected to grow into a large ice sheet, since the ice body itself would function as a highland to induce the necessary precipitation. Now, we know that Pliocene and early Pleistocene times saw great tectonic and volcanic activity, resulting in growth of mountain ranges. Upwarping of large parts of the more stable continental shields also occurred at this time, so it is reasonable to suppose that general uplift of the continents on a worldwide basis created topographic conditions favorable to the growth of Pleistocene ice sheets. Assuming a favorable global topography, what mechanisms can be invoked for repeated cycles of atmospheric cooling and warming?

One hypothesis requires that the sun's output of radiant energy underwent a reduction, causing a lowering of average air temperatures. Another hypothesis relies upon changes in the tilt of the earth's axis and changes in the earth's orbit to reduce the amount of solar energy falling upon the high-latitude zone. Other hypotheses of glaciation require a reduction in the content of atmospheric carbon dioxide, or an increase in the content of atmospheric dust, either

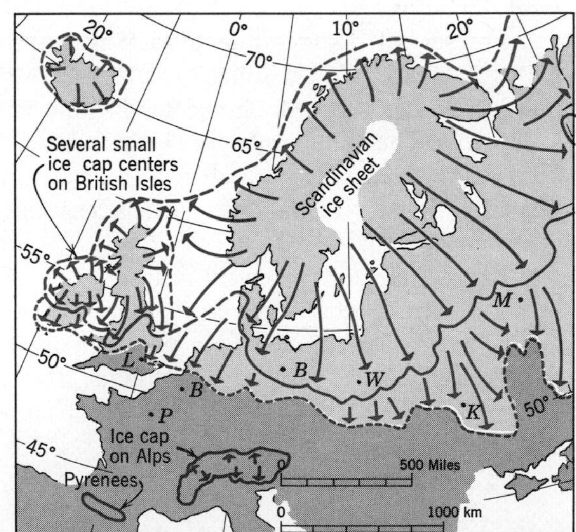

Figure 16.15 Limit of glacial ice of Europe in the last glaciation is shown by a solid line; maximum extent in the entire Pleistocene Epoch by a dashed line. (© 1973, John Wiley & Sons, New York.)

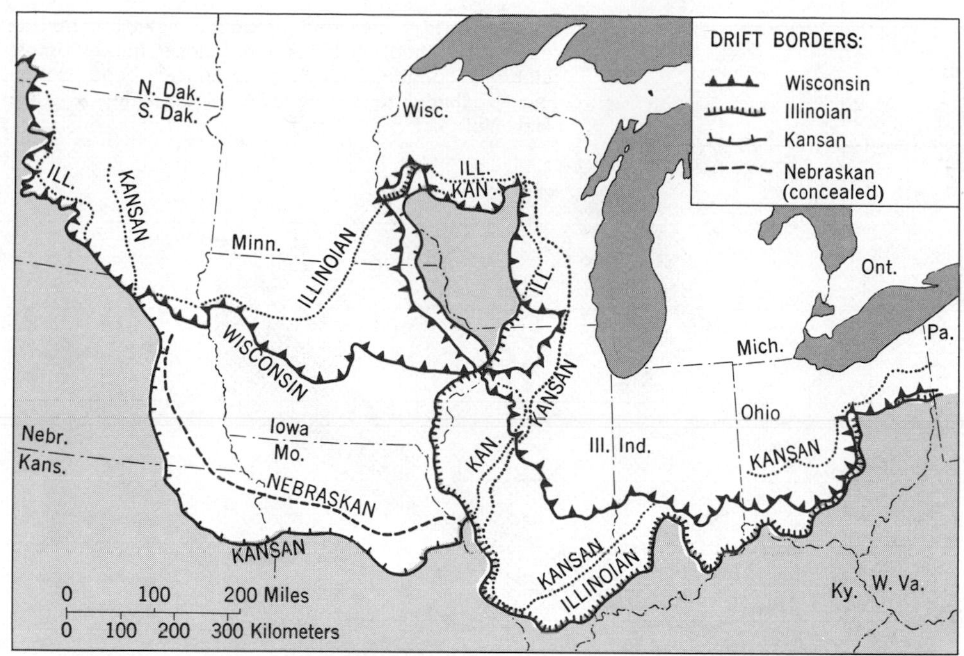

Figure 16.16 Ice limits in the north central United States for the four glaciations. (© 1973, John Wiley & Sons, New York.)

of which changes would tend to lower air temperatures. As yet, no single hypothesis of glaciation has gained general acceptance.

GLACIAL DRIFT

The term **glacial drift** applies to all varieties of rock debris deposited in close association with Pleistocene ice sheets. Drift consists of two major classes of materials: (1) **Stratified drift** is made up of sorted and layered clay, silt, sand, or gravel. This material was deposited as bed load or delta sediment from streams of meltwater, or settled from suspension into bodies of quiet water adjoining the ice. (2) **Till** is an unsorted mixture of rock and mineral fragments of a wide range of sizes—from clay to large boulders—deposited directly from glacial ice. One form of till consists of material dragged along beneath the moving ice and plastered upon the bedrock or upon other glacial deposits. Another form consists of debris held within the ice but dropped in place as the ice wasted away. Large rock fragments within glacial till are called **erratic boulders.** Some have been transported many tens of miles.

Within the area of the United States once covered by Pleistocene ice sheets, glacial drift has a thickness averaging from 20 ft (6 m) over mountainous terrain of the northeast to perhaps 50 ft (15 m) or more over the plains of the northcentral states. Drift is generally 50 to 200 ft (15 to 60 m) thick in Iowa and over 100 ft (30 m) in Illinois. As you would expect, the greatest thicknesses occur in preglacial valleys and lowlands. The thinnest deposits are over plateaus and hill summits. Consequently, the gross effect of continental glaciation has been to subdue the landscape to a more plainlike aspect.

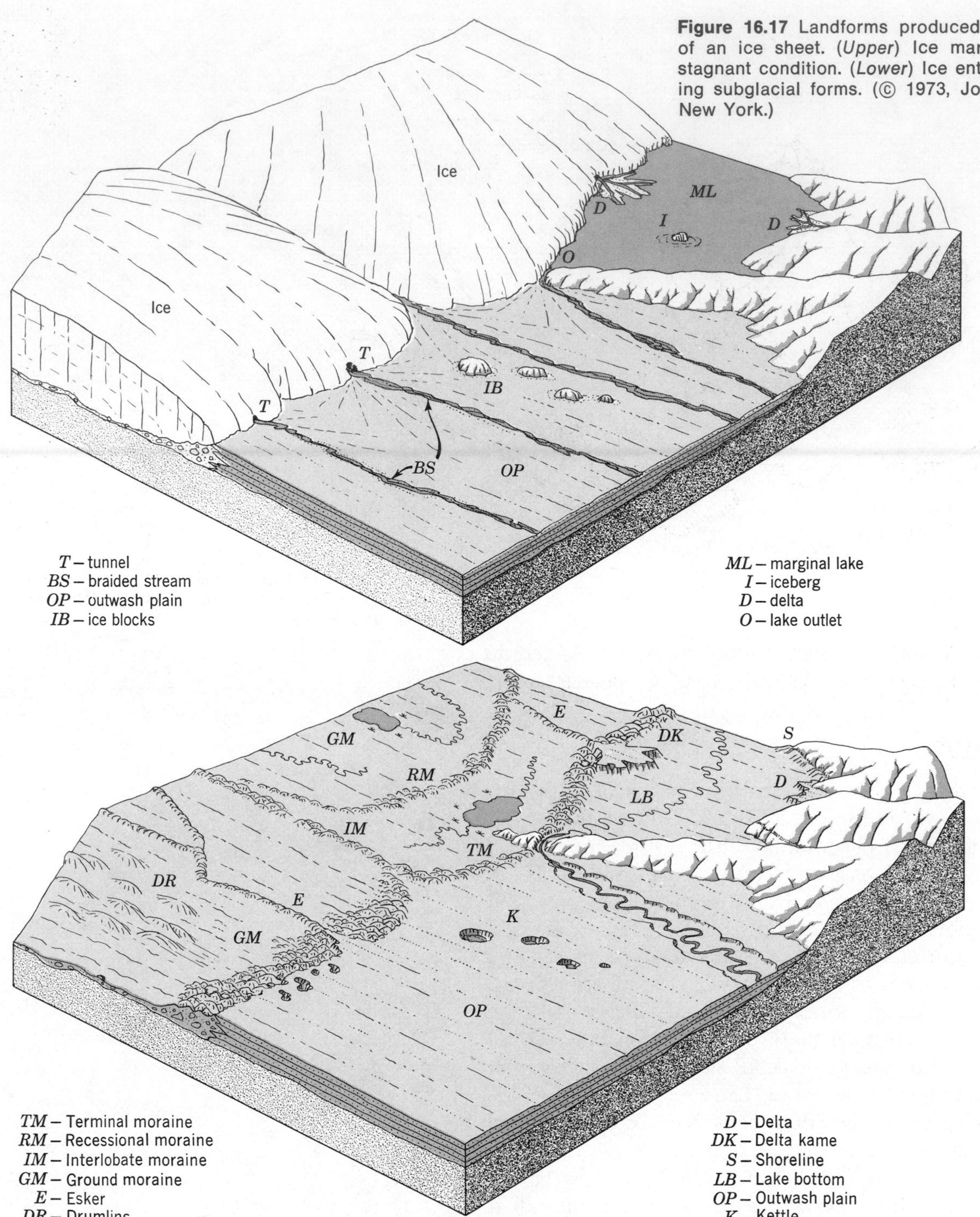

Figure 16.17 Landforms produced near the margin of an ice sheet. (*Upper*) Ice margin in an almost stagnant condition. (*Lower*) Ice entirely gone, exposing subglacial forms. (© 1973, John Wiley & Sons, New York.)

LANDFORMS OF CONTINENTAL GLACIATION

The advance and wastage of a great ice sheet leaves a host of distinctive minor landforms. They can best be understood and described with the aid of two block diagrams. One block shows the ice sheet in place; the other shows the landscape after the ice has disappeared (Figure 16.17). In the upper block the ice has reached a stable line of advance and is beginning to waste away in place. Previous forward movement of the ice has created a terminal moraine at the ice margin.

Farther back, beneath the ice, is a till layer of variable thickness, the **ground moraine.**

The ice has now become almost stagnant. Meltwater issues from the ice front in numerous streams. Some of these are discharging from tubes within and beneath the ice, others are flowing down the ice surface itself. Sand and gravel, carried as bed load in the meltwater streams, is spread in sheets in a zone in front of the ice, forming alluvial fans. These coalesce into a continuous alluvial sheet, the **outwash plain.** A series of ice blocks, left behind during a previous episode of advance and rapid wastage, are surrounded and perhaps buried in the outwash layers. At the rear right-hand side of block A the ice sheet has dammed the runoff system to produce a temporary **marginal lake.**

Disappearance of the ice sheet reveals more landforms shaped beneath the ice (lower block). The terminal moraine now appears as a belt of hilly ground with many deep closed depressions. This type of terrain is called **knob and kettle** (Figure 16.18). Sloping away from the terminal moraine is the smooth surface of the outwash plain. The plain is pitted here and there by steep-sided depressions, called **kettles,** formed where ice blocks were buried. Lakes and ponds are often held in the kettles (Figure 16.19).

Behind the moraine is an expanse of poorly drained, marshy surface underlain by ground moraine. Rising from this low ground are groups of smoothly rounded hills, called **drumlins.** They are oval in outline and composed of till shaped by the ice. The long axes of the hills are roughly parallel to one another and at right angles to the trend of the terminal moraine (Figure 16.20). Drumlins usually occur in groups.

Disappearance of the ice also uncovers long, narrow ridges of coarse sands and gravels. They extend in a sinuous course for miles, roughly parallel with the direction of ice movement (Figure 16.21).

Figure 16.18 Rugged topography of small knobs and kettles characterizes this interlobate moraine in Sheboygan County, Michigan. Many erratic boulders litter the surface. (U.S. Geological Survey.)

Figure 16.19 A deep kettle pond on Cape Cod. (A. N. Strahler.)

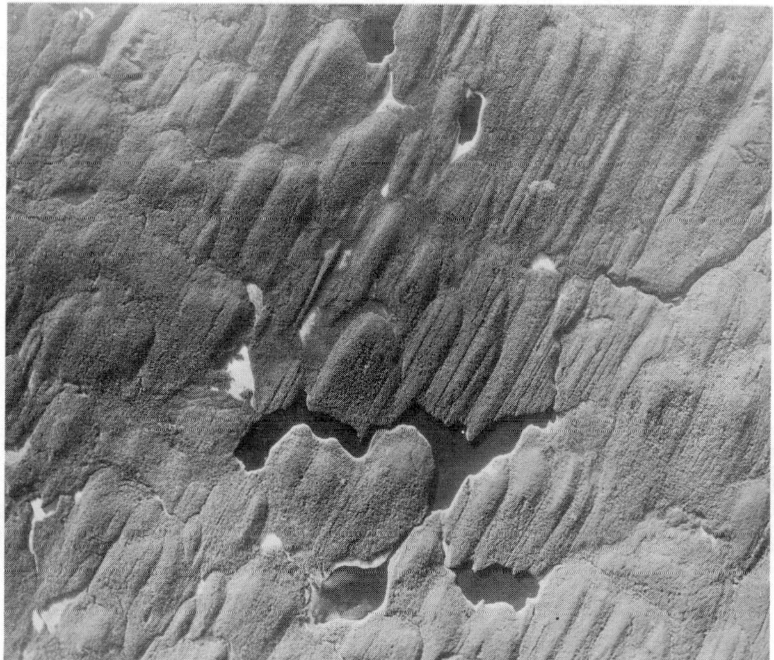

Figure 16.20 (*Left*) Drumlins of glacial till are seen in this vertical air photograph. Ice moved from upper right to lower left. The area shown is about 2 mi (3.2 km) wide; the drumlins are up to 75 ft (120 m) high. The locality is in northern British Columbia. (U.S. Army Air Force.)

Figure 16.21 (*Above*) This esker, near Boyd Lake in Canada, crosses irregular hills of glacially eroded bedrock and rock basin lakes. (Canadian Department of Mines, Geological Survey.)

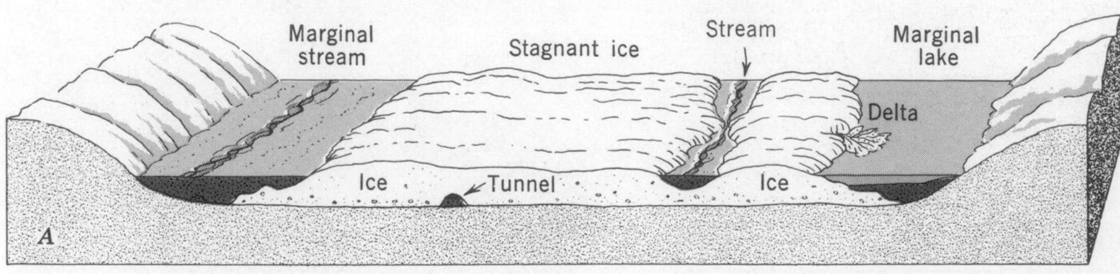

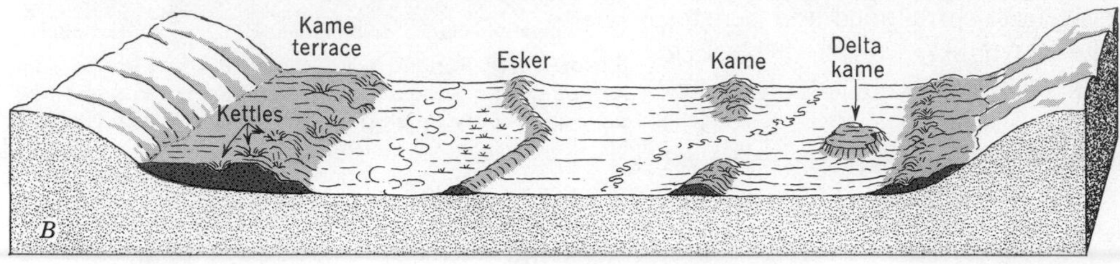

Figure 16.22 Glacial deposits are associated with a mass of stagnant ice in the axis of a broad valley. (© 1973, John Wiley & Sons, New York.)

These features are **eskers,** the bed-load deposits of meltwater streams that emerged from tunnels at the ice margin. Some eskers are traceable for tens of miles with few interruptions and may receive tributary eskers.

Most large lobes of an ice sheet, during the period of ice recession, underwent temporary halts and perhaps minor readvances. These halts resulted in the formation of additional moraines, termed **recessional moraines.** Between adjacent ice lobes are **interlobate moraines** (Figure 16.17).

DEPOSITS BUILT INTO STANDING WATER

A number of distinctive landforms are found in and around the basins of former temporary marginal lakes. These lakes drained away following disappearance of the ice (Figure 16.17). Meltwater streams emerging from the ice built deltas in the marginal lakes. After both ice and lake disappeared, deltas were left standing as flat-topped **delta kames.** A **kame** is any steep-sided hill of well-sorted glacial sand and gravel (Figure 16.22*B*). Delta kames contain the steeply sloping foreset beds typical of small, simple deltas (Figure 16.23). Layers of fine silt and clay were deposited upon the floors of the marginal lakes. Where an aggrading stream ran between the ice and the valley wall, a **kame terrace** was formed, resembling in some respects the alluvial terrace (Chapter 14), but commonly pitted with kettles (Figure 16.22*B*).

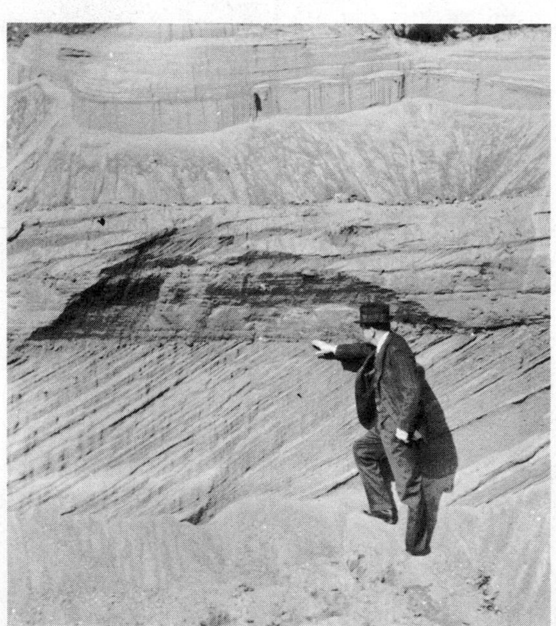

Figure 16.23 Sloping foreset beds of well-sorted sand in a delta kame near North Haven, Connecticut. This deposit and most like it have since disappeared, consumed for use in concrete in urban expansion. (R. J. Lougee.)

GLACIAL LANDFORMS IN REVIEW

This chapter has covered three major glacier topics. First is the alpine glacier. Sculptors of high mountain scenery, alpine glaciers have given us national parks of great beauty in the Rocky Mountains and the Sierra Nevada; they have shaped spectacular fiord coasts. Living alpine glaciers are abundant in Alaska; some are very close to coastal cities.

The second of our glacial topics is the great ice sheets existing on Greenland and Antarctica. Holding most of the world's fresh water in storage, these thick ice masses are beyond hope of yielding new living space for the growing global population. Perhaps someday, icebergs from these ice sheets will be towed to our major port cities to provide fresh water.

The third major topic of our chapter is continental glaciation of the Pleistocene Epoch. For people living in the northcentral and northeastern United States and in southern Canada the glacial deposits of the Pleistocene ice sheets make up the entire land surface they farm and on which they build their cities.

Man evolved to his present state as the supreme species of mammal while the Pleistocene ice sheets were advancing and retreating. Very likely, the rigorous glacial climate and the migrations required by the ice advances stimulated his amazing evolution in a way that might not have occurred otherwise.

YOUR GEOSCIENCE VOCABULARY

glacier
ablation
firn
alpine glaciers
ice sheets
cirque
firn field
ice fall
crevasse
zone of accumulation
zone of ablation
glacier terminus
glacier equilibrium
glacier advance
glacier recession
glaciologist
basal slip
glacial plucking
grinding
glacial striations
chatter marks
glacial grooves
glaciated rock knob
lateral moraine
medial moraine
terminal moraine
rock flour
horn
glacial trough
hanging trough
rock steps
rock basins
tarn
fiord
outlet glacier
ice shelf
glaciation
interglaciation
snow line
glacial drift
stratified drift
till
erratic boulder
ground moraine
outwash plain
marginal lake
knob and kettle
kettle
drumlin
esker
recessional moraine
interlobate moraine
delta kame
kame
kame terrace

SELF-TESTING QUESTIONS

1. Under what conditions of climate and topography do glaciers form? Describe the transformation of snow into glacial ice. What is unique about the form of an alpine glacier?

2. Explain the principle of glacier equilibrium. How do glaciers respond to climate changes?

3. Describe the flowage of a glacier, both at the surface and at depth. How are flowage rates measured?

4. How does a glacier perform its geologic work? What rock-surface features are produced by glacier erosion? How do they indicate the direction of ice movement? What forms of glacial deposition are associated with alpine glaciers?

5. Describe the landforms produced by alpine glacier action.

6. Compare the Greenland Ice Sheet with the Antarctic Ice Sheet as to the extent, volume of ice, and thickness. What is the nature of the rock floor beneath these ice sheets?

7. What parts of North America and Europe were covered by Pleistocene ice sheets? How far south did the ice extend in the United States? in Europe? Name in order the four glaciations and three interglaciations of the Pleistocene Epoch in North America.

8. What basic conditions of climate and topography are required to bring on glaciation? List several hypotheses intended to explain glaciations.

9. Describe the various kinds of glacial drift and the landforms they construct. Make a distinction between landforms built at or beyond the ice limit and those found beneath the ice. What landforms are built into standing water near the ice margin?

(*Above*) An air view of the Long Beach, California, harbor area. The amount of subsidence in feet during 1928–1960 is shown by contour lines. By 1971, when this picture was taken, subsidence had been halted by injection of brine under high pressure. (Port of Long Beach.)

(*Below*) Salt water, backing up through storm drains, flooded these municipal buildings in Long Beach, even though they were protected by dikes. Equipment at the left is pumping out the flood water. (Port of Long Beach.)

17 LANDFORMS SHAPED BY WAVES

A TALE OF TWO CITIES

The death of cities is usually attributed to the exodus of the affluent to the suburbs. But death by drowning? That seems a bit far out. Well, two cities have tried it and almost succeeded. One is Venice, on the Adriatic Sea; the other is Long Beach, California, on the Pacific Ocean. At first thought, you might suppose that these two cities have precious little in common. What they share is a recent history of sinking to the point of being engulfed in ocean water. For both, sinking is due to the satisfaction of a great thirst. For the Venetians, it has been a thirst for water; for Long Beach residents, a thirst for oil.

Venice was built on some hundred-odd tiny islands in the middle of a coastal lagoon on the Adriatic coast. A barrier beach, the Lido, protects the lagoon from storm waves, but not from high water. Although the islands of Venice rise barely higher than the tide waters of the lagoon, they offered safety to Romans fleeing the invading barbarians in the fifth century. By the ninth century, Venice was a thriving city handling a large volume of ocean trade. Queen of the Adriatic, Venice became a city of splendid churches and palaces.

City planning in the eleventh century did not include an engineering study of the geologic formations at depth, and this is where Venice started off on a poor footing, so to speak. The Adriatic Coast in this area is underlain by over 3000 feet of soft, unconsolidated sediments; these are layers of sand, gravel, clay, and silt, and a few layers of peat. The sinking of Venice, as the weak sediments beneath the city settled under the load, has probably been going on very gradually for many centuries, although there are no measurements available. Since precise surveys were begun 40 years ago, sinking has totaled 6 inches. This does not seem like much, but it has greatly increased the damage from flooding during high water of tides and Adriatic storms. The great churches and palaces are threatened by these floodings, and the future of Venice looks grim. As a footnote, we add that Venice's canals are also her sewers.

Water withdrawal from the sediments beneath Venice has been the cause of the recent sinking. The sediments become more compact as the water is pumped out. Most of the water has been withdrawn from industrial wells at Porto Marghera, the modern port of Venice, located on the mainland shore a few miles distant. Recent engineering studies report that if all pumping in the area is held constant in the future, Venice will sink only another inch or so. The city will of course continue to be plagued with inundations of sea water.

Long Beach has been blessed with a rich oil pool beneath the city. Withdrawal of petroleum since 1936 has resulted in a sinking of the land totaling 20 feet in the harbor area. Protective dikes were built to keep out the sea, but even so, flooding from rainwater was a problem because of poor storm drainage. To solve the problem, water was injected into the strata beneath the city. The sinking has not only been halted, but even reversed.

Living below sea level is no novelty to the Dutch people. Much of the agricultural land of the Netherlands is reclaimed land—polders—lying more than 8 feet below sea level. Dikes and dune ridges keep out the North Sea. Vulnerable, too, are the inhabitants of the English fenlands, along the southeastern coast of Great Britain. Here reclaimed tidal lands have subsided many feet following ditching and cultivation; only a barrier of dune ridges and dikes keeps out the North Sea. A great storm in January 1963 flooded hundreds of square miles of English fenlands and Dutch polders and brought death by drowning to more than 300 persons.

Man's struggle against the encroaching sea is one topic of this chapter, which covers the shaping of coasts by waves and currents. The predatory role of the sea as a consumer of continents has been played out during all of recorded geologic history; it is little wonder that man has not yet learned how to live in harmony with the monster.

THE SURF ZONE

The shores of all continents and islands and of all inland lakes are shaped by the unceasing work of waves. Energy derived from wind is carried forward by waves, and as they reach the shallow waters of a coastline, their energy is transformed into currents and powerful surges able to erode rock and transport sediment. In this chapter we turn first to the zone of breakers and surf, a strip of land alternately covered by water and exposed to the air.

After a breaker has formed and collapsed, the wave is transformed into a landward-moving sheet of highly turbulent water, the **swash** (Figure 17.1). Most shores have a rather gently sloping surface of sand or rock. The swash of large waves can surge up this slope to reach a point several feet higher than the still-water level. The swash is quickly slowed by friction and finally stopped. The water then begins to pour seaward down the slope in a reverse flow called the **backwash.** Return flow is shallower and less turbulent than the swash.

Swash and backwash together make an alternating water current that exerts a frictional drag against the surface over which it moves. The reversing flow is capable of moving rock particles of a wide range of sizes, from fine sand to cobblestones.

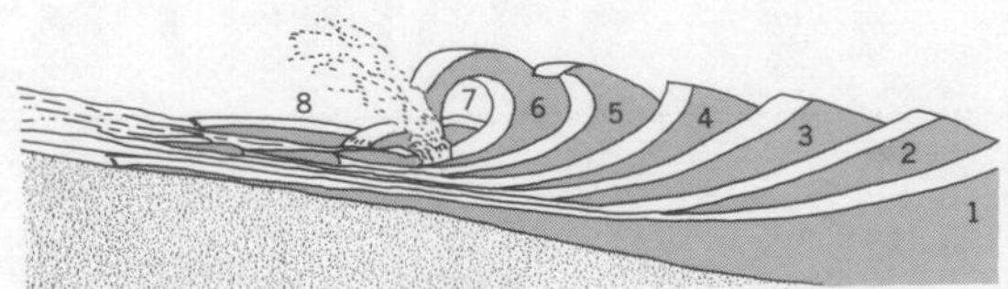

Figure 17.1 A breaking wave sets in motion the swash and backwash. (© 1973, John Wiley & Sons, New York.)

STORM WAVE EROSION

Storm waves breaking against a natural rock cliff or a man-made seawall deliver tremendous pistonlike blows. The attack is capable of undermining the cliff and dislodging boulder-sized masses of rock or masonry (Figure 17.2). The force of a breaking storm wave can run as high as several thousand pounds per square foot. Blocks of stone and concrete weighing from 2 to 10 tons have been moved about and lifted several feet vertically by storm waves striking a sea wall.

Figure 17.2 Storm waves breaking against a sea wall at Hastings, England. (Photographer not known.)

Storm wave erosion affects man adversely on exposed coasts. It is particularly severe in weak regolith. Glacial drift is highly susceptible to erosion (Figure 17.3). The outer shore of Cape Cod, facing

Figure 17.3 Storm waves rapidly undermined this marine cliff composed of weak glacial sands and gravels. Suffolk, England. (H. M. Geological Survey.)

Figure 17.4 This wide abrasion platform lies exposed at low tide. Notice the pocket beach at lower left. Pacific coast, south of Cape Flattery, Washington. (Photographer not known.)

easterly Atlantic storms, is a particularly good example. Formed of glacial sands and clays, this coastline has retreated at a rate of about 3 ft (1 m) per year for at least a century. The Highland Light, a Coast Guard beacon located at the brink of a 300-ft marine scarp of glacial deposits, has had to be moved back three times since it was first installed.

MARINE CLIFFS

Along a coast of hard bedrock, a gently inclined rock surface is gradually carved to accommodate the swash and backwash. This surface is the **abrasion platform.** It is usually completely covered at high tide, but exposed in a broad zone, often many yards wide, at low tide (Figure 17.4). Fragments the size of pebbles and cobbles, serving as tools for abrasion, litter the surface of the wave-abraded platform. Fine particles are moved seaward into deeper water.

A shoreline being carved into bedrock possesses a steep **marine cliff,** rising abruptly from the inner edge of the abrasion platform. The swash of storm waves thrusts rock fragments with great violence against the cliff base, eroding the cliff base and developing a **wave-cut notch.** Undermining leads to the fall of masses of bedrock from the cliff face, furnishing blocks for fragmentation into sediment.

The development of an abrasion platform and marine cliff are shown in Figure 17.5. Initially (block A), the sea has come to rest against a steeply sloping landmass of resistant bedrock. Waves have carved a small notch. Gradually the abrasion platform is developed and widened (block *B*). The detritus is swept seaward to accumulate in deeper water. Weaker places in the bedrock are more rapidly excavated, resulting in the formation of crevices and sea caves. Remnants of bedrock projecting seaward are sometimes cut through to produce arches; these may collapse, leaving columnar rock stacks (see Figure 4.9).

Block C illustrates a more advanced stage. The broad expanse of shallow water over the abrasion platform absorbs much of the energy of the breaking waves, leaving little energy to be expended in attack upon the cliff base. Weathering has subdued the cliff. In this stage a broad beach is normally present.

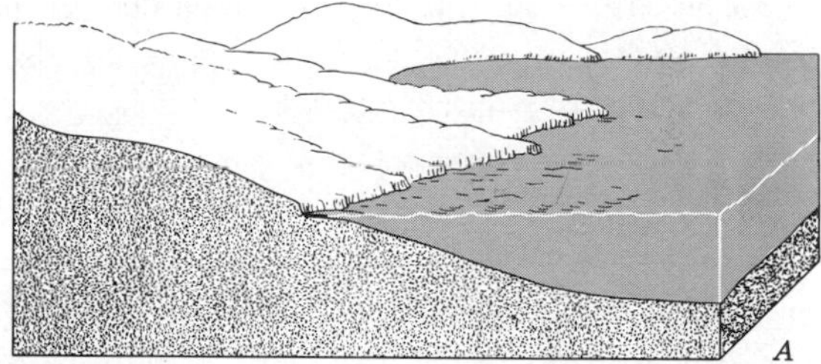

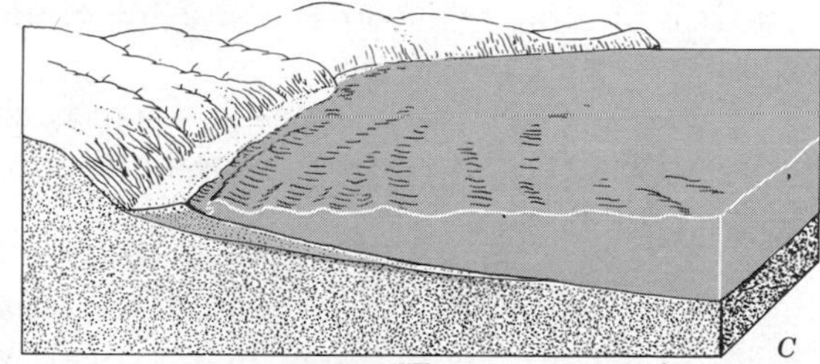

Figure 17.5 Stages in the evolution of a marine cliff. (*Block A*) A small cliff is first cut. (*Block B*) As wave erosion continues, the cliff grows higher, and many details appear. A: arch; B: beach; C: cave; N: notch; P: abrasion platform; R: crevice. (*Block C*) A broad beach is established.

BEACHES

A **beach** is an accumulation of sand, gravel, or cobbles in the zone of breakers and surf. A beach is a depositional landform formed of sediment produced by wave erosion or brought to the shore at various points by streams draining the land. Clay and silt, readily held in suspension, do not remain in the zone of breakers and surf; they diffuse seaward to settle upon the continental shelf in deeper water.

The first beach to form on a coastline having an abrasion platform and cliffs is the **shingle beach,** composed of well-rounded fragments the size of pebbles, cobbles, or, rarely, small boulders. Shingle beaches are narrow and have a very steep seaward slope (Figure 17.6). Shingle beaches form in the most sheltered locations —in bays between rocky promontories—and are called **pocket beaches** (Figure 17.4). They are typically crescent-shaped and are concave toward the sea. Where large quantities of sediment are available, broad beaches of sand are continuous along the coast.

Figure 17.6 Cobblestones form this multiple-crested beach at Smith Cove, Guysboro, Nova Scotia. The beach forms a crescent between rocky headlands. (Maurice L. Schwartz.)

Small summer waves tend to move sand inshore from deeper water, causing a summer widening of the beach. Large waves of winter storms tend to cause a sand movement in the seaward direction, depleting the beach and increasing the water depth.

Throughout this chapter we will use the term **shoreline** to mean the shifting line of contact between water and land. The broader term **coastline,** or simply **coast,** includes not only all of the beach zone, but also marine cliffs (where present), dunes, and such related features as may parallel the beach.

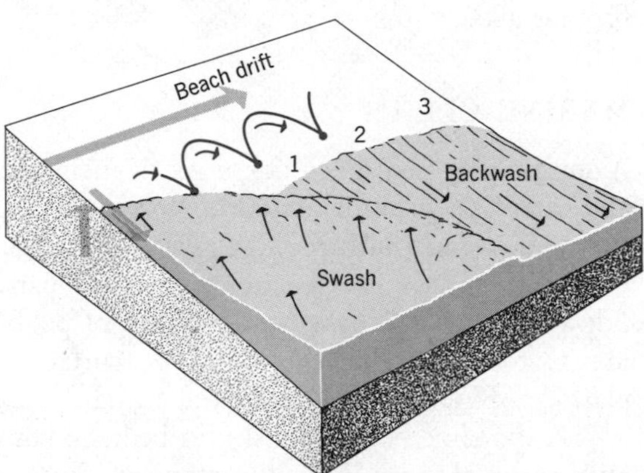

Figure 17.7 Beach drift of sand, caused by oblique approach of swash. (© 1973, John Wiley & Sons, New York.)

LITTORAL DRIFT

Rarely do waves approach a beach exactly head-on. Most of the time the wave crests approach obliquely. Breaking begins at a point on the wave close to shore and travels along the wave crest. Surfers make use of this effect, moving sidewise and keeping just ahead of the breaking crest. Because of the oblique approach, the swash is usually directed diagonally up the beach slope, as shown in Figure 17.7.

Particles carried in the swash ride obliquely up the beach face, but tend to be brought back directly downslope by the backwash. With each cycle of movement the particles are moved along the beach as much as several feet. Countless repetitions of this lateral movement account for transport of vast quantities of sediment. **Beach drifting** is the term applied to this form of sediment transport. It is of primary importance in shaping various kinds of beach deposits along a coast.

Waves approaching a shoreline obliquely also set up a current in the breaker zone. Known as a **longshore current,** it flows parallel with the beach, in the same direction as sediment being moved by beach drifting (Figure 17.8). At times when the longshore current is strong, it is capable of transporting sediment parallel with the beach. Beach drifting and longshore current work together; their combined action is called **littoral drift.**

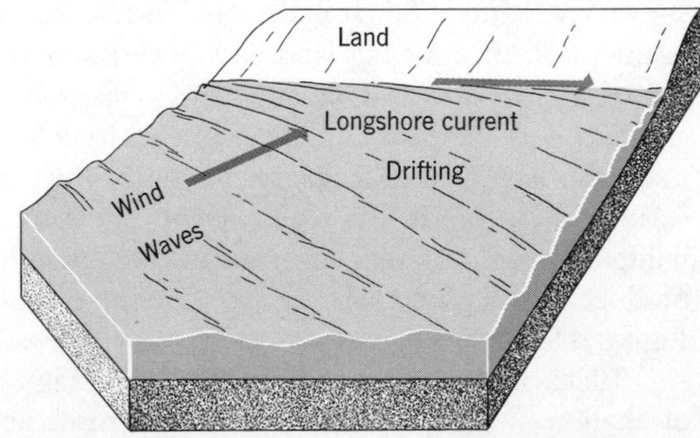

Figure 17.8 Longshore currents set in motion by oblique approach of waves. (© 1973, John Wiley & Sons, New York.)

Along a straight coast, littoral drift carries sediment continuously along the beach, often for many tens of miles, much as bed load is carried by a river (Figure 17.9). However, where the coastline undergoes an abrupt change in direction (as where a bay occurs), sediment is carried out into open water to form a **sandspit.** It is a fingerlike extension of the beach. Bending of waves around the end of the spit causes the end to curve landward in a characteristic spiral: The sandspit is said to be recurved.

WAVE REFRACTION AT HEADLANDS

As waves approach a shoreline indented with bays and headlands (promontories), the wave crests are turned toward the headlands. This process of wave bending is called **wave refraction.** Figure 17.10 shows how refraction affects the forms of the wave crests. Wave refraction causes a concentration of wave energy upon the headlands. Surf there is higher, and the breakers have more energy, so they erode more vigorously. A cliff and abrasion platform form quickly on the headland. Waves entering the bays are bent so as to be convex toward the mainland. At the same time, the wave is weakened and diminishes in height in this region. By the time the wave reaches the head of the bay it may have little energy left.

As Figure 17.10 shows, the refracted waves approach the bay sides at an oblique angle. Littoral drift along the bay sides will be landward, toward the bayhead, as the arrows indicate. Consequently, sediment torn from the headlands is carried to the bayheads, where it accumulates as a pocket beach.

WAVE EROSION AND SHORE PROTECTION

Postglacial rise of sea level has brought the powerful erosive action of storm waves to bear high upon continental coasts. Great lengths of coastline consist of easily erodible materials, particularly glacial drift and alluvium. Once man has occupied these coastlines with expensive buildings and highways, he feels obligated to resist the wave action that cuts away his land.

One example of man-made interference with the natural littoral drift is seen in the use of groins to collect beach sand. The widened beach forms a buffer zone that can absorb the short-lived but intense attacks of storm waves. A **groin** is simply a broad wall of huge rock masses built perpendicular to the shoreline (Figure 17.11). The groin interferes with the littoral drift of sand, causing a crescentic beach deposit to accumulate on the updrift side of the groin. In the case of the single groin shown in the diagram, the trapping of sand results in a deficiency of sand along the downdrift side. Here the beach is depleted, and the shoreline recedes.

ELEVATED SHORELINES AND MARINE TERRACES

The development of a shoreline is sometimes interrupted by a sudden rise of the coast. When this occurs, an **elevated shoreline** is formed. The marine cliff and abrasion platform are abruptly raised above the level of wave action. The former abrasion platform is now a **marine terrace** (Figure 17.12). Denudation processes begin to destroy the terrace and eventually will obliterate it entirely.

Elevated shorelines are common along the continental and island

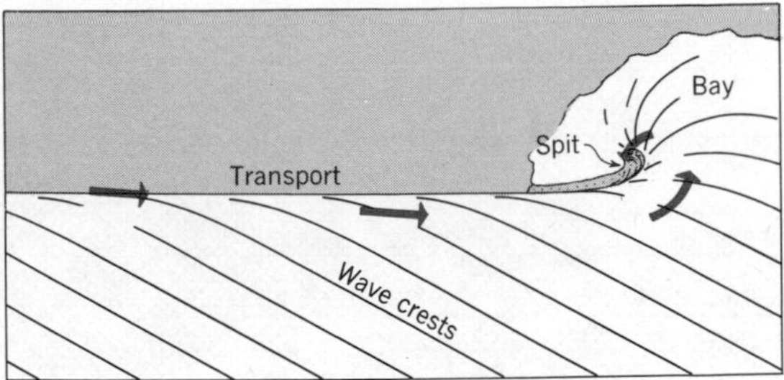

Figure 17.9 Littoral drift along a straight section of coastline, ending in a bay. A recurved sandspit has been built into the bay.

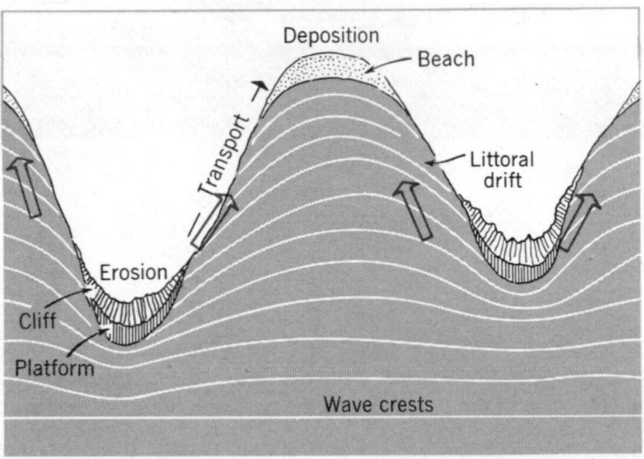

Figure 17.10 Littoral drift along the sides of a bay carries sediment from the headlands toward the bay head.

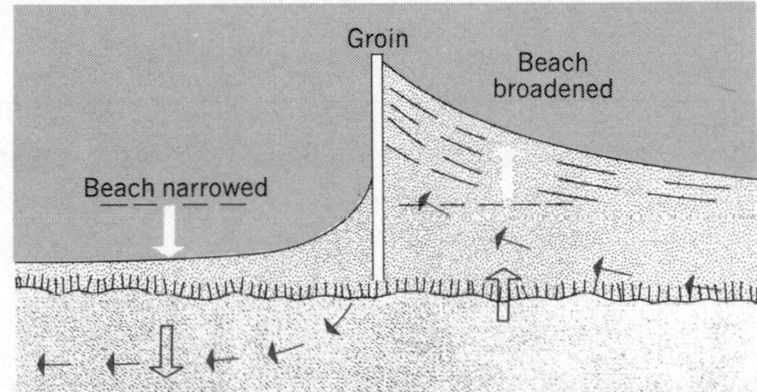

Figure 17.11 Construction of a groin causes marked changes in configuration of the sand beach.

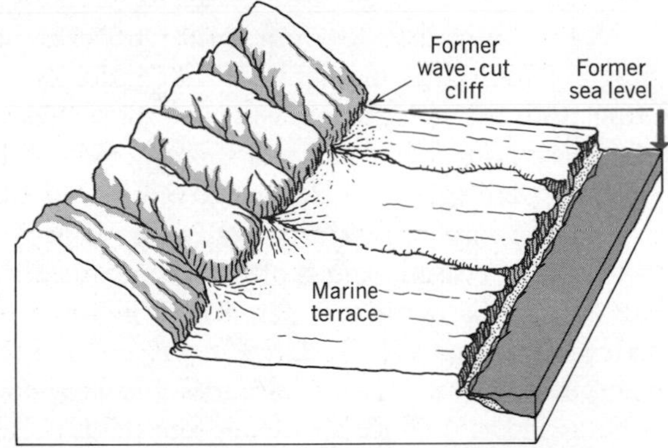

Figure 17.12 A marine terrace is the old abrasion platform of an elevated shoreline. Alluvial fan deposits are beginning to form a cover, while recession of the new marine cliff is narrowing the terrace.

Figure 17.13 These elevated marine terraces form a staircase on the western slope of San Clemente Island, off the southern California coast. There are more than 20 terraces in this series; the highest is 1300 ft (400 m) above sea level. (John S. Shelton.)

coasts of the Pacific Ocean, because here rapid uplift of a crustal block along an active fault is a common event. Repeated uplifts result in a series of marine terraces in a steplike arrangement. Fine examples of these multiple terraces are seen on the western slope of San Clemente Island, off the California coast (Figure 17.13).

COMMON TYPES OF COASTLINES

There are many kinds of coastlines. Each variety is unique because of the distinctive landmass against which the shoreline has come to rest. A rising of sea level or a sinking of the crust results in partial drowning, or **submergence,** of landforms shaped by such terrestrial agents as running water or glacial ice. On the other hand, **emergence,** caused by either a lowering of sea level or a crustal rise, results in exposure of submarine landforms. Emergence produces coastlines distinct from coastlines of submergence. Still another class of coastlines results when new land is built out into the ocean by volcanoes and lava flows, or by growth of deltas, or by coral reef growth.

We will describe a few important types of coastlines as examples. These are illustrated in Figure 17.14. A deeply embayed coast, resulting from submergence of a fluvially dissected landmass, is called a **ria coast** (*A*); it has many offshore islands. A **fiord coast** results from the submergence of deeply carved glacial troughs (*B*); it has long, narrow bays with steep rock walls. Emergence of a smooth, gently sloping coastal plain is often accompanied by the construction of an offshore barrier island, paralleling the inner shoreline. This is a **barrier-island coast** (*C*). Large rivers build elaborate deltas, producing **delta coasts** (*D*). Volcanoes and lava flows, built up from the ocean floor, produce a **volcano coast** (*E*). Reef-building corals create new land and make a **coral-reef coast** (*F*). Downfaulting of a landmass can bring the shoreline to rest against a fault scarp, producing a **fault-coast** (*G*).

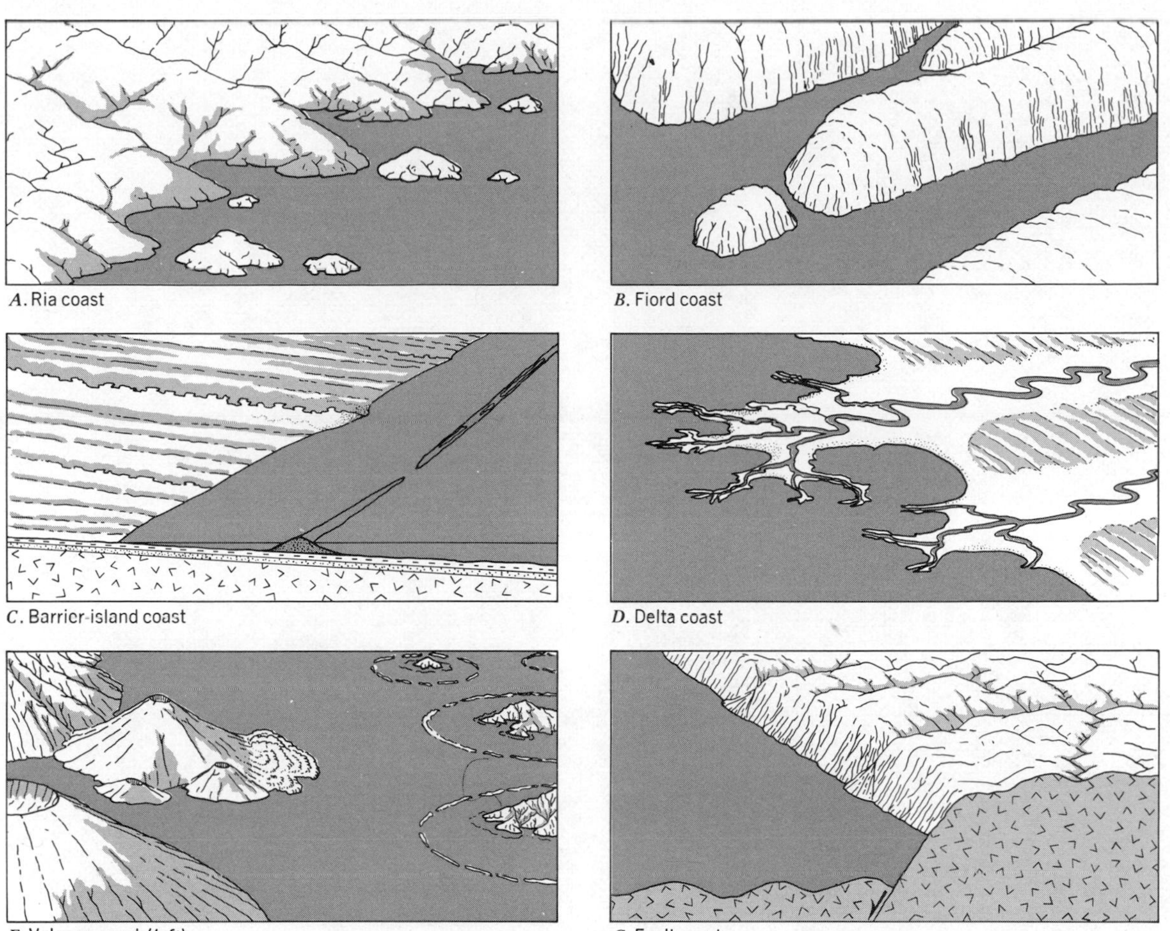

Figure 17.14 Some common types of coastlines.

TIDAL FLATS AND SALT MARSHES

Bays and lagoons shut off from the open ocean are gradually filled by layers of clay and silt brought into the quiet water by streams draining the land. The sediment is distributed over the bay by tidal currents.

Organic matter, both that carried in suspension in streams and that produced by growth of plants and animals on the bottom, makes up a substantial proportion of the sediment. Gradually, the sediment is built upward, until the upper surface is approximately at the level of low tide. The result is a mud flat, or **tidal flat**, exposed at low tide but covered at high water.

Salt-tolerant vegetation takes hold upon the tidal mud flats, eventually forming a resistant mat of plant roots. More sediment is trapped by plant stems, and the deposit is built up to the level of high tide by a layer of peat. The completed surface is described as a **salt marsh** (Figure 17.15). Tidal flats and tidal marshes encroach in succession upon the open water until the entire bay or lagoon is filled, except for a system of tidal channels.

Figure 17.15 This late winter scene shows stubble of salt-marsh grass over a peat layer. A small tidal channel in the foreground lies empty at low tide. South Wellfleet, Massachusetts. (A. N. Strahler.)

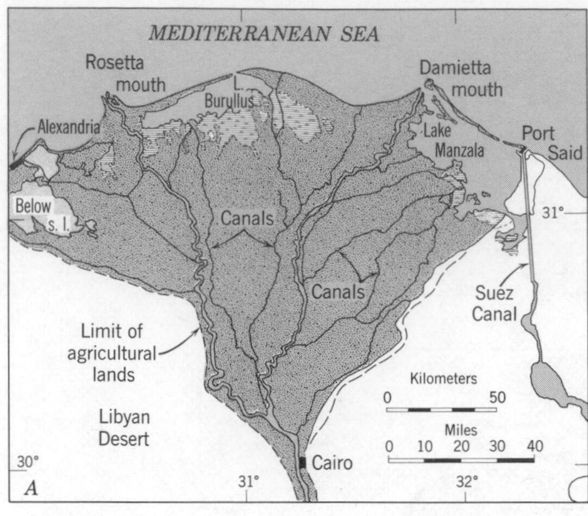

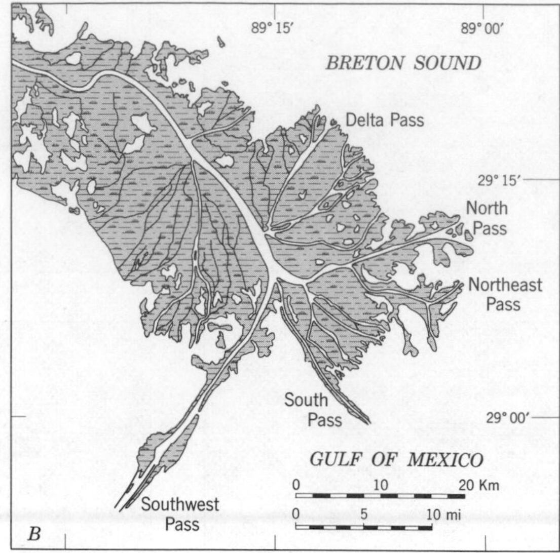

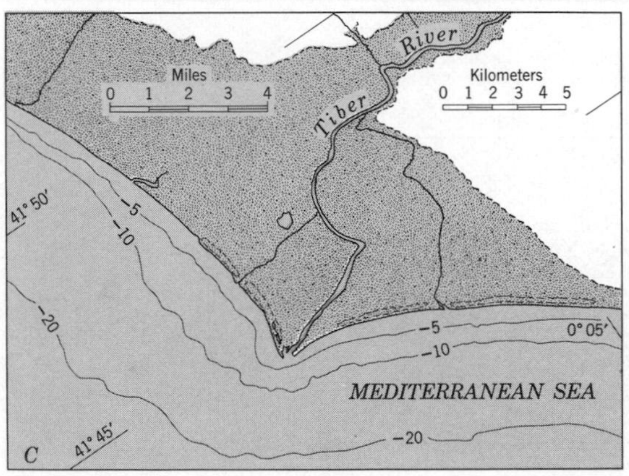

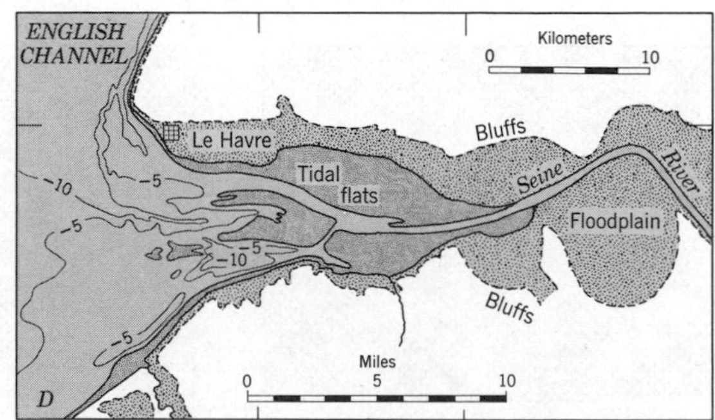

Figure 17.16 Four varieties of coastal delta configuration have been created by these four well-known rivers. (© 1973, John Wiley & Sons, New York.)

DELTA COASTS

Deltas built into the ocean by large streams create a complex coastline, because the outbuilding action of the stream is countered by wave and current action. The Nile delta is a classic case; from its triangular outline, resembling the Greek letter *delta*, we have inherited the name of this coastal landform (Figure 17.16A). The Nile branches into two major distributaries, starting about 100 mi (60 km) from the ocean, and these carry sediment to the Mediterranean shore in a broad fan-shaped front. Wave attack and littoral drift have spread the sediment from the two major mouths along the coast to form barrier beaches.

The Mississippi River delta is a highly specialized type with a long and complex history of growth. The modern, active delta is of the bird-foot type, with long narrow fingers built out by many distributaries (Figure 17.16*B*). Salt marsh forms at the surface between the natural levees of distributaries.

The delta of the Tiber River is cuspate (tooth-shaped) in outline, because of strong littoral drift of sediment away from a single mouth (Figure 17.16C). The delta of the Seine River of France illustrates a fourth type of delta (Figure 17.16*D*). Postglacial submergence of the coast allowed the sea to form a narrow estuary,

Figure 17.17 A wide fringing coral reef on the south coast of Java. The reef is best developed off the promontory where surf is strongest. The white band is a beach of coral sand, bordering tropical rainforest. (Luchtvaart-Afdeeling, Ned. Ind. Leger, Bandoeng.)

Figure 17.18 Barrier coral reef, reef islands, and lagoon. Tahaa Island, Society Islands. (Official U.S. Navy photograph.)

occupying the former floodplain of the river. Today, the delta is filling the estuary with sediments and these are creating a tidal flat.

CORAL-REEF COASTS

Coral reefs are massive rock structures built by living corals and algae in the surf zone of tropical oceans. The organisms secrete calcium carbonate to build new structures upon old. In this way, reefs are extended seaward into deeper water. Wave attack pulverizes exposed coral structures to form a calcareous sediment, which adds to the bulk of the reef.

Coral reefs are largely limited to the latitude zone between 30° N and 25° S, where water temperatures are mostly between 77° and 86° F (25° and 30° C). Reef growth is most rapid along exposed coastal locations, such as off headlands and along island shores facing into the prevailing direction of wave approach.

Coral reefs take three basic forms. First and simplest is the **fringing reef,** a shelflike attachment to the land, pictured in Figure 17.17. The reef surface is exposed at low tide, but covered by surf at high tide.

A second reef form is the **barrier reef.** This is a long narrow coral embankment lying offshore and enclosing a lagoon between reef and mainland (Figure 17.18). At intervals along the barrier reef, there are passes through which excess water brought into the lagoon by breaking waves is returned to the sea.

A third reef form is the **atoll,** a ringlike reef of coral enclosing only a lagoon of open water (Figure 17.19). Here and there on the atoll reef are low islands of coral sand. These are sometimes sufficiently large and high to be habitable but they are vulnerable to inundation in tropical storms. Atolls appear in isolated groups, far from any islands of noncoralline rock, in the vast expanses of the western Pacific Ocean.

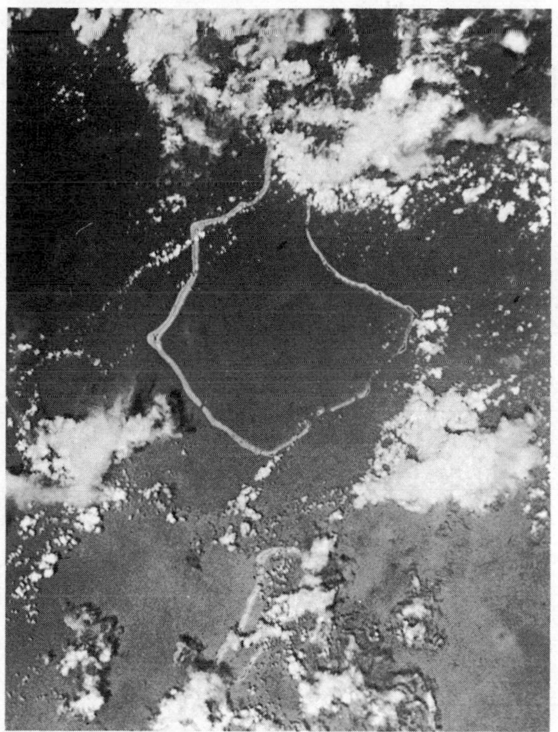

Figure 17.19 Rongelap Atoll in the Marshall Islands of the Pacific appears as a slender string of pearls beneath cloud masses in this photograph taken by astronauts aboard *Gemini V* spacecraft. Their altitude was about 150 mi (240 km). (NASA)

Several plausible but divergent theories of origin of the fringing reefs and atolls have been proposed. The earliest and one of the most successful is the **subsidence theory** proposed by Charles Darwin in 1842, as a result of that great naturalist's observations during the voyage of H.M.S. *Beagle*.

According to Darwin, a fringing reef is first formed during the slow subsidence of a volcanic island (Figure 17.20). Coral growth, continuing uninterrupted during subsidence, builds the reef upward, maintaining the reef surface close to sea level. Because the reef is built directly upward while the island shore is gradually being inundated, a lagoon is formed and becomes wider as subsidence continues. Note that in the barrier-reef stage, shown in sector *B* of Figure 17.20, the island shows the characteristic embayed coast to be expected from submergence of a fluvially dissected landmass. With continued subsidence the volcanic island is diminished to a small remnant, then finally disappears, leaving an atoll lagoon in its place.

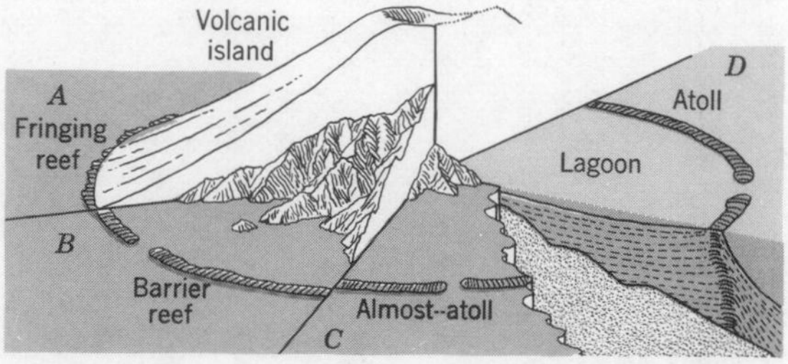

Figure 17.20 Stages in the development of an atoll, according to the subsidence theory. (© 1971, John Wiley & Sons, New York.)

SHORELINE PROCESSES IN REVIEW

Waves and coastal currents are not only major agents of landform development of the present, but also of the past. Sedimentary deposits transported and deposited along ancient shorelines are now part of the record of stratigraphy. Many important sandstone and conglomerate formations prove to be lithified beaches. Many ancient deltas have been recognized in the geological record, and even coral reefs can be identified within limestone formations. Some of these ancient shoreline deposits form valuable petroleum reservoirs. Wave erosion has been responsible for the planation of continents in the final stages of denudation. Often a peneplain proves to be a platform abraded by waves and then buried in the advancing deposits of marine sediments.

The great variety in coastlines is due to the fact that we live at a time when the crust has been moving up in one place and down in another. Today volcanism and faulting are widespread. The growth and disappearance of Pleistocene ice sheets set off wide swings of sea level, ending in a final rise to the present level.

YOUR GEOSCIENCE VOCABULARY

swash
backwash
abrasion platform
marine cliff
wave-cut notch
beach
shingle beach
pocket beach
shoreline
coastline (coast)
beach drifting
longshore current
littoral drift
sandspit
wave refraction
groin
elevated shoreline
marine terrace
submergence
emergence
ria coast
fiord coast
barrier-island coast
delta coast
volcano coast
coral-reef coast
fault coast
tidal flat
salt marsh
coral reef
fringing reef
barrier reef
atoll
subsidence theory

SELF-TESTING QUESTIONS

1. Explain the action of swash and backwash in shaping a beach and transporting sediment. Describe the action of storm waves in cutting back a coastline. What kinds of coastal materials are most susceptible to storm erosion?

2. Describe the stages in evolution of a sea cliff carved in bedrock. What landforms are produced in each stage?

3. What processes of transport are involved in littoral drift? What shoreline forms are shaped by littoral drift?

4. Explain how wave refraction alters the form of an embayed coastline. What effect has refraction upon waves entering a bay?

5. What engineering structures are used to induce widening of a beach? How is cutting away of the beach induced by man-made structures?

6. How are marine terraces formed? What eventually happens to a marine terrace?

7. Explain the significance of both submergence and emergence in terms of changes of sea level and crustal movement.

8. Describe six common kinds of coastlines. Which are caused by submergence? which by emergence? which by outbuilding of new land?

9. What processes are involved in shaping a large marine delta? Describe four varieties of delta configurations.

10. Where are coral reefs found? What forms do coral reefs take? Explain atolls according to the subsidence theory.

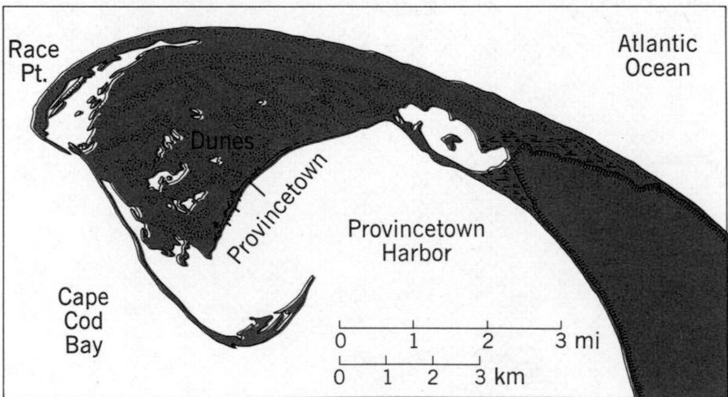

(*Left*) A sketch map of the Provincelands showing Provincetown as it was in the 1880s. Dunes, shown in stipple pattern, cover most of the peninsula. (*Above*) Moving dunes of the Provincelands. The view is toward Provincetown Harbor. (Harold L. Cooper, Cape Cod Photos.)

18 LANDFORMS SHAPED BY WIND

ORDEAL BY SAND

Provincetown, Massachusetts, was once a city with sand troubles. A famous fishing port on Cape Cod, thriving since colonial days, Provincetown is safely nestled in a quiet harbor within the curved fist of Cape Cod. A fingerlike sandspit encloses the harbor, while the wrist and forearm of the Cape protect it from vicious Atlantic storms attacking from the north and east. It was these same gale-force winds that brought the sand into the streets of Provincetown.

The northern tip of Cape Cod, called the Provincelands, is almost completely covered with sand dunes. Shaped into wavelike ridges, many of the dune summits rise to heights over 80 feet above sea level. At the time the first colonists settled here a forest of pitch pines kept the dunes pretty well in check, while beachgrass protected the higher dune surfaces.

The first settlers rapidly destroyed the protective plant cover of the Provincelands. Pine trees were cut for fuel and for pitch and turpentine needed for the fishing fleet. Sheep and cattle grazed the dune summits, destroying the beachgrass. Now winds from the north and east began to carry the loose sand south toward Provincetown Harbor. Early in the 1700s legislation was enacted to forbid grazing and tree-cutting, but the laws were poorly enforced or simply ignored.

The invasion of Provincetown by sand began in earnest about 1725. Steep dune slopes began to bank up against houses. Driving sand frosted the window panes. Sand drifts accumulated in the streets. Only the constant removal of sand by carts kept the streets open. By the late 1700s and early 1800s the dune ridges were advancing upon Provincetown Harbor along a front over 4 miles long. Commissioners appointed by the Commonwealth reported upon the situation in 1825, recommending protective legislation and the extensive planting of beachgrass. Their advice was heeded, and beachgrass grown from seed imported from Holland was set out in rows. Brush was piled between the rows to make a baffle for trapping the moving sand.

When Henry David Thoreau, the naturalist, visited Provincetown in 1849, he was told by natives that the sand had made no progress in the previous 10 years, but evidence of its invasion was still about. This is what Thoreau observed:

> The sand is the great enemy here. . . . The sand drifts like snow, and sometimes the lower story of a house is concealed by it, though it is kept off by a wall. The houses were formerly built on piles, in order that the driving sand might pass under them. . . . There was a school-house just under the hill on which we sat, filled with sand up to the tops of the desks, and of course the master and scholars had fled. Perhaps they had imprudently left the windows open one day, or neglected to mend a broken pane. . . . In some pictures of Provincetown the persons of the inhabitants are not drawn below the ankles, so much being supposed to be buried in the sand. . . . I saw a baby's wagon with tires six inches wide to keep it near the surface. The more tired the wheels, the less tired the horses.*

Invasion by coastal dunes is an old story in Europe. In the Landes region of western France, bordering the Bay of Biscay, deforestation set off the landward migration of great dunes; their speed was clocked at from 60 to 80 feet per year. One village was twice moved back from the advancing wave. Plantings of dunegrass and pines were made in the late 1700s, halting the sand invasion. To this day it is forbidden even to cross the dunes on foot, so sensitive are the grasses to any disturbance.

Perhaps a new era of sand invasions is at hand, now that the dune buggy is riding rough-shod over our coastal dunes, senselessly churning the sensitive roots of the beachgrass. What price recreation?

* From Henry D. Thoreau, *Cape Cod,* © 1961 by Thomas Y. Crowell Company, New York, pp. 256–258.

WIND EROSION

Wind is the fourth active agent of landform development. Although landforms shaped by wind cover only small parts of the continents, they are locally dominant on desert plains and along coastlines. Wind-transported silt is widespread in a thin layer over parts of the middle-latitude continents, where it forms the parent matter of highly fertile soils. Like a water stream, wind blowing over the land surface performs its geologic work by erosion, transportation, and deposition.

One form of wind erosion is **deflation,** the lifting of loose particles of clay and silt, collectively referred to as **dust.** The process is much like that of suspension of fine sediment in stream flow. Grains are carried up by vertical currents. The dust is diffused upward into the atmosphere and forms a **dust storm** (Figure 18.1). The smaller particles may quickly diffuse to heights of thousands of feet and will travel for hundreds of miles before settling to earth in less turbulent air.

Figure 18.1 This rapidly moving cloud is the leading edge of a dust storm. The dust is suspended within turbulent air of a cold front. Coconino Plateau, Arizona. (D. L. Babenroth.)

In the United States the passage of a rapidly moving cold front, bringing a turbulent mass of colder air southward over the Great Plains region, is a common cause of severe dust storms. As the front approaches, a dark dust cloud, representing the leading edge of the front, moves over the plain. When the cloud strikes it may bring semidarkness and reduce visibility to only a few yards. The fine dust, penetrating into all open spaces, makes breathing difficult.

Deflation occurs where clays and silts in a thoroughly dried state are exposed on barren land surfaces. Such conditions exist in deserts generally, and locally on dried-out floodplains, tidal flats, and lake beds. Even upon actively forming glacial outwash plains, deflation is active in times of cold, dry weather.

Deflation forms shallow depressions, called **blowouts** (Figure 18.2). Hollows develop where the natural vegetative cover of shrubs or grasses is broken down, exposing the bare soil to wind. Once formed, deflation hollows hold water after heavy rains. These puddles

Figure 18.2 A large blowout on the plains of Nebraska. A remnant column of the original material provides a natural yardstick for the depth of material removed by deflation. (U.S. Geological Survey.)

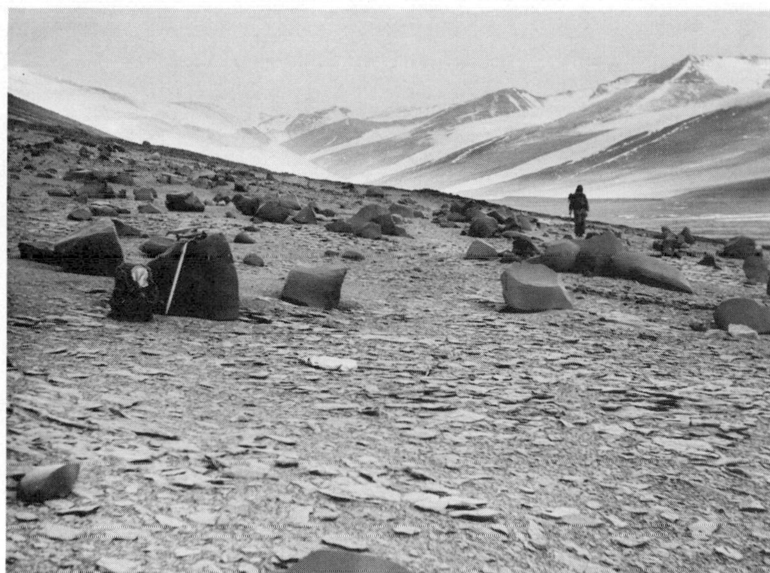

Figure 18.3 Eroded by sand driven on high winds, these strange sharp-edged boulders litter the floor of Wright Valley, McMurdo Sound, Antarctica. Ventifacts are formed in the cold deserts as well as in the hot deserts. (Robert L. Nichols.)

attract grazing animals whose trampling further loosens and disrupts the soil, preparing it for deflation when the soil is again dried out. Storm runoff attacks the sloping margins of the depressions, washing sediment into the bottom, ready for later removal by deflation. In this way deflation hollows deepen and enlarge, sometimes growing to widths of 1 mi (1.6 km) or more and to depths of 10 to 50 ft (3 to 15 m). Wind action over surfaces of alluvium removes the finer silts and sands, leaving behind an armoring layer of pebbles. This surface is called a **desert pavement.**

A second form of wind erosion is **sand-blast action,** in which hard sand grains (usually of quartz) are driven against exposed rock surfaces projecting above a plain. Wind-driven sand grains travel close to the ground, so that their erosive action is limited to surfaces no more than a few feet above the ground. Sand-blast action is responsible only for minor features, such as notches and hollows at the base of a cliff or a boulder.

Sand-blast erosion creates curiously shaped pebbles and boulders called **ventifacts.** These objects have curved faces intersecting in sharp edges (Figure 18.3).

LOESS

Thick deposits of wind-transported dust have accumulated in some large areas in middle latitudes. This material is **loess,** a porous friable, yellowish sediment. Loess consists of finely divided mineral fragments, mostly in the size range of silt. From 5 to 30% of the loess may be of clay sizes.

Loess usually shows no layered structure or stratification. On the other hand, it commonly has a natural vertical parting, or cleavage, along which masses readily break away from an undercut bank or cliff (Figure 18.4).

In terms of mineral composition, loess is commonly a mixture of mechanically pulverized fragments, dominantly of quartz, but with some feldspar, mica, hornblende, and pyroxene. Carbonate

Figure 18.4 Deflation of loess pulverized by heavy traffic caused this roadway to become deeply sunken between perpendicular walls, Shensi Province, China. (Frederick G. Clapp, *The Geographical Review.*)

mineral matter is also present. Small nodules and tubes of calcite (calcium carbonate) are found distributed through the loess. Fossil snail shells occur in large numbers in some thick loess layers.

The grain size and mineral composition of loess of the middle latitudes in the United States and Europe show that loess is wind-blown dust of Pleistocene age carried from alluvial valleys and outwash plains lying south of the limits of the retreating ice sheets. Most North American loess is found in a blanket over the north central states, with an important extension southward along the east side of the Mississippi alluvial valley (Figure 18.5). Most of this area lies within and immediately adjacent to the glaciated region.

Loess forms a nearly continuous sheet in European Russia, blanketing the plains topography. In northern China, loess reaches thicknesses commonly over 100 ft (30 m), and in some places as much as 300 ft (90 m) (Figure 18.4).

The great importance of loess to man is that it forms the parent material of some of the richest agricultural regions on earth. Soils developed on loess are largely the brown soils, rich in nutrients needed for grain crops. Corn and wheat thrive on soils of loess origin in the United States. Loess is easily cultivated, but is also subject to severe soil erosion and deep gullying when abused.

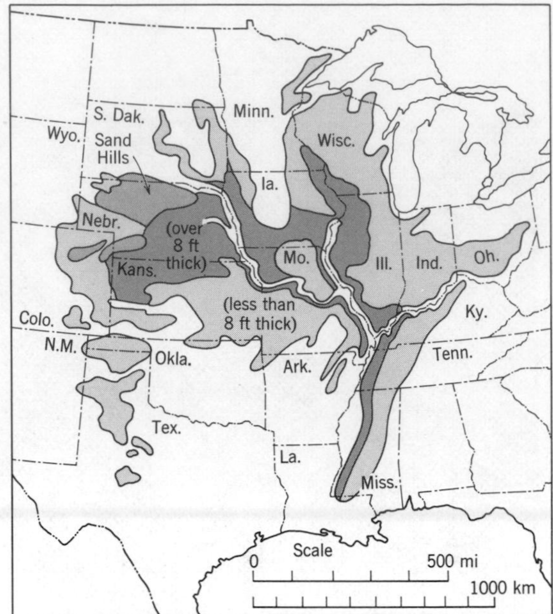

Figure 18.5 Map of loess distribution in the central United States. (From A. N. and A. H. Strahler, *Environmental Geoscience,* 1973, Santa Barbara, Calif., Hamilton Publ. Co.)

DUNES OF FREE SAND

Sand grains are carried by wind over a loose sand surface by numerous long, low leaps (Figure 18.6). The process is called **saltation.** Individual grains rebound upon impact, and at the same time push the impacted surface grains forward in the same direction. The creeping surface grains are formed into small ripples (Figure 18.7). During a **sand storm,** the motion of countless thousands of grains forms a low gliding carpet. Most of the sand moves in a layer only a few inches off the ground.

Transport of sand in saltation builds two forms of accumulations. One is a **sand drift,** attached to the lee of an obstacle, such as a boulder or a bush. The other is a **sand dune,** a mound or hill of sand rising to a single summit and not attached to any surface object. Most dunes are capable of movement in the downwind direction, but some kinds remain fixed in position for long periods of time.

One of the simplest of dune forms is the **barchan dune.** It is an isolated dune of crescent shape (Figure 18.8). The two points, or "horns" of the barchan point downwind. Between the points is a

Figure 18.6 Sand moves by saltation and surface creep. (© 1973, John Wiley & Sons, New York.)

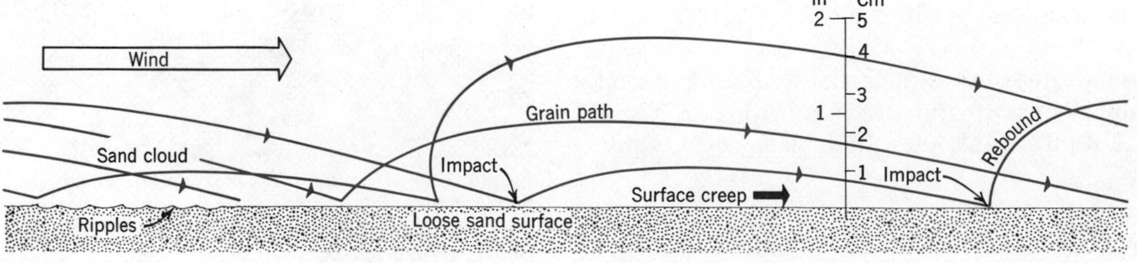

Figure 18.7 Sand ripples on dune surfaces. Quatif, Saudi Arabia. (Arabian American Oil Company.)

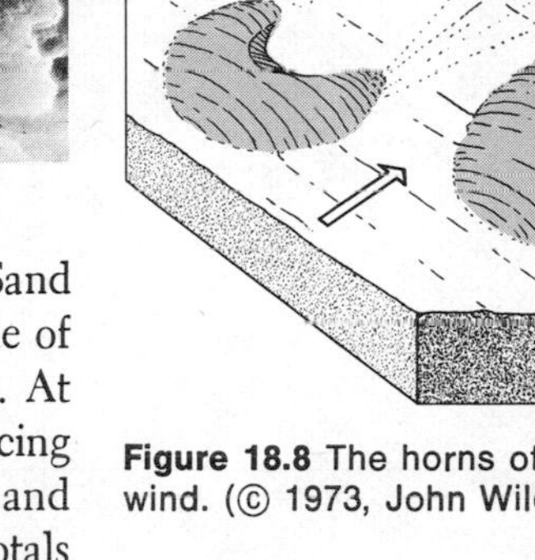

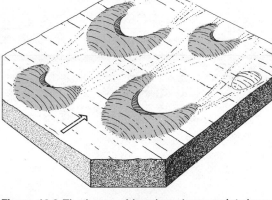

Figure 18.8 The horns of barchan dunes point downwind. (© 1973, John Wiley & Sons, New York.)

steep, concave sand slope, called the **slip face** (Figure 18.9). Sand grains carried in saltation over the smoothly rounded upwind side of the dune fall upon the slip face, gradually steepening the slope. At intervals, a thin layer of sand slides down the slip face, advancing the dune. Barchans grow to heights as great as 100 ft (30 m) and widths up to 1200 ft (360 m). Movement of the dune often totals several inches per day. Individual barchans have been found to travel as far as 50 ft (15 m) in a year.

In deserts where a great supply of sand is present, dune sand covers the entire surface. The sand surface is formed into wavelike ridges separated by troughs and hollows, much as if sea waves in time of storm were frozen into place (Figure 18.10). Such sand waves are described as **transverse dunes,** and the entire assemblage as a **sand sea.** Slip faces of a sand sea resemble those of barchans and face downwind. The origin of transverse dunes is similar to barchans.

Figure 18.9 These barchan dunes formed on a terrace bordering the Columbia River, near Briggs, Oregon. (U.S. Geological Survey.)

Figure 18.10 A great sea of transverse dunes in Imperial County between Calexico, California, and Yuma, Arizona. Some barchan dunes can be seen at the lower right. (Spence Air Photos.)

Toward the edge of a sand sea, where the sand layer thins out, individual transverse ridges become separated by bare ground. The ridges then become segmented into individual barchans, as we see in the lower-right portion of Figure 18.10.

Transverse dunes commonly occur in a coastal belt situated inland from beaches. They are best developed where sand supply is large but vegetation is absent.

DUNES CONTROLLED BY VEGETATION

Many common dune forms develop while bearing a scanty cover of grasses or small shrubs. The plants serve as sand traps. Here and there are plant-free hollows and slopes of loose sand. Simplest of these vegetation-controlled dunes are the coastal **foredunes.** They are

Figure 18.11 Beachgrass protects coastal foredunes on the Provincelands of Cape Cod, Massachusetts. (A. N. Strahler.)

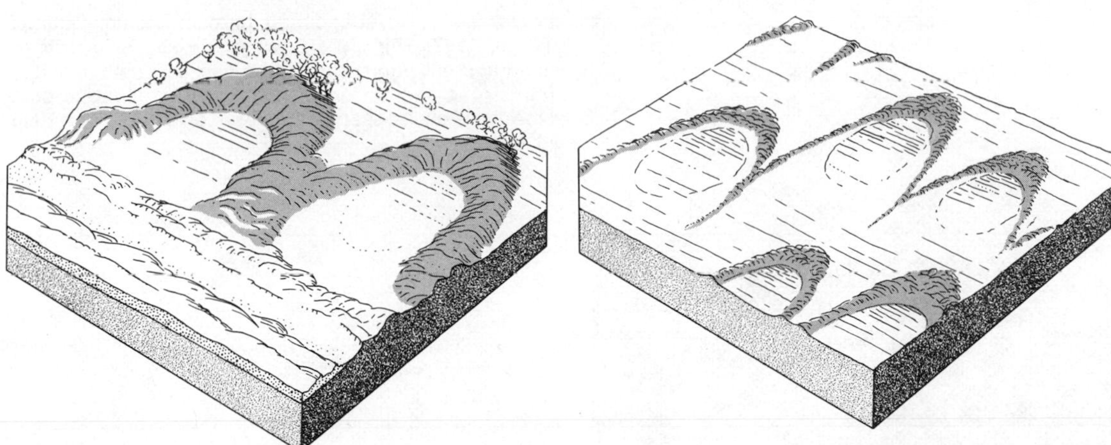

Figure 18.12 Dunes controlled by vegetation. Wind direction is from lower right. (*Left*) Coastal blowout dunes. (*Right*) Parabolic dunes of a semiarid steppe region. (© 1973, John Wiley & Sons, New York.)

found adjacent to the inner edge of a sand beach (Figure 18.11). Beachgrass upon these dunes is capable of maintaining itself as the sand accumulates. Foredunes are irregular in form, with numerous blowout hollows from which sand is excavated and carried to the dune summits.

Some other vegetation-controlled dune types have distinctive shapes, easily recognized when seen from the air. The two block diagrams in Figure 18.12 illustrate a distinctive dune family characterized by parabolic outlines. These **parabolic dunes** are convexly bowed in the downwind direction (opposite to the curvature of the barchan). One variety of parabolic dune forms along coasts, landward of beaches, where strong onshore winds are supplied with abundant sand (upper block). The sparse cover of protective grasses and shrubs is locally broken, permitting a deep deflation hollow to form. Sand is carried out of the hollow, building a curved embankment, or rim, which grows higher as the blowout is enlarged. When well developed, this **coastal blowout dune** may rise to heights of 100 ft (30 m). On the landward side it has a slip face that can override an inland forest, killing the trees (Figure 18.13).

Low parabolic dunes are formed in great numbers on semiarid plains where the regolith is sandy (lower block in Figure 18.12). The dune ridge, which is only a few feet high, is covered by grasses and shrubs. These serve to trap sand (or coarse silt) derived by deflation from a shallow depression.

MAN-INDUCED DEFLATION AND DUNE ACTIVATION

Man's activities have aggravated the work of wind where modern farming practices and intensified livestock grazing have been extended into semiarid climate zones. Here the plant cover is at best sparse and is easily broken down, baring the soil to deflation. The exposed soil of a harvested wheat field has little protection except the stubble of the previous crop. On rangelands, cattle are often allowed to graze in numbers that exceed the capacity of the native grasses to sustain growth. Trampling of the soil under the hooves of heavy animals induces easy deflation.

Now a new environmental impact has come into the picture. Use of the off-the-road vehicle (ORV) is increasing in popularity as

Figure 18.13 The steep landward slope of a coastal dune is advancing over a forest at Cape Henry, Virginia. (Douglas Johnson.)

Figure 18.14 An abandoned farm in Dallam County, Texas, 1937. This Dust Bowl scene shows a drift of sand and silt, too coarse to be carried in suspension, built up along a fence row during repeated dust storms. (Soil Conservation Service, U.S. Department of Agriculture.)

a form of recreation. Trail bikes and four-wheel-drive vehicles churn the desert surfaces, destroying both the plant cover and the protective pebble coating of the desert pavement. These activities allow clouds of dust to rise. Dune buggies are driven heedlessly over coastal foredunes, destroying the beachgrass and inducing the rapid landward movement of dunes.

Man-induced deflation and dust storms are not a recent phenomenon. In the Great Plains region, including part or all of New Mexico, Texas, Oklahoma, Kansas, Colorado, Nebraska, and the Dakotas, deflation rose to disastrous levels during the mid-1930s, when the region achieved recognition as the Dust Bowl. This resulted when a series of drought years followed agricultural expansion which had bared vast expanses of rich brown soils formed on loess. A particularly damaging effect of the repeated dust storms in the Dust Bowl was the accumulation of coarser fractions of the soil as drifts over fence rows and buildings (Figure 18.14).

As long as large expanses of this region are in cultivation, the dust storm will be a recurrent problem of drought years. Some of the deflation can be controlled through improved farming practices. For example, deeply plowed furrows (listed furrows) can act as traps to drifting soil. Tree belts can be effective in reducing wind stress at ground level.

In the opening pages of this chapter we related a case of man-induced dune activation. The impending burial of Provincetown was from sand moving toward the shoreline of the bay. Usually the case is the reverse, with activated coastal dunes moving inland to overwhelm houses and roads.

The coastal foredune ridge plays an important role in coastal protection. It absorbs much of the energy of storm surf, which reaches to heights many feet above the usual limit of the swash. The dune ridge may be cut back many yards in a single storm. It will normally be rebuilt in longer intervening periods of normal wave levels, receiving sand from beach deposits built by low waves during summer periods.

WIND AS A GEOLOGIC AGENT

The intensive work of wind is limited to those specific places where unattached mineral particles are available in quantity, such as desert plains and coastlines. Where they take place, wind erosion and transportation are significant both in terms of environment of today and in the stratigraphic records of the geologic past.

During the Pleistocene Epoch, when ice sheets were widespread and large expanses of unprotected detrital materials were exposed to the air, the geologic role of wind was much more important than we find it today. Going far back into geologic time the great fossil dune beds of Permian and Jurassic age in the Colorado Plateau bear witness to a scale of wind transport and deposition of sand that dwarfs anything found on earth today. Perhaps our judgment that wind is a geologic agent of secondary importance needs to be tempered by the evidence of the past.

YOUR GEOSCIENCE VOCABULARY

deflation
dust
dust storm
blowout
desert pavement
sand-blast action
ventifact
loess
saltation
sand storm
sand drift
sand dune
barchan dune
slip face
transverse dunes
sand sea
foredunes
parabolic dunes
coastal blowout dunes

SELF-TESTING QUESTIONS

1. Describe the processes involved in wind erosion. What landforms are produced by deflation? How is a ventifact shaped?

2. What is the origin of loess? Where is it found? What is the mineral composition of loess? How is loess related to agricultural soils?

3. How is sand transported by wind? Describe a barchan dune. How does it advance? How fast does it advance? How are transverse dunes related to barchan dunes?

4. How does vegetation exert control over dune development? Describe the parabolic dune. How is its shape oriented with respect to the wind? How would you be able to distinguish a parabolic dune from a barchan dune?

5. In what ways do man's activities promote deflation and dust storms? What role does the foredune belt play in modifying storm wave action?

6. What evidence is found in the geologic record of major activity by wind?

In a deep quarry at Crosby, Minnesota, a launch crew prepares to release a balloon carrying Maj. David G. Simons. On this day in August 1957, Major Simons reached an altitude of about 102,000 ft, enclosed in the spherical capsule you see just above the bed of the launch truck. (U.S. Air Force photo.)

19 THE ATMOSPHERE AND OCEANS

GHOSTS AND BOOMERANGS

Man's urge to rise silently into the air has found expression in many ways since the Montgolfier brothers launched the first successful free balloon almost two centuries ago. The primitive hot-air balloon has seen a modern revival, thanks to efficient propane burners. These small free balloons are great for seeing the countryside and even for racing in competition. But for exploring the upper atmosphere, a new breed was needed. Two pioneers in the early 1930s were the Piccard brothers. Using a sealed gondola and a large rubberized fabric bag filled with helium, first Auguste and then his brother Jean rose in successive ascents to heights just over 55,000 feet.

In the mid-1940's, the United States Navy became interested in using high-altitude balloons. Various kinds of scientific observations can best be made from slow-moving balloons remaining aloft for many hours. The new tough plastics now came into use. Mylar in particular has been used for high-altitude balloons. It is not only tough and leakproof but has little stretch. Many unmanned balloons were tested by the Office of Naval Research, and in 1956 one of them rose to 145,000 feet, a new altitude record. Balloons with a helium capacity of 3 million cubic feet and a payload of 400 pounds of instruments were used in one study of cosmic rays.

Next, the Navy began to develop manned high-altitude balloons in order to study physiological effects of man's entry into space. The Air Force was also getting into the act, and 1957, Maj. David G. Simons reached a height of over 100,000 feet while crammed into a small sealed aluminum capsule. In 1960, another Air Force officer, Capt. Joseph W. Kittinger, Jr., wearing a space suit but seated in an open gondola, ascended to an altitude of 102,800 feet. His return to earth was by parachute. He made a record free fall of over 84,000 feet, reaching a speed of over 600 miles per hour, before opening his chute. Then, in 1961 two Navy Officers set a new altitude record of 113,740 feet, which stands today. Cmdr. Malcolm D. Ross, USNR, a physicist, and his companion, Lt. Cmdr. Victor A. Prather, ascended in a sealed aluminum gondola equipped with elaborate research instruments. After drifting some hours, they descended for a landing in the Gulf of Mexico. Here, Prather met his death by drowning when he slipped and fell from the rescue helicopter that had lifted him from the floating gondola.

Manned stratospheric balloons have not been used since the early 1960s, but the unmanned Mylar balloons have been developed into new breeds of research vehicles. First came the GHOST (Global Horizontal Sounding Technique) family. These helium-filled Mylar bags were first launched in 1966 from Christchurch, New Zealand. Designed to stay at a constant altitude of 50,000 to 60,000 feet, GHOST balloons carried instrument packages for measuring weather data in the stratosphere. Drifting with the upper-air westerly winds, GHOST balloons were able to circle the globe repeatedly over the Southern Ocean. One GHOST balloon remained aloft 441 days and completed 35 circuits around the world.

The successor to GHOST is Boomerang, a NASA-funded research balloon designed to drift at an altitude of 80,000 feet. Carrying 90 pounds of scientific equipment, Boomerang I was launched in Australia in 1973. It completed two circuits and was recovered at the starting point. A "super-Boomerang" is planned for cruising at an altitude of 130,000 feet. One topic of this chapter is the density, pressure, and temperature of the atmosphere at these great heights.

ATMOSPHERE AND OCEANS

With this chapter we turn from the lithosphere to focus attention upon the atmosphere and oceans. The **atmosphere**, or gaseous realm, is the earth's envelope of gases, in which are suspended countless minute liquid and solid particles. The **oceans**, standing bodies of salt water, form the vast bulk of the earth's **hydrosphere**, or liquid realm. The oceans hold a large quantity of mineral matter, most of it dissolved as sea salts.

We introduce you to both the atmosphere and oceans in a single chapter to point out their similarities and differences. Both air and water are classed as **fluids**, substances that flow easily when subjected to unbalanced forces. Fluids also share the property of responding to the force of gravity so as to come to rest with the denser of two fluids forming a layer beneath the less-dense fluid. The two forms of fluids are liquids and gases.

As a **liquid**, the water of the oceans, lakes, and rivers is able to flow freely, but it cannot be appreciably compressed into a smaller volume. A liquid poured into a container will come to rest with a free horizontal surface and will not expand to fill the entire container. The atmosphere, on the other hand, behaves as a true **gas** in that it is easily compressed into a smaller volume when pressure is applied. Gas will also expand rapidly by diffusion to occupy any given small container uniformly. Consequently, the oceans and atmosphere have common flowage properties but different ways of occupying space.

Both the atmosphere and the oceans are vast fluid layers in ceaseless motion. Both have a common role in the physical processes of our planet. This role is to transfer heat received from the sun, from one part of the globe to another. Inequalities in the heating of the atmosphere and oceans are basically responsible for generating and sustaining the enormous flow systems of air and water which serve to regulate the earth's heat budget.

So far as we know, no other planet has both a dense atmosphere and extensive oceans of liquid matter. The earth possesses a unique combination of surface fluids, which together have provided a favorable environment for the development of life, sustained with remarkable constancy for hundreds of millions of years.

The science of **meteorology** deals with the physics of the atmosphere; the science of **physical oceanography** with the physics of the oceans. Because the atmosphere and oceans interact intensely, the sciences of meteorology and oceanography are closely interrelated.

COMPOSITION OF THE ATMOSPHERE

The earth's atmosphere consists of a mixture of various gases surrounding the earth to a height of many miles. Held to the earth by gravitational attraction, this spherical shell of gases is densest at sea level and thin rapidly upward. About 97% of the mass of the atmosphere lies within 18 mi (29 km) of the earth's surface.

The upper limit of the atmosphere cannot be drawn sharply because the density of gas molecules grades imperceptibly into the near-emptiness of interplanetary space. We can use 6000 mi (10,000 km) as a working figure for the thickness of the atmosphere. This thickness is almost as great as the diameter of the solid earth itself.

From the earth's surface upward to an altitude of about 60 mi (100 km), the chemical composition of the atmosphere is highly uniform throughout, in terms of the proportions of its gases. Above 60 mi (100 km), the very thin gases become separated into layers of different composition. We shall focus attention only on the lower, uniform layer, since it contains all important weather phenomena.

The uniform atmospheric layer consists of (1) a mixture of gases referred to collectively as the **pure dry air,** (2) water vapor, (3) dust particles, and (4) water droplets or ice particles, in the form of clouds and fog. The first two components are true gases composed of individual molecules. Dust consists of solid particles much larger than molecules, but still so tiny as to mix freely with the gases and to stay aloft almost indefinitely.

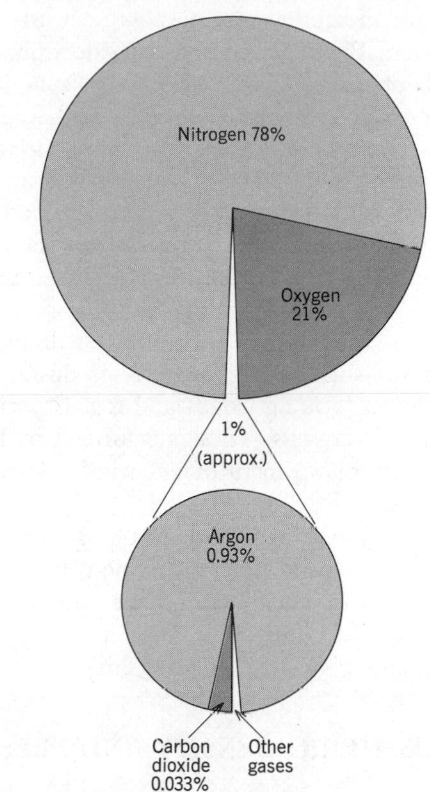

Figure 19.1 Components of the earth's lower atmosphere.

Consider first the individual gases of pure dry air (Figure 19.1). The largest part by far is nitrogen, about 78%, or more than three-fourths, of the pure dry air by volume. Nitrogen can be thought of as an inactive gas, or space filler, for the most part, although it is extracted from the air by certain bacteria which form nitrogen compounds vital to plant life.

Oxygen, the second largest component, makes up about 21%, or one-fifth, of the air by volume. Oxygen is chemically very active, combining readily with rock-forming minerals in rock decay, with metals in rusting, with fuels in burning, and with food to provide heat energy in animals. Despite its chemical activity the quantity of oxygen in the air remains constant from year to year because the amount used is exactly balanced by oxygen given back to the atmosphere by plants.

Nitrogen and oxygen together make up about 99% of the air; of the remaining 1% argon composes more than nine-tenths of 1%. Argon does not naturally combine chemically with other elements.

Carbon dioxide, although forming only 33/1000 of 1% of the air, is an extremely important gas. Climatically, carbon dioxide is important as an absorber of heat and as an insulating blanket. It helps to regulate air temperatures near the earth's surface. Biologically, carbon dioxide is essential for the growth of plant life.

Since man has begun to burn prodigious quantities of wood, coal, and oil, much more carbon dioxide is being released into the atmosphere now than half a century ago. Since 1900 the amount of carbon dioxide has increased more than 10% and is rising steadily.

WATER VAPOR AND DUST

The second major component of the lower atmosphere is **water vapor,** the gaseous state of water in which individual water molecules have the same freedom of movement as molecules of nitrogen or oxygen gas. The water molecules diffuse completely among gases of the pure dry air. Water vapor is not visible and should not be confused with fog and clouds, which are composed of liquid or solid particles.

The amount of water vapor in the air varies greatly from time to time and place to place throughout the earth's atmosphere. Water vapor supplies the water for all clouds and rain, and during condensation it releases heat that supplies the energy for storms.

Dust in the lower atmosphere consists of particles so tiny that 250,000 of them placed side by side would be needed to make a line about an inch long. Most atmospheric dust comes from the earth's surface. Smoke from grass and forest fires is an important source. Winds blowing over dry land surfaces of deserts raise mineral particles thousands of feet into the air. Volcanoes in eruption contribute dust clouds. Dust is also added by meteoroids that vaporize upon entering the upper atmosphere.

Water vapor condenses to form either water droplets or ice crystals, depending on how cold the air is. These forms of water appear as clouds and fog. Both water droplets and ice crystals form as deposits surrounding particles of dust. The dust particles serve as **nuclei** in forming clouds and fog. Especially important as nuclei are tiny salt crystals. These are formed by the evaporation of spray droplets swept up in turbulent winds blowing over crests of breaking waves.

Both water vapor and dust originate mainly from the earth's surface and depend upon air motions to be lifted vertically. As a result, these components tend to be most heavily concentrated in the lowest air layers and to diminish to nearly zero values in the upper atmosphere, above 60 mi (100 km).

ATMOSPHERIC DENSITY AND PRESSURE

The force of gravity acts on all matter, even the air, which we think of as having almost no substance or weight. Molecules of the atmospheric gases are attracted earthward and tend to crowd together. The crowding increases downward from the outer limits of the atmosphere to sea level, because any one layer of the atmosphere is being compressed by the weight of all the layers above it. About

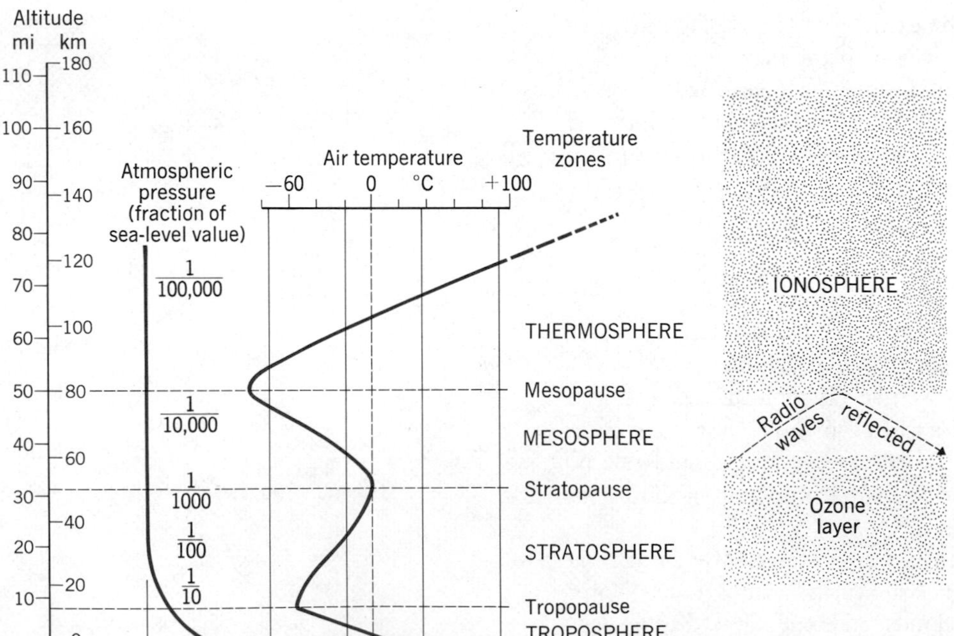

Figure 19.2 Some physical properties of the atmosphere.

three-fourths of the earth's air is contained in the lowest 6 mi (10 km) of the atmosphere, compressed by the weight of the overlying air in such a way that the density is greatest at sea level and decreases very rapidly upward.

Any surface exposed to the atmosphere is under a force represented by the weight of the gases lying above the surface. This force, when measured for a standard unit of area, is known as **atmospheric pressure.** It is the same on an exposed surface whether the surface is turned up, down, or in any other attitude. Atmospheric pressure, or simply **air pressure,** is measured routinely at all observing stations and is an essential part of all weather description and forecasts.

Atmospheric pressure has an average value at sea level of 14.7 lb/sq in. (about 1 kg/sq cm). The pressure graph in Figure 19.2 shows how rapidly air pressure decreases with increase in altitude above sea level. Taking the sea-level pressure as 100%, we see that air pressure has fallen to 10% at 11 mi (18 km) and to a mere 1% at 20 mi (32 km). At 70 mi (112 km) air pressure is only 1/100,000 of the sea-level value.

Air pressure can be demonstrated and measured by a very simple device, the **mercurial barometer** (Figure 19.3). The demonstration is often called Torricelli's experiment, after the man who first performed it, in 1643. A glass tube of very narrow bore and about 3 ft (90 cm) long is sealed at one end, filled with mercury, and inserted open end down into a dish of mercury. Instead of all pouring out, most of the mercury column remains in the tube with a height of about 30 in. (76 cm). A vacuum occupies the section of empty tube above it. External air pressure supports the mercury column.

We may imagine that the mercury column in the tube represents the balancing weight on the scales in Figure 19.3. In other words, a square column of mercury 1 in. thick and 30 in. high actually weighs as much as an entire column of atmosphere 1 in. sq in cross section extending from the solid earth to interplanetary space. The exact average height of the mercury column at sea level is 29.92 in.; this value is taken as the **standard sea-level pressure** of the atmosphere. In metric units the established value is 76 cm, or 760 mm.

In modern weather science the unit of air pressure is the **millibar.** One in. of mercury column is equivalent to 33.9 millibars (mb). Standard sea-level pressure is 1013.2 mb.

PRESSURE AND ALTITUDE

In Figure 19.4 is a detailed graph of air pressure in the lower part of the atmosphere, where most of man's activities take place. Pressure is given in inches of mercury at the top and in millibars at the bottom. Pressure decreases about 1 in. in the first 1000 ft of ascent, but this rate steadily becomes less at higher altitudes. A fairly accurate formula is that pressure decreases by about one-thirtieth of itself for each 950 ft of rise. The graph shows us that at an altitude of 3.5 mi (5.5 km) the pressure is half the sea-level value.

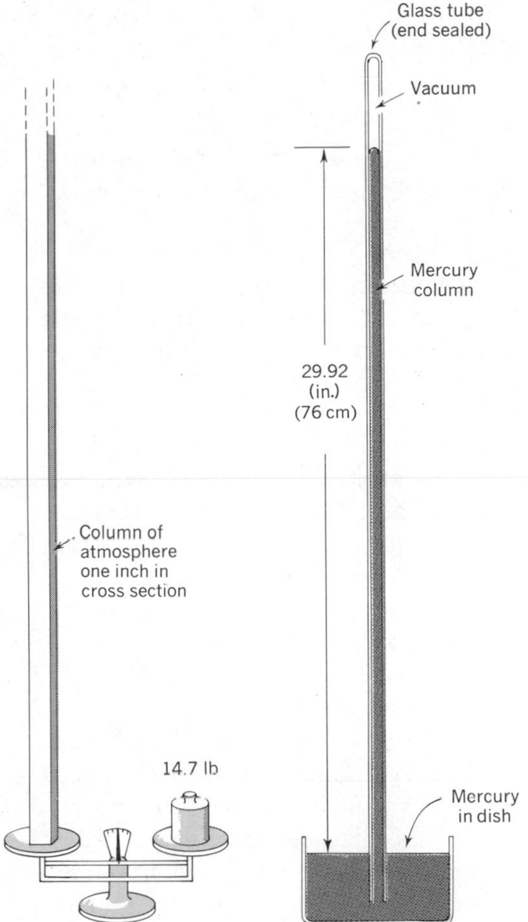

Figure 19.3 Atmospheric pressure can be visualized as the weight of a unit column of air. Torricelli's experiment demonstrates atmospheric pressure.

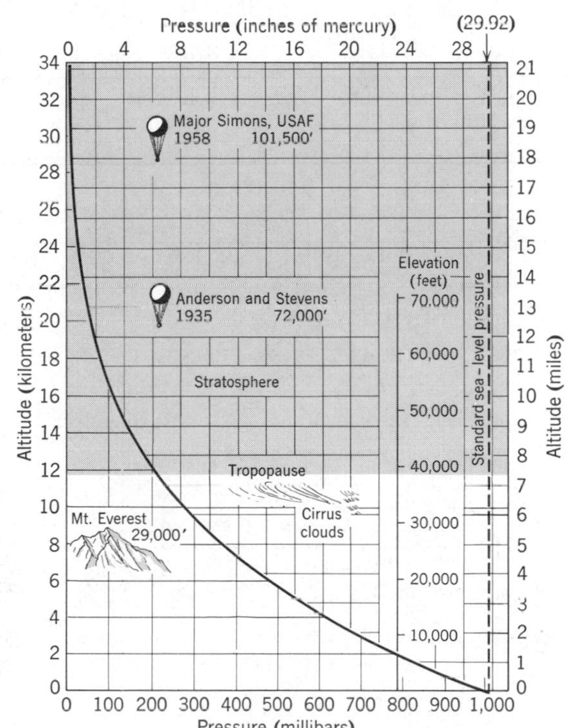

Figure 19.4 A smooth curve represents the decrease of atmospheric pressure with increasing elevation. No perceptible break occurs at the tropopause. (© 1973, John Wiley & Sons, New York.)

On the earth's highest point, Mt. Everest in the Himalayas, the pressure at 29,000 ft (8.8 km) is one-third that at sea level, and the mercury column stands only 10 in. (25 cm) high. Although mountain climbers who have spent weeks acclimating themselves to the rarefied air can survive on Mt. Everest, a flier needs oxygen to stay alive at this altitude. The 40,000-ft level (about 8 mi, 13 km) is approximately the highest level at which a person can survive while breathing oxygen but without being enclosed in a pressurized cabin or pressure suit, for here the pressure is a mere 6 in. (15 cm) of mercury. Above this level the human lungs could not absorb enough pure oxygen to stay alive because there would be insufficient pressure to force the gas into the lung tissues. Still worse difficulties would arise at the 60,000-ft (18-km) level, because here the human blood boils at its normal body temperature.

The limit of sustained, level jet-plane flight is somewhere in the range of 80,000–90,000 ft (15–17 mi, 24–27 km). Here the pressure is less than 1 in. (2.5 cm) of mercury. Air density above this level is so low that enough oxygen cannot be provided for fuel combustion and the lift force on the wings is too weak to keep the plane climbing at the available speed.

Free balloons filled with helium gas have carried men still higher. In 1961 Cmdr. Malcolm D. Ross, USNR, reached 113,740 ft (34.7 km), the current world record in a manned balloon. Man has, of course, ascended much higher into the atmosphere by means of rocket-powered aircraft, and he has gone into outer space by using multistage rockets. Unlike jet engines, rockets carry their own oxygen for combustion and are not hindered by lack of oxygen or reduced air pressure. In addition, rocket craft profit from the greatly reduced friction of the rarefied air.

TEMPERATURE LAYERS OF THE ATMOSPHERE

Temperature scales used in weather science are the **Fahrenheit** and **Celsius** (Centigrade) scales (Figure 19.5). On the Fahrenheit scale (designated as °F), the freezing temperature of water is 32° F, and the boiling point of water is 212° F. The Fahrenheit scale is used in the reports of surface weather conditions issued by the National Weather Service. The Celsius scale (designated °C) takes 0° C for the freezing point and 100° C for the boiling point of water.

Figure 19.2 includes a graph of temperature from ground level to more than 100 mi (160 km) height. Starting at the ground, let us examine the great range of temperatures one would encounter in a vertical ascent. As we climb a mountain or rise in an airplane, the temperature falls steadily. This rate of temperature drop, averaging

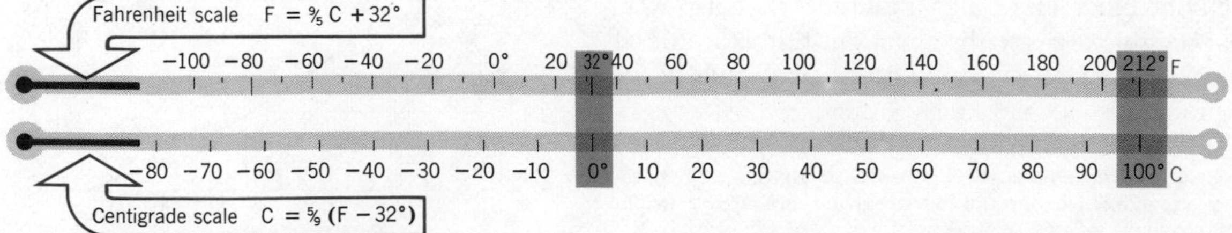

Figure 19.5 Comparison of Fahrenheit and Celsius scales. (© 1973, John Wiley & Sons, New York.)

about 3½ F° for each 1000 ft (6.4 C°/km), is called the **environmental temperature lapse rate.**

This almost constant decrease in air temperature is shown in more detail in Figure 19.6. Height is scaled on the vertical axis; temperature on the horizontal axis. The curve on this graph represents a typical sounding made by a balloon carrying up a thermometer and sending back its temperature values by radio. Notice that temperature falls steadily until a height of about 35,000 ft (11 km) is reached. At this point the curve breaks sharply and reverses itself. Here, at the curve break, we have reached the **tropopause.** It marks the top of the basal atmospheric layer known as the **troposphere.** Above lies the **stratosphere.**

The troposphere contains almost all the water vapor of the atmosphere. Consequently this basal layer contains nearly all clouds, precipitation, and storms. Today, jet aircraft can maintain flight in the stratosphere in middle and high latitudes and thus avoid many weather hazards.

Upward through the stratosphere, temperatures rise gradually up to a level of about 30 mi (50 km), where the **stratopause** is encountered (see Figure 19.2). Here, the temperature reaches a maximum of about 32° F (0° C).

Above the stratopause lies a zone of diminishing temperature, the **mesosphere.** At the **mesopause,** about 52 mi (85 km) up, temperatures reach a minimum value, averaging −120° F (−83° C).

Above the mesopause lies the **thermosphere,** a zone of rapid temperature increase, to extremely high values of over 1300° F (700° C) at an altitude of 125 mi (200 km). Keep in mind that at these great heights the air is extremely thin—a near-vacuum—so that little heat is actually held in the air.

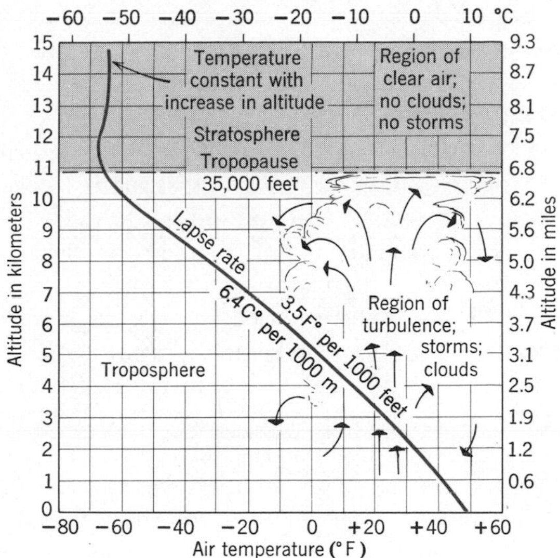

Figure 19.6 Air temperature falls steadily as altitude increases in the lowest atmospheric layer, or troposphere. (© 1973, John Wiley & Sons, New York.)

THE SOLAR RADIATION SPECTRUM

To understand the heating of the stratopause and thermosphere, as well as certain other phenomena of the upper atmosphere, we need first to analyze the nature of incoming solar energy.

Heated to incandescence at temperatures of many thousands of degrees, our sun radiates an enormous quantity of energy. This energy travels through space until it falls upon some gaseous, liquid, or solid matter. The receiving substance transmits or absorbs the radiation or turns it away by reflection. Solar radiation belongs to a class of energy flow called **electromagnetic radiation;** it can be thought of as being a wavelike motion. In some respects it moves like the waves which travel over the surface of a quiet pond from the point where a stone is dropped.

Using this concept, we can describe the waves in two ways: (1) by the distance separating successive wave crests, or **wavelength,** and (2) by the number of wave crests moving past a fixed point each second of time, or **frequency.** Figure 19.7 illustrates the point that long waves have low frequency, while short waves have high frequency, where all waves have the same speed of travel. Electromagnetic radiation travels at the constant rate of 186,000 mi (300,000 km) per second ("speed of light"). Solar radiation thus takes about 8⅓ min to reach the earth.

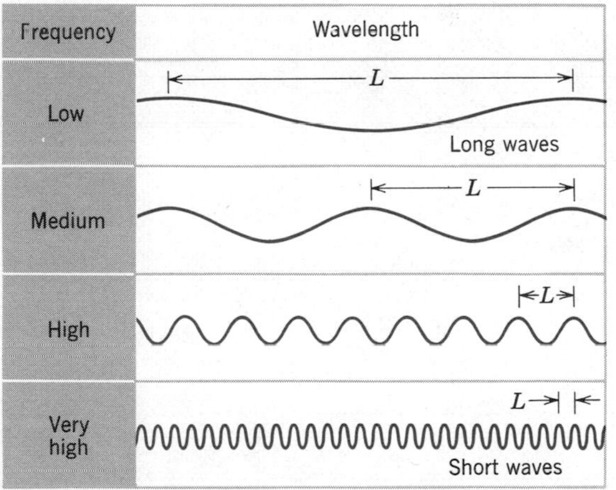

Figure 19.7 Long waves have low frequency; short waves have high frequency, when all travel at the same speed. (© 1975, John Wiley & Sons, New York.)

Figure 19.8 illustrates the component parts of the entire **electromagnetic spectrum.** The scales are logarithmic scales (constant ratio scales) such that each division has a value 10 times as great as the next lower division. The upper scale gives frequency in megahertz as powers of 10. One **megahertz** is a frequency of one million cycles per second. The wavelength scale is labeled in microns, centimeters, and meters. One **micron** is equivalent to 0.0001 cm (10^{-4} cm). The expanded scale of colors in the lower part of Figure 19.8 shows the visible light wavelength band.

At the very short wavelength (very high frequency) end of the spectrum are **gamma rays** and **X rays.** These are high-energy rays capable of deep penetration into opaque substances. Next comes the **ultraviolet** band, followed by the **visible light** band. Composed of still longer wavelengths is the **infrared** band, overlapping the even longer wavelength band of **microwaves.** The spectrum then continues as radar and radio waves, and these extend to lengths of many kilometers. All parts of the electromagnetic spectrum are radiated into space by the sun and other stars.

Electromagnetic radiation is a flow of energy outward into space in all directions from the spherical surface of the sun. The total energy of radiation is not equally distributed throughout the entire spectrum. The peak intensity of energy is within the visible part of the spectrum, which carries about 41% of the total energy. The shorter waves carry about 9%. The infrared and longer waves carry the other 50%. We will go into this energy distribution in more detail in Chapter 20.

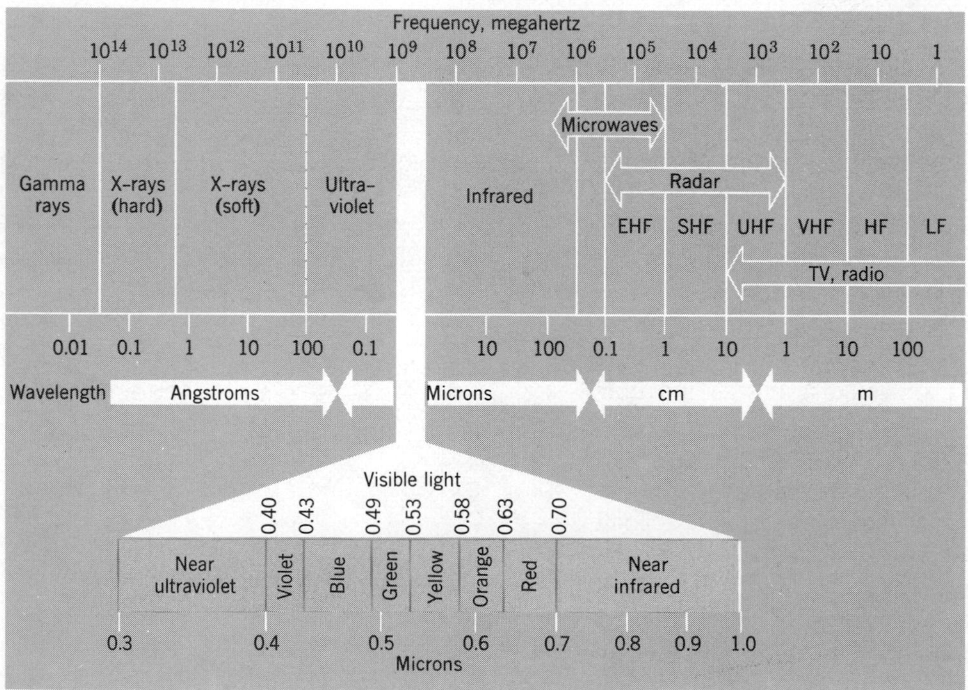

Figure 19.8 The electromagnetic radiation spectrum. (One angstrom unit is equal to 10^{-7} cm.) (© 1975, John Wiley & Sons, New York.)

THE IONOSPHERE

As the solar radiation penetrates our upper atmosphere, its rays strike atoms and molecules of the various gases.

First, molecules and atoms of nitrogen and oxygen of the upper atmosphere absorb the highly energetic gamma rays, X rays, and ultraviolet rays of the solar spectrum. In so doing, each affected molecule or atom loses an electron and becomes a positively charged molecule or atom known as an ion. The process is known as **ionization.** It begins at a height of about 600 mi (1000 km) and is effective down to about 30 mi (50 km). There is a region from about 50 to 250 mi (80 to 400 km) above the earth where the concentration of positive ions is most dense; this region is known as the **ionosphere** (see Figure 19.2).

Absorption of radiation in the ionosphere causes the atmosphere to become heated, and this phenomenon accounts for the high temperatures found in the thermosphere. Notice in Figure 9.2 that the mesopause, a very cold layer, coincides with the base of the ionosphere. Upward within the ionosphere temperatures rise rapidly to produce the thermosphere.

The ionosphere is extremely important in radio-wave transmission. Figure 19.9 shows how the paths of radio waves sent out from a transmitter on the earth are reflected back to earth by the lower part of the ionosphere. Before the nature of the ionosphere was understood, two early radio experimenters independently discovered the radio-wave reflection principle. The **Kennelly-Heaviside layer,** named in their honor, was the term first applied to the lower part of the ionosphere, at about the 50- to 60-mi (80- to 97-km) level. Extremely long radio waves, with lengths over 1000 ft (300 m), travel by reflection with very little loss and are used in transoceanic radiotelephone transmission.

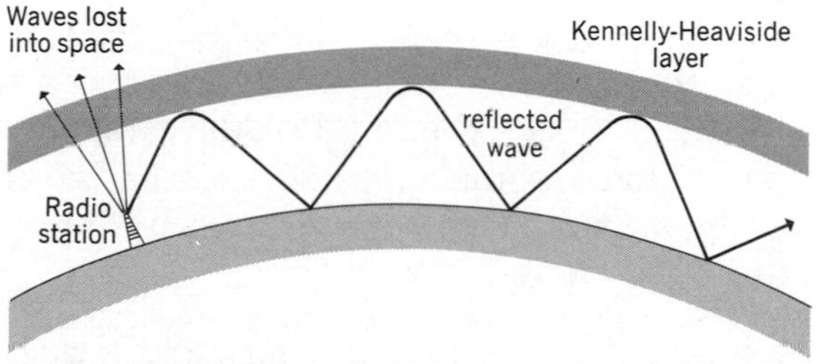

Figure 19.9 Radio waves are reflected from the ionosphere. In this way, long-distance radio communication is possible.

THE OZONE LAYER

Let us continue to trace the effects of solar radiation as it penetrates deeper into the earth's atmosphere. Solar X rays and the shorter ultraviolet rays are almost completely absorbed within the ionospheric region. However, the longer ultraviolet rays pass readily into lower levels of the atmosphere.

The region known as the **ozone layer** is largely concentrated in the stratosphere in the altitude range of 12 to 31 mi (20 to 35 km) (Figure 19.2). Here, the absorption of the longer ultraviolet rays produces a chemical effect upon the atmospheric oxygen that is essential in creating the gas for which the ozone layer is named. The gas **ozone** is a form of oxygen in which the molecule consists of three oxygen atoms instead of the usual molecule of two atoms. In the ozone layers of the atmosphere, ultraviolet rays cause ordinary oxygen molecules to split into single atoms. When a single oxygen atom collides with an ordinary two-atom oxygen molecule, an ozone molecule is formed. The process is reversible, so that ozone eventually breaks down and re-forms into ordinary oxygen molecules.

Ozone is a deadly poison to life when present in large concentrations. Fortunately it is almost completely absent in the troposphere except in the polluted air of cities. The ozone layer is essential to all life on earth, because it fully absorbs the shorter ultraviolet rays that would otherwise destroy all exposed bacteria and severely damage animal tissues.

The ozone layer also absorbs much of the longer ultraviolet radiation and some of the visible and infrared wave lengths as well. This absorption heats the ozone layer, causing the temperature maximum of the stratopause, reached at about 30 mi (50 km). It is interesting to note that if all atmospheric ozone were brought down to sea level, at these pressures it would form a layer only 0.1 in. (0.25 cm) thick.

How much we owe to this small amount of ozone! There is concern among scientists that jet engine exhausts of supersonic transport (SST) aircraft might reduce this protective layer and expose us to the deadly ultraviolet radiation.

THE MAGNETOSPHERE

The earth has a magnetic atmosphere—the magnetosphere—as well as an atmosphere of gases, but the two are entirely different.

Recall from Chapter 2 that the earth acts as a great magnet. Lines of force, shown in Figure 2.4, extend far out into space, making up the magnetosphere. This field extends many times farther from earth than the gaseous atmosphere.

A magnetic field is a form of energy and has no matter, but it can trap and hold matter in the form of particles. These particles are electrons and protons which stream from the sun and pass by the earth and other planets in the form of a **solar wind.** When a particle from the solar wind reaches the earth's magnetic field, it can become entrapped between lines of force, as shown in Figure 19.10. Countless billions of such particles are continually entrapped in the magnetic field, and these make up the magnetosphere.

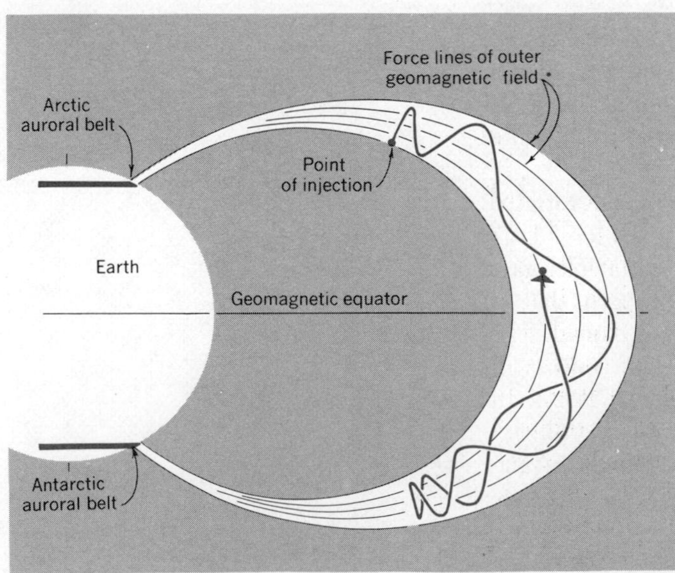

Figure 19.10 The sinuous line shows the typical path of a particle trapped within the lines of force of the earth's outer magnetic field.

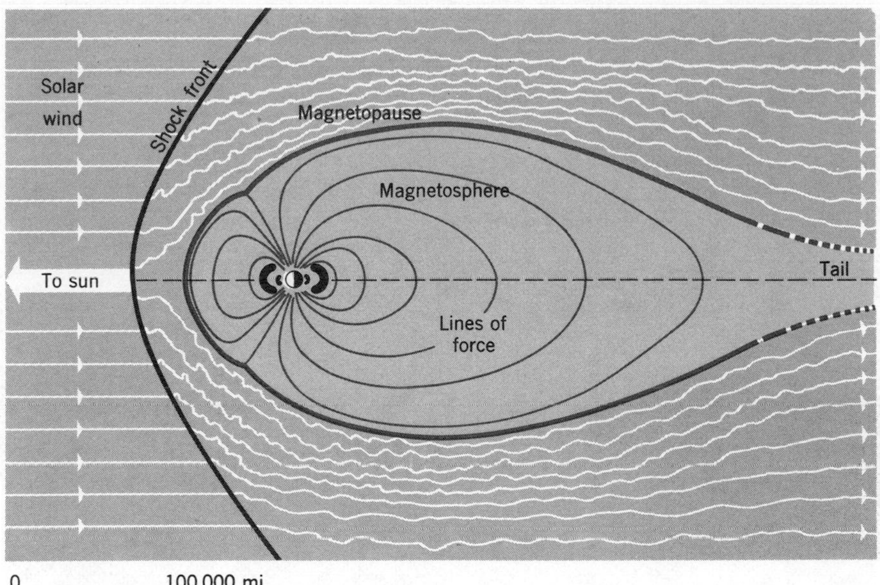

Figure 19.11 Cross section of magnetosphere, showing magnetopause and shock front. (© 1973, John Wiley & Sons, New York.)

The earth's magnetic field would be symmetrical and doughnut-shaped were it not for the pressure exerted by the solar wind. As Figure 19.11 shows, the magnetosphere is deformed into a teardrop shape, with a long, tapering tail pointed away from the sun. Streamlines of the solar wind are sent into wavy ripples along a shock front, where the wind first encounters the magnetic field. The streamlines of the solar wind are then forced to flow around the magnetosphere, enclosing it in a sharp boundary called the **magnetopause.**

Trapped particles in the magnetosphere are said to be energetic, meaning that they produce **ionizing radiation,** much like that generated by an X-ray machine or a radioactive isotope, like radium. Early space probes carried Geiger counters, which as you may know are used to detect radiation coming from radioactive isotopes. The Geiger counters detected a belt of heavy radiation located several thousand miles out. Within a doughnut-shaped belt surrounding the earth the radiation from trapped particles is intense (Figure 19.12). This region was named the **Van Allen radiation belt,** after the scientist who first reported its existence. The Van Allen belt is shown to scale in Figure 19.11 by black crescents beside the earth.

THE AURORA

One of nature's great spectacles is the **aurora borealis,** or **northern lights,** of the northern latitudes, and its southern hemisphere counterpart, the **aurora australis.** The aurora borealis is seen at night in the northern sky. It takes the form of bands, rays, or draperies of light, continually shifting in pattern and intensity (Figure 19.13).

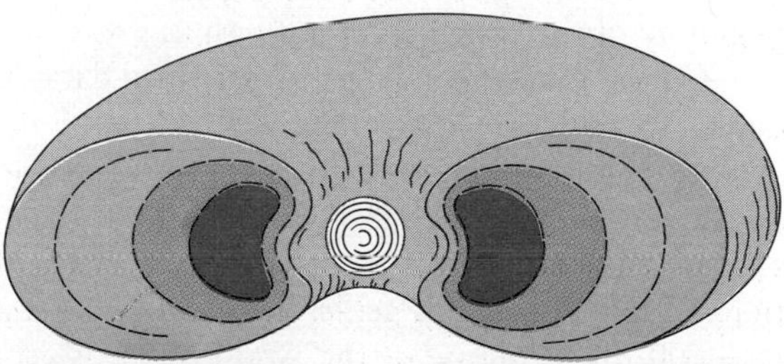

Figure 19.12 The Van Allen radiation belt forms a doughnut-shaped ring surrounding the earth. The outer boundary is actually a zone of gradation to lower radiation intensity, not a sharp boundary. Radiation is most intense where the color is deepest.

The aurora emanates from the ionosphere in a region 50–175 mi (80–280 km) above the earth. The auroral light continues during

Figure 19.13 This photograph of the aurora borealis was taken during the winter night in Alaska. (The American Museum of Natural History—Hayden Planetarium.)

the hours of darkness, after the solar radiation is gone, with maximum intensity one to two hours before midnight.

The aurora is caused by highly energetic electrons and protons trapped within the lines of force of the external magnetic field. As Figure 19.10 shows, activity is concentrated in two belts, one in the Northern Hemisphere, the other in the Southern Hemisphere. The energetic particles excite molecules of oxygen and nitrogen, causing them to emit light. The aurora is most commonly greenish in color, but red and violet colors are seen on rare occasions. The aurora is associated with exceptionally intense bursts of electrons and protons from the sun. The arrival of such a burst is called **magnetic storm** because the earth's magnetic field shows a strong disturbance at such times. One effect of a magnetic storm is to disrupt long-distance radio communication.

THE OCEANS

Throughout the remainder of this chapter we turn to the oceans of the globe. Our concern is with various phases of physical and chemical oceanography. We covered in Chapter 8 certain aspects of geological oceanography—the size, depth, distribution, and bottom topography of the principal ocean basins.

Consider some comparative statistical data, using the term **world ocean** to designate the combined oceans of the earth. The world ocean occupies about 71% of the earth's surface and has a mean depth of about 12,500 ft (3800 m). This figure includes shallow seas as well as the main basins. The average depth of the main portions of the Atlantic, Pacific, and Indian oceans is roughly 13,000 ft (4000 m). Volume of the world ocean is estimated as about 1.4 billion cu km, which is equivalent to about 317 million cu mi and constitutes 97.2% of the world's free water. Most of the remaining 2.8% is locked up in ice sheets.

Comparing masses of atmosphere and world ocean with the

total earth mass (including atmosphere and oceans) the estimates given at the right apply. (The unit used here is 10^{21} kg.)

Atmosphere	0.005
World ocean	1.4
Entire earth	6,000

On a fractional basis, these figures show that the world ocean has a mass about 1/4000 that of the entire earth, while the atmosphere has a mass about 1/275 that of the world ocean.

In what basic ways does the water of the ocean behave similarly to the air of the atmosphere? In what ways is their behavior different? How is their behavior linked? In the introduction to this chapter we noted that both air and water are fluids in the strict sense: both tend to come to rest in a state of fluid equilibrium; both respond freely to unequal forces by flowage.

One important difference is that the atmosphere, being gaseous, has no distinct upper boundary. In contrast, the fluid of the oceans has a clearly defined upper surface.

As regards the atmosphere, our primary concern is the study of the lower boundary, which is essential to an understanding of the earths' atmospheric circulation and storms. In contrast, the upper boundary is the most vital region of the oceans. The direct absorption and radiation of heat occurs through this upper ocean boundary; the loss of moisture by evaporation and the gain by precipitation take place here; the dragging force of the wind to produce waves and surface currents is exerted upon this surface.

Another basic difference is that oceans occupy basins, in which the continental margins act as side walls to restrict the motion of the ocean water. On the other hand, the atmosphere has no such restriction and is free to circulate on a global scale.

Air, a substance of low density and viscosity, moves freely, and quickly attains high velocities; it also quickly comes to rest. In contrast, ocean water, with its much greater density and viscosity, is comparatively sluggish in motion. It changes velocity only very slowly and to a much lesser degree than air.

Perhaps the most striking physical difference between the atmosphere and oceans occurs when the confining pressure is changed. The atmosphere, a gas, expands readily when pressure is reduced (as in the case of rising air) and contracts when pressure is increased (as in the case of descending air). Such volume changes are accompanied by large changes in temperature. The water of the oceans, in contrast, is nearly incompressible and experiences only very slight temperature changes when it is raised or lowered through many thousands of feet of vertical distance.

COMPOSITION OF SEA WATER

Sea water is **brine,** a solution of salts. Certain of the elements in sea water have come from deep within the earth's interior throughout geologic time by the process of outgassing, mentioned in Chapter 10. The elements chlorine, sulfur, fluorine, and bromine—all important constituents of sea water—are products of outgassing.

Another group of elements has been derived from the lands, through processes of rock weathering. These include sodium, magnesium, calcium, and potassium. While streams can bring these elements in solution to the oceans, evaporation of sea water cannot remove them through the atmosphere. However, the various elements of sea water can enter the solid state as sedimentary deposits on the ocean floors. In this way they can be removed from sea water. It is believed that the rate at which the various elements are added to the oceans is closely balanced by the rate of removal through precipitation as sediment. As a result, the composition of sea water has held quite constant for much of the geologic time in which complex life forms have existed.

Table 19.1 lists the five most important chemical constituents of sea water. Notice that in these five salts the first element is always a metal: sodium, magnesium, calcium, or potassium. If you refer back to Chapter 1 on the composition of igneous rocks, you will find that these elements are common in the feldspars and dark minerals making up the bulk of igneous rocks. Notice also that in four of the five salts, chlorine (as chloride) is the second element.

Chemical analysis of sea water shows that chlorine makes up 55% of the total weight of all matter dissolved in sea water; sodium is next with 31%.

Less abundant but important elements not appearing in the table are bromine, carbon, strontium, boron, silicon, and fluorine. A complete list of elements known to be present in sea water would include at least half of all the naturally occurring elements.

Table 19.1 **Principal constituents of sea water**

Name of salt	Grams of salt per 1000 grams of water
Sodium chloride	23
Magnesium chloride	5
Sodium sulfate	4
Calcium chloride	1
Potassium chloride	0.7
With other minor ingredients to total	34.5

TEMPERATURE LAYERS OF THE OCEANS

We observed that within the troposphere, air temperatures are typically highest at sea level, diminishing with altitude to very cold values at a rather uniform rate. In contrast, ocean water is warmest at the sea surface, because it is here that solar radiation is received. Temperature of the ocean body does not change uniformly with depth.

The temperature structure of the oceans over middle and low latitudes can be described as a three-layer system (Figure 19.14). The surface water is subjected to intense solar radiation—year-round in low latitudes and in summer in middle latitudes. The heated water takes the form of an upper layer of quite uniform temperature, a result of mixing within the layer. This warm surface layer attains a thickness of 1600 ft (500 m) and a temperature of 70° to 80° F (20° to 25° C) or higher in equatorial latitudes.

Immediately below the warm layer, water temperatures drop sharply. This layer of rapid temperature change is called the **thermocline** (Figure 19.14). It is 1600 to 3300 ft (500 to 1000 m) thick. Below the thermocline, temperatures decline gradually from about 40° F (5° C) immediately below the thermocline to about 34° F (1° C) close to the bottom. This statement applies to low latitudes. Figure 19.15 shows the situation at arctic and antarctic latitudes. Here, surface water temperatures are close to 32° F (0° C), and the temperature changes with increasing depth are very slight.

In contrasting the atmosphere and oceans, remember that the entire atmosphere is penetrated by solar radiation, and consequently receives an input of energy at all levels, as well as receiving heat from the earth's surface. In contrast, the ocean receives its heat only at the upper surface. Depth of penetration of solar radiation in

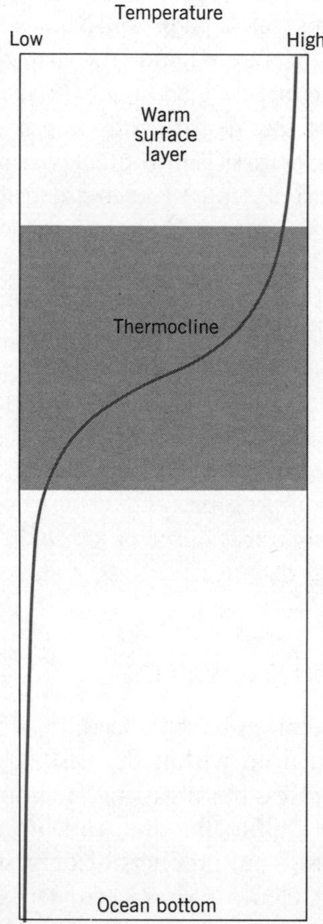

Figure 19.14 The oceans of middle and low latitudes show a warm surface layer, but very cold water at depth.

the oceans is very small, and warm water tends to remain on top. Only a very small flow of heat takes place from within the earth and is of practically no importance in warming the oceans.

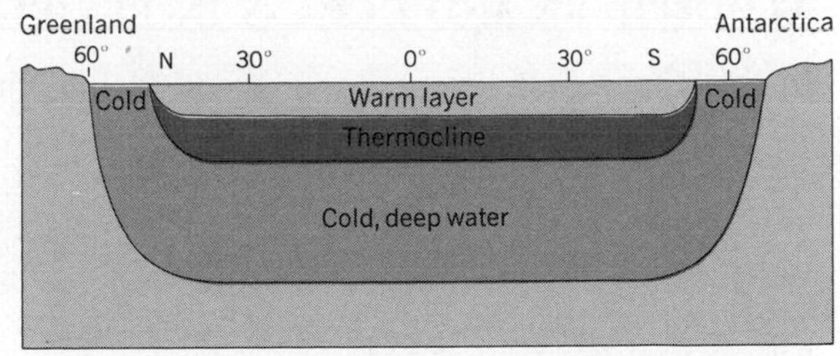

Figure 19.15 This schematic north-south cross section of the oceans shows that the warm surface layer and thermocline disappear in arctic latitudes, where the water is uniformly cold.

OCEAN SALINITY AND DENSITY

Salinity of sea water is the ratio of the weight of dissolved salts to the weight of water. In Table 19.1, we showed a salinity of 34.5 gm of salt for 1000 gm of water (34.5 gm/kg). This is a good average figure, but salinity differs from place to place over the oceans. The full range in salinity is from as low as 33 gm/kg to as high as 40 gm/kg. High values occur in certain bays or gulfs largely shut off from the ocean in tropical desert areas. The Red Sea is an example of a gulf with very high salinity. Low salinity is found over the equatorial oceans where rainfall is very heavy and the salt water tends to be diluted.

Sea water, because of the presence of dissolved salts, is slightly denser than pure fresh water. Compared with a density of 1.000 gm/cc for pure fresh water, sea water has a density of about 1.026–1.028 gm/cc. Both temperature and salinity affect the density.

Sea water becomes increasingly dense as it becomes colder, until the freezing point is reached, at about 28° F (−2° C). This is an important principle, because it means that sea water cooled near the surface will tend to sink, displacing water of less density.

Density also becomes greater as salinity increases, so that where surface evaporation is great, the water near the surface may become slightly denser than that below it, and will sink to a lower level. Because temperature is the stronger of the two controls of density, the densest sea water is formed in the cold arctic and polar seas. This very cold water sinks to the bottom and tends to remain close to the floor of the deep ocean basins.

PRESSURE IN THE OCEAN

Pressure in the oceans is governed by laws of fluids. In descending into the ocean, the pressure which water exerts equally in all directions upon any exposed surface increases in direct proportion to depth. This rule applies because water is practically incompressible (as compared to air). We simply multiply the weight of a cubic foot of water (62.4 lb) by the depth in feet to obtain the pressure in pounds per square foot (lb/sq ft). Of course, we must add the weight of the atmosphere upon the ocean, which is about 2000 lb/sq ft. At a depth of 1000 ft the pressure of the ocean is 64,400 lb/sq ft. At a depth of 20,000 ft in a deep ocean trench, the pressure is over 1 million lb/sq ft! Constructing a deep-diving submarine to reach these depths under such enormous pressures is a major engineering problem.

The important point is that ocean pressure increases uniformly with depth. In contrast, in the atmosphere gases are easily compressed into a much smaller volume by the weight of overlying layers. As a result, the pressure of the atmosphere falls off much faster near the ground than at high altitude.

ATMOSPHERE AND OCEANS IN REVIEW

In this chapter we have taken a broad look at the atmosphere and oceans as global layers of gas and of liquid water, respectively. We have compared these layers in several respects. One is the contrast in densities of air and water, and the rapidly diminishing density and pressure of the atmosphere upward from the surface. Someone has remarked that we humans live at the bottom of an ocean of air. All terrestrial life can be thought of as bottom-dwelling organisms thickly concentrated in a shallow layer. In this context, man is like a crawling marine crustacean—perhaps the counterpart of a lobster or a crab. Birds can swim in the sea of air; they play the role of fishes.

Solar radiation, penetrating the upper atmosphere, causes physical and chemical changes in the atoms and molecules of the gases it encounters. The ionosphere and ozone layer have no counterpart in the oceans. Another important point we have made is that the oceans occupy basins, separated by continents. This containment strongly controls the motions of ocean waters, whereas the atmosphere is free to move globally.

In the next three chapters we will investigate the atmosphere in much more detail. Our study of the oceans will be brief in comparison and will be concerned largely with the ways in which the sea surface interacts with the lower atmosphere.

YOUR GEOSCIENCE VOCABULARY

atmosphere
oceans
hydrosphere
fluid
liquid
gas
meteorology
physical oceanography
pure dry air
water vapor
nuclei
atmospheric pressure (air pressure)
mercurial barometer
standard sea-level pressure
millibar
Fahrenheit scale
Celsius scale
environmental temperature lapse rate
tropopause
troposphere
stratosphere
stratopause
mesophere
mesopause
thermosphere
electromagnetic radiation
wavelength
frequency
electromagnetic spectrum
megahertz
micron
gamma rays
X rays
ultraviolet rays
visible light
infrared radiation
microwaves
ionization
ionosphere
Kennelly-Heaviside layer
ozone layer
ozone
solar wind
magnetopause
ionizing radiation
Van Allen radiation belt
aurora (borealis, australis)
northern lights
magnetic storm
world ocean
brine
thermocline
salinity

SELF-TESTING QUESTIONS

1. How do liquids and gases differ in their behavior as fluids?

2. What gases comprise the pure dry air? What additional components make up the atmosphere? Of what importance in earth science are atmospheric oxygen, carbon dioxide, and water vapor?

3. Why does atmospheric density decrease from the surface upward? At what rate does atmospheric pressure decrease upward? Describe the mercurial barometer.

4. Describe the temperature layers of the atmosphere from the ground up. What is the cause of warming of the thermosphere?

5. Describe the electromagnetic radiation spectrum. How much of the total energy is carried in each part of the solar radiation spectrum? What changes take place in the radiation spectrum as solar energy penetrates the atmosphere?

6. Describe the ionosphere. What effect does the ionosphere have on radio-wave travel?

7. What is the ozone layer? How is ozone formed? What is the importance of the ozone layer to life on earth?

8. Describe the magnetosphere. How does the magnetosphere originate? What causes the Van Allen radiation belt? What causes the light of the aurora?

9. Describe the world ocean in terms of surface area and depth. Compare the masses of the world ocean, the atmosphere, and the solid earth. What are the most striking differences in physical properties of the oceans and atmosphere?

10. What elements comprise the important salts in sea water? Which are metals? What is the origin of the major elements in these sea salts? How is salinity measured? Where is salinity greatest? Where least?

11. Describe the three-layer system of ocean temperatures. What causes this layering? Why is layering absent in the arctic and polar oceans?

12. Describe conditions of density and pressure from the ocean surface to the deep ocean floor. Compare the rate of change of pressure in the ocean with that in the atmosphere.

Form No. 1009—Met'l.

U. S. Department of Agriculture, Weather Bureau.

COOPERATIVE OBSERVERS' METEOROLOGICAL RECORD:

Month of July, 1913; *Station,* Greenland Ranch *County,* Inyo.
Hour of Observation, 5 Pm
State, California *Latitude,* [illegible]; *Longitude,* 116-50W; *Time used on this form,*

DATE.	TEMPERATURE. Maximum.	Minimum.	Range.	§ Set Max.	PRECIPITATION. Time of Beginning.	Time of Ending.	† Amount.	Snowfall, in Inches.	Depth of Snow on Ground at Time of Observation.	Prevailing Wind Direction.	Character of Day.	‡ Miscellaneous Phenomena.
1	117	70		115							Clear	
2	121	71		116							Clear	
3	124	83		118							Clear	
4	119	77		114							Clear	
5	126	79		121							Cloudy	
6	125	89		122							Cloudy	
7	127	89		125								
8	128	90		125								
9	129	93		129								
10	134	85		128								
11	129	86		122								
12	130	85		126								
13	131	85		107								
14	127	86		118								
15	118	86		115								
16	118	88		116								
17	118	95		115								
18	108	94		106								
19	106	83		105								
20	105	70		96			1/10					
21	98	71		95			2/10					
22	100	70		98			3/10					
23	105	87		106								
24	106	81		106								
25	105	77		104								
26	104	74		98								
27	106	75		104								
28	108	74		102								
29	108	75		104								
30	111	73		110								
31	111	73		110								
Sum	3609	2506					0.60					
Mean	116.4	80.8										

MONTHLY SUMMARY.

TEMPERATURE.

Mean maximum, 116.4

Mean minimum, 80.8

Mean, 98.6

Maximum, 134; date, 10th

Minimum, 70; date, 1st-22nd

Greatest daily range, 53

PRECIPITATION.

Total, 0.60 inches.

Greatest in 24 hours, 0.30; date, 22nd

SNOW.

Total fall, ______ inches; on ground 15th, ______ inches; at end of month, ______ inches.

NUMBER OF DAYS.

With .01 inch or more precipitation, 3

Clear, 18; partly cloudy, 0; cloudy, 13

† Including rain, hail, sleet, and melted snow.
‡ Thunderstorms, halos, auroras, etc.
§ Reading of maximum thermometer immediately after setting.

(IN TRIPLICATE.) 8—253

______, *Cooperative Observer.*

Post-Office address, Ryan, Calif

The controversial record high air temperature for the United States occurred on the tenth day in August 1913, during a 10-day period in which every daily maximum exceeded 120° F. Figures in the third column show the reading made by the observer at the time he reset the maximum thermometer. (National Weather Service.)

20 THE SUN'S ENERGY AND AIR TEMPERATURES

THE THERMOMETER WATCHERS

In the American colonies the thermometer was a rare and precious instrument, imported from England and reverently installed inside the homes of some distinguished persons. A few of the instruments began to arrive in the early 1700s. One of the first persons to keep a record of air temperatures was John Winthrop, a professor of natural philosophy in Harvard College. His daily record is complete, with scarcely a break, from 1742 until the year of his death, 1779. Evidently he first used an instrument hung in the home of Harvard's President Holyoke. These early thermometers were too costly and fragile to hang outdoors, so they were placed in an unheated room.

Not surprisingly, Benjamin Franklin obtained a thermometer in 1745, for he was an active scientific investigator. Thomas Jefferson, also an amateur scientist of distinction, obtained his thermometer in Philadelphia and paid for it on July 4, 1776.

What seems to have been the first authentic record of an unusual thermal event was made on "Hot Sunday," a day in June 1749. The Harvard thermometer registered 98° F in its indoor location in front of an open window in the President's house; while Ben Franklin's instrument ascended to the 100-degree mark in Philadelphia.*

Record high air temperatures invite just about as much skepticism as any sports record and most have to be discarded because they were not taken under the strictest of rules. First, the thermometer must be certified as accurate. Second, it must be housed in a very special shelter, shaded from the sun but well ventilated by louvers, and set at a specified height above the ground surface. Third, the observer must be properly certified and supervised, and his records kept in an approved manner.

Many textbooks cite as an all-time record for maximum observed air temperature, the value of 58° C (136.4° F) taken at El-Azizia, Libya, on September 13, 1922. Exactly the same value has been reported from San Luis, Mexico, for August 11, 1933. The experts are unwilling to accept either reading as authentic. In the United States, the official highest reading of all time is 134° F, recorded on July 10, 1913, in a standard U.S. Weather Bureau thermometer shelter at Greenland Ranch in Death Valley, California. This location is 178 feet below sea level. Even though this record has official status, there are some weather scientists who remain unconvinced.

The hottest city in the United States may well be Yuma, Arizona. Their maximum reading of record is 123° F. Their hottest summer was in 1937, when the mercury climbed above 100° F on each of 101 consecutive days. Actually, some states we think of as cool in summer have come off with some pretty high maximum readings: Maine, 105° F; Vermont, 105° F, and New Hampshire, 106° F. Even Alaska can boast of 100° F, recorded at Fort Yukon in June of 1915.

The hottest place on earth, year-in and year-out, no holds barred, may prove to be Bou-Bernous, Algeria, in the heart of the great Sahara Desert. According to the data published in tables of Her Majesty's Meteorological Office of Great Britain, the average daily maximum reading for July, the hottest month, is a whopping 116° F. The average value of the highest reading taken in each July between 1925 and 1942 is 120° F. The absolute maximum during this same period was 123° F, but this is well below the record at Death Valley and many other United States localities.

For the world's hottest large city, our nominee is Baghdad, Iraq, with its 1½ million sweltering souls. During the four hottest months, June through September, the average daily maximum temperature in Baghdad is 104° F or above; while for both July and August it is 100° F. For each of these months, the corresponding values are approximately four degrees higher than at Yuma.

In this chapter we will investigate the heating of the atmosphere and earth under the sun's rays.

* Much of the information in this essay comes from an article by David M. Ludlum titled "Extremes of Heat in the United States," published in the June 1963 issue of *Weatherwise.*

THE SUN-EARTH-SPACE RADIATION SYSTEM

Planet Earth intercepts radiant energy from the sun. This energy powers the motions of the atmosphere and oceans and is the source of energy for all precipitation and storms. But the earth is also a radiator of energy, emitting energy into outer space. As Figure 20.1 shows, the input of energy from the sun and the earth's output of energy into space go on simultaneously and constantly. As a result there is a **radiation balance** between incoming energy and outgoing energy.

If the earth were to intercept more radiant energy than it emits to outer space, the planetary temperature would rise indefinitely, and our planet would melt and finally vaporize. If, on the other hand, the earth were to lose more energy than it gains, the planetary temperature would fall, and earth would become a solidly frozen mass close to absolute zero.

Geological evidence shows that our earth's surface temperature has remained in a middle range. Temperatures of the oceans and atmosphere have probably been close to present values for a long time. Some modern invertebrate marine animals—clams, snails, and worms—are quite similar to forms of the Cambrian period. This evidence suggests that water temperatures of the shallow seas half a billion years ago where not greatly different from today.

SOLAR RADIATION AND EARTH RADIATION COMPARED

Of the solar energy we receive, about half is within the shorter wavelengths of the electromagnetic spectrum—X rays, ultraviolet rays, and visible light. On the other hand, energy radiated by the earth is entirely of the longer waves, or infrared radiation. This is obvious from the fact that the earth's surface at night emits no light.

The intensity of energy radiation increases rapidly with an increase in surface temperature of a radiating object. In this case temperature is given in **degrees Kelvin** (° K). (A single Kelvin degree matches a single Celsius degree in heat value, but the Kelvin scale starts with **absolute zero,** which is −273° on the Celsius scale.) The total energy radiated by a unit of surface area (such as 1 square centimeter) increases as the fourth power of the temperature in

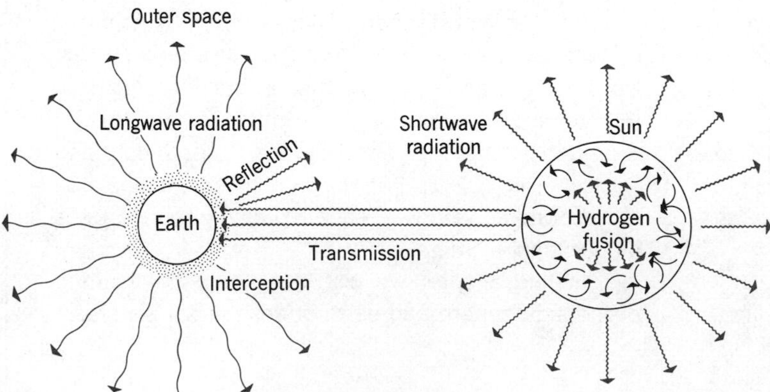

Figure 20.1 The sun-earth-space radiation system. (© 1973, John Wiley & Sons, New York.)

Kelvin degrees. We can approximate the sun's energy output as a perfect radiator with a temperature of 6000° K.

Figure 20.2 shows the curve of radiation intensity for our ideal sun and earth. The meaning of the units of radiation intensity on the vertical scale doesn't need to concern us here; we're only interested in comparing sun with earth. The scale is a logarithmic scale, based on powers of ten. On the horizontal scale of the graph is wavelength, using microns as our unit (see Figure 19.8). The scale of microns is also a logarithmic scale. The peak intensity of our model sun is right in the middle of the visible light band. About 41% of the solar energy lies in the relatively small visible light portion, about 9% in the ultraviolet portion, and about 50% in the infrared portion. Very little energy is carried by infrared rays longer than 50 microns.

Now let us consider the earth as a radiator. The ideal model for our earth is an object with a temperature of 300° K. This is equal to 80° F or 27° C, a pleasant outdoor temperature. The radiation intensity curve for the earth model is shown at the right side of Figure 20.2. Practically all of the energy is emitted in longer wavelengths than the solar spectrum, most of it between 5 and 50 microns. The peak is at about 10 microns. However, the peak for earth radiation intensity has a value of about 5/100 of a unit, whereas the peak for the sun is about 4 units. This means that the peak intensity of the sun's radiation is about 80 times that of earth for each square centimeter of surface.

Scientists often refer to solar radiation as **shortwave radiation** and to the earth's radiation as **longwave radiation.** Figure 20.2 makes clear the reason for using these terms.

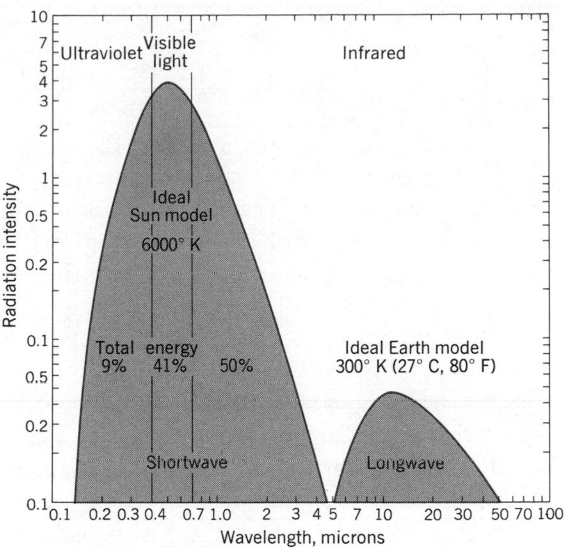

Figure 20.2 Radiation intensity curve of the ideal sun peaks at a much higher lever and shorter wavelength than the curve for an ideal earth. (Radiation units are langleys per minute per micron width.)

THE SOLAR CONSTANT

Because the sun's rays diverge radially outward from its surface, the intensity of the sun's energy diminishes rapidly with distance. At the average distance of the earth, solar radiation has an intensity of about 2 calories of heat energy per square centimeter per minute on a surface held at right angles to the sun's rays. A radiation unit long used by scientists is the **langley;** it has a value of 1 gram-calorie per square centimeter. (The gram-calorie is the quantity of heat needed to raise the temperature of 1 gram of pure water through 1 Celsius degree at 15° C.) We can express the intensity of solar radiation reaching earth at the average distance from the sun as 2 langleys per minute (2 ly/min). This quantity is called the **solar constant.** Only minor fluctuations of 1% or so can be observed in the solar constant, and there is no reason to believe that it has changed significantly in many decades of observation.

SOLAR RADIATION OVER A SPHERICAL EARTH

Because the earth is almost a true sphere, at any given instant only one point on earth, the **subsolar point,** presents a surface at right angles to the sun's rays (Figure 20.3). In all directions away from the subsolar point, the earth's surface curves away from the sun and forms an angle with the rays that decreases with distance from the

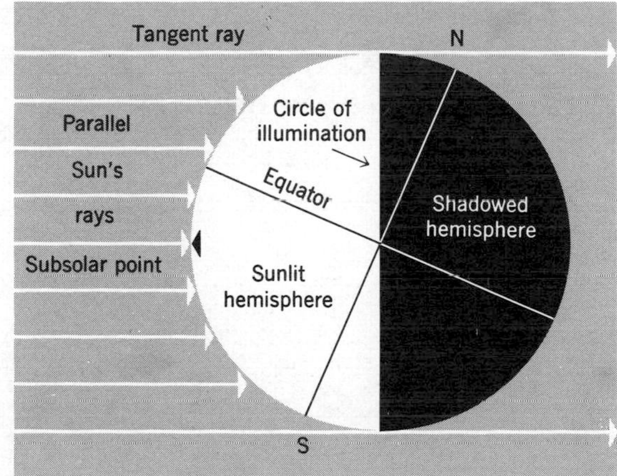

Figure 20.3 Relation of the sun's rays to latitude at a date intermediate between equinox and solstice.

subsolar point. At the **circle of illumination,** the earth's surface parallels the rays (see tangent ray in Figure 20.3). The circle of illumination is a line separating the sunlit hemisphere from the hemisphere in shadow.

Assuming for the moment that the earth has no atmosphere, the total quantity of solar energy received by 1 sq cm of horizontal surface in one day will depend upon two factors: (a) the angle at which the sun's rays strike the earth, and (b) the length of time of exposure to rays. These factors vary by latitude and by the seasonal changes in the height of the sun above the horizon.

It is important to understand that the earth's axis of rotation is not oriented perpendicular to the plane of the earth's orbit around the sun. Instead, as shown in a perspective drawing in Figure 20.4, the earth's axis is inclined at an angle of 23½° away from the perpendicular. The earth's axis constantly maintains this angle, while always aimed at the same point in space. As a result, there are two points in the orbit at which the axis is inclined neither toward nor away from the sun: These are the **equinoxes.** There are also two other points in the orbit in which the full value of axis inclination is directed toward the sun: These are the **solstices.** Names and dates of equinoxes and solstices are given in Figure 20.4.

If the earth's axis were perpendicular to the plane of the orbit (that is, if there were no axial tilt), the conditions of equinox would prevail throughout the entire year. Equinox conditions at noon are shown in Figure 20.5. Radiation at the equator is 100%, with a value of 2.0 ly/min. At 30° N and S the percentage is reduced to 87%; at 90° N and S, values are zero. These facts lead us to conclude correctly that the earth receives its greatest total solar radiation at the equator and the least at the poles, considered on a yearly average basis.

The earth's annual cycle of change of axial tilt relative to the sun's rays must next be taken into account. Persons living in the middle latitude portions of the United States and Canada (35° to

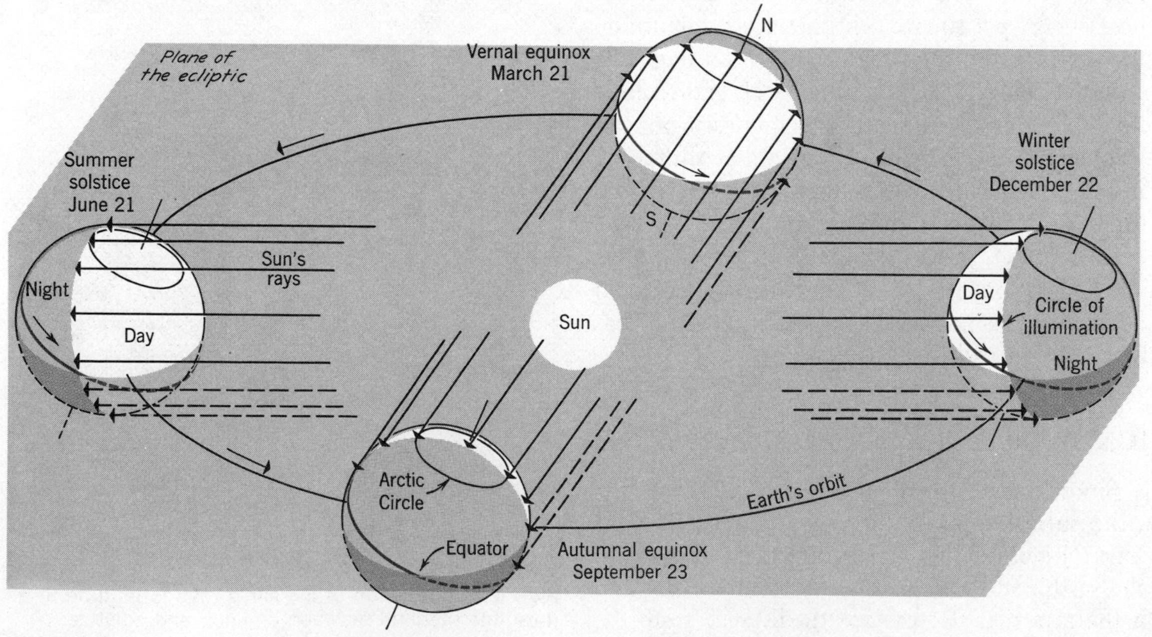

Figure 20.4 Orientation of the earth's axis remains fixed in space as the earth revolves about the sun, producing the seasons. (© 1973, John Wiley & Sons, New York.)

50° north latitude) are well aware that in midsummer the sun rides high in the sky and remains above the horizon for some 14 to 16 consecutive hours. For this reason the quantity of solar radiation accumulated during one day at summer solstice is much greater than at either equinox. In contrast, in midwinter the sun takes a low path in the sky and is above the horizon for only 8 to 10 consecutive hours. Thus, at winter solstice, the day's accumulation of radiation is much less than at either equinox. Plotted as a curve throughout the year, the changing value of incoming solar radiation makes a strong annual cycle, illustrated in Figure 20.6 by the curve for 40° north latitude. Units are langleys per day (ly/day).

Additional curves for various other latitudes in the Northern Hemisphere are shown in Figure 20.6. Notice that the equator (0°) has a cycle with two high points (maxima) and two low points (minima), because the sun passes overhead twice each year. Nevertheless, radiation is very strong throughout the entire year at the equator. At 20° N latitude there is only a slight dip at summer solstice. At all latitudes between 23½° and 66½° the radiation curve shows one maximum and one minimum, the amplitude increasing with higher latitude. Poleward of 66½°, for part of the year there is no incoming radiation. At the poles (90° N and S), this period of zero value is six months long and spans the entire period between equinoxes.

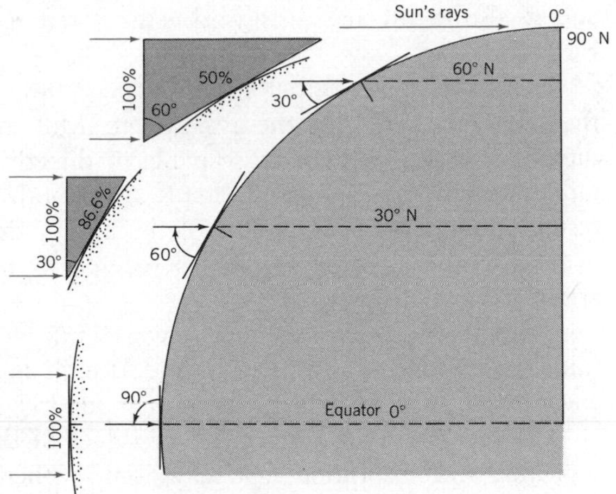

Figure 20.5 Radiation intensity decreases rapidly with increasing latitude.

ENERGY LOSSES IN THE LOWER ATMOSPHERE

As solar radiation penetrates into deeper and denser atmospheric layers, gas molecules cause the visible light rays to be turned aside in all possible directions. This process is known as **scattering.** Where dust particles are encountered in the troposphere, further scattering occurs. The total process can be called **diffuse reflection.** On a global basis about 5% of the incoming radiation is lost into space by diffuse reflection. At the same time, some scattered shortwave

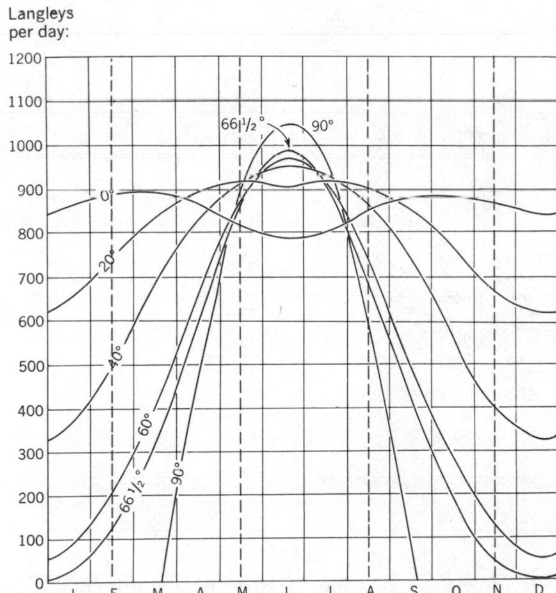

Figure 20.6 Annual cycles of change in daily quantities of solar radiation at selected latitudes in the Northern Hemisphere.

energy also is directed earthward. This is referred to as **diffuse sky radiation.**

Absorption is another form of energy loss that takes place as the sun's rays penetrate the atmosphere. Molecules of both carbon dioxide and water vapor are capable of directly absorbing infrared radiation. Dust particles and clouds also absorb energy. Absorption results in a rise of temperature of the air. In this way some direct heating of the lower atmosphere is caused by incoming solar radiation.

Direct absorption can be as low as 15% of incoming radiation under conditions of clear, dry air and as high as 35% when a cloud cover exists. A world average figure is about 18%.

Figure 20.7 shows the range of values of the various forms of reflection and absorption that may occur. When skies are clear, reflection and absorption combined may total about 20%, leaving as much as 80% of the incoming rays to reach the ground.

We must now bring into the picture another form of energy loss. The upper surfaces of clouds are extremely good reflectors of shortwave radiation. As an air traveler you are well aware of how painfully brilliant the sunlit upper surface of a cloud deck can be when seen from above. Cloud reflection can account for a direct turning back into space of from 30% to 60% of total incoming radiation (Figure 20.7). A world average of 21% for cloud reflection is reasonable.

Clouds can also absorb large amounts of energy—5% to 20% of the total. Under conditions of a heavy cloud layer, the combined reflection and absorption from all causes can account for a loss of 55% to 100% of the incoming radiation, allowing from 45% to 0% to reach the ground.

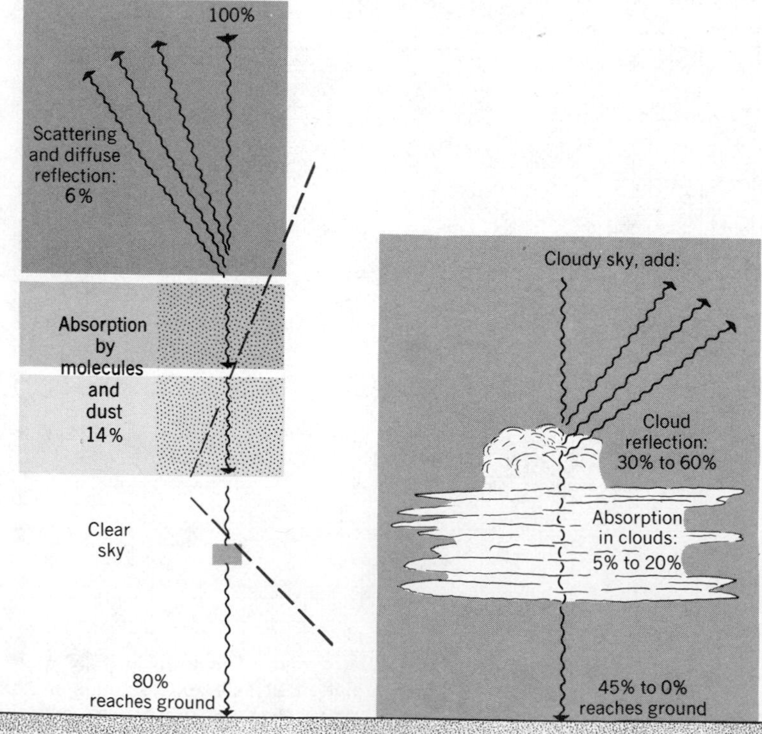

Figure 20.7 Losses of incoming radiation can be very much greater on cloudy days than on clear days.

The surfaces of the land and ocean reflect some shortwave radiation directly back into the atmosphere. This quantity, which is very small, averages about 6% on a global basis. It may be combined with cloud reflection and diffuse reflection to give total average reflective losses of about 32% for the globe as a whole.

We have summarized the losses of incoming radiation from all causes in Figure 20.8. Keep in mind that these figures are good estimates, but that small changes can be expected as our scientific knowledge becomes more exact. Orbiting earth satellites are continually collecting data on atmospheric phenomena. As these data accumulate, our average figures will be refined.

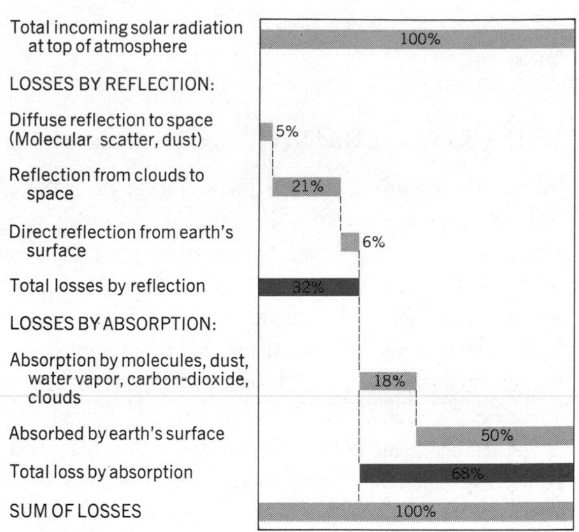

Figure 20.8 The percentage bars show what average proportion of total incoming solar radiation is lost by reflection and absorption.

LONGWAVE RADIATION

The surfaces of the continents and oceans possess heat derived originally from absorption of the sun's rays. Land and water surfaces continually radiate this energy back into the atmosphere. The process is known as **ground radiation.** This radiation is of the infrared wavelengths, longer than 4 microns, as shown in Figure 20.2. All of this energy is what we have called longwave radiation.

The atmosphere also radiates longwave energy both toward the earth and outward into space, where it is lost. Be sure you understand that longwave radiation is quite different from reflection, in which the rays are turned back directly without being absorbed. Longwave radiation from both ground and atmosphere continues during the night, when no solar radiation is being received.

Longwave energy radiated from the ground is easily absorbed by the atmosphere. Absorption of longwave radiation by water vapor and carbon dioxide takes place largely in wavelengths from 5 to 8 microns and 12 to 20 microns. However, radiation in the range of wavelengths between 8 and 11 microns passes freely through the earth's atmosphere and into outer space. About 8% of the longwave radiation directed outward leaves the atmosphere in this manner.

So we see that the atmosphere receives much of its heat by an indirect process in which the incoming energy in shortwave form is allowed to pass through, but the outgoing energy in longwave form cannot easily escape. For this reason the lower atmosphere, with its water vapor and carbon dioxide, acts as a warm blanket. Longwave radiation from the atmosphere back to the earth's surface is called **counter-radiation.** It returns heat to the earth and helps to keep surface temperatures from dropping excessively during the night or in winter at middle and high latitudes.

Somewhat the same principle is used in solar heating of greenhouses. Here the glass acts like the lower atmosphere and permits entry of shortwave energy. However, accumulated heat cannot escape by mixing with coolor air outside. The expression **greenhouse effect** is used by meteorologists to describe this heating principle as it occurs in the atmosphere.

Our earth as a planet must radiate as much energy into space as it absorbs. Orbiting satellites, which operate above the atmosphere, are ideally suited to measuring the earth's longwave radiation. Obviously, the greatest longwave output is from low latitudes, where the lower air and surface are warm year-round. Output is least from the cold arctic and polar surfaces, covered much of the year with snow and ice.

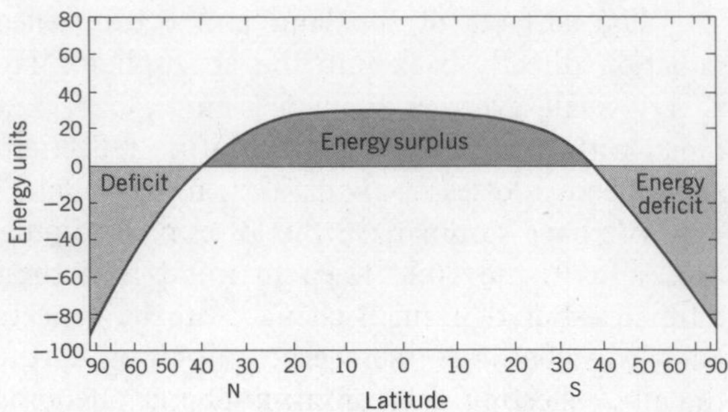

Figure 20.9 The yearly average energy balance shows a surplus between 40° N and 40° S, but a deficit in both hemispheres at high latitudes.

THE GLOBAL RADIATION BALANCE

We have already evaluated the losses of incoming shortwave solar energy; now we must evaluate the outgoing longwave energy on a global basis. Recall that the earth's land and water surface aborbs 50% of the total incoming shortwage energy as an annual average percentage. The amount of energy received by the earth from the sun each year must be balanced by an equal amount of energy radiated into outer space. This situation represents an **energy balance.**

Of the total energy leaving the earth's land and water surface, about half leaves as longwave radiation and the remainder by other forms of heat transfer. One such mechanism is by evaporation of water of ocean and land surfaces. This concept will be explained in Chapter 21. A second mechanism of heat transfer is by conduction from ground to air, accompanied by mixing with higher air layers.

THE ENERGY BALANCE AND LATITUDE

The global energy balance does not apply to particular regions of the globe. At low latitudes, much more energy is received than lost; over the two polar regions, much more energy is lost than received. Figure 20.9 shows this effect. An **energy surplus** exists between about 40° N and 40° S; an **energy deficit** exists poleward of the 40th parallels. (Philadelphia and Salt Lake City lie on about the 40th parallel of latitude.) The surplus equals the combined deficits of the two hemispheres.

Why aren't the low latitudes getting hotter and hotter and the two polar regions getting colder and colder? Temperature records show that the yearly averages stay more or less constant at any given observing station no matter where it is located. The answer must be that heat is being transported by some means other than radiation from the belt of surplus to the regions of deficit. We will find in the next chapter that heat transfer mechanisms exist in the circulation of the atmosphere and oceans.

DAILY CYCLES OF RADIATION AND AIR TEMPERATURE

Because the earth turns on its axis, there is a daily cycle of incoming solar radiation. This cycle determines the daily cycle of rising and falling air temperatures with which we are all familiar. Let us see how radiation and air temperature are linked in this cycle.

The upper graph of Figure 20.10 shows incoming shortwave radiation on a clear day at a city in the United States located about 40° N latitude, such as Chicago or New York. At equinox, solar radiation begins at sunrise, 6 AM local time, and reaches a peak at noon, when the sun is highest in the sky. Radiation then falls off and ceases at sunset, 6 P.M.

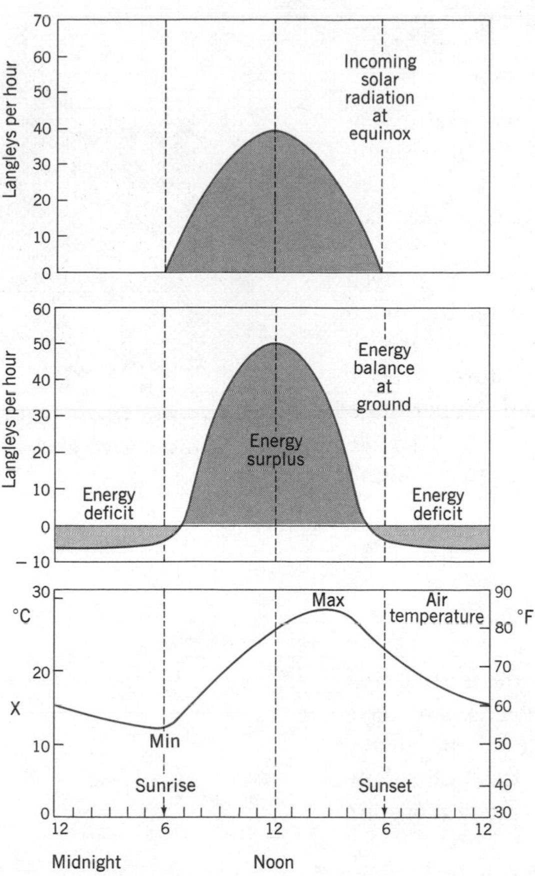

Figure 20.10 The daily cycle of solar radiation controls the cycle of air temperature.

The middle graph of Figure 20.10 shows the energy balance at the ground. Throughout the night there has been an energy deficit, because outgoing longwave radiation continues during darkness when there is no incoming solar radiation. Just after sunrise the situation is reversed, and an energy surplus sets in. This surplus builds as the sun rises higher, reaching a maximum at noon. More energy is entering the lower air layer than is leaving, so the air temperature rises rapidly, as the lower graph shows.

In the afternoon the energy surplus declines, and just before sunset the surplus gives way to a deficit. The air temperature keeps rising into the middle afternoon, then begins to fall in late afternoon and continues down through sunset and into the night. All night long, air temperature falls, reaching its minimum just before sunrise.

Figure 20.11 A standard thermometer shelter. Inside is a maximum-minimum thermometer (*upper left*). (National Weather Service.)

MEASUREMENT OF AIR TEMPERATURES

Air temperature is one of the most familiar bits of daily weather information. This information comes from standard United States National Weather Service observing stations and is taken following a carefully standardized procedure. Thermometers are mounted in a standard instrument shelter, shown in Figure 20.11. The shelter shades the instruments from sunlight, but louvers allow air to circulate freely past the thermometers. The instruments are mounted from 4 to 6 ft (1.2 to 1.8 m) above ground level, at a height easy to read.

At most stations only the highest and lowest temperatures of the day are recorded. To save the observer's time, the **maximum-minimum thermometer** is used. This instrument uses two thermometers; one to show the highest temperature since it was last reset; the other to show the lowest.

When the maximum and minimum temperatures of a given day are added together and divided by two, we obtain the **mean daily temperature.** The mean daily temperatures of an entire month can be averaged to give the **mean monthly temperature.** Average daily means for the whole year give the **mean annual temperature.** Usually such averages are compiled for many years at a given observing station. These averages are used in describing the climate of the station and its surrounding area.

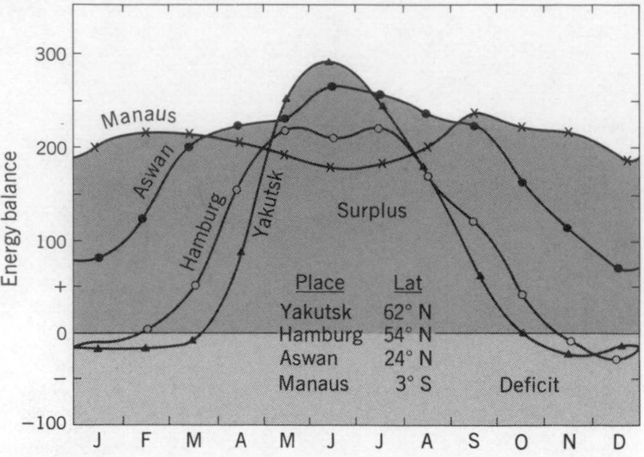

Figure 20.12 The energy balance throughout the year at four representative stations.

ANNUAL CYCLES OF RADIATION AND AIR TEMPERATURE

As the earth revolves about the sun, the tilt of the earth's axis causes an annual cycle of incoming solar radiation, as we have learned. This cycle is also felt in an annual cycle of the energy balance, which in turn causes an annual cycle in the mean daily and monthly air temperatures. In this way the climatic seasons are generated.

Figure 20.12 shows average monthly energy balances for four stations, ranging in latitude from the equator almost to the arctic circle. Figure 20.13 shows mean monthly air temperatures for these same stations. Starting with Manaus, a city on the Amazon River in Brazil, let us compare the net radiation graph with the air temperature graph. At Manaus almost on the equator, there is a large energy surplus every month. A look at the temperature graph of Manaus shows monotonously uniform air temperatures, averaging about 81° F (27° C) for the year. The **annual temperature range,** or difference between the highest-month temperature and lowest-month temperature, is only 3 F° (1.7 C°). In other words, one month is about like the next, temperaturewise. There are no temperature seasons.

We go next to Aswan, United Arab Republic (Egypt), on the Nile River at altitude 24° N. The energy curve has a strong annual cycle, and the energy surplus is large in every month. The temperature graph shows a corresponding annual cycle, with an annual range of about 30 F° (17 C°). The months of June, July, and August are terribly hot, averaging over 90° F (32° C).

Moving further north, we come to the German city of Hamburg, 54° N latitude. The energy cycle is strongly developed. The energy surplus lasts for 9 months, and there is a deficit for 3 winter months. The temperature cycle reflects the reduced total energy at this latitude. Summer months reach a maximum of just over 60° F (16° C); winter months reach a minimum of just about freezing (32° F; 0° C). The annual range is 30 F° (17 C°).

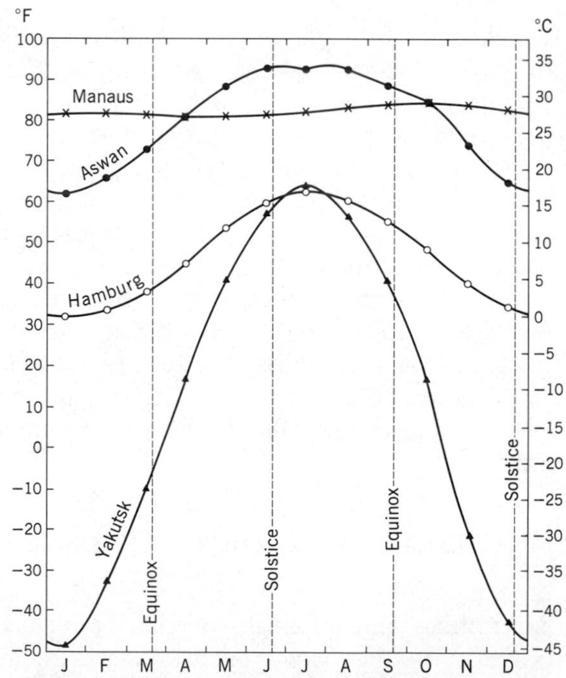

Figure 20.13 Mean monthly air temperatures for the same four stations shown in Figure 20.12. Compare the annual temperature cycles with the annual energy cycles.

Finally, we travel to the Siberian city of Yakutsk, lat. 62° N. During the long dark winters there is an energy deficit; it lasts about 6 months. During this time air temperatures drop to extremely low levels. For three of the winter months, monthly mean temperatures are between −30° and −50° F (−35° and −45° C). Actually, this is one of the coldest places on earth. In summer, when daylight lasts most of the 24 hours, the energy surplus rises to a strong peak. This is a value higher than any of the other three stations. As a result, air temperatures show a phenomenal spring rise to summer-month values of over 55° F (13° C). In July the mean temperature is about the same as for Hamburg. The annual range at Yakutsk is enormous—over 110 F° (61 C°). No other region on earth, even the south pole, has so great an annual range.

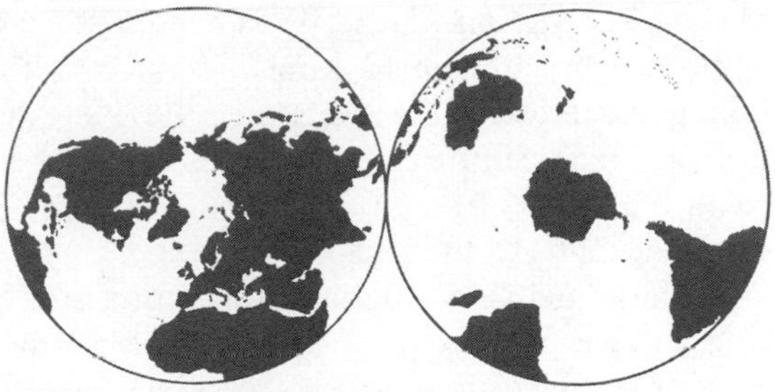

Figure 20.14 Northern and Southern Hemispheres—a contrast in land-ocean distributions. (© 1973, John Wiley & Sons, New York.)

LAND AND WATER TEMPERATURE CONTRASTS

What gives our planet its great variety of climates is the odd distribution of continents and ocean basins. Take a look at the two global hemispheres—northern and southern—outlined in Figure 20.14. The Northern Hemisphere displays a polar sea surrounded by massive continents; the Southern Hemisphere shows the very opposite—a pole-centered continent surrounded by a vast ocean. The Americas form a north-south barrier between two oceans—Atlantic and Pacific. The continents of Eurasia and Africa together form another great north-south barrier. Oceans and continents have quite different properties when it comes to absorbing and radiating energy. Consequently, when it comes to air temperatures, land surfaces behave differently from water surfaces.

The important principle is this: The surface of any extensive deep body of water heats more slowly and cools more slowly than the surface of a large body of land, when both are subject to the same intensity of incoming radiation.

The effect of land and water contrasts is seen in two sets of daily air temperature curves (Figure 20.15). El Paso, Texas, exemplifies the temperature environment of an interior desert in middle latitudes. Responding to intense heating and cooling of the ground surface, air temperatures show an average daily range of 20 to 25 F°

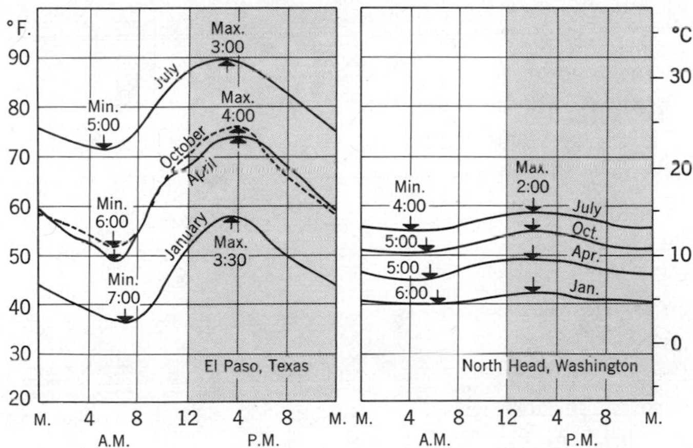

Figure 20.15 Average temperatures throughout the day at El Paso, Texas, a desert station of the continental interior, and at North Head, Washington, a coastal station. Notice how much larger the daily range is at El Paso than at North Head. (© 1973, John Wiley & Sons, New York.)

(11 to 14 C°). North Head, Washington, is a coastal station strongly influenced by air brought from the adjacent Pacific Ocean by prevailing westerly winds. Consequently, North Head exemplifies a maritime temperature environment. The average daily range at North Head is a mere 5 F° (3 C°) or less. Persistent fogs and cloud cover also contribute to the small daily range.

The principle of contrasts in heating and cooling of water and land surfaces also explains the contrasts in the seasonal or annual cycle of temperature of such places. Notice in Figure 20.15 that the annual range at El Paso is about 35 F° (20 C°), while that at North Head is only about 15 F° (8 C°). So we see that at these middle latitudes the seasons are very weak close to the ocean, but very strongly developed in the continental interior.

TEMPERATURE INVERSION AND FROST

During the night, when the sky is clear and the air calm, the ground surface rapidly radiates longwave energy into the atmosphere above it. Soil temperatures drop rapidly, and the overlying air layer becomes colder. When we plot temperature on such a night against altitude, as in Figure 20.16, the straight, slanting line of the normal environmental lapse rate becomes bent to the left in a J-hook. In the case shown, the air temperature at the surface, point A, has dropped to 30° F (−1° C). As we move up from ground level, temperatures become warmer up to about 1000 ft (300 m). Here the curve reverses itself and the normal lapse rate takes over. The lower, reversed portion of the lapse rate curve is called a **low-level temperature inversion.**

In the case above, temperature of the lowermost air has fallen below the freezing point (32° F, 0° C). This condition is called a **killing frost** when it occurs during the growing season. Killing frost can be prevented in citrus groves by setting up air circulation to mix the cold basal air with warmer air above. One method is to use oil-burning heaters; another is to operate powerful motor-driven propellers to circulate the air.

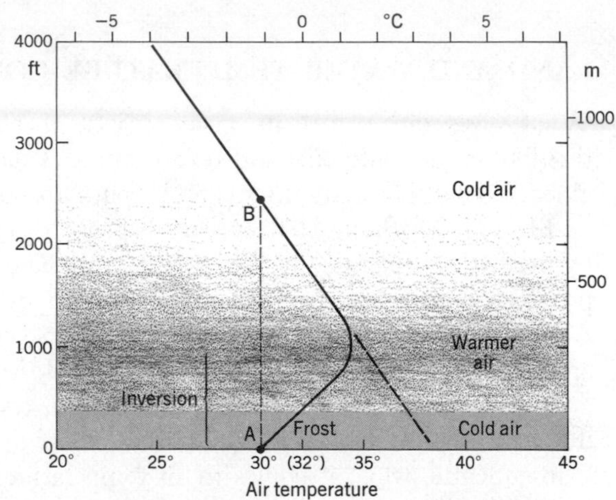

Figure 20.16 In low-level temperature inversion, a colder air layer develops at night close to the ground.

ALTITUDE AND AIR TEMPERATURES

Everyone knows that places located at high altitudes have cooler air temperatures than nearby low areas. Two effects of altitude are conspicuous. (1) The mean monthly temperature decreases progressively with altitude. (2) The daily range becomes much greater as altitude increases, because the clearer, more rarefied air at higher altitude permits more intense solar radiation to reach the ground. As a result, the ground is more intensely heated during the day. At night the loss of heat by longwave radiation into outer space is much more rapid at high altitude because of the smaller amount of water vapor and carbon dioxide in a given air volume.

As far as average yearly temperature is concerned, then, the effect of increase in altitude is very much the same as a great poleward shift in latitude. You may reach the region of permanent snowbanks and glaciers by climbing above 15,000 ft (4.5 km) in the equatorial belt just as surely as by traveling to the arctic regions.

MAN'S IMPACT UPON THE ATMOSPHERE

Industrial processes and urbanization in modern times have had a strong impact upon the atmosphere. Man has made changes in four categories:

1. He has changed the proportions of the component gases of the lower atmosphere through combustion of fuels.
2. He has changed the water-vapor content of the lower atmosphere.
3. He has introduced dust particles in suspension in the lower atmosphere, along with gaseous compounds not normally present in measurable quantities in preindustrial times.
4. He has changed the characteristics of the ground surface and its plant cover by agriculture and urbanization.

The above changes have affected the radiation balance and air temperatures, both locally over cities and over the globe as a whole.

CHANGE IN THE ATMOSPHERIC CARBON DIOXIDE LEVEL

Consider first the change in the proportion of carbon dioxide in the atmosphere. We have seen that carbon dioxide is an important absorber (along with water vapor) of longwave radiation, converting it into sensible heat.

Increase in atmospheric carbon dioxide during the past 120 years is reasonably well documented. The carbon dioxide percentage has risen from about 0.029% to about 0.032% during this period.

What effects can we anticipate from increase in the atmospheric carbon dioxide level? These effects follow logically from the role of carbon dioxide as an absorber of longwave radiation. Average temperature of the lower atmosphere will tend to rise, as longwave radiation into space is reduced. Observations seem to lend some support to this predicted temperature rise; but there is also conflicting evidence, and the situation proves to be far more complex than we might at first suppose.

Figure 20.17 is a graph of change in average hemispherical air temperature based upon observation for about the past century. From 1920 to 1940 the temperature increased by about 0.6 F° (0.4 C°), but since then it has been falling. Evidently, other causes of temperature change working in the opposite direction have been more important than rising carbon dioxide in the past three decades. It seems likely that a temporary increase in the content of volcanic dust in the upper atmosphere has had a cooling effect. In the long run the warming trend should assert itself.

The increased production of waste heat by fuel combustion is another cause of rise in global air temperatures to be added to the effect of increased carbon dioxide. At present, the amount of heat released by combustion and nuclear reactors is only an extremely tiny fraction of the total longwave energy radiated from the earth. But because our use of fuels is increasing rapidly, this added source of heat may become important. An estimate has been made that

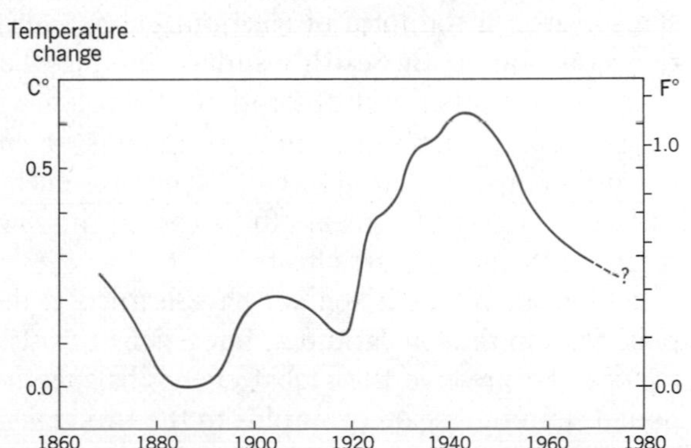

Figure 20.17 Trend of global mean annual air temperatures for latitudes 0° to 80° N was upward for many decades, until about 1940, when it started down.

about 90 years from now the added heat of combustion will have raised the earth's average air temperature by about 2 F° (1 C°).

EFFECT OF URBANIZATION UPON THE ENERGY BALANCE

The study of climate near the ground shows clearly the impact of man on the radiation balance and upon air temperatures. The effect of urbanization is particularly strong, because buildings, paved streets, and parking lots replace a preexisting surface clothed in trees, shrubs, and grass. A completely new set of surface properties is substituted for the natural surface properties. In addition, the construction of buildings produces a large number of vertical masonry surfaces. These affect the energy balance and also influence the structure of winds close to the ground. These changes, of course, must be viewed in conjunction with emission of heat and air pollutants through combustion.

Consider the changes in the heat environment. The absorption of solar radiation produces higher ground-surface temperatures in urban areas for two reasons.

First, there is no shading by foliage and no cooling effect of evaporation from plants. As in a desert environment, ground-surface temperatures can rise to high levels under direct solar heating. This heat is conducted readily into the ground, and a very large quantity is stored there.

Second, evaporation may be nil on pavement and masonry surfaces, and there can be no cooling such as that which occurs by evaporation from a moist soil surface. Vertical masonry surfaces also absorb direct and reflected radiation, and this is radiated back into the air between buildings.

Maximum daytime air temperature in summer usually averages several degrees higher in the center of a city than in surrounding suburbs. This fact is shown by the map of Washington, D.C., for a day in August (Figure 20.18). The city is said to produce a **heat island.**

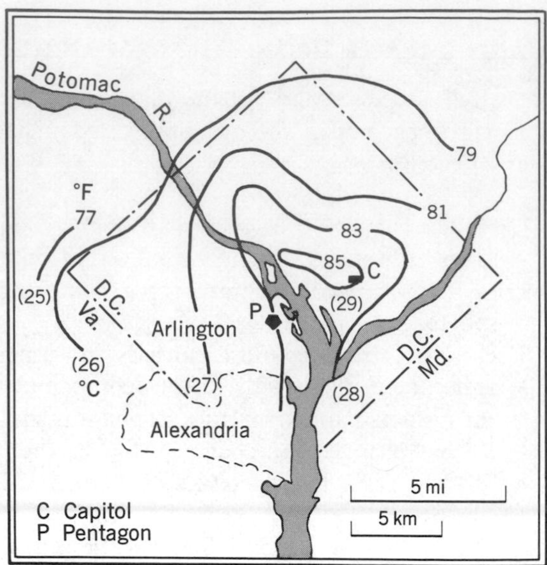

Figure 20.18 Heat island over Washington, D.C., is revealed by lines of equal air temperature at 10 PM local time on a day in early August. (Data of H. E. Landsberg.)

RADIATION AND AIR TEMPERATURES IN REVIEW

This chapter began with an evaluation of the flow of energy from sun to earth in the form of electromagnetic radiation. We followed this radiation to the earth's surface, noting the various kinds of losses by reflection and absorption. To balance the earth's energy budget, our planet must emit an amount of energy by longwave radiation equal to that absorbed. Longwave energy from the earth's surface is largely responsible for warming the lower atmosphere, by means of the greenhouse effect.

One important lesson we have learned is that a radiation surplus exists in the low latitudes, but a deficit exists over the two polar regions. To preserve the global energy balance heat must be transported from the region of surplus to the two regions of deficit. Circulation of both atmosphere and oceans accomplishes this transport (Figure 20.19). Our objective in the next chapter is to analyze this circulation system.

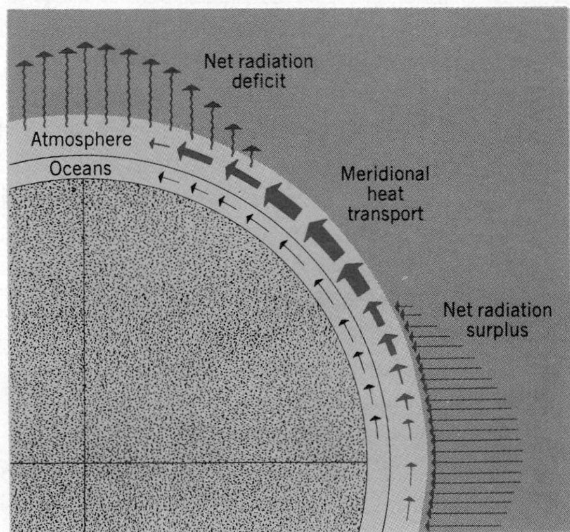

Figure 20.19 Poleward heat transport by motions of the atmosphere and oceans is necessary to maintain the earth's energy balance.

Earth rotation and revolution cause daily and seasonal cycles of radiation and air temperature. These and the effects of latitude create a wide range in temperature environments from equator to poles. The contrast in physical properties of ocean and land surfaces is still another factor that lends variety to the earth's temperature environments.

Man's combustion of fossil fuels and the spread of his cities are beginning to make changes in the radiation balance of the earth as a whole, and in that of urban areas in particular. If we are wise, we will monitor these effects very closely and be guided by changes that today are small, but tomorrow may be much larger.

YOUR GEOSCIENCE VOCABULARY

radiation balance
degrees Kelvin
absolute zero
shortwave radiation
longwave radiation
langley
solar constant
subsolar point
circle of illumination
equinoxes
solstices
scattering
diffuse reflection
diffuse sky radiation
absorption of radiation
ground radiation
counter-radiation
greenhouse effect
energy balance
energy surplus
energy deficit
maximum-minimum thermometer
recording thermometer
mean daily temperature
mean monthly temperature
mean annual temperature
annual temperature range
low-level temperature inversion
killing frost
heat island

SELF-TESTING QUESTIONS

1. What evidence can you cite that our planet has enjoyed a fairly uniform temperature environment for the past half-billion years?

2. How is the intensity of electromagnetic radiation related to temperature? Compare the sun's radiation with that of the earth. How constant is the solar constant?

3. How does the earth's spherical shape affect the input of solar radiation? How does tilt of the earth's axis affect the input of solar radiation?

4. Describe the various forms of energy losses affecting incoming solar radiation as it passes through the atmosphere.

5. Which wavelengths of ground radiation escape readily into space? Explain the greenhouse effect. If the lower atmosphere were only half as dense as it is, would air temperatures be higher or lower?

6. Describe the global energy balance. In what ways is energy returned from the earth's land and water surfaces to the atmosphere?

7. For the globe as a whole, where is there an energy surplus? Where a deficit? How can surface temperatures be kept uniform, on the average, at all places on the globe?

8. Describe the daily cycles of incoming solar radiation, energy balance, and air temperature. At what hour is air temperature lowest? When is it highest? Explain. Under what conditions would you expect the daily air temperature range to be very great? Very small?

9. Explain the techniques of measurement of air temperature and the derivation of the mean daily temperature. How are the mean monthly and mean annual temperatures calculated?

10. Describe the annual cycles of energy and air temperature for stations near the equator, in middle latitudes, and in the arctic zone. Where is the annual temperature range highest?

11. In what ways does a water surface influence the air temperatures immediately above it? In what ways does a land surface influence air temperatures?

12. What causes a low-level temperature inversion? What conditions favor such an inversion? What measures can be taken to alleviate a killing frost?

13. How does air temperature change with increasing altitude of places on the ground? How is the daily range affected?

14. In what ways does man have an impact on the atmosphere? What kinds of changes are man-made?

15. Explain in detail the effect of increased atmospheric carbon dioxide levels on air temperatures. What evidence have we of such effects? What effect will the increased output of waste heat of combustion have upon global air temperatures?

16. How does urbanization alter the energy balance? How are air temperatures affected? Cite evidence for your answer.

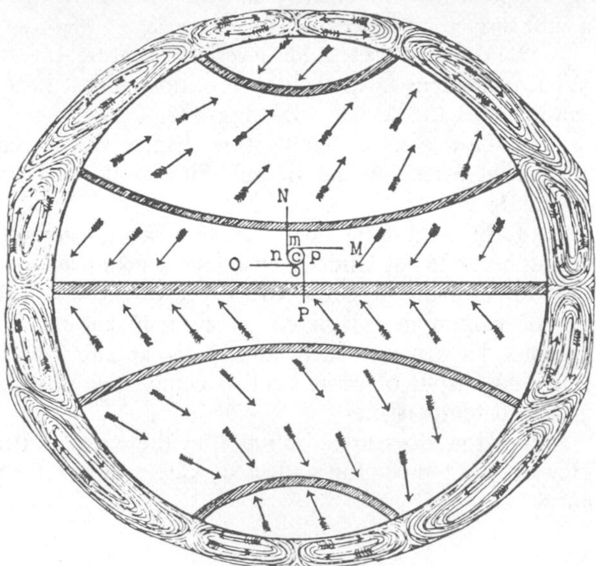

(*Above*) A dying cyclonic storm located about 1200 mi north of Hawaii in the Pacific Ocean. Cloud bands form a tightly wound spiral about the core of the disturbance, which was photographed by astronauts of *Apollo 9* space flight. (NASA) (*Left*) William Ferrel's original illustration of the planetary wind system, published in 1856, shows three cells of atmospheric circulation in each hemisphere.

21 CIRCULATION OF THE ATMOSPHERE AND OCEANS

MR. FERREL'S VEERING WINDS

There will always be someone ready to bet that, without fail, water in a washbowl drains in an anticlockwise spiral in the Northern Hemisphere, but always turns clockwise in the Southern Hemisphere. A letter to the London *Sunday Times,* written in 1962, made the most fantastic claim of all. The correspondent had verified the fact that precisely at the moment his ship crossed the equator heading south, the exiting bath water in his tub drain switched from a left-hand swirl to a right-hand swirl.

The alleged washbowl phenomenon is supposed by its chroniclers to be produced by the Coriolis force, which is generated by the earth's rotation. Physicists assure us that the direction of spiraling of the washbowl vortex is not controlled by the earth's rotation. Nevertheless, the Coriolis force plays a significant role for all life on earth, since it governs the spiraling flow of winds and ocean currents on a planetary scale, and so controls the global climate.

Prior to the Civil War, few salient contributions to science were being made by Americans, for science was still firmly in the grip of Old-World scholars, manning the fortress walls of centuries-old universities. One person who overcame the intellectual wilderness of the New World was William Ferrel, a schoolteacher in Nashville, Tennessee. In his day, the prevailing winds of the earth were already well charted by mariners. Their long voyages made them closely familiar with the prevailing sou'westerlies and the trades. But it remained for Ferrel, a landlubber, to discover that the earth's rotation controlled these planetary winds. Publishing in 1856 in the *Nashville Journal of Medicine and Surgery,* Ferrel gave an accurate analysis of the forces which control the global winds.

To this day, we recite Ferrel's law of winds, to the effect that winds in the Northern Hemisphere are deflected toward the right; those in the Southern Hemisphere, to the left. But the Europeans were not far behind. Only a year later, a similar analysis was published by C. H. D. Buys Ballot, Chief of the Dutch Meteorological Service. There is a law named after him, too. Ballot's law states that in the Northern Hemisphere, if you stand with the wind at your back, the lower pressure will be at your left. Ferrel's model of the global winds required three circulation cells in each hemisphere, as his original illustration shows (*opposite page*).

Where did the name of Coriolis get into the act? The name belonged to a Frenchman, G. G. de Coriolis, who died in 1842, long before Ferrel and Buys Ballot came on the scene. A brilliant mathematician, Coriolis had worked out the theoretical aspects of the mysterious force that pulls sideways upon moving objects. His work was based upon Newton's laws and, until Ferrel came along, largely an abstraction so far as natural science was concerned.

Ferrel's law applies to water as well as to wind, and even to solid objects in motion. Henry M. Eakin, a geologist working in Alaska, wrote in 1910 that he observed the Yukon River to be cutting only against the right bank throughout all of the 600-mile course on its lower floodplain. The entire floodplain lies on its left, so surely this meant that some mysterious force kept urging the river toward its right side. What's more, driftwood hugged the right bank, with almost none to be found on the left. The natives were well aware of this, for driftwood was their fuel. Artillerymen have known about Ferrel's law for a long time, too. A shell from a battleship veers as much as 200 feet to the right before it reaches a target 20 miles away. Famed Big Bertha, Germany's enormous rifle of World War I, fired upon Paris from a distance of 70 miles. During the 3-minute flight, Big Bertha's huge shell drifted a full 3 miles to the right of the straight line on which the barrel was aimed.

In this chapter you will become expert at applying Ferrel's law to currents of air and water, as you investigate the global circulation systems.

WINDS AND THE GLOBAL CIRCULATION

Two great global circulation systems exist on our planet. In these systems air and water—both of which are fluids—move in closed circuits, transporting heat and distributing it more uniformly over the planetary surface. If there were no such mechanisms of heat transport, the equatorial region of the earth would be hotter and the polar regions colder than they actually are. The gaseous atmosphere and liquid oceans flow across the parallels of latitude, taking heat from equatorial regions of surplus to polar regions of deficit (Figure 21.1). This equalizing process is of major importance in determining the environment of life of the lands.

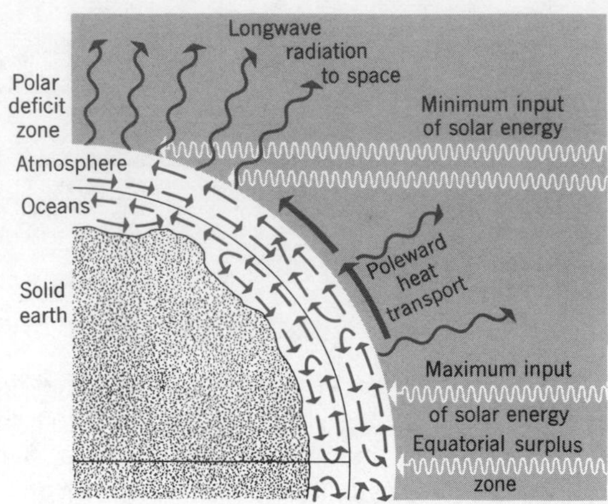

Figure 21.1 The global system of poleward heat transport. (© 1973, John Wiley & Sons, New York.)

The flow of the atmosphere is powered by unequal heating of large masses of air. Movement of the air in turn sets surface layers in motion. We can say that the oceanic circulation is largely wind-driven.

Wind is broadly defined as air in motion. In weather science, wind refers to air motions that are dominantly horizontal. Vertical motions are designated by other terms, such as updraft or downdraft. Forces that cause winds will be the first subject of inquiry in this chapter.

To describe a wind you must tell both its speed and direction. Speed is given in miles or kilometers per hour for public weather bulletins, in knots (nautical miles per hour) on the weather map, and in meters per second for scientific analysis. Direction of a wind is always stated as the direction from which the wind is blowing. For example, a west wind travels from west to east. For public weather bulletins, direction is given as one of the eight compass points. For scientific and navigational purposes, direction is given as a compass bearing, starting with 0° at geographic north and working clockwise throughout 360°.

WINDS AND THE PRESSURE GRADIENT

Recall from Chapter 19 that atmospheric pressure, or barometric pressure, decreases from the ground up. For an ideal atmosphere at rest, the barometric pressure will be the same throughout a given horizontal surface at any height. This condition is shown in the upper part of Figure 21.2. Surfaces of equal barometric pressure, called **isobaric surfaces,** appear in cross section in this diagram as horizontal, parallel lines. Pressures in millibars have been assigned to these surfaces.

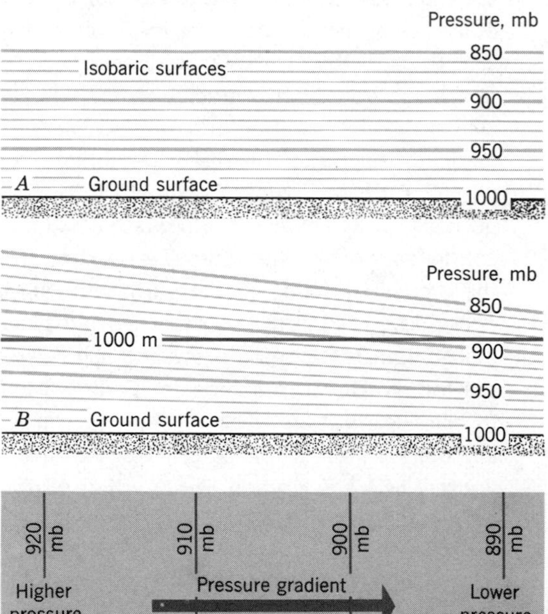

Figure 21.2 Isobaric surfaces are sloping where a pressure gradient exists. (© 1973, John Wiley & Sons, New York.)

Most of the time, isobaric surfaces are not horizontal. Instead, as shown in the middle diagram of Figure 21.2, the isobaric surfaces slope down or up from one area to another. Consequently, at some given height, such as 1000 m, pressure is higher in one place and lower in another. In the diagram, pressure at 1000 m is higher at the left than at the right. Specifically, pressure is 920 mb at the left; at the right it is 890 mb. We describe this situation as a **pressure gradient,** in this case sloping down from left to right.

Next, we make a map of the situation. As shown in the bottom diagram of Figure 21.2, a map of the 1000-meter level consists of lines of equal pressure, called **isobars.** These lines cut across the map from top to bottom and are labeled with their pressure values. A broad arrow shows the direction of the pressure gradient.

Where a pressure gradient exists, air tends to move in the same direction as the gradient, that is, from higher to lower pressure. For simplicity, we say that a **pressure-gradient force** acts upon the air, urging it to move. The stronger (steeper) the pressure gradient, the stronger is the force, and so the stronger will be the wind.

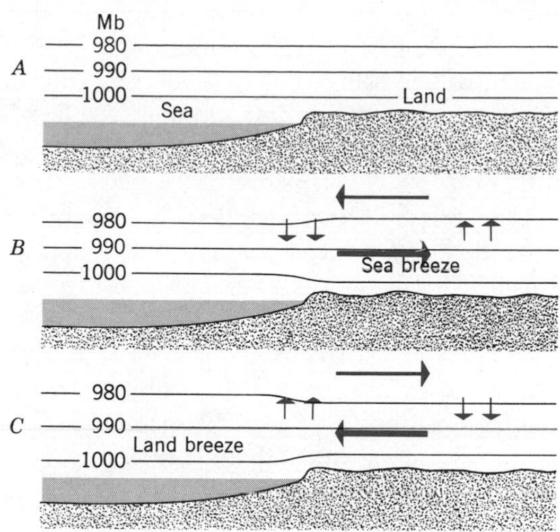

Figure 21.3 A sea breeze is set up during the day, when the land is warmed. The land breeze at night reverses the flow. (© 1973, John Wiley & Sons, New York.)

SEA BREEZE AND LAND BREEZE

A simple illustration of winds and the pressure gradient is the summer sea breeze along a coast. In diagram A of Figure 21.3 it is early morning and calm. The isobaric surfaces are horizontal, and no gradient exists. By afternoon, the air layer over the land has become warmed by longwave radiation from the heated ground surface. The air layer expands, and the isobaric surfaces are raised. As shown in the diagram *B*, low pressure now exists over the land, as compared with higher pressure over the adjoining ocean. A landward pressure gradient is developed, and air moves toward the land as a **sea breeze.** At upper levels, the gradient is the reverse, and a weak air flow moves seaward.

At night, the reverse situation occurs (diagram C). The lower air layer over the land cools down by longwave radiation. The isobaric surfaces are compressed, and pressure becomes higher over the land than over the ocean. Now the gradient is from land to sea, and a **land breeze** sets in.

THE CORIOLIS EFFECT

If the earth did not rotate on its axis, wind would move exactly in the direction of the pressure gradient. Wind arrows would cut across the isobars at right angles. However, because of earth rotation, a second force comes into play. This is the **Coriolis force.** In the Northern Hemisphere, the Coriolis force pulls toward the right upon any substance in horizontal motion, whether it be a solid object, a liquid, or a gas. In response to the Coriolis force, air in motion is deflected toward the right. As shown in Figure 21.4, it does not matter in what compass direction the air is moving—east, west, north, or south—the deflection is always toward the right.

The Coriolis force is nonexistent precisely on the equator; it increases in strength to a maximum at each pole, as the widening arrows in Figure 21.4 suggest. At the 30th parallel the force is 50% of the polar value; at the 60th parallel it is about 87%. The Coriolis force also becomes stronger as the speed of motion of the substance increases.

A full explanation of the Coriolis force would be too lengthy and complex to present here. However, the principle is easy to illustrate with a rotating disk, such as a record turntable (Figure 21.5). Imagine this disk to represent a small region centered on the north pole. A straightedge is mounted above the table. Using this straightedge as a guide, move a pencil poleward over the turning disk. The mark you have made will be a curving path, deflected toward the

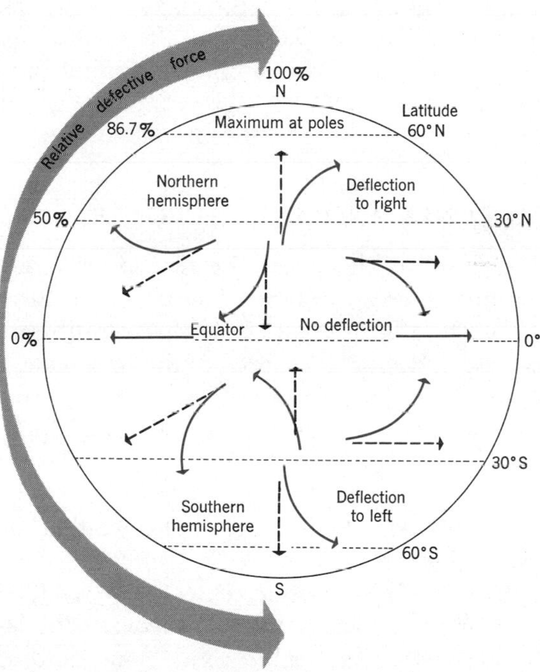

Figure 21.4 The Coriolis force deflects moving air to the right in the Northern Hemisphere and to the left in the Southern Hemisphere. (© 1973, John Wiley & Sons, New York.)

right of the direction of motion when the disk is turned counterclockwise (Northern Hemisphere). Your pencil point followed a straight path in space, but its track was curved on the turning surface. If you reverse direction and draw the line from the pole toward the outer edge of the disk, the track will again be curved toward the right.

The principle involved here is Newton's first law of motion, stating that any body in motion will follow a straight line unless compelled to change its path by some external force. The straight line mentioned in the law is a line fixed in space with respect to the stars. Of course, objects moving on the earth's surface are held to that surface by gravity and also subjected to friction, so that they cannot move freely in a straight space path. The deflecting force is nevertheless acting upon them. Air and water respond very readily to the Coriolis force.

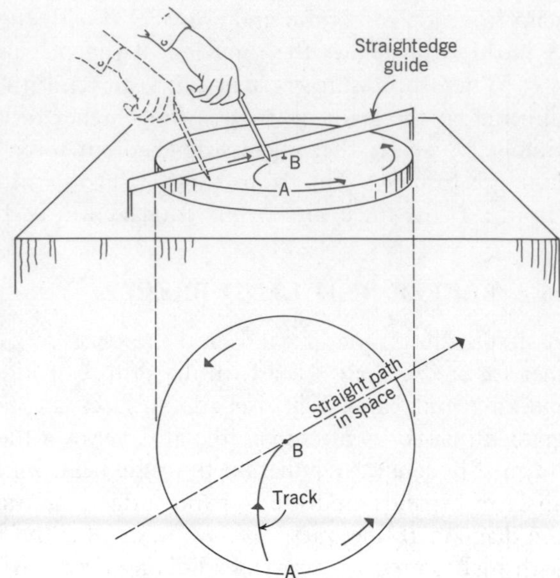

Figure 21.5 A record turntable can be used to illustrate the Coriolis effect in the polar region of the Northern Hemisphere.

THE GEOSTROPHIC WIND

Let us now apply the Coriolis effect to winds. Visualize the motion of a small parcel of air starting at rest from point A of Figure 21.6. Isobars show the pressure gradient to be uniformly lower toward the left. The pressure-gradient force acts toward the left with a constant value at all times. The pressure-gradient force tends to set the air parcel in motion at right angles across the isobars. However, as soon as motion starts, the Coriolis force acts at right angles to the path of motion, as shown by very small arrows. The air parcel responds by turning toward the right. As speed of motion increases, the Coriolis force also increases. Finally the path is turned to achieve a direction at right angles to the pressure gradient but parallel with the isobars (point *B* of Figure 21.6). Here the Coriolis force exactly balances the pressure-gradient force, and no further turning ensues. The resultant flow of air at point *B* is termed the **geostrophic wind,** which occurs where isobars are straight and parallel.

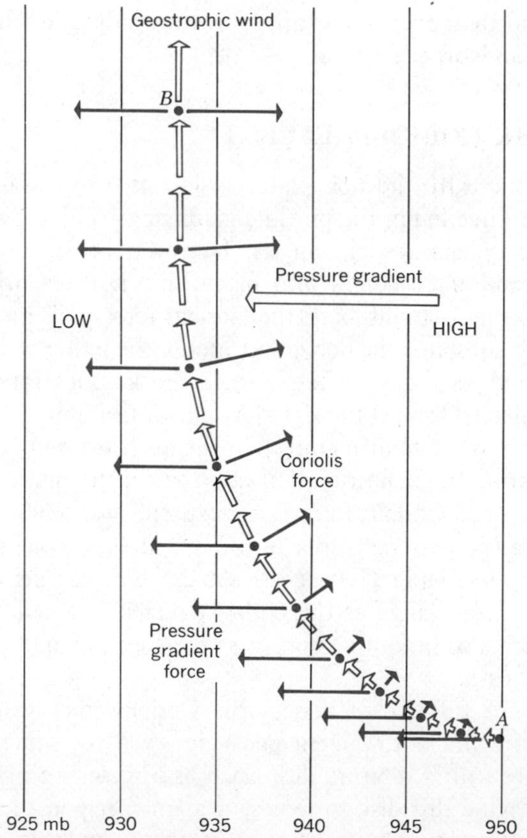

Figure 21.6 A parcel of air starting from a position of rest at *A* is deflected until it is moving parallel with the isobars at *B*.

CYCLONES AND ANTICYCLONES

Where isobars are curved, the air also moves in a curved path. Motion in a curved path brings into play centrifugal force. We will not go into details, except to note that centrifugal force acts either to increase or decrease the speed of the air motion, depending upon the geometry of the given situation. So far as we are concerned here, wind at high altitudes will closely follow the curving isobars along whatever configurations they happen to take.

Using the Coriolis principle, analyze a simple map showing isobars and winds (Figure 21.7). This weather map shows conditions several thousand feet above the surface. A low pressure center lies at the left; a high pressure center at the right. The lower diagram is an enlarged portion of the map between the high and the low. At this point the geostrophic wind blows northward, paralleling the isobars. Arrows on the map show the air flow paralleling the isobars and circling the pressure centers.

Winds circling a low pressure center constitute a **cyclone;** winds

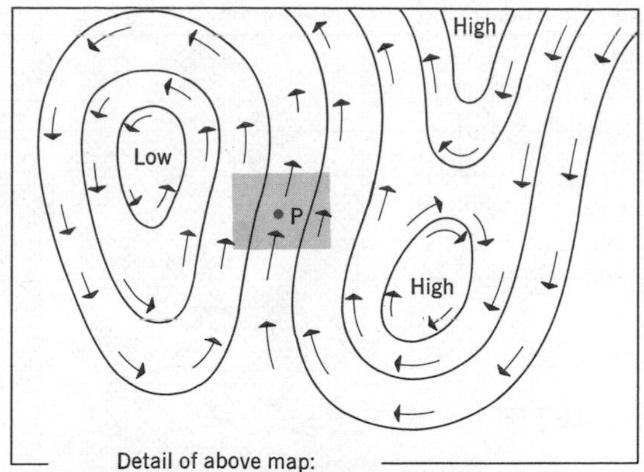

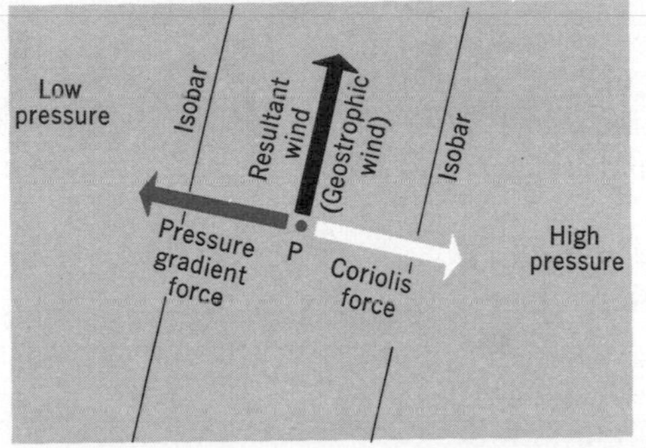

Figure 21.7 Wind follows isobars at high levels.

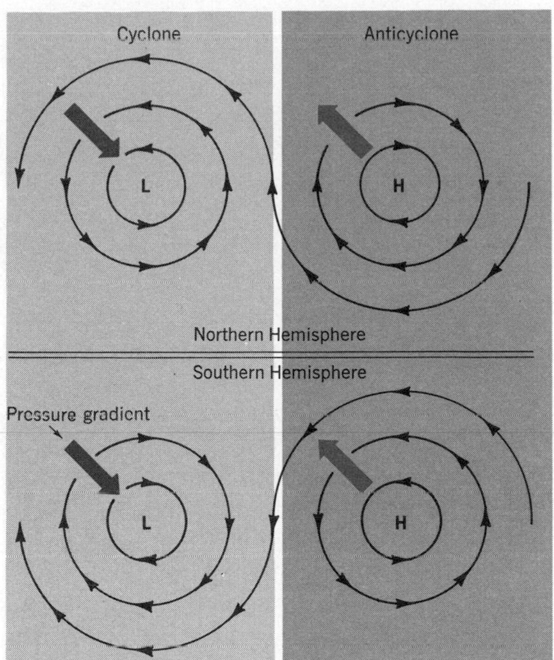

Figure 21.8 Air flow around cyclones and anticyclones in the Southern Hemisphere is exactly the reverse from that in the Northern Hemisphere.

circling a high pressure center constitute an **anticyclone.** Figure 21.8 illustrates cyclones and anticyclones. The broad arrows show the pressure gradient—inward toward the center of the cyclone but outward from the center of the anticyclone. In the Northern Hemisphere, winds move counterclockwise in the cyclone; clockwise in the anticyclone. Reverse directions apply to the Southern Hemisphere, as you see in the lower half of the figure.

If you were an airline pilot, plotting your course so as to have a tailwind at all times, your rule would be "Keep the lows on your left and the highs on your right" (in the Northern Hemisphere).

SURFACE WINDS

Air moving close to the surface encounters friction with the land or water surface beneath it. This drag not only retards the air flow, reducing the wind speed, but also changes the angle between wind and isobars. Because of friction, deflection by the Coriolis force is not enough to bring the wind to a path parallel with the isobars. Instead, the wind crosses the isobars at an angle, as shown on a surface weather map (Figure 21.9). The wind arrows typically make an angle of 25° to 45° with the isobars.

Figure 21.9 also shows that winds are stronger where the isobars are more closely crowded; weaker where they are spaced farther apart. This is what we would expect, since close crowding of isobars means a steep pressure gradient and a strong pressure-gradient force. Calm prevails precisely at the center of both a cyclone and an anticyclone.

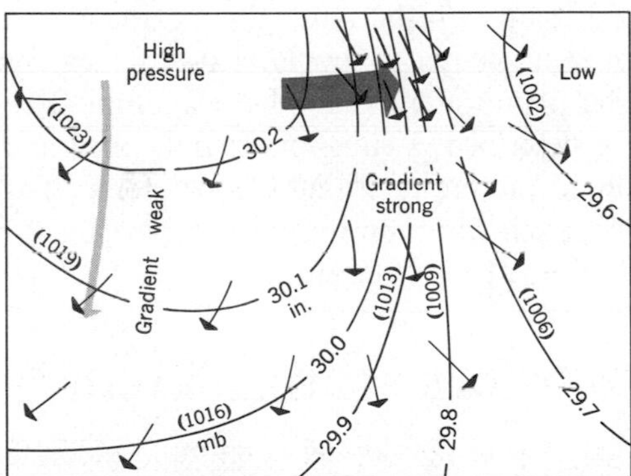

Figure 21.9 On this simplified weather map, surface winds are shown by short arrows, crossing the isobars obliquely. (© 1973, John Wiley & Sons, New York.)

Surface winds within cyclones and anticyclones are illustrated in Figure 21.10. Cyclones have inspiraling winds; anticyclones have outspiraling winds. Directions of spiraling are reversed in the Southern Hemisphere. Now, it must be obvious that where air is spiraling into the center of a cyclone it is converging and must be disposed of by rising to higher levels. Air spiraling out from an anticyclone is diverging and must be replaced by air subsiding from higher levels. Therefore, **convergence** characterizes a cyclone; **divergence** an anticyclone.

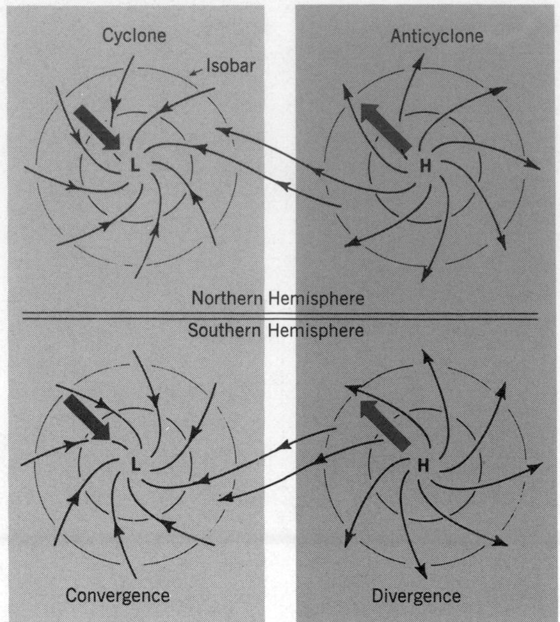

Figure 21.10 Inspiraling surface winds characterize a cyclone; outspiraling winds, an anticyclone.

WINDS ON AN IMAGINED NONROTATING EARTH

To begin our analysis of the global winds, we set up an imaginary situation in which the earth does not rotate on its axis (or rotates very slowly). In so doing, we are eliminating the effect of the Coriolis force, but will introduce it later. The nonrotating earth is imagined to receive solar radiation most intensely around its equatorial girdle, but none at either pole.

The atmosphere is heated most in the equatorial zone and expands, making an equatorial low pressure belt at the surface (Figure 21.11). By comparison, the atmosphere over the poles is colder and produces high pressure centers at the surface. In this way, a pressure gradient is set up from each pole toward the equator. Air moves with this gradient and the result is surface winds blowing along the meridians. Such winds are described as **meridional winds.** In the Northern Hemisphere these are north winds; in the Southern Hemisphere they are south winds.

Air converging upon the equator rises and spreads out at high levels; then travels poleward. Over the poles, air from high levels converges and descends, completing the circulation. As a result, we have a single **circulation cell** in each hemisphere. Altogether, the two cells make up a global **heat engine,** which is simply a machine driven by an input of heat. Heat leaves largely by longwave radiation over the polar zones.

The imaginary global circulation illustrates a principle that is valid and serves to explain certain of the observed features of the earth's circulation. There is, in fact, an equatorial belt of low barometric pressure. Furthermore, there exists a strongly developed polar high pressure center at low level over ice-covered Antarctica.

The planet Venus may have a simple two-cell system such as we have described. Venus rotates only once in 243 days, and this is too slow to produce a strong Coriolis force. There is now some evidence for such winds on that planet, developed most strongly on the side of Venus facing the sun.

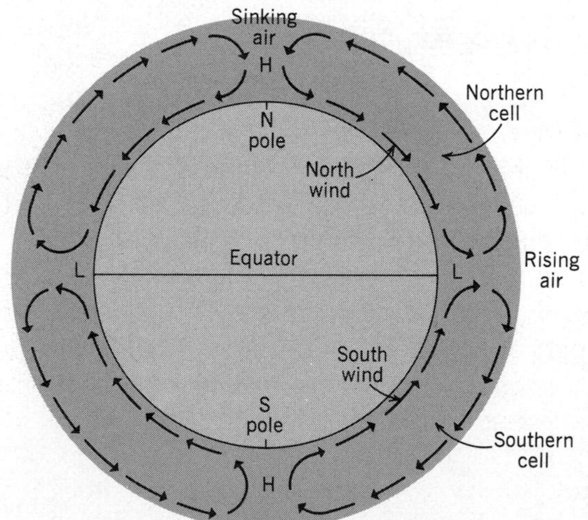

Figure 21.11 On an imagined nonrotating earth a simple two-cell system of meridional winds would be maintained.

WINDS ON OUR ROTATING EARTH

The next step in developing our simple model of atmospheric circulation is to introduce the Coriolis force present on a rotating globe (Figure 21.12). As the high-level air over the equatorial belt begins to move poleward, it is deflected to the right (in the Northern Hemisphere) to become a system of upper-air **westerly winds** paralleling the isobars at about the 30th parallel. Correspondingly, in

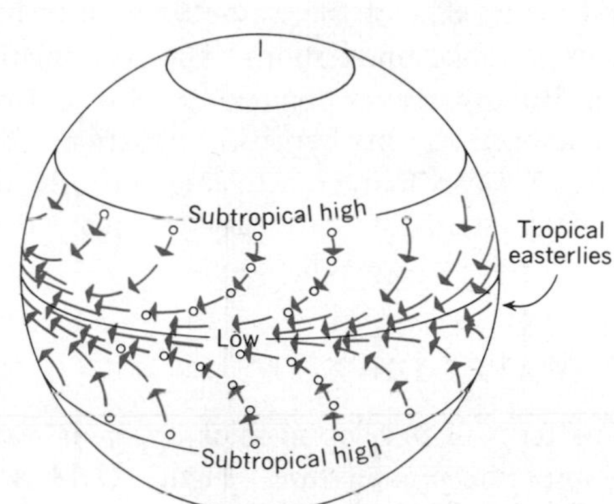

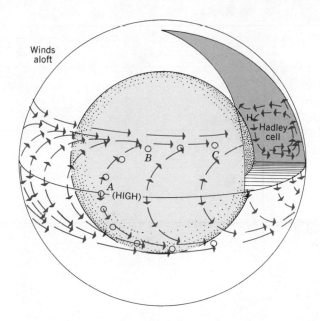

Figure 21.12 At high levels, a parcel of air starting northward from a position at rest at *A* over the equator would be deflected to the right until moving due eastward at *B* (*left figure*). Near the earth's surface, air starting to travel equatorward from the subtropical highs is turned westward to produce the tropical easterlies (*right figure*).

the Southern Hemisphere, deflection to the left also turns the poleward flow at high level into a westerly wind system.

But now we anticipate still another phenomenon. Because air moving at high levels poleward from the equatorial belt has been turned into westerly flow, following the earth's parallels of latitude, it tends to accumulate more rapidly than it can escape poleward. This accumulation, or banking up, of air takes place in zones between lat. 20° N and 30° N and between lat. 20° S and 30° S. Here, accumulation aloft produces at the surface two belts of high barometric pressure known as the **subtropical high pressure belts,** one in each hemisphere.

Part of the air subsiding within the subtropical high pressure belt starts to spread equatorward. As this air follows the barometric pressure gradient from subtropical high to equatorial low, it is deflected westward to create a system of prevailing winds known as the **tropical easterlies** (Figure 21.12, *right*). The tropical easterlies form a broad, steady, and deep air stream moving around the earth over the equatorial regions and extending to high altitudes.

The atmospheric circulation system of equatorial and tropical latitudes thus consists of two cells, one in each hemisphere. Seen in cross section, and neglecting east-west components of motion, the meridional circulation within each cell consists of horizontal and vertical motions, together forming a complete circuit. The existence of such a circulation system was first postulated by George Hadley in 1735, and is now called the **Hadley cell** by meteorologists.

THE UPPER-AIR WESTERLIES

Poleward of the subtropical high pressure belts, circulation in the troposphere takes the form of a prevailing system of **upper-air westerlies.** These winds are shown schematically in Figure 21.13. Air moving northward (Northern Hemisphere) is deflected by the Coriolis force to the right (toward the east) and becomes a west wind. The flow constitutes a great vortex moving counterclockwise around

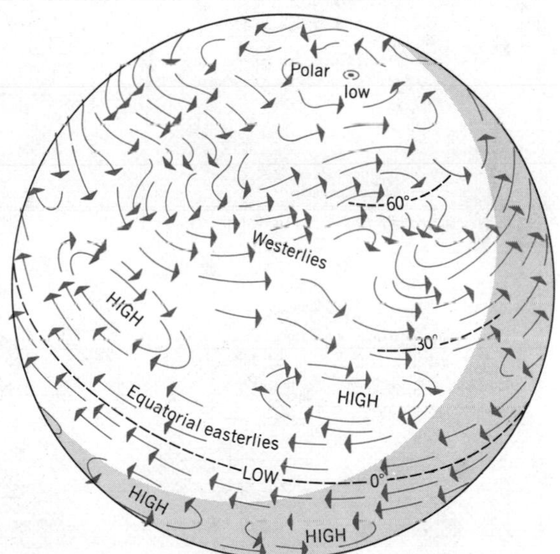

Figure 21.13 Global circulation at high levels in the troposphere takes the form of westerlies and easterlies, with a belt of high pressure centers between.

a prevailing center of low barometric pressure, the **polar low.** A corresponding system of upper-air westerlies exists in the Southern Hemisphere.

The simple west-to-east flow of the westerlies is disturbed by a ceaseless succession of wavelike undulations. These undulations are **upper-air waves,** or **Rossby waves,** named for C.-G. Rossby, a meteorologist who developed the mathematical equations governing the waves. The upper-air waves may grow, change in form, and dissolve. They may remain essentially stationary for many days, and may also drift slowly in the east-west direction.

HOW UPPER-AIR WAVES DEVELOP

Let us analyze the patterns of development of upper-air waves and their relation to atmospheric temperatures. Figure 21.14 is a schematic diagram showing wave evolution in four stages. Long, heavy arrows show the location of a high-speed air stream called the **jet stream.** The jet stream defines the position of the waves. Conditions shown are those existing near the top of the troposphere, which ranges in height from about 30,000 ft (9 km) over the poles to about 50,000 ft (17 km) over the equator.

The troposphere lying poleward of the jet stream consists of cold polar air, whereas that on the equatorward side consists of warm tropical air. Such large bodies of the atmosphere are referred to as **air masses.** Air masses are identified on the basis of both temperature and water-vapor content. We will explain in greater detail in Chapter 22 that polar air masses, because they are cold, can hold

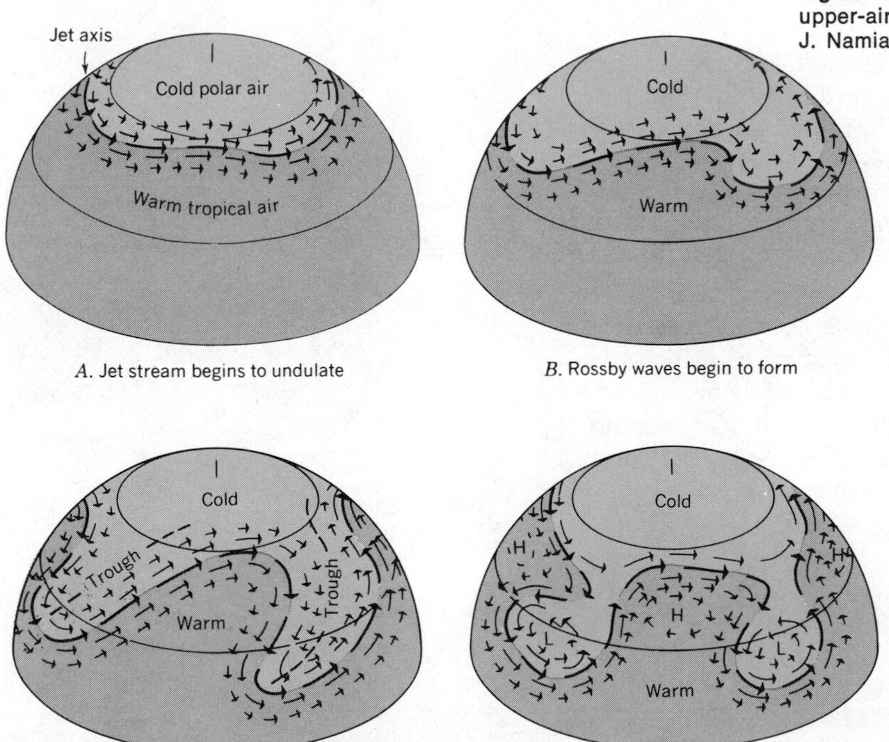

Figure 21.14 Four stages in the development of upper-air waves in the Northern Hemisphere. (After J. Namias, National Weather Service.)

little moisture. In contrast, warm tropical air masses can hold comparatively large quantities of moisture.

The jet stream in middle latitudes occupies a position at the contact between the polar air mass and the tropical air mass. A contact surface between adjacent air masses is known as a **front.** In the case we are examining, the front lying beneath the jet stream is known as the **polar front.**

Diagram A of Figure 21.14 shows the jet stream lying over the high latitudes and with only small undulations. As waves form (diagram *B*), the polar air pushes south at one place, and the tropical air moves north at another. Soon great tongues of air form an interlocking pattern, with the jet stream taking a sinuous path between them (diagram C). Finally, a wave constricts at the base, and a mass of cold or warm air is detached, forming an isolated **pressure cell.**

A cell of stranded cold air aloft at subtropical latitudes forms a low pressure center with counterclockwise circulation. An isolated cell of warm air aloft at the higher latitude becomes a high pressure center with clockwise air flow. At the close of the wave development cycle, which takes four to six weeks to complete, the isolated cells dissolve. The jet stream then resumes its simple course over the high latitudes.

The cycle of upper-air wave development explains how great quantities of heat are transferred from equatorial regions to polar regions. North-moving tropical air carries heat to the high latitudes, where the heat is lost. South-moving tongues bring cold air to the low latitudes. Here the cold air mass absorbs part of the excess heat. Although this form of heat and moisture transfer fluctuates in intensity and location, the average effect, year in and year out, is to maintain a balance in the earth's heat budget.

Figure 21.15 This three-dimensional sketch will help you to visualize a jet stream with its high-speed core. (National Weather Service.)

THE JET STREAM

We now turn to details of the jet stream phenomenon. During World War II, American pilots flying B-29 aircraft at high altitude in bombing missions over Japan occasionally reported extremely strong westerly winds above 20,000 ft (6 km). These winds were of such high speed as to equal the forward air speed of the planes and so reduce the aircraft's ground speed to zero. Transport planes bound eastward at high altitudes over the Pacific were sometimes aided by tail winds which doubled the ground speed of travel. These were the first indications of the nature of the jet stream as a powerful but narrow stream of fast-moving air in the westerlies of the upper troposphere.

You can think of this jet stream as resembling the high pressure flow of water from a hose nozzle submerged and pointed horizontally in the direction of flow of a slowly moving stream (Figure 21.15). The jet stream is shaped like a tube which lies roughly horizontally. The tube may curve from side to side to change direction from, say, northwest to west to southwest.

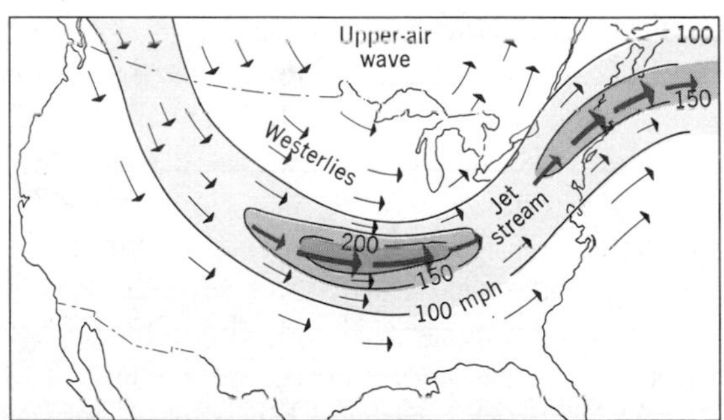

Figure 21.16 This map shows a typical jet stream in action. Numbered lines give the wind speed, while color arrows show the jet stream core. (© 1973, John Wiley & Sons, New York.)

Figure 21.16 is a map of the United States showing wind speeds at the 30,000-ft (9-km) level on a particular day. The heavy arrows show the jet core as making a great loop southward over the Middle

West. Wind speed is over 200 mi (320 km) per hour in the fastest section of the stream.

Most Americans who travel cross-country frequently in commercial jet aircraft are already quite familiar with the jet stream. The captain will usually announce when a jet stream is being crossed, because it is accompanied by **clear air turbulence** (CAT). This form of turbulence is difficult to anticipate, because the air is normally quite clear. The plane is subjected to shaking that is sometimes severe enough to place dangerous stresses on the aircraft. The turbulence results from adjacent air layers moving at different speeds, setting up small but intense eddies between the layers.

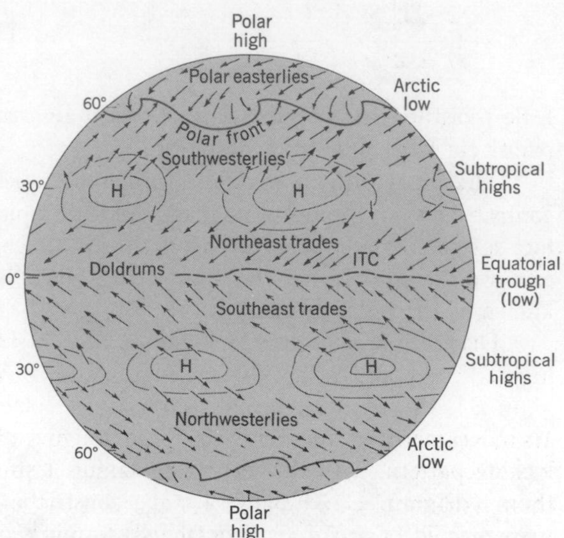

Figure 21.17 A schematic diagram of the global pattern of surface winds.

GLOBAL PATTERNS OF SURFACE WINDS

So far, our study of global circulation has been about very large flow patterns affecting the atmosphere above the level at which surface friction is effective in modifying winds. Recall that near the ground, surface friction prevents the air from moving parallel with isobars. Instead, wind blows obliquely across isobars (Figure 21.9).

The characteristic global pattern of surface winds is shown in Figure 21.17. Notice first that the subtropical high pressure belt is actually composed of a number of centers, or pressure cells. Typically one or two cells are centered over each ocean, while other cells are centered over tropical portions of the continents. Calm prevails much of the time in the centers of these **subtropical high pressure cells.** Each cell is a persistent site of subsiding air, which is dry. The world's great tropical deserts lie beneath the high pressure cells.

Air movement equatorward from the high pressure cells constitutes the **trade winds.** These surface winds are remarkably persistent over the tropical oceans, being almost entirely from the northeast quarter in the Northern Hemisphere. Near the equator, the trades converge in the **equatorial trough.** Here barometric pressure is somewhat lower than normal. Locally, the line of trade-wind convergence is sharply defined and is called the **intertropical convergence zone** (abbreviated as ITC in Figure 21.17). Elsewhere there is a zone of stagnant air, the **doldrums,** in which winds can be described as being light and variable, or absent in extended periods of calms. It is within the equatorial trough that the rise of heated air takes place in the Hadley-cell circulation.

Poleward of the subtropical cells of high pressure the lower air tends to move toward higher latitudes as a southwest or northwest wind. This wind pattern is intensified in summer of the respective hemisphere. It has given rise to the mariner's labels **southwesterlies** (Northern Hemisphere) and **northwesterlies** (Southern Hemisphere) within the latitude belt from about 30° to 50°. The facts are that this latitude belt experiences winds from all quarters, depending upon local storms and seasonal effects. In the sense that west winds are most frequent, this latitude zone can aptly be described as a belt of **prevailing westerlies.**

At the earth's surface, the contact between cold air of polar origin and warm air of tropical origin is usually a sharply defined line, the polar front. As shown in Figure 21.17, the polar front has a deeply indented pattern. The front is constantly shifting. It may

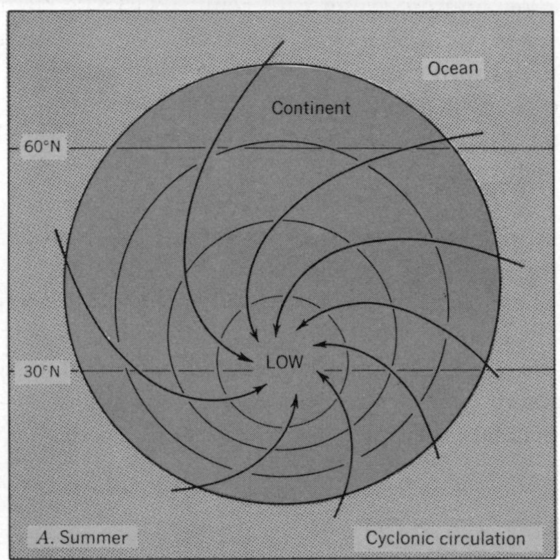

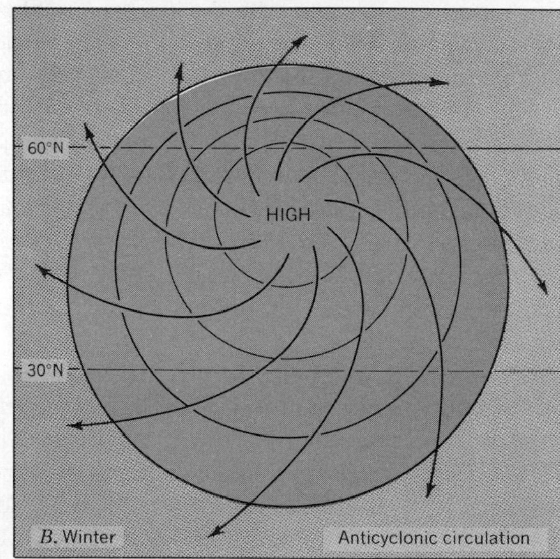

Figure 21.18 A schematic diagram of yearly alternations in surface pressure and air flow over a large middle-latitude continent in the Northern Hemisphere.

at one time sweep far equatorward; at another time it is formed in arctic latitudes. The cold air masses are associated with winds from a northerly quarter. Northwest, north, and northeast winds occur from time to time from middle latitudes to polar regions. The **polar easterlies,** labeled on the illustration, should not be thought of as persistent winds, such as are the trades, but rather as frequent winds associated with traveling disturbances of the lower atmosphere. We shall investigate these disturbances in Chapter 23.

MONSOON CIRCULATION PATTERNS

Barometric pressures at low levels in the atmosphere reflect air temperatures. For example, over a highly heated continent, low pressure prevails. Over a cold continent in winter, high pressure prevails. Adjacent oceans, slow to warm and cool, at the same season show an opposite barometric tendency. Consequently, a pressure gradient is set up from high to low pressure. Winds develop in response to these gradients.

The result is a seasonal flow of air into a continent from the surrounding ocean, alternating with a reversal of this flow at the opposite season. Such seasonal patterns go by the general term **monsoon systems.** Actually, the monsoon circulation of southwestern Asia is by far the most important example. Much weaker displays of the monsoon effect can be observed in northern Australia and in the central United States.

Figure 21.18 is a schematic diagram of the monsoon system over a continent in the Northern Hemisphere. The summer cyclone tends to be displaced toward the southern side of the continent, while the winter anticyclone tends to occupy a position north of the continental center.

In Figure 21.19, the surface winds of eastern Asia are shown for July and January. Notice the placement of the July low pressure region over northern India, Pakistan, and Afghanistan. In contrast, the strong winter high lies far to the north in Siberia and coincides with a center of intense cold.

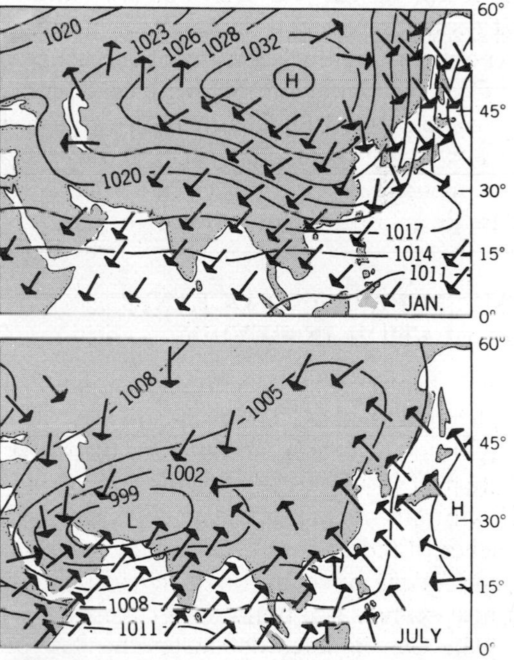

Figure 21.19 These surface maps of pressure and winds over southeastern Asia in January and July show the monsoon system. Pressure in millibars.

WINDS AND OCEAN CURRENTS

Winds blowing over the ocean surface transfer vast quantities of energy from the atmosphere to the oceans. One can think of the atmospheric circulation systems, such as the Hadley cell and the prevailing westerlies, as great gear wheels. Meshed with the sea, these wheels turn related systems of surface water motion.

All horizontal motions of the surface ocean layer are included in the general term **ocean currents.** As wind blows over a smooth water surface, a frictional drag is exerted by the air upon the water. This drag force sets the surface water layer in slow motion. As the uppermost water layer is dragged in the downwind direction, it drags along the next lower layer. In this way the motion is propagated toward increasing depth. The moving water is also subjected to the Coriolis effect. In the Northern Hemisphere, Coriolis acts as a force pulling toward the right in a direction perpendicular to the direction of motion. The action is the same as for air in motion.

The very slow water motion caused by winds is called a **drift current.** The Coriolis effect upon a drift current was observed by Fridtjof Nansen, the arctic explorer, as his vessel, the *Fram,* was drifting upon the polar sea, firmly held in pack ice. Nansen noted that the pack ice was moving in a direction some 20° to 45° toward the right of the wind direction. He correctly attributed this difference in direction to the Coriolis effect. When we compare drift direction with prevailing winds, we find that direction of water motion is approximately at 45° with respect to wind direction.

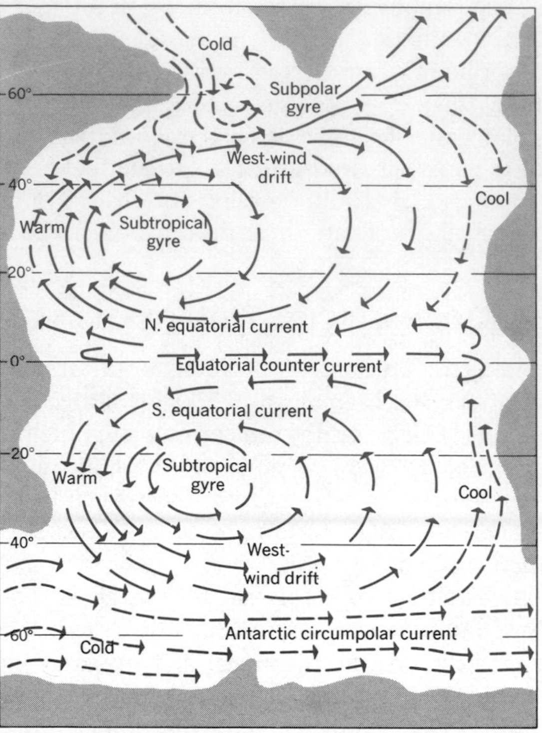

Figure 21.20 This schematic diagram of an ocean shows an idealized system of surface currents dominated by gyres.

THE GENERAL PATTERN OF OCEANIC CIRCULATION

Within each ocean a characteristic pattern of surface currents and drifts is repeated. Figure 21.20 shows schematically the major elements of the flow system. The major features are two great **gyres,** or circular flow systems. One gyre is located in each hemisphere, centered approximately upon the subtropical cell of high barometric pressure. Water motion is clockwise about the gyres.

The trade winds set in motion a west-moving **equatorial current,** paralleling the equator. The prevailing westerlies set in motion the east-moving **west-wind drift** in middle latitudes. The gyre is completed by a strong poleward flow at the western side of the gyre.

Because of the Coriolis effect, the gyres are pushed toward the west side of the ocean. Consequently, the poleward currents on the west sides are intensified. Examples are the Gulf Stream off North America, and the Kuroshio Current off Japan. These currents are relatively warm and serve to transport heat from low to high latitudes.

On the eastern sides of the gyres, the drift is turned equatorward, bringing cool water of arctic and antarctic origin into low latitudes. These equatorward currents bring unusually cool air temperatures to continental west coasts at low latitudes. The effect of the great oceanic gyres upon the earth's heat budget is most important, for enormous quantities of water are exchanged by the flow.

Configuration of the Antarctic continent, a pole-centered landmass completely surrounded by the great belt of southern ocean, results in a simple circulation system. Strong prevailing westerly winds set in motion an **antarctic circumpolar current,** which encircles Antarctica in a continuous stream.

ATMOSPHERIC AND OCEANIC CIRCULATION IN REVIEW

In this chapter we have examined a heat engine within the atmosphere and oceans. The main drivings wheels of this engine are in the atmosphere. Solar energy turns the wheels. Air heated by solar radiation in low latitudes expands and becomes less dense, setting up low barometric pressure in contrast with high barometric pressure at higher latitudes. The pressure gradient sets winds in motion. Rising and sinking air completes the circulation system. The Hadley cell is a nice example of a heat-driven circulation system.

The Coriolis force dominates the flow paths of both air and water. Because of Coriolis, we have tropical easterly winds in low latitudes and prevailing westerly winds in middle and high latitudes.

The secondary wheels of the planetary heat engine are the gyres of the oceans. Winds set the surface water in motion, while the Coriolis force deflects the flowpaths into east-flowing and west-flowing currents. These currents are connected into complete gyres. In this way heat is carried into high latitudes to be radiated away, helping to maintain the earth's heat balance.

YOUR GEOSCIENCE VOCABULARY

wind
isobaric surface
pressure gradient
isobar
pressure-gradient force
sea breeze
land breeze
Coriolis force
geostrophic wind
cyclone
anticyclone
convergence
divergence
meridional winds
circulation cell
heat engine
westerly winds
subtropical high pressure belts
tropical easterlies
Hadley cell
upper-air westerlies
polar low
upper-air waves
Rossby waves
jet stream
air mass
front
polar front
pressure cell
clear air turbulence (CAT)
subtropical high pressure cells
trade winds
equatorial trough
intertropical convergence zone (ITC)
doldrums
southwesterlies
northwesterlies
prevailing westerlies
polar easterlies
monsoon system
ocean currents
drift currents
gyres
equatorial current
west-wind drift
antarctic circumpolar current

SELF-TESTING QUESTIONS

1. Explain how circulation of the atmosphere and oceans play a key role in maintaining the earth's heat balance.

2. Explain fully the relationship between barometric pressure and winds. How is wind speed related to pressure gradient?

3. Describe the action of the Coriolis force in terms of direction, latitude, and hemisphere. What device can be used to illustrate the deviation caused by the Coriolis force?

4. Apply the Coriolis principle to the geostrophic wind. Show by a simple sketch map the direction of air flow around upper-level centers of low and high pressure in both hemispheres.

5. How does friction affect the direction of surface winds in relation to isobars? Illustrate with sketch maps of surface winds in cyclones and anticyclones in both hemispheres. Where does convergence occur? Where does divergence occur?

6. Describe an idealized global circulation system on an imaginary nonrotating earth. In what way is this system a heat engine? How many cells has the system?

7. Describe the system of high-level winds on a rotating earth. How does the Coriolis force take part in controlling the air flow in this system? How are the tropical easterlies related to the Hadley cell?

8. Describe the development of upper-air waves (Rossby waves). Where is the polar front situated with respect to these waves? Where are warm and cold air masses situated with respect to the polar front? Describe the flow of air in a jet stream.

9. Name and describe the major surface winds of the globe. Where do calms prevail much of the time? Where is the polar front situated with respect to wind belts?

10. What conditions are required for the development of a monsoon wind system? Describe the seasonal alternation of winds in a monsoon system. Where is the monsoon system best developed? Why?

11. What causes ocean currents? How does the Coriolis force influence the direction of a drift current? Describe the general pattern of ocean currents in an ocean basin confined by continents.

(*Left*) Wilson Bentley and his snowflake camera. This picture is thought to have been taken in 1918, when his photographs were widely known. Although he posed outdoors in front of his farm home in Vermont, Bentley used his camera in a coldroom at the rear of the house. (Courtesy of Dr. Duncan Blanchard and National Weather Service.) (*Above*) A selection from Bentley's best snowflake photographs. (National Weather Service.)

22 MOISTURE IN THE ATMOSPHERE

THE SNOWFLAKE MAN

Within the past half-century a rare and interesting subspecies of *Homo sapiens* has become extinct. He was the self-educated dilettante researcher who worked alone and yet made a notable contribution to science. In the year 1900 one might have counted on the fingers of a hand the surviving individuals of this subspecies. One of them was Wilson Alwyn Bentley, a farmer of Jericho, Vermont. Born during the Civil War, Bentley grew up on his father's farm at the foot of Bolton Mountain, located far north in the state. His mother had been a schoolteacher before her marriage, and she taught her son at home up to his fourteenth year.

Among the family possessions was a small microscope his mother had used as a schoolteacher. Bentley wrote:

> When other boys my age were playing with popguns and slingshots, I was absorbed in studying things under the microscope: drops of water, tiny fragments of stone, a feather dropped from a bird's wing, a delicately veined petal from some flower. But always, from the very beginning, it was snowflakes that fascinated me most. The farm folks, up in this north country dread the winter; but I was supremely happy, from the day of the first snowfall . . . until the last one.

At first young Bentley tried to draw the tiny hexagonal snow crystals he observed, but their intricacy defied him. Then he read in his encyclopedia about a camera for taking photographs through a microscope. Nothing would do but to possess this device. He and his mother finally badgered his father into purchasing the camera, and so began a long and discouraging series of attempts to record the crystals on film. Finally, after a year's work, he mastered the technique, and in 1885 produced his first successful photograph. He wrote: "The day that I developed the first negative made by this method, and found it good, I felt almost like falling on my knees beside that apparatus and worshipping it! It was the greatest moment of my life."

During the next 13 years Bentley worked on in seclusion, obtaining over 400 photographs of snow crystals, and in 1898 he published his first paper in the *Popular Scientific Monthly*. After further study, he published several technical papers in the *Monthly Weather Review*. His work consisted of much more than mere photography. He was able to classify snow crystals into several basic types and offer hypotheses as to their origin in various parts of a storm and the changes they underwent as they fell.

During the summers, Bentley turned to the study of raindrops. This he did while working long hours as a successful dairy and potato farmer. He pioneered in the measurement of the sizes of raindrops. Using a pan of sifted flour, he would catch some drops of falling rain. Each drop formed a pellet of dough, and when these hardened they could be removed from the flour and measured for diameter. He had carefully ascertained by indoor experiments that the dough pellet was about the same diameter as the drop that made it. Again he published his results, along with hypotheses of different ways that drops were produced in clouds. Certain of these ideas are considered valid today.

In all, Bentley accumulated some 4500 photographs of snow crystals. The best of these were published in 1931 under auspices of the United States Weather Bureau. It was none too soon, for in December of that year, at the age of 66, Bentley died of pneumonia in the same farmhouse where he had lived and worked all his life. By then his fame was widespread, not as a scientist, but as The Snowflake Man. Recognition of his contributions as a scientist were to come much later, when his pioneer writings were restudied by elegantly trained Ph.D.s in search of the elusive origins of snow crystals and raindrops high up in the clouds.*

* Information in this essay is based on "Wilson Bentley, The Snowflake Man," by Duncan C. Blanchard, *Weatherwise,* vol. 23, no. 6, 1970.

THE HYDROLOGIC CYCLE COMPLETED

In Chapter 13 we introduced the hydrologic cycle. Now we return to the hydrologic cycle to deal with the movement of water into the atmosphere as vapor and back to earth as precipitation in the form of water or ice.

To understand how clouds are formed, how rain and snow occur, and how storms are generated, we must use some elementary concepts of physics. These concepts relate to the way in which energy exists in different forms in the atmosphere.

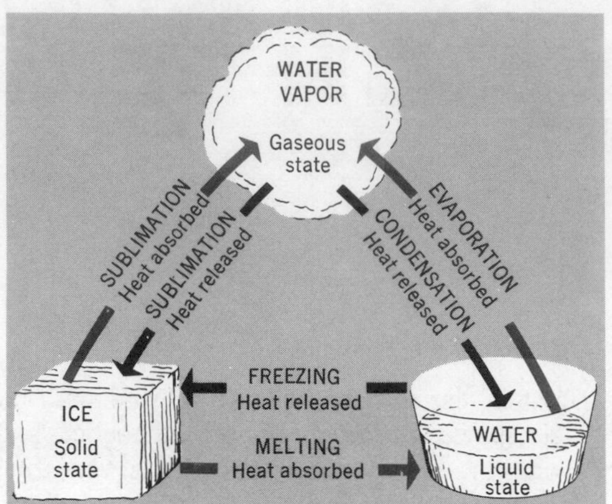

Figure 22.1 Water exists in three states: gaseous, liquid, and solid. Changes of state are accompanied by the release or absorption of heat energy, depending upon the direction of the change. (© 1973, John Wiley & Sons, New York.)

WATER STATES AND HEAT

Water occurs in three states: (1) **solid state,** frozen as ice, a crystalline substance; (2) **liquid state,** as water; and (3) **gaseous state,** as water vapor (Figure 22.1). From the gaseous vapor state, molecules may pass into the liquid state by **condensation.** When temperatures are below the freezing point, water vapor can, by **sublimation,** pass directly into the solid state to form ice crystals. By **evaporation,** molecules leave a water surface to become gas molecules of water vapor. The change from ice directly into water vapor is also called sublimation. Then, of course, water may pass from liquid to solid state by **freezing,** and from solid state to liquid state by **melting.**

Of utmost importance in weather science are the exchanges of heat energy accompanying changes of state. **Sensible heat** is heat we can feel and can measure by thermometer. When water evaporates, sensible heat is absorbed and passes into a hidden form held by the water vapor. This hidden form is known as the **latent heat of vaporization.**

Change from sensible heat to latent heat during evaporation results in a drop in temperature of the remaining liquid. The cooling effect produced by evaporation of perspiration from the skin is an obvious example. For every gram of water that is evaporated, about 600 calories of sensible heat pass into the latent form.

In the reverse process of condensation, an equal amount of energy is released to become sensible heat. This energy release from latent form tends to increase the temperature of the air in which condensation is taking place.

The freezing process releases heat energy in the amount of about 80 calories per gram of water. Melting absorbs an equal quantity of heat, referred to as the **latent heat of fusion.**

When sublimation occurs, the heat absorbed by vaporization or released by crystallization is even greater for each gram of water, because the latent heats of vaporization and fusion are added together.

RELATIVE HUMIDITY

The amount of water vapor that may be present in the air at a given time varies widely from place to place. It ranges from almost nothing in the cold, dry air of arctic regions in winter to as much as 4% or 5% of the volume of the atmosphere in the warm equatorial belt.

Water vapor enters the atmosphere by evaporation from exposed water surfaces such as oceans, lakes, rivers, or moist ground. Some is supplied by plants which transpire water (a form of evaporation).

We use the term **humidity** to refer generally to the degree to which water vapor is present in the air. For any specified temperature there is a definite limit to the quantity of moisture that can be held by the air. When this limit is reached the air is said to be **saturated.**

The proportion of water vapor present relative to the maximum quantity is the **relative humidity,** expressed as a percentage. For saturated air, relative humidity is 100%. When half of the total possible quantity of vapor is present, relative humidity is 50%, and so on.

A change in relative humidity of the atmosphere can be caused in one of two ways. If ar exposed water surface is present, the humidity can be increased by evaporation. This is a slow process, requiring that the water vapor diffuse upward through the air.

The other way relative humidity can change is through a change of air temperature. Even though no water vapor is added, a lowering of temperature results in a rise of relative humidity. This is an automatic change. It is due to the fact that the capacity of the air to hold water vapor has been reduced by cooling, so that the existing amount of vapor represents a higher percentage of the capacity of the air.

Likewise, as air temperature rises, relative humidity decreases, even though no water vapor has been taken away. The principle of relative-humidity change caused by temperature change is illustrated by graphs of these two properties throughout the day (Figure 22.2). As air temperature rises, relative humidity falls and vice versa.

A simple example illustrates these principles (Figure 22.3). At a certain place, the midmorning temperature of the air is 60° F (16° C); the relative humidity (RH) is 50%. In midafternoon the air, warmed by infrared radiation from the sun and ground surface, reaches 90° F (32° C). The relative humidity has now dropped to 20%, which is very dry air. By outgoing longwave radiation the air becomes chilled during the night, and by early morning its temperature falls to 40° F (5° C). Now the relative humidity has automatically risen to 100%.

Any further cooling of saturated air will usually cause condensation of the excess vapor into liquid form. As the air temperature continues to fall, the humidity remains at 100%, but condensation continues, taking the form of minute droplets of dew or fog. If the temperature falls below freezing, condensation usually occurs as frost upon exposed surfaces.

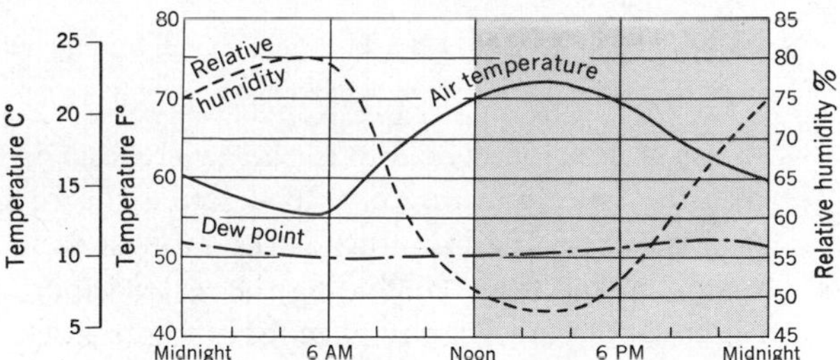

Figure 22.2 Curves of mean hourly air temperature, dew point, and relative humidity for the month of May at Washington, D.C. (Data of National Weather Service.)

DEW POINT

The **dew point** is that critical temperature at which the air is fully saturated. Below this temperature condensation normally occurs. We have an excellent illustration of condensation due to cooling in summertime when beads of moisture form on the outside surface of a pitcher filled with ice water. Air immediately adjacent to the cold glass surface is chilled enough to fall below the dew-point temperature. Moisture then condenses on the surface of the glass.

Figure 22.2 shows that the dew-point temperature throughout the day remains about constant, despite large swings in both air temperature and relative humidity. The small fluctuations shown on the graph are caused by minor gains or losses of water vapor from or to the ground surface and its plant cover.

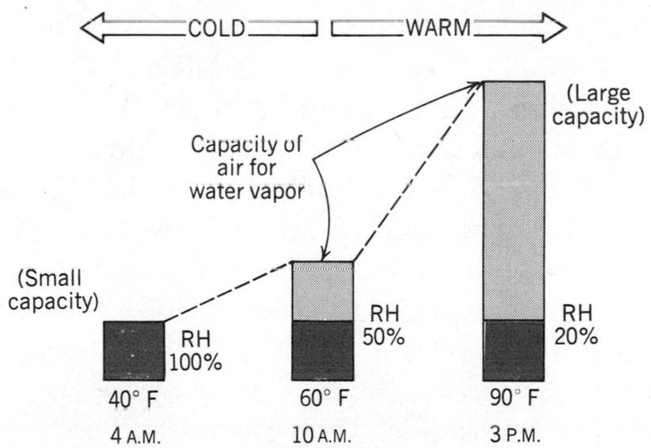

Figure 22.3 A change of air temperature changes the capacity of the air to hold water vapor. The cold air is saturated (RH = 100%). The warm air holds only one-fifth of its capacity (RH = 20%).

SPECIFIC HUMIDITY

Relative humidity, being a percentage, does not tell how much moisture is actually held in the air. Meteorologists use a more definite measure, specific humidity, to describe the moisture content of an air mass. **Specific humidity** is the ratio of weight of water vapor to weight of moist air (including the water vapor). Units are grams of water vapor per kilogram of moist air (gm/km). When a given mass of air is lifted to higher altitudes without gain or loss of moisture, the specific humidity remains constant even though the air expands to occupy a larger volume.

Specific humidity is used to describe the moisture characteristics of a large mass of air. For example, extremely cold, dry air over arctic regions in winter often has a specific humidity as low as 0.2 gm/kg, whereas extremely warm, moist air of tropical regions may hold as much as 18 gm/kg. The total natural range on a worldwide basis is such that the largest values of specific humidity are from 100 to 200 times as great as the least.

In a sense, specific humidity is a yardstick of a basic natural resource—fresh water—to be applied from equatorial to polar regions. It is a measure of the quantity of water that can be extracted from the atmosphere as precipitation. Cold air can supply only a small quantity of rain or snow; warm air is capable of supplying large quantities. This is a very important concept to keep in mind, and we shall apply it in our study of storms.

AIR MASSES

Important in your understanding of weather phenomena—such as rain and snow, thunderstorms, and hurricanes—is the concept of the air mass. We referred briefly to air masses in the previous chapter. Recall that a cold polar air mass lies poleward of the polar jet stream, while a warm tropical air mass lies on the equatorward side.

In its dimensions, a single **air mass** is a body of air extending horizontally over a substantial part of a continent or ocean. The air mass also extends vertically through a major fraction of the troposphere. A given air mass possesses nearly uniform temperature and water-vapor content in all horizontal directions at any given altitude. An important fact to know when describing an air mass is the rate at which its temperature changes with altitude.

Typically, a given air mass has a sharply defined boundary in contact with a different air mass adjacent to it. As we said in Chapter 20, this sharp line of contact between two air masses is a **front.** Some fronts are nearly vertical; others are inclined almost to horizontality. In the latter case, one air mass will overlie another. Fronts are often in rapid motion, so that one air mass displaces another.

An air mass is a product of its **source region,** the ocean or land surface from which it derives its physical properties. For example, over a warm ocean, the air mass derives a large water-vapor content and high temperature. Over a cold continent in winter, an air mass is not only intensely cold, especially in the lower layer, but its water-vapor content is extremely small as well.

Although an air mass may be stagnant over long periods over the source region, it can travel into other regions. It follows the movement of regional winds in response to the pressure gradient. Modification of the air mass gradually takes place during migration. Heat may be gained or lost to the ground surface through longwave radiation. Mixing of warm lower air layers with colder air aloft may occur. Water vapor may be added by evaporation from a sea surface below and find its way upward through such mixing.

The meteorologist's classification of air masses is based upon the global position of the source area and the nature of the underlying surface, whether continent or ocean. Latitude relates closely with air temperature. Nature of the surface below strongly influences the water-vapor content.

An air mass can be identified by a two-letter symbol. The types of air masses based on latitude are listed in Table 22.1 (capital letter). They are arctic, antarctic, polar, tropical, and equatorial, in order from poles to equator. Temperatures of these air masses range from extremely cold for the arctic types to very warm for the tropical types.

Two further subdivisions, based on source region surface, are recognized: maritime and continental (Table 22.1, lower-case letter). As you might expect, maritime air masses have high water-vapor content; continental air masses have low water-vapor content.

Examples of the principal combinations of latitudinal divisions and source-region divisions are shown in Table 22.2. Some typical examples illustrate the conditions of temperature and specific humidity that may be expected. Notice that the moisture content of the maritime equatorial (*mE*) air mass is almost 200 times as great as that of the continental arctic (*cA*) air mass. The example of a continental tropical (*cT*) air mass, with 11 gm/kg specific humidity, represents air of the tropical deserts. Although this desert air has low relative humidity when highly heated during the daytime, it actually holds substantial quantities of water vapor. The maritime tropical (*mT*) air mass has a very high specific humidity of 17 gm/kg and can yield heavy rainfall. The maritime polar (*mP*) air mass, shown in the example with a specific humidity of 4.4 gm/kg, is also capable of yielding large amounts of precipitation under favorable conditions.

Table 22.1 **Kinds of air masses**

Air mass	Symbol	Source region
Arctic	A	Arctic Ocean and fringing lands
Antarctic	AA	Antarctica
Polar	*P*	Lands and oceans, 50° to 65°, N and S latitudes
Tropical	*T*	Lands and oceans under subtropical high pressure cells, 20° to 35°, N and S latitudes
Equatorial	*E*	Oceans close to the equator, in equatorial trough and doldrums
Maritime	*m*	Oceans
Continental	*c*	Continents

Table 22.2 **Typical examples of air masses**

Air mass	Symbol	Properties	Temperature (°F)	Temperature (°C)	Specific humidity (gm/kg)
Continental arctic and continental antarctic	*cA* (*cAA*)	Very cold, very dry (winter)	−50°	−46°	0.1
Continental polar	*cP*	Cold, dry (winter)	12°	−11°	1.4
Maritime polar	*mP*	Cool, moist (winter)	39°	4°	4.4
Continental tropical	*cT*	Warm, dry	75°	24°	11
Maritime tropical	*mT*	Warm, moist	75°	24°	17
Martime equatorial	*mE*	Warm, very moist	80°	27°	19

In contrast, the continental polar (cP) air mass has a specific humidity of only 1.4 gm/kg and can produce little precipitation.

The subtropical high pressure cells form source regions of mT air masses over the oceans but of cT air masses over continents. The source regions of mP air masses coincide with great centers of low pressure over the Aleutian region in the north Pacific, the Icelandic region in the north Atlantic, and the Southern Ocean surrounding Antarctica.

THE ADIABATIC PROCESS

Precipitation is the general term applying collectively to actively falling rain, snow, sleet, or hail. Precipitation can result only where large masses of air are experiencing continued drop in temperature below the dew point. This required drop in temperature cannot be brought about by the simple process of chilling of the air through loss of heat by radiation during the night. Instead, it is essential that a large mass of air be rising to higher altitudes. To understand why this must be so requires us to investigate an important concept of physics.

One of the most important laws of meteorology is that rising air experiences a drop in temperature, even though no heat energy is lost to the outside (Figure 22.4). The drop of temperature is a result of the decrease in air pressure at higher altitudes, which permits the rising air to expand. Individual molecules of the gas become more widely diffused and do not strike one another so frequently. Less frequent collisions result in a lower sensible temperature of the gas.

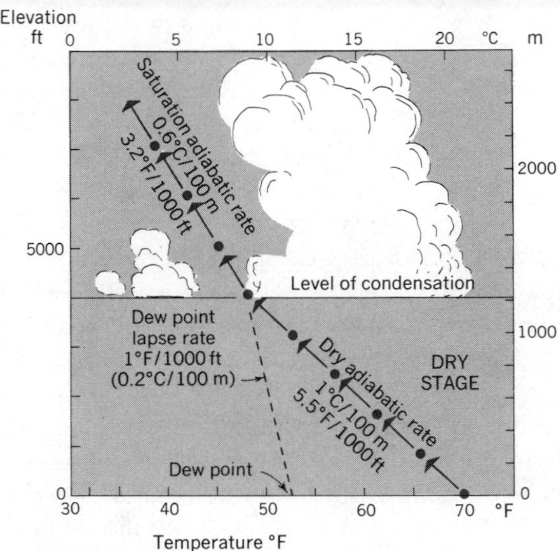

Figure 22.4 Rising air is cooled by the adiabatic process, but the rate of cooling is reduced when condensation is occurring.

Physicists refer to this spontaneous temperature change accompanying volume change as the **adiabatic process.** The word adiabatic means "taking place without gain or loss of heat energy from or to the outside." The adiabatic process is nicely illustrated when you inflate an automobile tire with a hand pump. You will find that the forced reduction of air volume makes the pump very hot. When you let the air escape from the tire valve the air feels cool, because it is expanding in volume.

When no condensation is occurring, the rate of drop of air temperature is termed the **dry adiabatic lapse rate;** it has a value of about 5½° F per 1000 ft of vertical rise of air. (In metric units the rate is 1 C° per 100 m.) The dew point also declines as air rises; the rate is 1 F° per 1000 ft (0.2 C° per 100 m). In Figure 22.4 the drop in dew-point temperature is labeled **dew-point lapse rate.**

When water vapor in the air is condensing, the adiabatic rate is less; a typical value is 3.2 F° per 1000 ft (0.6 C° per 100 m). The reduced rate results from the liberation of latent heat during the condensation process. The production of latent heat partially compensates for the adiabatic cooling. This modified rate is referred to as the **wet** (or **saturation**) **adiabatic lapse rate** (Figure 22.4).

Don't confuse the adiabatic lapse rate with the environmental temperature lapse rate, explained in Chapter 19. The environmental lapse rate applies to still air whose temperature is measured at successively higher levels by a thermometer carried upward. As Figure 22.4 shows, the rise of air according to the dry adiabatic rate falls within the **dry stage.** When the converging dew-point lapse rate and dry adiabatic lapse rate meet, the air is saturated and the **level of condensation** is reached. With further rise, condensation sets in and a cloud forms.

CLOUDS

Clouds are dense concentrations of suspended minute water droplets or ice crystals. These particles have diameters in the range of 0.0008 to 0.0024 in. (0.02 to 0.06 mm). Under a microscope, cloud droplets look like tiny spheres. Each droplet forms by condensation of water upon a tiny nucleus. Usually the nucleus is a bit of salt which is **hygroscopic,** that is, which has an affinity for water. Common table salt is hygroscopic, unless specially treated, and turns damp and sticky in moist weather.

Salt particles are abundant in the atmosphere because the turbulent winds blowing over ocean waves lift bits of salt spray into the air. Evaporation of these spray particles leaves salt particles that travel easily into all parts of the troposphere and make excellent cores upon which cloud particles can be formed. Growth of cloud droplets begins while air is still not fully saturated, and the droplets become rapidly larger when the saturation point is reached.

Minute particles of water can remain in the liquid state at temperatures far below the normal freezing point of 32° F (0° C). Such liquid water is said to be **supercooled.** At temperatures down to about 14° F (−10° C) cloud particles are almost entirely in the liquid state. In colder air the cloud is a mixture of liquid and ice particles. Below −40° F (−40° C), the cloud consists entirely of ice crystals. Generally speaking, the highest forms of clouds—above about 4 mi (6½ km) high—are composed entirely of ice crystals, because air temperatures are very low at high altitudes.

Clouds have such a very high capacity to reflect sunlight that reflection of the entire visible spectrum occurs, accounting for their brilliant snowy appearance when lighted by the sun. Dense cloud masses appear gray or black on the underside because sunlight is unable to pass through. Thin layers transmit enough sunlight to appear gray, whereas some of the thinnest, veillike clouds seem scarcely to weaken the intensity of direct sunlight.

Apart from their major role as producers of rain, snow, sleet, and hail, clouds are excellent indicators of the general weather situation, the direction and speed of air movement, and the moisture state of the air.

CLOUD FORMS

Clouds formed into blanketlike layers are described as **stratiform.** Obviously such layers could not exist with rapid, large-scale vertical motions of the air. However, stratiform clouds may indicate that air is moving in layers, one sliding over the other, with a gradual rate of rise.

Flat-based clouds of massive globular shape, often higher than wide, are described as **cumuliform.** These forms generally indicate strong rising air currents, or updrafts, carrying moist air rapidly to higher levels and causing continued adiabatic cooling and condensation.

Clouds are named according to height and general form (stratiform or cumuliform). An international system of classification recognizes four **cloud families:** high clouds, middle clouds, low clouds, and clouds of vertical development (Figure 22.5).

The high-cloud family above 23,000 ft (7 km) includes individual types named cirrus, cirrocumulus, and cirrostratus. All are composed of ice crystals. Cirrus is a wispy, featherlike cloud, commonly forming streaks or plumes named "mares' tails" (Figure 22.6). Cirrus clouds are so thin as to make no barrier to sunlight. Streaked cirrus bands usually indicate the presence of a high-altitude jet stream, with the wind direction paralleling the long lines of the cloud.

The middle-cloud family, extending from 6500 to 23,000 ft (2 to 7 km) in height, includes two cloud types—altocumulus and altostratus. The low-cloud family, found from ground level to a height of 6500 ft (2 km) above the earth's surface includes three types: stratus, nimbostratus, and stratocumulus. Stratus is a uniform cloud sheet at low height and usually completely covers the sky. The gray undersurface is foglike in appearance. Where stratus thickens to the point that rain or snow begins to fall from it, the cloud becomes nimbostratus, the prefix nimbo meaning "rain." Nimbostratus is usually dense and dark gray, shutting out much daylight. The cloud may extend upward many thousands of feet into the middle-cloud height range. Stratocumulus consists of low individual masses of dense cloud, often in large cigar-shaped rolls, the masses forming a distinct layer with an approximately uniform base altitude.

Clouds of the fourth family, those of vertical or upright development, are all of the cumuliform type. The smallest and most pleasant are the simple cumulus of fair weather. These are snow-white cottonlike clouds, generally with rounded tops and rather flattened bases (Figure 22.7); their shaded undersides are gray. The accompanying weather is fair, with much sunshine.

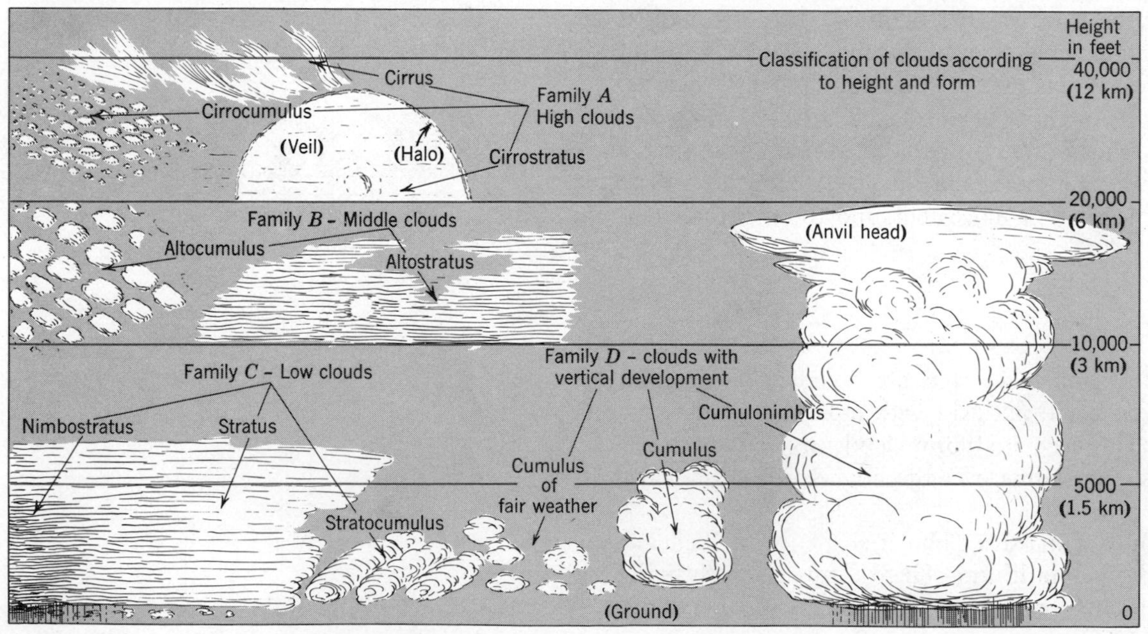

Figure 22.5 Cloud forms are classified into families based on height and vertical development.

Figure 22.6 This fibrous form of cirrus cloud is often called "mares' tails." (National Weather Service.)

Small cumulus can grow larger and denser to form congested cumulus with rounded tops resembling heads of cauliflower and flat, dark gray bases. These larger cumulus in turn sometimes grow into gigantic **cumulonimbus,** or thunderheads. From this cloud we get heavy rain, hail, wind gusts, and thunder and lightning (Figure 22.8). Cumulonimbus clouds on occasion extend upward to heights of 60,000 ft (18 km) in the tropics and can occupy low-, middle-, and high-cloud zones simultaneously.

Figure 22.7 Cumulus of fair weather. (National Weather Service.)

FOG

Fog is simply a cloud at the earth's surface. Dense fog is an indication that the air is at or close to the dew-point temperature and that sufficient moisture has condensed to produce abundant cloud droplets or ice particles. Perhaps the simplest type of fog is **radiation fog,** produced at night when a cold land surface conducts heat away from the lowest layer of the atmosphere.

On calm clear nights, because of downslope air drainage, the coldest air tends to collect in valley bottoms. It is here that the radiation fog is usually seen, often appearing as a "lake" of fog when viewed from higher ground. Radiation fogs disappear by evaporation when air temperature increases soon after sunrise, when they are said to "burn off."

Another type of fog is **advection fog.** Advection simply means horizontal transfer of air. One type of advection fog is formed where moist warm air blows over a colder surface, whether land or water. Air passing close to the surface loses heat by conduction to the colder surface beneath and its temperature is brought to the dew point. Among the most famous of advection fogs is that over the Grand Banks off Newfoundland, where warm, moist air overlying the Gulf Stream is close to the cold Labrador Current.

Figure 22.8 Cumulonimbus. This isolated thunderstorm has heavy rain falling from the central region. (U.S. Navy.)

FORMS OF PRECIPITATION

Precipitation includes all forms of water particles that fall from the atmosphere and reach the ground. Excluded from precipitation are dew and hoarfrost, which are produced when moisture condenses directly upon soil or plant surfaces.

Commonly recognized forms of precipitation are rain, snow, hail, and sleet. **Rain** consists of water droplets larger than 1/50 in. (0.5 mm) in diameter. The droplets form by rapid condensation and grow by joining with other droplets in frequent collisions. The average raindrop contains roughly one million times the quantity of water found in a single cloud particle and may grow as large as 1/5 in. (5 mm) in diameter. Above this size, the drop is unstable and will break apart as it falls. **Drizzle** is simply precipitation composed of tiny droplets, each less than 1/50 in. (0.5 mm) in diameter. Drizzle falls from low-lying nimbostratus clouds.

Snow is a form of ice in tabular or branched hexagonal (six-sided) crystals, such as those pictured on the first page of this chapter. These crystals mat together to form snowflakes. **Sleet** (American usage) consists of small grains or pellets of ice formed by the freezing of raindrops falling through a cold air layer.

Hail consists of rounded pieces of ice, often made up of concentric ice layers much like the layers of an onion (Figure 22.9). Hail is formed only in cumulonimbus clouds, when powerful updrafts within the clouds carry raindrops above the freezing level. Ice layers are formed from coats of supercooled water during repeated lifting and delayed fall of hailstones within a moist air layer where subfreezing temperatures exist.

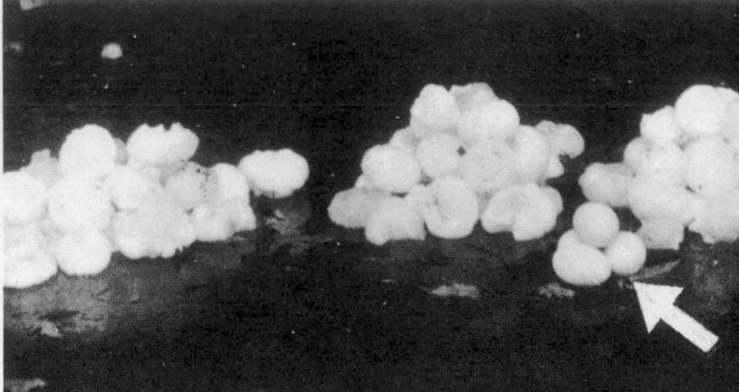

Figure 22.9 These hailstones are larger than hen's eggs, shown in the right foreground. (National Weather Service.)

Hailstorms can cause severe crop damage (Figure 22.10). They are particularly prevalent over the Great Plains region of the United States. The national crop damage by hail averages over $200 million annually in value.

Related to precipitation is the **ice storm,** or **glaze,** a coating of clear ice that forms on branches, wires, pavements, and all exposed surfaces (Figure 22.11). Glaze forms when rain falls through a cold air layer lying close to the ground. The droplets freeze as they touch exposed surfaces. Ice storms cause great damage in the form of broken wires and tree branches and make travel by foot or automobile extremely hazardous.

Quantity of precipitation is stated in terms of depth in inches or millimeters. This value represents the depth of water caught in a straight-sided, flat-bottomed pan from which there is no loss by evaporation.

In describing precipitation, not only must the depth be given, but the period of time must be given as well. The total depth in a month or year constitutes the monthly precipitation or yearly (annual) precipitation, respectively.

Figure 22.10 A severe hailstorm devastated this corn crop in the Middle West. (NCAR)

CONDITIONS THAT PRODUCE PRECIPITATION

Precipitation can occur only if large masses of moist air are cooled rapidly below the dew-point temperature. Condensation must continue until large droplets or ice particles are formed. Only through the vertical rise of large air masses can such continued cooling take place. So we must investigate the ways in which large masses of air are made to rise through several thousands of feet in altitude.

The rise of large masses of air may be either spontaneous or forced. Commonly a forced rise triggers spontaneous rise. Precipitation resulting from spontaneous rise of air is described as **convectional precipitation.** Forced ascent of air produces precipitation of two quite different types. **Orographic precipitation** is caused by the forced ascent of air in crossing a mountain barrier. **Frontal precipitation** is caused by the forced rise of air occurring when unlike air masses meet. We shall explain convectional and orographic precipitation in this chapter, saving the subject of frontal precipitation for the next chapter, in which fronts are covered in detail.

Figure 22.11 The weight of this heavy glaze of ice caused many tree limbs and power lines to break at Cherry Valley, N.Y., December 1942. (National Weather Service and N.Y. Power & Light Co., Albany.)

CONVECTION AND THUNDERSTORMS

Convection is dominantly vertical motion of air. Convectional rainfall forms from rising columns of moist, warm air. The rainfall usually takes the form of the torrential downpour of a **thunderstorm,** with its massive cumulonimbus cloud and associated lightning, thunder, and, occasionally, hail. You can think of the rising convection column as similar to the updraft in a chimney caused by the rise of the less dense heated air produced in a fireplace.

In their simplest form, small convection units are formed by uneven heating of an air layer near the ground (Figure 22.12). A mass of warmer air detaches itself from the ground layer, to rise as a bubble of air might rise in a vessel of water. Cooled adiabatically, the rising air soon arrives at the level at which the dew-point temperature is reached, and a cumulus cloud begins to form. The cloud is usually small and will dissolve after being carried some distance downwind. In other cases, the condensation triggers the growth of a shower-producing cloud, or even the great cumulonimbus storm cloud.

How can convection, once begun, intensify itself into a thunderstorm? Actually, heating of the ground layer is only one of several triggering mechanisms involved in spontaneous growth of powerful convection columns. Another source of energy needs to be called upon to explain the continued rise of air for thousands of feet. This energy source is the latent heat of condensation present in the moist air and spontaneously released. You can think of the energy as the heat of a fire obtained by release of energy stored in another form as fuel. A tiny match can start a great conflagration; similarly, a small rising air current can grow into a violent storm.

When condensation begins, the wet adiabatic rate takes over, and liberated heat is added to the rising air. Because of this added heat the rising air continues to be warmer and less dense than the surrounding air. Consequently the convection grows stronger, and the upward currents intensify. The result is a violent storm with heavy bursts of rain. Finally, at a high altitude condensation in the rising air becomes greatly reduced, and its temperature falls to the same level as the surrounding air. The rising air is now of the same density as the surrounding air. The upward movement of the convection system ceases, and the air spreads sideways.

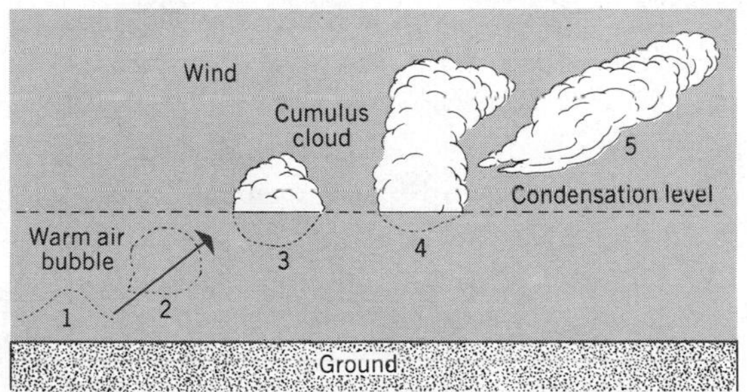

Figure 22.12 When a rising bubble of warm air penetrates the condensation level, a cumulus cloud is formed.

Thunderstorms consist of individual parts, called **cells.** Within each cell, air rises in a succession of bubblelike masses, rather than in a single, continuous column (Figure 22.13). At all levels, air is brought into the cell from the sides in the wake of the rising bubble. Vertical air speeds up to 3000 ft (900 m) per minute commonly develop. Condensation goes on rapidly. The particles so produced range from droplets of water in the lower and middle parts, through mixed rain and snow, then wet snow, to dry snow in the extremely cold upper part.

As the rising air bubble travels upward to high levels, heavy precipitation occurs. The top of the cloud, above the freezing level, spreads laterally to form an anvil top. Falling ice particles cool the cloud. These particles also serve as nuclei for condensation. Their action is described as **cloud seeding,** a process that causes rapid condensation. Falling rain drops actually drag the air downward to produce a strong downdraft of cold air which strikes the ground at the time of the heavy initial burst of rain (Figure 22.13). This gusty squall wind spreads out horizontally along the ground.

Air turbulence within a thunderstorm is violent and can destroy light aircraft that venture into the cell. Raindrops caught in the updraft may grow into hailstones.

A familiar effect of the thunderstorm is **lightning,** an electrical discharge. It is simply a great spark, or arc, from one part of a cloud to another, or from the cloud to the ground. **Thunder** is the sound produced by lightning. What we hear is the shock wave of sound sent out by the lightning stroke.

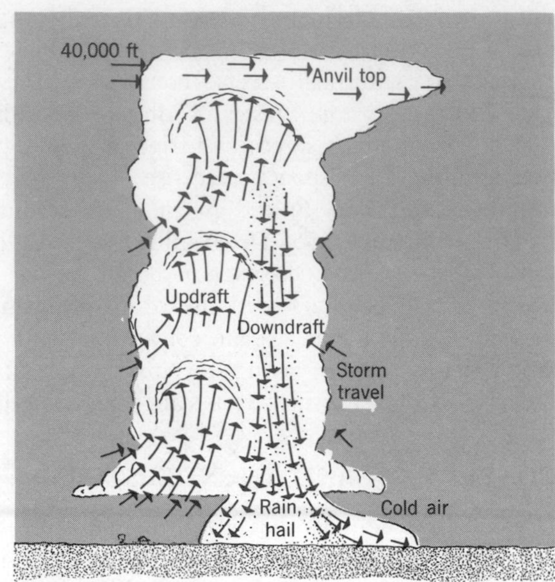

Figure 22.13 Air rises in a thunderstorm in a succession of bubbles, drawing in air from the sides. Falling rain drags the air down to produce the downdraft and squall winds. (© 1973, John Wiley & Sons, New York.)

OROGRAPHIC PRECIPITATION

The word orographic is an adjective meaning "relating to mountains." Orographic precipitation is therefore related to the existence of mountainous terrain. The principle of orographic rainfall is explained in Figure 22.14. Where prevailing winds blow from an ocean across a mountainous coast, air is forced to rise, often through many thousands of feet. If the moisture content is high, as it normally is in maritime air masses, the dew-point temperature is quickly reached. Further ascent of the air produces rain or snow, which falls on the windward slopes and crest of the range.

Orographic precipitation can take two forms: (1) a persistent rain or snow resulting from the steady lift of air, or (2) heavy convectional showers or thunderstorms set off by the forced rise.

In equatorial and tropical regions orographic rainfall is from maritime equatorial (*mE*) and maritime tropical (*mT*) air masses with large water-vapor content. This rainfall is usually convectional and brings frequent violent downpours when the rainy monsoon is on. The world record for rain probably goes to Cherrapunji, India, a station at 4300 ft (1300 m) altitude on the southern slopes of the Khasi Hills in Assam. Here, in the single month of August, 1841, 241 in. (612 cm) of rain fell. Of this, 150 in. (380 cm) fell on five consecutive days, and 40.8 in. (104 cm) in one 24-hour period!

Orographic rainfall is also heavy along the western coasts of both North America and Europe wherever mountains lie close to the sea and are exposed to prevailing westerly winds. For example, the

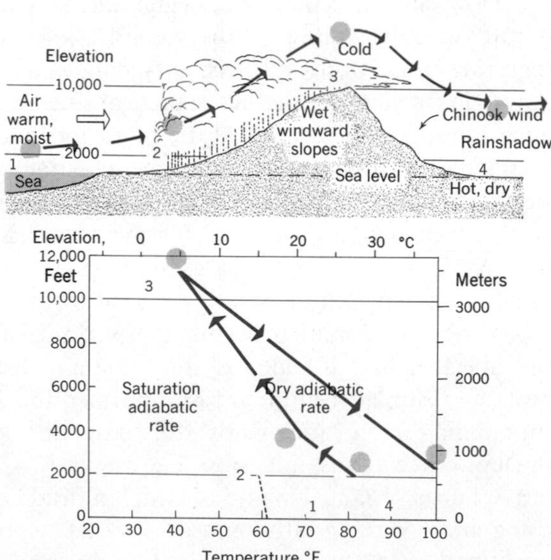

Figure 22.14 Cooling, condensation, and precipitation occur within a mass of air forced to ascend a mountain barrier.

Klamath Mountains of northern California receive over 100 in. (254 cm) of precipitation annually, much of it in the form of snow.

RAINSHADOW DESERTS

Subsiding air is warmed at the dry adiabatic rate. As Figure 22.14 shows, air descending the lee slope of a range warms rapidly, and the relative humidity also declines rapidly. The remaining water droplets of the clouds quickly evaporate, clearing the air. Since no more moisture is available, the air temperature continues to increase at the dry adiabatic lapse rate of 5½ F° per 1000 ft (1 C° per 100 m). Notice that the graph informs us that at any given level the descending air is considerably warmer than the rising air at the same level on the windward slope of the mountain range.

By the time the air has reached the valley floor at sea level on the lee side of the range, it is hot and very dry. A desert, called a **rainshadow desert,** will exist here if prevailing winds blow from the ocean toward the continent throughout the year.

ARIDITY IN HIGH PRESSURE CELLS

Another important example of dryness associated with sinking air is the high pressure cell, or anticyclone, explained in Chapter 21. Air in a high pressure center is slowly sinking and spreading outward toward areas of lower barometric pressure. Such air is warmed adiabatically and becomes drier as it descends. This process tends to produce fair skies and sunny weather.

The subtropical high pressure cells centered on tropical oceans are especially dry in their central and eastern portions, producing nearly rainless conditions that extend over large expanses of the ocean.

AIR POLLUTION AND SMOG

Over highly urbanized areas, where large amounts of fossil fuels are burned, the lower atmosphere undergoes important changes in composition and quality. **Air pollution** is the term applied to these changes, which are on the whole detrimental to the health of man and plants. Air pollution also causes corrosion, staining, and other forms of deterioration to such man-made materials as buildings, clothing, and automobiles.

Substances responsible for air pollution go by the general name of **air pollutants.** These belong to two classes. First, there are solid and liquid particles designated as a group by the term **particulates.** Most of the particles comprising smoke from fuel combustion are in this category, as are droplets of clouds and fog. Second, there are compounds in the gaseous state which are not normally found in appreciable quantities in clean air over uninhabited regions. These can be covered under the heading of **chemical pollutants.** For the most part, industrial and urban pollutants consist of carbon monoxide, ozone, sulfur dioxide, hydrocarbon compounds, and oxides of nitrogen. Chemical pollutants and particulates may be combined within a single suspended particle. Many dust particles are hygroscopic and readily absorb a film of water. Certain of the pollutant gases enter into solution in the water films.

Sulfur dioxide may combine with oxygen to produce sulfur trioxide, which in turn reacts with water of suspended droplets to produce sulfuric acid. This acid is a particularly irritating and corrosive ingredient of polluted air over industrial areas. Ozone is formed in the presence of sunlight during reactions between nitrogen oxides and organic compounds. These reactions are described as **photochemical.** Another class of pollutants is the hydrocarbons, among them ethylene, a compound highly toxic to plants.

The entire contaminant mixture in its dense state is well known by the name of **smog** and is familiar to every city dweller. Its appearance and effects upon the eyes and respiratory system scarcely need to be mentioned. Lesser concentrations of particulates and pollutants produce **haze.** Keep in mind that atmospheric haze can also develop through the natural lifting of mineral dusts, crystals of sea salts, and hydrocarbon compounds exuded by plants, in slow-moving air masses of the troposphere.

Major air pollutants and their sources are shown in Figure 22.15. Practically all of the carbon monoxide, two-thirds of the hydrocarbons, and one-half of the nitrogen oxides are contributed by exhausts of gasoline and diesel-powered vehicles. Industrial processes and the generating of electricity furnish most of the sulfur oxides through the combustion of coal and lower-grade fuel oil, comparatively rich in sulfur. These sources also supply most of the particulates. Heating of homes and commercial buildings is a relatively small producer of contaminants, because the fuel is largely natural gas or oil of higher grade and is efficiently burned. Refuse disposal appears as a minor contributor in all categories.

The smog of cities, in addition to the compositions mentioned above, contains special contributions from automobile and truck engines and exhausts. These include solid particles containing lead, chlorine, bromine, and carbon. Another contribution by the automobile is finely divided rubber worn from tires.

The lifting of particulates and chemical pollutants into the lower troposphere takes place by rising air currents. The reverse process of settling out under gravity is **fallout.** Only the larger solid

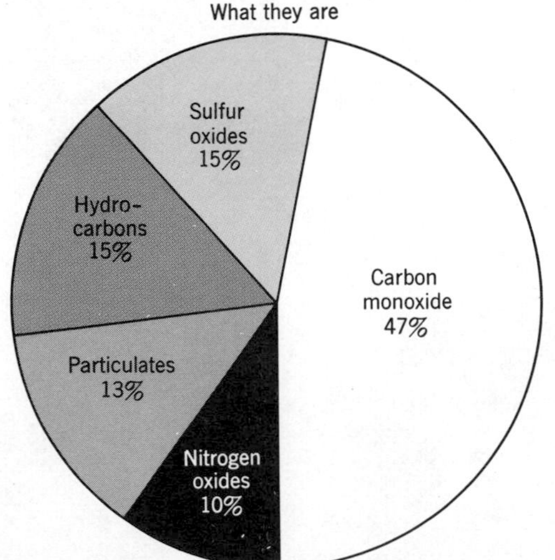

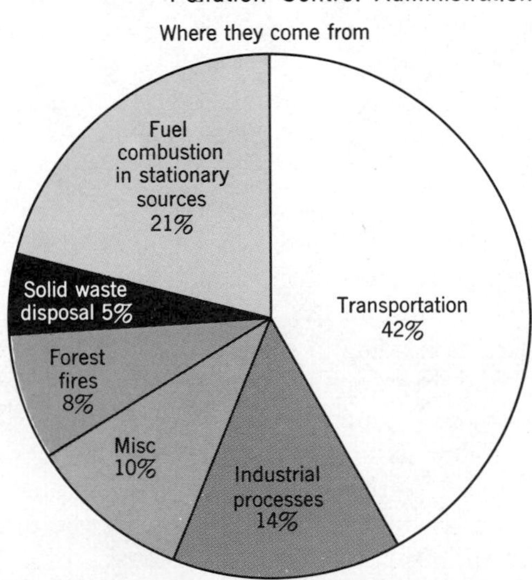

Figure 22.15 Air pollution emissions in the United States, 1968. Percentage by weight. (National Air Pollution Control Administration, HEW)

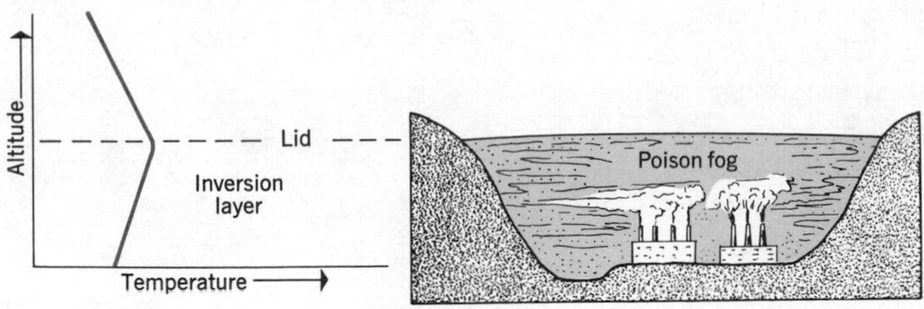

Figure 22.16 A low-level temperature inversion was the predisposing condition for poison fog accumulated at Donora, Pennsylvania, in October 1948.

particles engage in fallout. No city dweller needs to be told what fallout his cleanly washed automobile acquires when left parked outdoors for a few hours. **Washout** by precipitation sweeps down particles too small to fall readily by gravity. Fallout and washout tend to rid the atmosphere of its contaminants, so that there is achieved in the long run a balance between recharge and elimination of air pollutants. Pollutants are also removed from urban areas of concentration by strong, turbulent winds; these mix the contaminated air within a large air mass.

INVERSIONS AND SMOG

Air pollution is strongly favored by temperature inversions. The low-level inversion occurring at night was explained in Chapter 20 and illustrated in Figure 20.16. Such inversions occur when air masses remain stagnant and winds are very light or a calm prevails. The low-level inversion is developed up to a height of perhaps 500 to 1000 ft (150 to 300 m) above the ground and forms the layer of contamination. An **inversion lid** coincides with the level at which a normal lapse rate sets in (Figure 22.16).

A particularly good example of a low-level inversion is that which developed at Donora, Pennsylvania, in late October 1948. As shown on the right side of Figure 22.16, the city lies in a river valley hemmed in by rising valley walls several hundred feet high. Pollution of air trapped in the inversion layer increased for five days. The high concentration of trapped industrial smoke and gases, together with high humidity, resulted in a poisonous fog that caused illness to several thousands of people and death to 20.

AIR POLLUTION AND URBAN CLIMATE

Man-made air pollution has an important atmospheric effect upon visibility and illumination. Although the sun may be shining through a cloudless sky, the smog layer cuts illumination appreciably—by some 10% in summer and 20% in winter. Ultraviolet radiation is at times almost completely prevented from reaching the ground, and this depletion may be important in allowing increased bacterial growth at ground levels.

Horizontal visibility can be reduced by city smog to only 10% to 20% of normal values for uncontaminated air. The hygroscopic suspended particles can readily acquire condensed water films to result in fogs, which are estimated to be from two to five times more frequent over cities than in the surrounding countryside.

The rise of water vapor, particulates, and chemical pollutants produced by burning of fuels within the heat island (Chapter 20) causes a marked increase in cloudiness and precipitation over a city. It is thought that certain of the pollutants rising above centers of intense industrial activity act as cloud-seeding agents, effective in increasing rapid condensation. Over the city of London thundershowers produce 30% more rainfall than over the surrounding country area. The general increase of precipitation over urban areas is estimated at about 10%.

ATMOSPHERIC MOISTURE IN REVIEW

Variations in moisture present in the lower atmosphere make the weather, so to speak. Excess moisture leads to cloud formation, and then to precipitation. Two basic principles must be applied if we are to understand clouds and precipitation: First is the principle of latent heat as a form of stored energy. Latent heat provides the energy for convectional precipitation; and we shall find in the next chapter that latent heat is the energy source for intense cyclonic storms. Second is the adiabatic principle that air is cooled by volume expansion upon rising, but warmed by contraction into smaller volume upon sinking. Without adiabatic cooling, there could be no significant production of clouds and precipitation.

Cloud forms can tell us much about weather conditions at various levels. The stratiform clouds indicate air layers gliding over one another; the cumuliform clouds show spontaneous rise of air in vertical convection columns. The concept of the air mass has been important in this chapter, but we will make more use of it in the next, as we investigate the interaction of unlike air masses along fronts and in cyclonic storms.

Another important concept in this chapter has been that precipitation requires the large-scale lifting of air masses, for otherwise the adiabatic process could not operate. While under some conditions air rises spontaneously in convection cells, much of the global precipitation depends upon forced lift, whether by mountain barriers or by interaction of two air masses. Adiabatic warming over high pressure cells produces aridity, and thus our great deserts are sustained. So far as precipitation as a natural resource is concerned, it is feast or famine for much of the globe.

We have also touched upon the role of man through his industrial processes as a modifier of climate over cities. The implications of air pollution for global climate change remain to be determined, but changes there will surely be if the rising rate of fuel combustion continues unchecked.

YOUR GEOSCIENCE VOCABULARY

solid state
liquid state
gaseous state
condensation
sublimation
evaporation
freezing
melting
sensible heat
latent heat, of vaporization, of fusion
humidity
saturated air
relative humidity
dew point temperature
specific humidity
air mass
front
source region

precipitation
adiabatic process
dry adiabatic lapse rate
dew-point lapse rate
wet (saturation) adiabatic lapse rate
dry stage
level of condensation
clouds
hygroscopic nuclei
supercooled water
stratiform clouds
cumuliform clouds
cloud families
cumulonimbus
fog
radiation fog
advection fog
rain
drizzle
snow
sleet
hail
ice storm (glaze)
convectional precipitation
orographic precipitation
frontal precipitation
thunderstorm
cell of thunderstorm
cloud seeding
lightning
thunder
rainshadow desert
air pollution
air pollutants
particulates
chemical pollutants
photochemical reactions
smog
haze
fallout
washout
inversion lid

SELF-TESTING QUESTIONS

1. Describe the three states of water and the changes of state that can occur. How does sensible heat differ from latent heat? What amount of heat is absorbed by evaporation? by melting?

2. How does relative humidity change as air temperature changes? Describe the typical daily cycle of relative humidity change. How is the dew-point temperature related to relative humidity?

3. What advantage has specific humidity over relative humidity as a means of describing the content of moisture in the air?

4. What properties serve to define a given air mass? What part does the source region play in determining air mass properties? Give examples. How are air masses classified? Name the types of air masses based on latitudinal position; based upon moisture content. Name six air masses, and give the approximate temperature, moisture content, and source region of each.

5. Explain how the adiabatic process works. Why is the wet adiabatic rate different from the dry adiabatic rate?

6. Of what form of water are cloud particles composed? How does air temperature determine the makeup of a cloud? Name the four cloud families. What significance have stratiform and cumuliform clouds in terms of weather?

7. How is fog formed? Explain how two important types of fog are produced. Where and when does each type occur most frequently?

8. Name the forms of precipitation. What are the basic conditions necessary for precipitation to occur? How does spontaneous rise of air occur? How does forced rise of air occur?

9. Describe the development of convectional rainfall. What supplies the energy for a thunderstorm? Describe the internal structure and activity within a thunderstorm. How is hail formed? What causes lightning?

10. Explain the production of orographic rainfall as a moist air mass rises over the windward slope of a mountain range. What changes occur as the air descends the lee slope? Why are high pressure cells normally dry?

11. What kinds of pollutants are found in smog? What is the major source of each type of pollutant? Explain how temperature inversions are conducive to buildup of smog. In what ways is the climate of a city changed by fuel combustion and the rise of pollutants?

(*Above*) Survivors of the cyclone which struck the coast of Bangladesh in 1970 file past the body of a drowning victim. The great storm surge which accompanied the storm inundated these flat paddy fields under nearly 20 feet of water in a few minutes, leaving no escape for man or beast. (United Press International.) (*Below*) A West Indian hurricane. (National Geographic Magazine, 1896.)

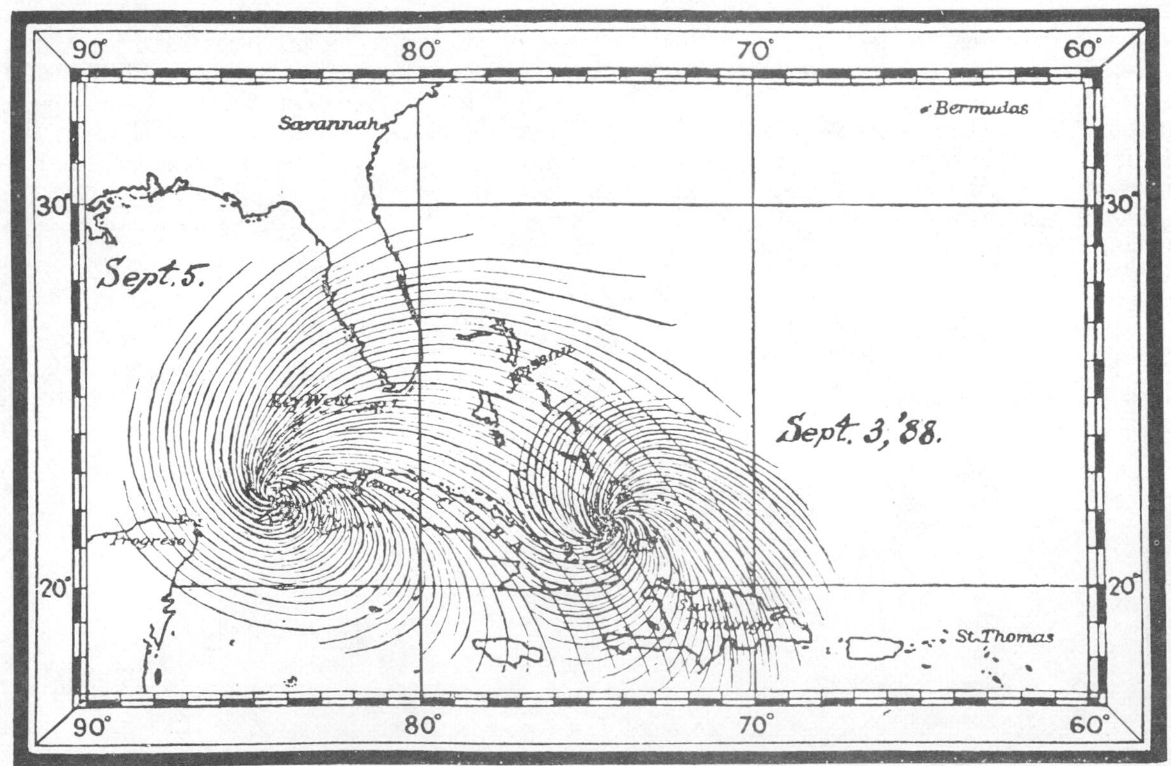

23 WEATHER FRONTS AND CYCLONIC STORMS

THE DEADLIEST STORM IN HISTORY?

What may have been the greatest killer storm in history rode north out of the Bay of Bengal on November 12, 1970. It was aimed at the most vulnerable coastline in the world. Millions of persons inhabit the low-lying deltaic coast of what was then East Pakistan and the adjoining Bengal Province of India. Here the distributaries of the Ganges and Brahmaputra rivers branch repeatedly, dividing the coastal fringe into numerous low islands. Fishing and rice cultivation support a population as dense as in any agricultural region on earth, as well as supplying food to millions dwelling on higher ground farther inland.

The record of deaths in this portion of East Pakistan (now Bangladesh) from tropical cyclones is almost unbelievable: 1822, 40,000; 1876, 100,000; 1897, 175,000; 1970, 300,000. Most deaths were by drowning, as the ocean water moved landward in a great wave, called a storm surge. This kind of wave can cause the water to rise 20 feet or more within minutes. Few places on the coastal islands are as high as 20 feet, and there is little refuge save in the upper branches of some larger trees.

A district official living on one of the islands described his experience on the night of November 12, 1970:

> At midnight we heard a great roar growing louder from the southeast. I looked out. It was pitch black, but in the distance I could see a glow. The glow got nearer and bigger and then I realized it was the crest of a huge wave. I was lucky because I live in a solidly built house and we went upstairs. But thousands were just swept away. The wave came as high as the first floor of my house. We were not poor people on this island but prosperous fishermen and paddy (rice) farmers. Now we are all street beggars. Everything has gone. All the cattle are dead, all the sheep and goats and most of the buffaloes. All the fishing vessels have been lost and all the nets. We are shy to beg from you, but please, I do beg you to get help for us. We have no drinking water—that we need above all. But we must have vaccines and other medicines too, and we need food.*

On the basis of later fact-finding investigations, the toll was assessed as follows: There were 200,000 confirmed burials; the missing were estimated as between 50,000 and 100,000 persons. Crop losses were valued at $63 million. Nearly 300,000 cattle perished, and 400,000 homes were destroyed or damaged. Of the offshore fishing fleet, 9000 boats were destroyed; of the inland-water fleet, 90,000 boats. Because most of the protein in the diet of the 73 million persons of East Pakistan came from fish, this loss of 65% of the total fishing capacity pointed to a disastrous famine.

Storm surges accompanying hurricanes (our American word for tropical cyclones) are also well known along the Gulf coast of Louisiana and Texas. The Galveston, Texas, disaster of September 8, 1900, was undoubtedly the worst of its kind on record. Here the rapid rise of water conspired with the wind. About 3000 houses were destroyed by the force of the wind, hurling the people inside them into water 10 to 15 feet deep. Of the 6000 who died that night, most succumbed by drowning, but many were killed by flying planks and timbers as they clung to floating masses of wreckage.

The delta plain of Louisiana has had a history of hurricane disasters paralleling that of the Bay of Bengal in storm intensity, but with far fewer casualties. Audrey, a hurricane of June 1957, was a notable storm in this area. She came ashore about halfway between New Orleans and Galveston. The number of dead in Louisiana was estimated at about 550 persons; the damage in that state alone was valued at $120 million.

In this chapter we will investigate the inner workings of severe storms, as well as weather disturbances of lesser magnitude.

* Quoted from a dispatch of the *Daily Telegraph,* London, as reprinted in *The New York Times,* November 17, 1970.

WEATHER FORECASTS

With the introduction of the telegraph, which made possible instantaneous communication between cities, it first became possible for meteorologists to prepare current weather maps, showing the surface distribution of pressure, temperature, winds, and weather. Such weather maps are called **synoptic maps,** because they give a synopsis, or general overview, of weather elements at a given moment over a large area.

The first American synoptic maps and forecasts were prepared in 1857 by Joseph Henry of the Smithsonian Institution in Washington, D.C. Systematic weather reporting was later undertaken by the U.S. Signal Corps and led to organization in 1890 of the U.S. Weather Bureau, today named the National Weather Service.

The examination of maps drawn at 12- or 24-hour intervals demonstrated that bad weather—involving cloudiness, precipitation, and strong winds—was associated with centers of low barometric pressure, or cyclones, which move cross-country, generally from west to east. Most cyclones are not dangerous, and few would be called severe storms by the average person. Clear skies and fair weather were seen to be associated with traveling centers of high barometric pressure, or anticyclones.

By anticipating the speed and direction of movement of lows and highs, meteorologists developed the principles of weather forecasting. Initially, forecasting was a mixture of science and art, based on correlations of weather and pressure patterns. At the time of World War I, rapid advances in weather theory were made as a result of much new information on conditions in the upper air provided by the first aircraft. Not until World War II, however, did the nature of large-scale weather systems become really apparent. At that time, military aircraft operations were extended into the equatorial, tropical, and arctic regions on a vast scale. A great network of weather-observing stations accompanied the construction of military bases and airfields. Synoptic charts of the entire Northern Hemisphere became possible, not only for surface weather conditions, but for upper levels as well.

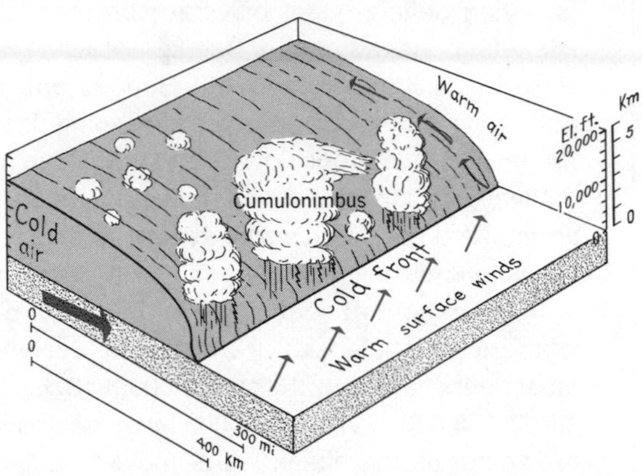

Figure 23.1 Along a fast-moving cold front, warm air is forced to rise off the ground. Violent weather often results. (© 1973, John Wiley & Sons, New York.)

WEATHER FRONTS

In Chapter 22, when describing air masses, we noted that adjacent air masses of unlike properties lie in contact along fronts. To understand weather phenomena of middle and high latitudes we must go into further detail about the various kinds of fronts.

A weather front is the boundary separating two air masses of unlike properties. Air masses do not mix easily. Instead, because of their different properties, unlike air masses tend to have distinct boundaries between them, just as oil and water tend to remain in separate layers or drops without mixing.

In most weather fronts one mass of air is invading a region occupied by an unlike air mass. Not only is the air on both sides of the front in motion, but also the front itself is moving over the earth's surface beneath it. The primary rule of weather fronts is that the air of the colder mass, being the denser, stays close to the ground, forcing the warm (or less cold) air to slide over it and rise upward.

In other words, the interaction of fronts leads to forced rise of air masses.

Putting these principles to use, consider the three basic types of fronts that may develop. The **cold front,** shown in Figure 23.1, is formed by a cold air mass invading the region occupied by a warm (or less cold) air mass. Staying close to the ground, the cold air forms a wedge, pushing the warm air upward from its advancing edge. Ground friction slows the advancing cold air close to the ground, so that it may develop a steep or blunt leading edge. The lifting of the warm air is therefore abrupt and violent on occasions.

Of course, the vertical scale in Figure 23.1 is very greatly exaggerated. Actually the cold air forms a thin wedge. For example, the cold air layer might be 1 mi thick at a distance of 40 mi back from the line where the front touches the ground. In meteorological terms, this slope is considered steep. The front itself is the entire surface of contact between the two air masses, not just the trace of this surface on the ground.

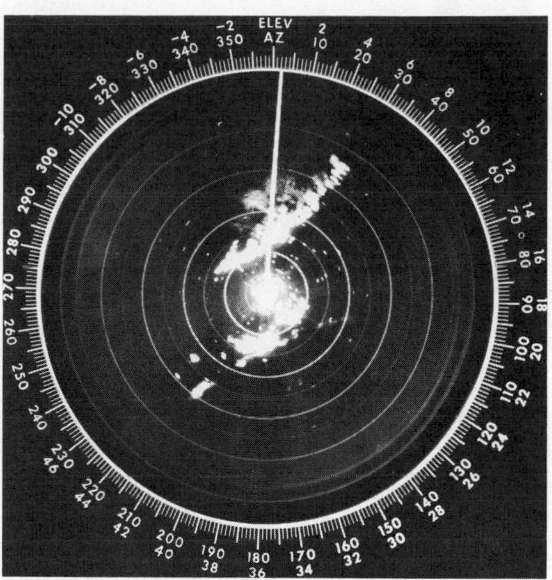

Figure 23.2 Lines of thunderstorms show as light patches on this radar screen. The heavy circles are spaced 50 nautical mi (70 km) apart. (National Weather Service.)

If the warm air is of maritime tropical (*mT*) type, as is most often the case, it may break into spontaneous convection. Warm, moist air masses are said to be **unstable,** because a small amount of forced lift sets off spontaneous rise. Convection produces dense cumulus and cumulonimbus clouds (thunderstorms) extending to extreme heights. Often a cold front produces a long line of thunderstorms 200–500 mi (300–800 km) in length. Such fronts are important to pilots, for the storms often rise too high to be surmounted and extend too low to be flown beneath. The pilot must try to pass between individual convection cells. A radar screen is most useful for this purpose, for it will show the active convection cells (Figure 23.2). Not all cold fronts are accompanied by violent weather, but cloudiness will normally be present, and the passage of the front will bring a marked drop in temperature and humidity, together with a quick shift in wind direction.

The **warm front,** shown in Figure 23.3, is formed by a relatively warm air mass moving into a region occupied by colder air. The cold air remains close to the ground, while the warm air slides up over it on a broad, gently sloping front that may have a slope 1 mi in 100 mi, or smaller. If the warm air mass is stable, as is often so, stratiform clouds mark the overriding air layer, for the forced ascent of air causes steady adiabatic cooling and condensation. The highest fringe of advancing warm air is marked by cirrus and cirrostratus clouds. As the front advances, these clouds are replaced by altostratus, then by a dense stratus, and finally by nimbostratus, with a broad zone of light, steady, and prolonged precipitation. On the other hand, if the warm air is unstable, it will break into heavy showers or into thunderstorms rising above the stratus layer.

If the cold air beneath a warm front is below freezing, rain originating in the warm air above may freeze to form ice pellets (sleet) as it falls through the cold air layer, or the drops may freeze upon contact with the ground, producing a glaze or ice storm.

In contrast to cold fronts, which are usually fast-moving and narrow, a warm front may move slowly and cover a belt 200 to 400 mi (300 to 600 km) wide. As a result, a warm front brings a long period of precipitation and cloudiness. As the warm front passes, air

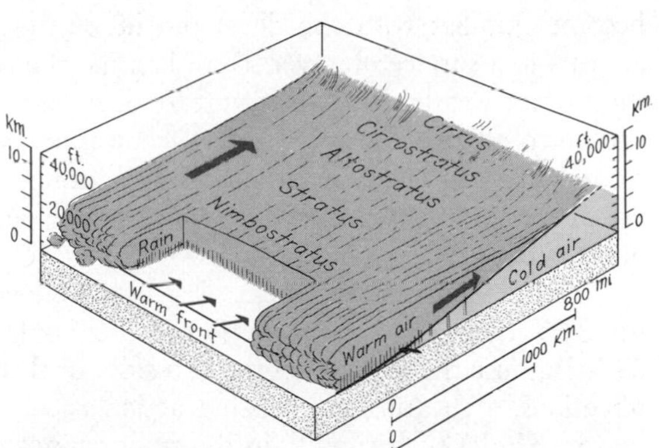

Figure 23.3 A broad band of stratiform clouds and rain lies along a gently sloping warm front. (© 1973, John Wiley & Sons, New York.)

temperatures rise gradually, and winds will normally shift to a southerly or southwesterly direction.

A third type of front is the **occluded front,** diagramed in Figure 23.4. Here a fast-moving cold front has caught up with and pushed into a slower-moving warm front, completely lifting the warm air off the ground. The invading cold air remains close to the ground and then comes in contact with the air under the warm front, which is colder than the warm air mass but less cold than the invading cold air. The warm front is now said to be "occluded," that is, cut off from contact with the ground. An occluded warm front would not show on a weather map of surface conditions, but is found at higher altitudes. The warm air mass aloft produces precipitation reaching the ground. Again, as with both cold and warm fronts, an occluded front may contain relatively stable warm air, which forms a dense stratus cloud, with steady rain or snow. In other cases the occluded air is unstable and breaks into cumulonimbus clouds as the cold air pushes under, as illustrated in Figure 23.4.

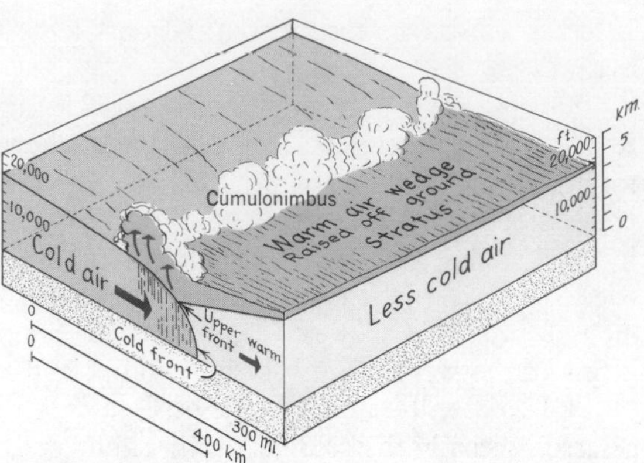

Figure 23.4 In this occluded front, a fast-moving cold front has overtaken a warm front. (© 1973, John Wiley & Sons, New York.)

Cold, warm, and occluded fronts form in the middle and high latitudes wherever unlike air masses meet in conflict. Fronts are most frequent in the region of conflict between polar and maritime air masses.

WAVE CYCLONES OF THE MIDDLE AND HIGH LATITUDES

We have defined a cyclone as a center of low barometric pressure. In the Northern Hemisphere a cyclone has a counterclockwise flow of winds about its center. Near the ground the air spirals inward toward the center of the cyclone (see Figure 21.10).

Cyclones occur at all latitudes and vary greatly in size and intensity. Some cyclones of the tropics are of a violent type, such as the hurricane or typhoon. A special type of cyclone is the tornado, a very small but intense storm restricted in occurrence to a few special locations. Our first concern here is with cyclones in the middle and high latitudes. They are produced by wavelike kinks developing on the fronts separating cold from warm air masses. For this reason such disturbances are called **wave cyclones.**

We will first examine surface weather conditions, in order to become familiar with the development of the wave cyclone as it appears to a surface observer. Examine first the conditions shown on the surface weather map in Figure 23.5. A mass of cold polar air on the north adjoins a mass of warm tropical air on the south, with a front lying between. The front lies in a trough of low pressure between the two highs. Air flow is approximately opposite on either side of the front.

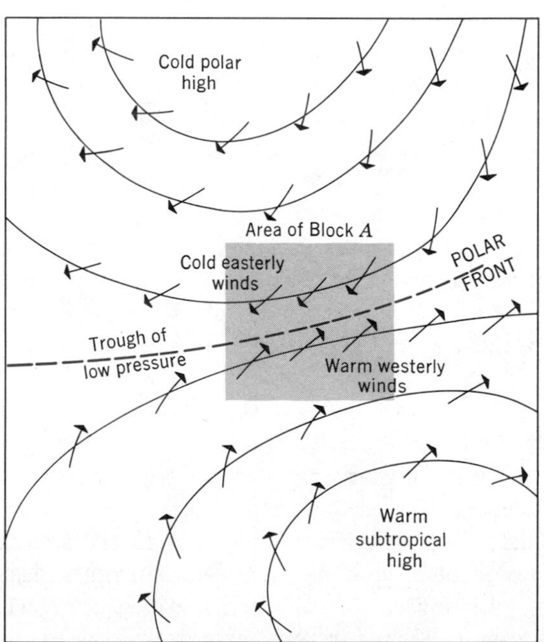

Figure 23.5 Surface weather conditions typically associated with the formation of a wave cyclone. (© 1973, John Wiley & Sons, New York.)

Stages in the development of a wave cyclone are shown in Figure 23.6. Block A shows a portion of the front described in Figure 23.5. Because the air flow on the two sides of this front is in opposite directions, a shearing or dragging action is set up between the air masses. Under such conditions the front cannot remain smooth, but will tend to develop a bend or kink, such as that shown beginning to form in block A.

As the frontal wave in block A develops, the flow of air is modified so that cold air begins to push into the region of the warm, and the warm begins to move into the region occupied by the cold. This motion resembles a revolving door, with one person going out of a building while another enters.

Block *B* shows the open stage of the wave cyclone. The frontal wave is well developed and forms a sharp point toward the region of cold air. To the left of this point, or crest, lies a cold front with cold air actively pushing south. To the right lies a warm front curved in a great arc bowed to the east.

Block *C* of Figure 23.6 shows that the wave has now steepened to the point that the crest has been cut off. Here the cold front has caught up with the warm front to produce an occluded front separating a layer of warm air from contact with the ground. The cyclone has reached the occluded stage.

After the occluded stage the cyclone enters the dissipating stage, shown in block *D*. The cyclone is no longer supplied with moist, warm air. Lacking condensation to supply latent heat energy, the cyclone dies out. Now the front is re-formed as a smooth line between polar and tropical air masses.

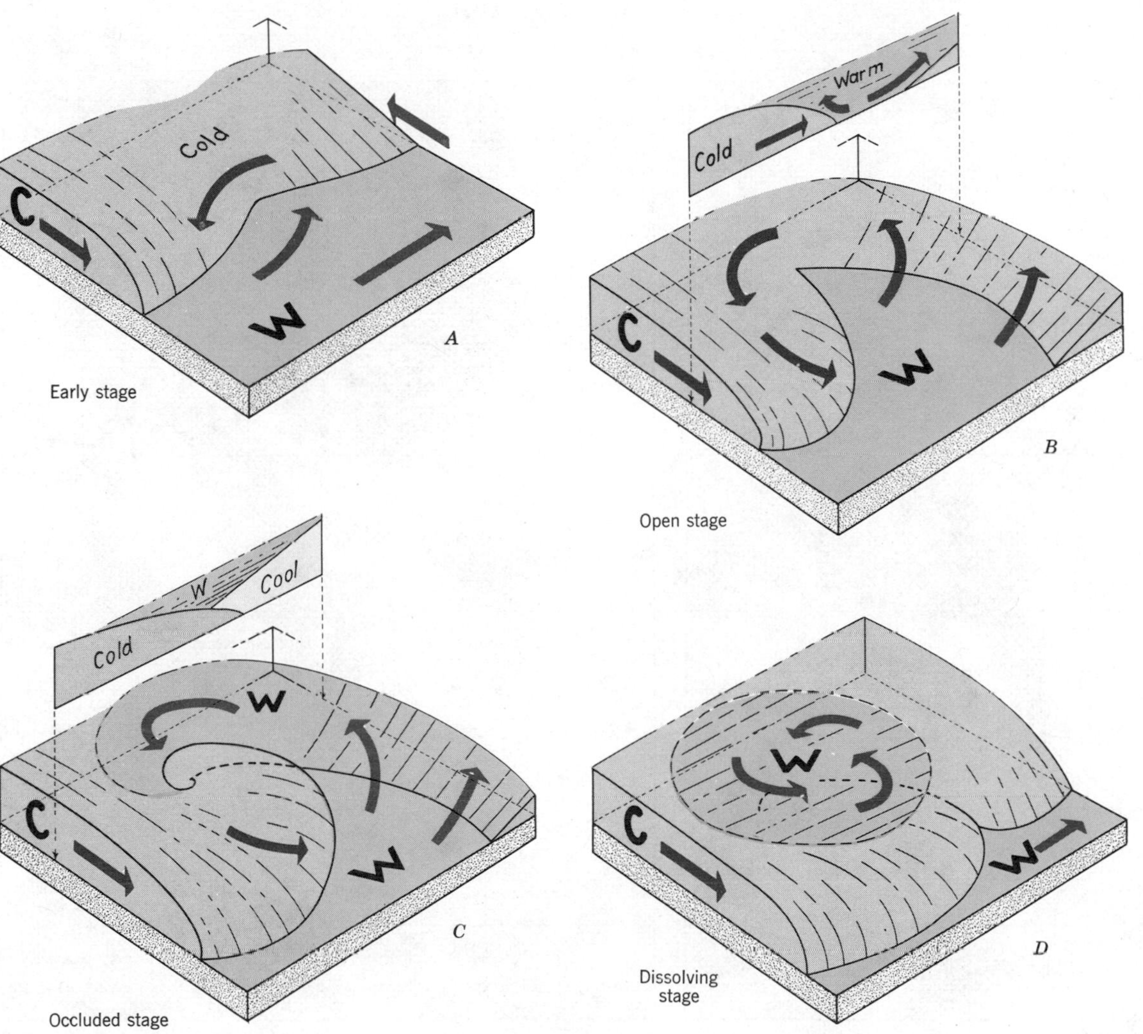

Figure 23.6 Four stages in the life history of a wave cyclone of middle latitudes. (© 1973, John Wiley & Sons, New York.)

WAVE CYCLONES ON THE DAILY WEATHER MAP

The anatomy of a wave cyclone is best studied by means of the surface weather map. Figure 23.7 is a pair of such maps, showing conditions on two successive days in the month of April. Data of each observing station are represented by station symbols. Fronts are shown by standard conventions, as labeled on the map. A dashed line crosses each map; it is an **isotherm,** or line of equal air temperature. In this case the isotherm is for the freezing point, 32° F (0° C).

Map A shows a cyclone in the open stage, approximately equivalent to block *B* of Figure 23.6. The storm is centered over western Illinois and is moving northeastward. Notice the following points: (*a*) Isobars of the low are closed to form an oval-shaped pattern; (*b*) isobars make a sharp V where crossing the cold front; (*c*) wind directions, indicated by arrows, are at an angle to the trend of the isobars and form a pattern of counterclockwise inspiraling; (*d*) in the warm-air sector, there is northward flow of tropical air toward the direction of the warm front; (*e*) there is a sudden shift of wind direction accompanying the passage of the cold front, as indicated by the widely different wind directions at stations close to the cold front, but on opposite sides; (*f*) there is a severe drop in temperature accompanying the passage of the cold front, as shown by differences

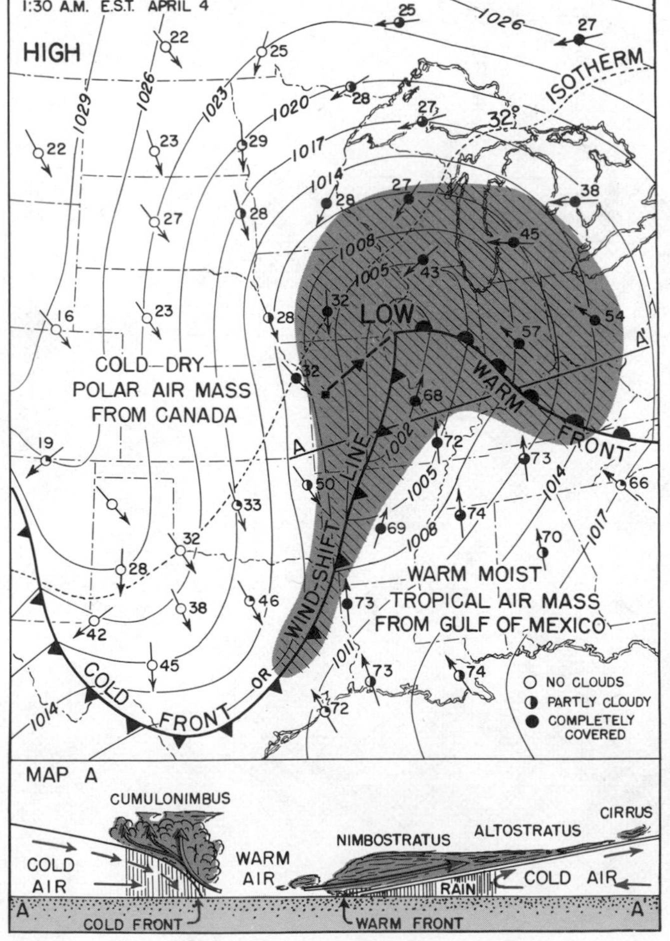

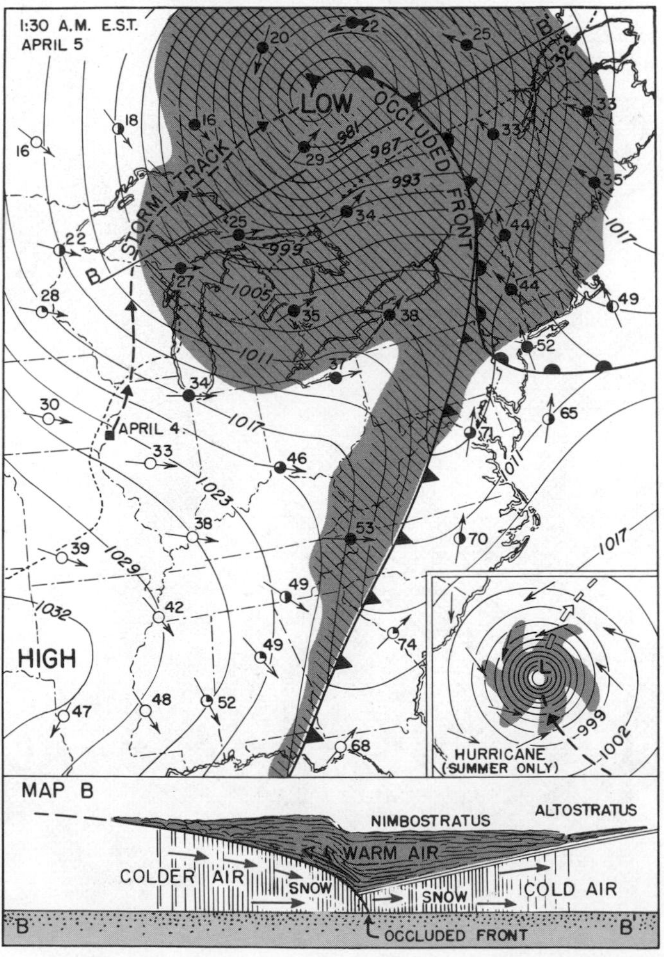

Figure 23.7 Surface-weather maps showing two stages in the development of a cyclonic storm. Isobars are labeled in millibars. The figure beside each station gives air temperature in degrees Fahrenheit. Arrows fly with the wind. Diagonal shading shows areas experiencing precipitation. These maps are modified from daily weather maps of the National Weather Service. (From *Climates of the World,* by A. K. Lobeck, reproduced by permission of the publisher, Hammond Incorporated.)

in temperature readings at stations on either side of the cold front; (*g*) precipitation, shown by shading, is occurring over a broad zone near the warm front and in the central area of the cyclone, but extends as a thin band down the length of the cold front; (*h*) cloudiness, shown by degree of blackness of station circles, is greatest in the warm sector and northeastern part of the cyclone, but the western part is mostly clear; (*i*) the low is followed on the west by a high (anticyclone) in which low temperatures and clear skies prevail; (*j*) the 32° F (0° C) isotherm crosses the cyclone diagonally from northeast to southwest, showing that the southeastern part is warmer than the northwestern part.

A cross section below map A, made along the line *AA′*, shows how the fronts and clouds are related. Along the warm front is a broad area of stratiform clouds. These take the form of a wedge with a thin leading edge of cirrus. Westward this thickens to altostratus, then to stratus, and finally to nimbostratus with steady rain. Within the warm air mass sector, the sky may partially clear, with scattered cumulus. Along the cold front are violent thunderstorms with heavy rains, but this is along a narrow belt that passes quickly.

The second weather map, map *B*, shows conditions 24 hours later. The cyclone has moved rapidly northeastward into Canada, its path shown by the line labeled **storm track.** The cyclone has occluded. An occluded front replaces the separate warm and cold fronts in the central part of the disturbance. The high pressure area, or tongue of cold polar air, has moved in to the west and south of the cyclone, and the cold front is passing over the eastern states. Within the anticyclone, the skies are clear and winds are weak. In another day the entire storm will have passed out to sea, leaving the eastern United States with cold but clear weather. A cross section below the map shows conditions along the line *BB′*, cutting through the occluded part of the storm. Observe that the warm air mass is being lifted higher off the ground and is giving heavy precipitation.

CYCLONE TRACKS AND FAMILIES

Cyclones travel in a generally easterly direction in the Northern Hemisphere, at speeds ranging from 20 to 40 mi (32 to 64 km) per hour. Some travel a distance of one-third to one-half of the way around the earth during their life cycle. The intensity of these cyclones is extremely varied. Considering that a cyclone is experienced every few days throughout the year by persons living in North America and Europe, it is obvious that most cyclones pass almost unnoticed as spells of cloudy or rainy weather.

On the other hand, a large, intense wave cyclone can be a powerful and devastating storm. The storm can have winds up to 70 mi (112 km) or more per hour and can bring flooding rains or deep snows. At sea, over the North Atlantic and Pacific Oceans and over the Southern Ocean in lat. 40° to 70°, cyclonic storms tend to intensify. These deep cyclones become extremely severe in the winter season, causing high seas and great peril to shipping. A cold air mass pushing with great strength into an occluded storm can build up mountainous seas which lash the western coasts of North America and Europe with great fury.

Long observation of the tracks of wave cyclones established that they tend to follow certain paths with greater frequency than others. Figure 23.8 shows the principal tracks of cyclones of the middle latitudes. In the Southern Hemisphere the storms are rather uniformly distributed around the Southern Ocean. In the Northern Hemisphere there is a definite concentration into two characteristic paths. One originates in easternmost Asia and the Japanese Islands and extends across the North Pacific to Alaska and the northwestern coast of North America. Another originates in eastern North America and crosses the North Atlantic to northern Europe. Many a storm begins in the southern plains of the United States, travels northeast, deepens and becomes occluded as it passes out over the Atlantic Ocean, and arrives in the British Isles in a late stage of development.

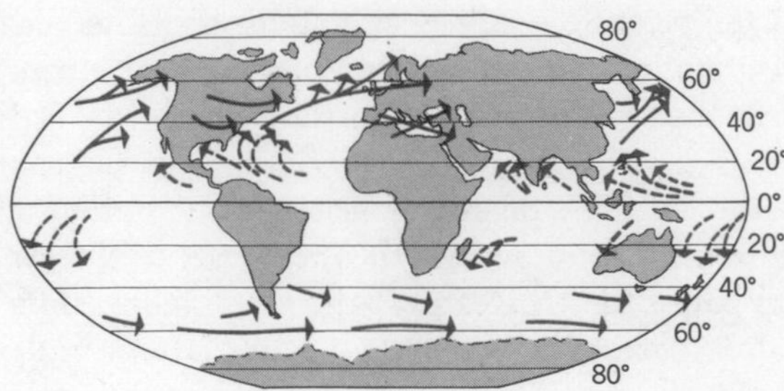

Figure 23.8 Solid lines on this world map show principal tracks of middle-latitude cyclones. Dashed lines show the characteristic tracks of tropical cyclones. (© 1960, John Wiley & Sons, New York.)

Along the prevailing tracks wave cyclones tend to form in succession as **cyclone families.** Within a single family, the stage of development ranges from open wave to occlusion, from southwest to northeast. Figure 23.9 is a weather map of the entire world, showing a number of wave cyclones in both the Northern and Southern hemispheres. Those of the Southern Hemisphere bear a mirror-image relation to those of the Northern Hemisphere, because all travel from west to east. Notice the cyclone families located over the North Pacific, North Atlantic, and western Asia.

Cyclones are closely tied in with the upper-air, or Rossby, waves described in Chapter 21. Cyclones typically form near the southerly base of a well-developed wave and are dragged along beneath the jet stream. This is the reason that accurate forecasting of the direction of storm travel depends upon a knowledge of air flow at high levels.

TORNADOES

The **tornado** is a small but very intense wind vortex. It extends down from a cumulonimbus cloud, taking the form of a tapering funnel (Figure 23.10). Although only a few hundred feet in diameter, the funnel cloud contains winds with speeds up to 250 mi (400 km) per

Figure 23.9 On a typical day in July or August, a surface-weather map of the entire world might have weather systems such as those shown here. (© 1960, John Wiley & Sons, New York.)

hour. Within the center of the funnel cloud is a vortex in which air pressure is but a fraction of normal pressure.

Tornadoes commonly move northeastward cross-country at 25–40 mi (40–65 km) per hour, following the general motion of thunderstorms and associated cold fronts. Where the funnel cloud touches ground, there is complete destruction along a swath about 1000 ft (300 m) wide and often many miles long. Not only does the great wind speed prove irresistible, but the sudden lowering of air pressure as the vortex passes may cause buildings to explode from the expansion of air entrapped within (Figure 23.11).

Tornadoes are the small eddies of intense turbulence generated by the mixing of dry polar air masses (*mP*) with unstable moist maritime tropical air (*mT*). The region most favorable for producing tornadoes is the Great Plains region and the Mississippi Valley. The states of Kansas, Oklahoma, and Texas have the highest frequency of tornadoes; Iowa, Missouri, and Arkansas rank next.

Tornadoes are most likely to occur in months in which the greatest contrast exists between polar and tropical air masses. These months are from early spring to late summer, with May leading, followed by June and April. Tornadoes are typically an American weather phenomenon. Although reported in other parts of the world, they are nowhere as frequent as in the United States.

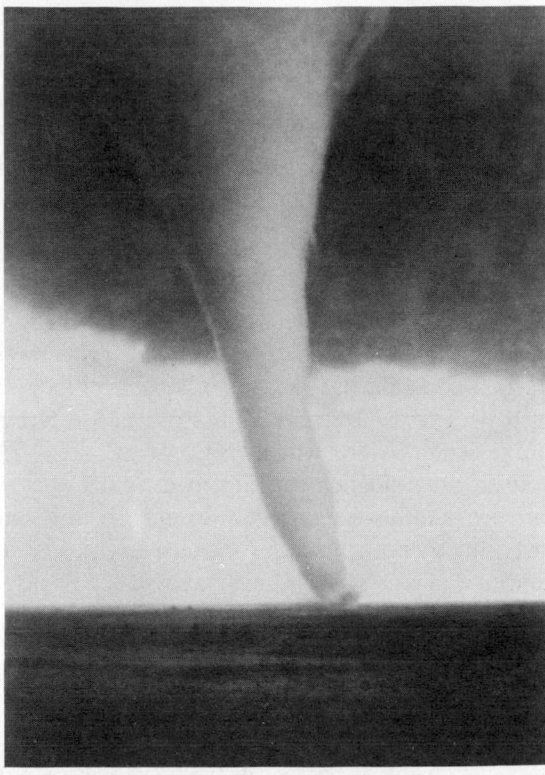

Figure 23.10 A tornado funnel cloud seen at Hardtner, Kansas. (National Weather Service.)

WEATHER DISTURBANCES OF LOW LATITUDES

In contrast to our detailed weather knowledge of the middle latitudes of the Northern Hemisphere, knowledge of weather systems in the low latitudes is not yet complete. Weather processes of the low latitudes are dominantly those of vertical rising and sinking air motions, or convection. The Coriolis force is weak and air masses stagnate for long periods of time over warm oceans, picking up large quantities of water vapor. Much of the rainfall in the trade-wind belt of low latitudes occurs in weak low pressure systems that travel slowly eastward across the oceans. Moist air masses converge into these depressions and numerous convective showers and thunderstorms break out.

Figure 23.11 Two persons died in the wreckage of this store at Ionia, Iowa, struck by a tornado. Under the vacuumlike effect in the tornado vortex, the building roof was lifted and the walls fell outward. The roof then fell back on top of the wreckage, trapping the people inside. (*Des Moines Register* and National Weather Service.)

TROPICAL CYCLONES

The otherwise peaceful belt of tropical easterlies breeds the most violent of all large cyclonic storms, the **tropical cyclone.** It is a nearly circular storm with extremely low pressure at the center, accompanied by high winds, dense clouds, and heavy precipitation (Figure 23.12). The names **hurricane** (West Indies) and **typhoon** (western Pacific) are other names for severe tropical cyclones.

Figure 23.12 Hurricane Gladys, located about 150 mi (240 km) southwest of Tampa, Florida, on October 8, 1968. The powerful storm was photographed from *Apollo 7* spacecraft at an altitude of about 110 mi (180 km). (NASA)

Tropical cyclones originate only over oceans. At first, a weak center of low pressure forms. If conditions are favorable, the low deepens rapidly, with the isobars taking on the form of nearly concentric circles. Not all such tropical cyclones deepen into severe storms—some die out quickly, others travel long distances as mild disturbances. However, if the pressure becomes extremely low in the storm center, winds will increase to speeds of 75 mi (120 km) per hour or much higher, accompanied by dense clouds, extreme air turbulence, and heavy rain. The storm is then designated as a hurricane or typhoon and is a serious menace to ships at sea and to islands or continental coasts over which it may pass. The storm normally travels westward or northwestward at a rate of 6–12 mi (10–20 km) per hour.

Figure 23.7*B* shows a surface-weather chart of a hurricane of the West Indies. The pressure in the storm center is often as low as 28 in. (72 cm) of mercury, a low value rarely found in even the deepest of middle-latitude cyclones. Surface winds blow in toward the center with a counterclockwise spiral (Northern Hemisphere). Wind speeds are commonly from 75 to 125 mi (120 to 200 km) per hour, but much higher gusts have been reported. These extreme winds may affect a circle of radius as great as 200 mi (300 km) for a very large storm.

Figure 23.13 is a schematic cross section through a tropical cyclone, showing cloud formations and rain bands. Upon reaching high levels, air flows outward, producing a cirrus cloud cap. The tropical cyclone has a **central eye,** a strange hollow vortex several miles wide surrounded by a dense cloud wall. In the eye, the air is almost calm, and the sky may clear.

Severe tropical cyclones occur in all oceans of the world except the South Atlantic. Most North Atlantic tropical cyclones (Figure 23.14) originate in lat. 10° to 20° N. and travel westward, following the tropical easterlies. The paths then turn more northwestward, bringing the storms into the region of westerlies. There the storms turn north, then northeast. In so doing, they broaden into middle-latitude cyclones, and are carried eastward across the ocean. Many storms stay entirely over the ocean, but others curve over the eastern margin of the adjoining continent.

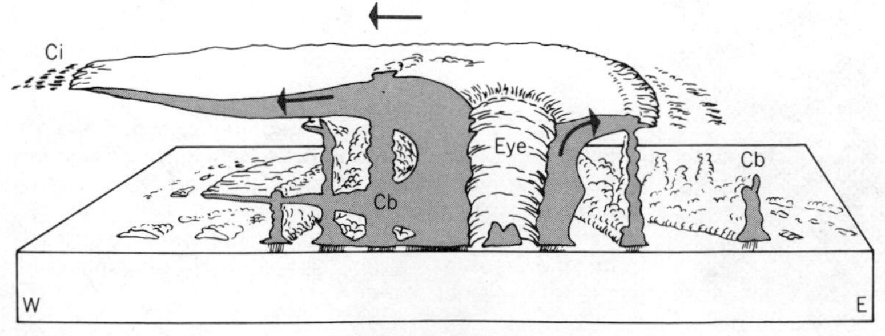

Figure 23.13 This cross section of a hurricane cuts through the eye. Cumulonimbus clouds in concentric rings rise through dense stratiform clouds in which air spirals upward. Width of the diagram represents about 600 mi (1000 km). Highest clouds are at elevations often over 30,000 ft (9 km). (After R. C. Gentry, NOAA, National Weather Service.)

Tropical cyclones occur in that part of the year including and immediately following the period of high sun, or summer season, in the hemisphere in question. For example, the hurricanes and typhoons of the Caribbean and northwestern Pacific occur from June through November; those of the western South Pacific and South Indian Oceans occur from December through March.

Tropical storms rank among the great natural catastrophes. Although of prime importance as a hazard to ships at sea, these storms do their greatest damage when passing over densely inhabited islands and coasts (Figure 23.15). Destruction of harbor facilities and small craft is especially great.

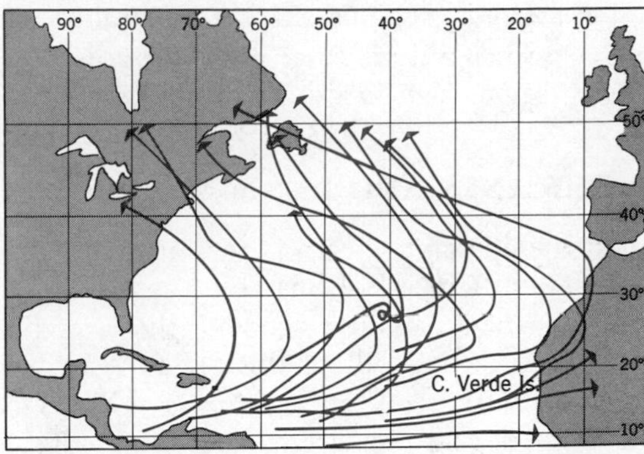

Figure 23.14 Hurricane tracks for the month of August illustrate typical storm paths in the western North Atlantic. (© 1973, John Wiley & Sons, New York. Data of U.S. Navy Oceanographic Office.)

THE STORM SURGE

Tropical cyclones approaching a continental coast are often accompanied by a sudden rise of water level amounting to many feet. This phenomenon is the **storm surge.** It has nothing to do with the seismic sea wave (or tsunami) described in Chapter 6. Instead, it is a combined effect of lowered barometric pressure and strong onshore winds.

Sea level rises in response to a fall of barometric pressure. In a severe tropical storm, the sea surface can be raised as much as 2 ft (0.6 m) by this cause alone. Strong winds also drag surface water toward the shore, causing it to pile up temporarily. The combined action of lowered barometric pressure and strong winds generates the storm surge. If a high tide occurs simultaneously, the effect is much greater.

In terms of loss of life, few natural disasters can match the destructiveness of the storm surge. The water level at a given coastal point may rise several feet in a minute or less. Extreme surges raise the water level 20 ft (6 m) or more, perhaps even as much as 40 ft (12 m), if accounts of great storms are to be believed. As described

Figure 23.15 Palm trees and storm surf along the Miami, Florida waterfront at the height of a severe hurricane. (National Weather Service.)

at the beginning of this chapter, vast numbers of humans may be drowned when a storm surge inundates a heavily populated low-lying coastal region, such as a deltaic plain.

HURRICANES AND RIVER FLOODS

Another important effect of tropical cyclones reaching land is the heavy fall of rain, which may total as much as 12 in. (30 cm) in a 24-hour period. Summer floods of the Gulf coast and eastern seaboard states are usually of this origin. A striking example was provided by hurricane Diane in August 1955. This storm took about 200 lives and did property damage amounting to $1.5 million as it swept over the eastern states. Torrential rains produced unprecedented floods on the rivers of Connecticut, Rhode Island, and Massachusetts. Similar flooding rains are brought to Japan and the coast of China by summer typhoons. An occasional tropical storm formed off the west coast of Mexico will pass inland over the Sonoran Desert region of southeastern California and southern Arizona, causing torrential rains. These bring floods to desert stream channels, washing out highways and railroad bridges.

CYCLONIC STORMS IN REVIEW

The theme of this chapter has been the stimulation of large-scale atmospheric activity through the release of latent heat, held as water vapor in moist air masses. In low latitudes, much of this activity is in the form of spontaneous rise of convection columns.

In contrast, middle and high latitudes are subjected to strong upper-air westerly winds, shaped into upper-air waves and swift jet streams. These vast horizontal air motions, aided by strong Coriolis force, favor the formation of wave cyclones, in which very unlike air masses interact and are forced to form fronts. Wave cyclones dominate the weather of these higher latitudes. Although convectional storms occur in substantial numbers in these same latitudes, they are mostly activated by swiftly moving cold fronts. A byproduct is the deadly tornado. However, the most severe threat of all major weather phenomena is the tropical cyclone, produced within the otherwise peaceful belt of tropical easterlies.

With this chapter we conclude our study of the atmosphere and oceans. Planet Earth possesses a unique combination of a thick oceanic layer and a dense atmosphere. In contrast, Mercury and Mars are almost totally devoid of both oceans and atmospheres. The same can be said of our Moon. Venus has a dense atmosphere but no ocean and practically no free water. The great outer planets have dense atmospheres, but their cold surfaces cannot maintain water in the liquid state.

Combined with the remarkable endowment of ocean and atmosphere are a rotation rate that strongly controls the circulatory systems of planet Earth. The oceanic gyres and the circulation systems of the atmosphere exist as they do because a particular rate of planetary rotation is combined with a particular degree of imbalance between incoming solar radiation in equatorial latitudes and outgoing longwave radiation in polar regions.

Our final investigation of planet Earth takes us out into the solar system. Then, in the closing chapter, we shall place our solar system in perspective within our galaxy; and that galaxy in its place among countless others.

YOUR GEOSCIENCE VOCABULARY

synoptic map
cold front
unstable air mass
warm front
occluded front
wave cyclone
isotherm
storm track
cyclone family
tornado
tropical cyclone
hurricane
typhoon
central eye
storm surge

SELF-TESTING QUESTIONS

1. Describe the three basic types of weather fronts, including slope of the front, air masses, cloud patterns, and precipitation types associated with each.

2. Describe the stages in formation of a wave cyclone of the middle latitudes. How are fronts situated and shaped within the cyclone. Where does precipitation take place? What cloud types are found in the cyclone? What changes in wind direction take place with the passage of a cold front?

3. Describe the development of wave cyclones within a cyclone family. Where can such families be found? Give the locations of the common tracks of cyclonic storms in both hemispheres.

4. Describe the tornado, giving information on its appearance, size, direction of travel, speed of travel, and wind speed. Where are tornadoes most abundant? In what months are tornadoes most frequent? Explain.

5. Describe the tropical cyclone, including its form, size, direction of travel, speed of travel, clouds, and wind speed. Where do tropical cyclones originate? In what season do tropical cyclones occur?

6. How is a storm surge formed? When and where does it occur? How do hurricanes affect the flow of rivers on continents?

The planet Venus, photographed at five different phases, appears much smaller at full phase, when it is farthest from Earth, than in crescent phase, when it is closer to Earth. (Lowell Observatory.)

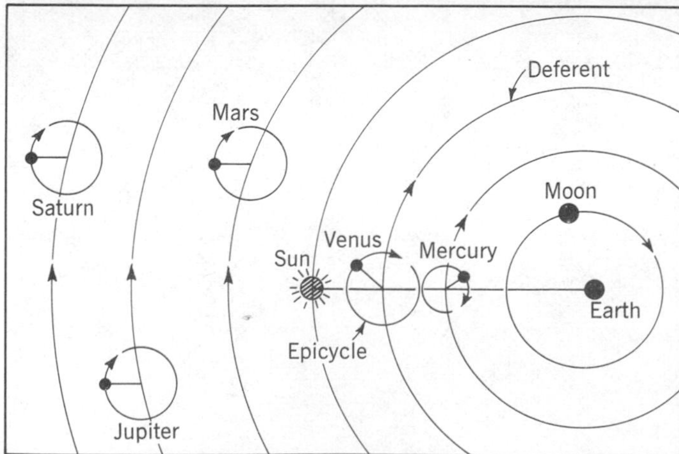

The Ptolemaic system of astronomy required small orbits, called epicycles, superposed on large orbits, called deferents. Strangely enough, in this system the Sun lies in an orbit between the planets Venus and Mars.

24 THE SOLAR SYSTEM

EARTH OR SUN AT THE CENTER?

The idea that observable members of the solar system are discrete bodies in space, moving in a systematic manner with respect to one another, was accepted by the earliest astronomers of Greece and Egypt. However, the question of which of two bodies, Earth or Sun, was the fixed object about which the others revolved was disputed for perhaps 2000 years.

The followers of Pythagoras, a Greek philosopher and mathematician who lived in the sixth century B.C., asserted as a principle of their philosophy that the Earth rotates upon its axis and revolves in an orbit around the Sun. Although they offered no scientific observations or reasoning to support their views, the Pythagoreans can be credited with first stating the **heliocentric theory** (from the Greek word *helios,* "Sun") of the solar system, that the Sun is at the center of planetary orbital motion. Not long thereafter, a Greek astronomer, Aristarchus (about 310–250 B.C.), advanced the heliocentric theory in a systematic manner based on careful astronomical observations and calculations, but this was to be the last time for many centuries that the theory was supported.

The **geocentric theory** (from the Greek word *geos,* "Earth"), places Earth at the center of the universe. This theory was put forward about the middle of the second century A.D. by Claudius Ptolemy, an astronomer and mathematician of Alexandria. Called also the **Ptolemaic system** of astronomy, the geocentric theory describes the individual apparent paths of Sun, Moon, planets, and stars as complex systems of cyclic motions relative to a stationary (nonrotating) Earth. It is possible in this way to represent the apparent motions of these bodies geometrically, as the illustration on the facing page shows. But a good deal of ingenuity is required to cope with each new motion discovered by more precise measurements. Moreover, the method fails completely to give any explanation of why the bodies move as they do.

Heliocentric theory was revived by the Polish astronomer Nikolaus Copernicus (1473–1543). He referred to Aristarchus' concepts of nearly 1800 years earlier to develop what is now generally called the **Copernican system** of the solar system. Arguing that it was a much simpler way to explain the known facts, Copernican theory placed the Sun at the center of the solar system, with the planets revolving about the Sun in a set of circular orbits. Bear in mind that the telescope was not yet invented, nor were the laws of gravitation and motion known. There was no way Copernicus could marshall absolute evidence of the Earth's motions. Copernicus' contemporaries and successors were strongly divided on the merits of a heliocentric system. The Copernican theory had some very tough sledding in store.

Great strides in establishing the heliocentric theory came through the use of the first astronomical telescope by its inventor, the Italian scientist Galileo Galilei (1564–1642). In 1610 Galileo discovered that there are moons revolving around the planet Jupiter.

Then Galileo discovered that Venus shows phases similar to those of the Moon and, moreover, changes apparent size with phase (photographs on opposite page). Because the changes of phase are clearly the different proportions which we see of a spherical surface illuminated by the Sun's rays, the only logical conclusion was that Venus revolves about the Sun.

Galileo was not free to support his own beliefs publicly, because Copernican theory was then regarded by Church authorities as a religious heresy. He was compelled to recant his views and to retire into seclusion for his remaining years.

Johannes Kepler (1571–1630) carried the baton passed to him by Galileo. Using a long series of observations made by the Danish astronomer Tycho Brahe, Kepler discovered the three fundamental laws of planetary motion which describe the Copernican theory.

In this chapter we will investigate the solor system, secure in the belief that it is a heliocentric system. How easy it is now to accept the laws of science without the slightest question!

MEMBERS OF THE SOLAR FAMILY

The **solar system** consists of the planets, their satellites, the asteroids, meteoroids, and comets. All of these objects move in the gravitational field of the Sun, the preeminent body and center of the entire system.

The term **planet** is limited in common usage to the nine largest bodies revolving about the Sun; they are also often called the **major planets.** In order of distance from the Sun they are: Mercury, Venus, Earth, Mars, Jupiter, Saturn, Uranus, Neptune, and Pluto. The asteroids, sometimes referred to as the minor planets, number in the thousands. All have diameters less than 500 mi (800 km), and most are less than a few miles in diameter. In general they are found between the orbits of Mars and Jupiter. The meteoroids are extremely minute solid particles traveling in swarms in orbits around the sun. A comet is a rather large, diffuse body of very small mass.

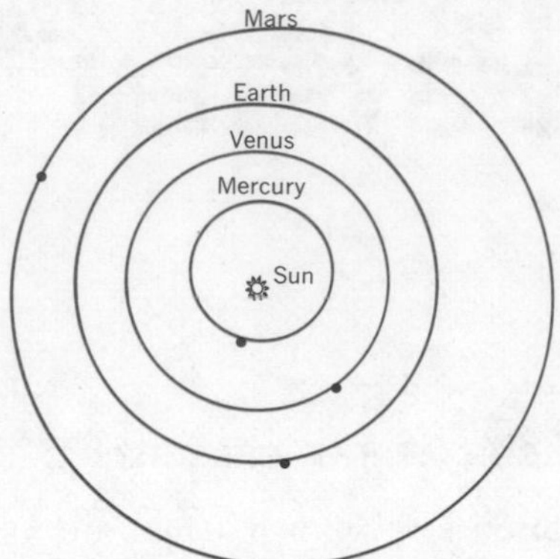

Figure 24.1 Orbits of the four planets near the Sun are shown here to scale. The black dots represent perihelion points.

INNER AND OUTER PLANETS

As we noted in Chapter 10, astronomers classify the nine major planets into two groups. The **inner planets,** also referred to as the **terrestrial planets,** are those four lying closest to the Sun: Mercury, Venus, Earth, and Mars (Figure 24.1). The four are grouped together because they are comparatively small and have orbits relatively close to one another and to the Sun.

The five **outer planets,** Jupiter, Saturn, Uranus, Neptune, and Pluto, move in orbits of vastly greater diameter than those of the four inner planets and, except for Pluto, are vastly greater in size. We can't show all nine planetary orbits to the same scale on one drawing, so the scale of Figure 24.2, in which the orbits of the outer planets are shown, is about one-twentieth that used in Figure 24.1.

In order to convey a stronger impression of the size differences among the planets and Sun, the diameters of these bodies are drawn to a common scale in Figure 24.3, although only a small part of the Sun's disk can be shown.

Table 24.1 gives information in several categories for the nine major planets. It is obvious from the figures giving distance from

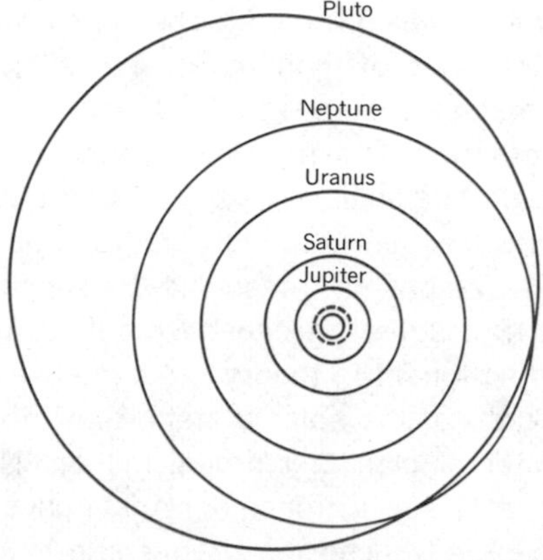

Figure 24.2 Orbits of the outer planets. The innermost circle represents Mars' orbit, and the dashed circle represents the zone of asteroid orbits.

Table 24.1 **The principal planets**

		Distance from sun, millions of miles*		Period of revolution, sidereal	Diameter, thousands of miles		Mass, relative to Earth	Mean density gm/cm³	Period of rotation	Number of moons
	Inner planets			Days						
The Terrestrial Planets	Mercury	36	(58)	88	3	(4.9)	0.06	5.4	58^d16^h	0
	Venus	67	(108)	225	7.6	(12.2)	0.80	5.2	243^d	0
	Earth	93	(150)	365¼	7.9	(12.7)	1.00	5.5	23^h56^m	1
	Mars	142	(228)	687	4.2	(6.7)	0.11	4.0	24^h37^m	2
	Outer planets			Years						
The Great Planets	Jupiter	484	(779)	12	89	(142)	315	1.3	9^h50^m	12
	Saturn	886	(1430)	29½	72	(115)	94	0.7	10^h14^m	10
	Uranus	1780	(2870)	84	29	(47.4)	15	1.6	10^h42^m	5
	Neptune	2790	(4500)	165	28	(44.6)	17	1.6	15^h48^m	2
	Pluto	3670	(5900)	248	4?	(6.4?)	0.11	5.0?	6^d	0

* Kilometer equivalents in parentheses

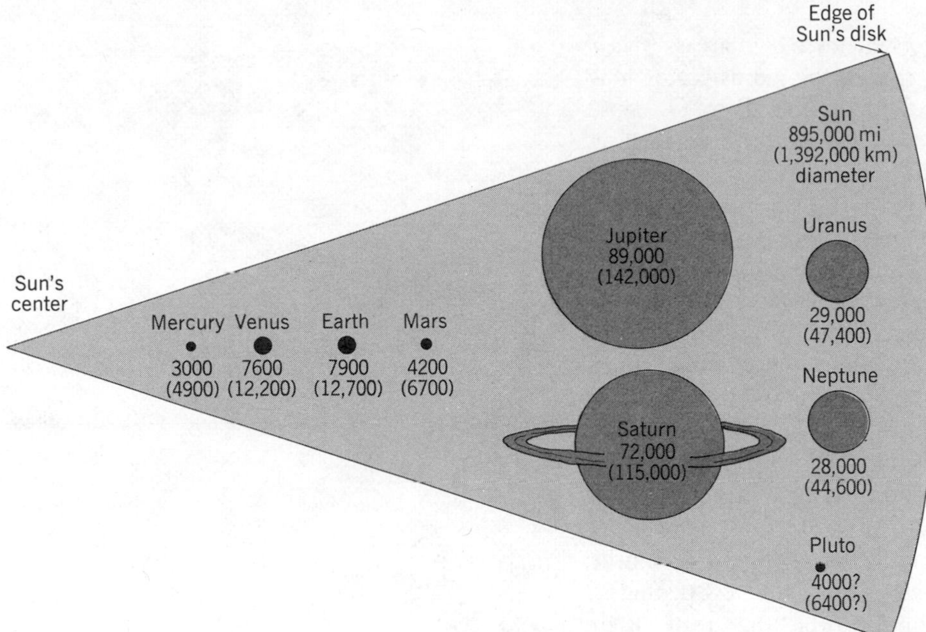

Figure 24.3 Relative diameters of the Sun and planets. Figures give diameters in thousands of miles, with thousands of kilometers in parentheses.

Sun, diameter, and mass, that the four inner, or terrestrial, planets form a group of quite similar bodies. Also, the first four outer planets —Jupiter, Saturn, Uranus, and Neptune—can appropriately be grouped as the **great planets.** In contrast, Pluto has a very much smaller size and is in a class by itself.

ORBITS OF THE PLANETS

The **orbit** of a planet is the path it follows around the Sun. The orbit of every planet is an **ellipse** (Figure 24.4). The Sun occupies one **focus** of the ellipse. An ellipse has a longest diameter, the **major axis,** and a shortest diameter, the **minor axis.** These two axes are at right angles to one another and intersect at the center of the ellipse. Along the major axis are two foci (plural of **focus**). From any point on the ellipse, two straight lines, known as **radius vectors,** can be drawn, one to each of the two foci.

A law of the geometry of the ellipse is that the sum of the two radius vectors remains a constant for all points on the ellipse. This law can be demonstrated by a device for drawing ellipses, using a loop of thread and two pins or thumbtacks on a drawing board (Figure 24.5). You can vary the degree of flattening of the ellipse by adjusting the spacing of the two foci in relation to the length of loop. As the two foci are brought closer to the center, the ellipse approaches a circle in form.

The German astronomer Johannes Kepler (1571–1630) discovered three laws of planetary motion, which bear his name. Kepler's first law simply states that the orbit of each planet is an ellipse, with the Sun located at one focus of the ellipse.

Kepler's second law states that a planet moves in its orbit about the Sun at a varying speed, such that the radius vector from the Sun sweeps over equal areas in equal times. This concept is illustrated in Figure 24.6, a greatly exaggerated ellipse. The circumference has been divided into 12 segments representing 12 months, assumed

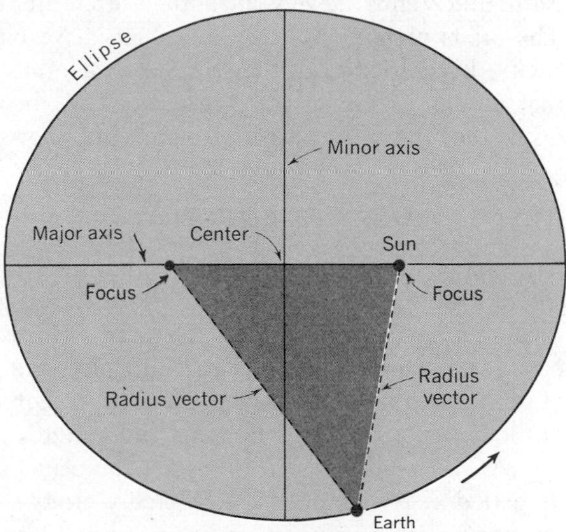

Figure 24.4 The orbit of every planet is an ellipse in which the Sun occupies one focus.

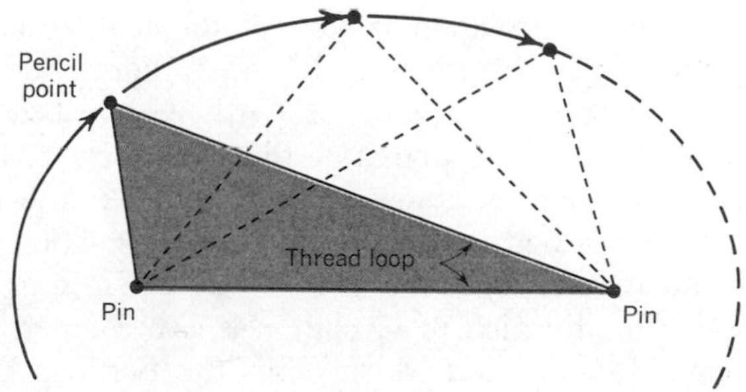

Figure 24.5 An ellipse is easy to draw, using a loop of thread.

exactly equal in time. The radius vectors enclose equal areas. Area A is equal to area B. In sweeping across area A, the radius vector must travel from M to N, a longer distance than from P to Q; consequently, a point on the orbit must travel at a greater average speed between M and N than between P and Q.

It is obvious from Figure 24.6 that the planet in its orbit is nearest the Sun when located at one end of the major axis of the ellipse. This closest point is termed **perihelion** (from the Greek words *peri*, "about or near"; *helios*, "Sun"); the most distant point is at the opposite end of the major axis, a position termed **aphelion** (from the Greek *ap*, "away from"; and *helios*). For example, Earth is at perihelion about January 3, and at aphelion about July 4 each year. At perihelion the radius vector from the Sun to Earth is about 91½ million mi, at aphelion about 94½ million mi, giving a mean value of about 93 million mi (150 million km) for the whole orbit.

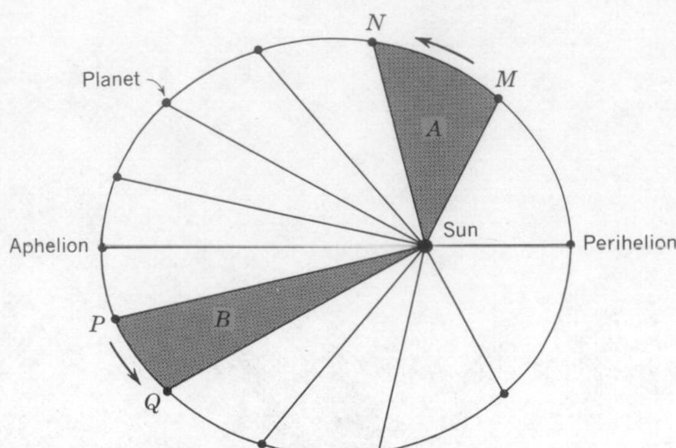

Figure 24.6 An ellipse divided into 12 equal areas.

It follows that the planet's speed in its orbit must be continuously changing. From a maximum at perihelion the speed diminishes to a minimum at aphelion, then increases again to the next perihelion.

The more flattened a planet's elliptical orbit, the greater is the eccentricity of its orbit. **Eccentricity** can be visualized as the extent to which the focus, represented by the Sun, lies to one side of the center point of the orbit ellipse. The eccentricities of the orbits of Earth and Venus are very small; their orbits are nearly perfect circles. The orbits of both Mercury and Mars have only moderate eccentricity. Eccentricities of the great planets' orbits are also small, but that of Pluto is abnormally great. In fact, the orbit of Pluto cuts inside the orbit of Neptune, as Figure 24.2 shows.

REVOLUTION AND ROTATION

The motion of a planet in its orbit about the Sun is called **revolution**, whereas the spinning of the planet on an axis is called **rotation.** One revolution in the orbit defines the **year,** whereas one complete turn on its axis defines the **day.** Table 24.1 gives the period of revolution of each planet in Earth days and Earth years; it also gives the period of rotation of each planet in hours and minutes of Earth time.

As you would expect, the larger the planet's orbit, the longer is its period of revolution. Little Mercury orbits the Sun in only one-quarter the time required by Earth, while Pluto requires 248 Earth years to make the circuit. When it comes to the periods of rotation, no such relationship to distance applies. Venus' day lasts for 243 Earth days, while the Jovian day is less than half an Earth day. Actually, the rotation periods of the four great planets are quite similar—all between about 10 and 16 hours.

Direction of rotation of Earth is described as eastward. This simply means that a point on the equator is traveling in an eastward direction. If we were out in space, situated at a point directly over the Earth's north pole, the direction of rotation would seem to be counterclockwise, as shown in Figure 24.7. From this same point of view, the direction of revolution is also counterclockwise about the Sun. Moreover, our Moon orbits the Earth in the same direction. Astronomers refer to this counterclockwise turning as **direct motion.** (The opposite direction of turning is called **retrograde motion.**)

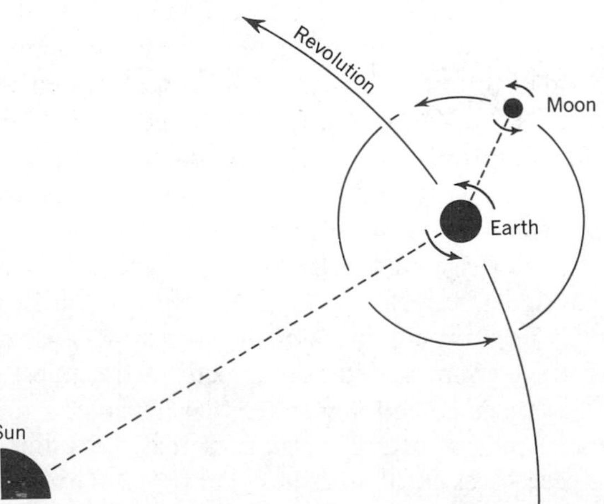

Figure 24.7 Moon and Earth both rotate and revolve in the same direction.

All of the planets and most of their satellites orbit in direct motion. The Sun also rotates with direct motion. These facts strongly suggest that all members of the solar system came into existence, through condensation of a single rotation nebula (Chapter 10). This conclusion is strengthened by the fact that all planetary orbits lie approximately in the same plane, which is about the same as the plane of the Sun's equator. We would expect this uniformity if the hypothetical solar nebula which gave rise to the present solar system were a thin, flat disk, spinning with the Sun at its hub.

GRAVITATION AND CENTRIPETAL FORCE

Kepler based his laws of planetary motion solely upon observations of the planets themselves. A valid physical explanation for elliptical orbits and varying speeds was not furnished until Sir Isaac Newton developed his laws of gravitation and motion, published in 1687.

Gravitation is the mutual attraction between any two masses. The key to planetary motion lies in the **law of gravitation:** Any two bodies attract each other with a force that is directly proportional to the product of their masses and inversely proportional to the square of the distance between them. Let M_1 and M_2 represent the two masses, R the separating distance, and F the gravitational force between them. Then

$$F \propto \frac{M_1 \times M_2}{R^2}$$

The symbol $\propto$ means "proportional to." Recall that R^2 is read "R-squared," which is the same as "R times R." What this law means in simple language is, first, that the larger the masses, the greater is the force with which they attract each other, and second, that the attractive force diminishes very rapidly as the masses increase their distance of separation.

Newton's **first law of motion** states: Every body continues in a state of rest or of uniform motion in a straight line unless made to change that state by some external force. This law is easily illustrated (Figure 24.8). The apparatus consists of a weight attached to a cord. The weight is swung by hand in a horizontal circle. Because of inertia the weight tends to follow a straight-line course tangent to the circle at any given instant. Should the cord break, the weight would fly off in a straight path, labeled "tangent path" in the illustration. Tangential flight is prevented by the **centripetal force** of tension exerted by your hand, acting through the cord. It is this force which changes the direction of the weight from a straight line and causes it to follow a circular path.

In the case of a planet, the force of gravitational attraction between planet and Sun is the centripetal force that constantly deflects the planet from its straight tangential path.

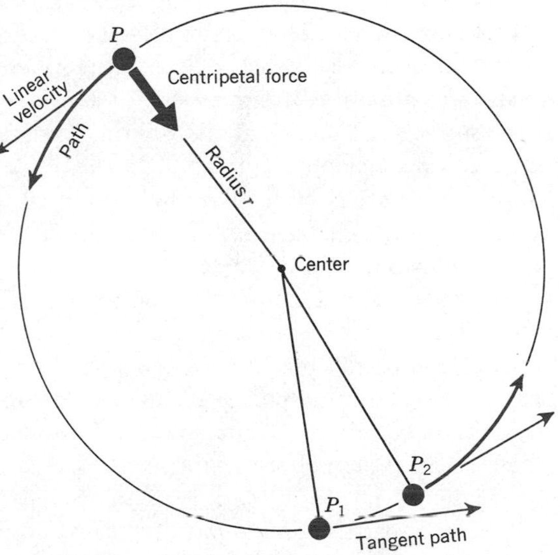

Figure 24.8 Centripetal force acts upon an object in a circular path of motion.

MERCURY

A brief description of the astronomical characteristics of each planet helps us to appreciate the unique environment of Earth. We covered the geological character of the inner planets in Chapter 10.

Mercury, fourth brightest of the planets, is the closest planet to the Sun. Its orbit has a radius only two-fifths that of Earth. In looking over the facts about Mercury in Table 24.1 we note that this planet is a pygmy compared to Earth.

Mercury's atmosphere is so rarefied as to be almost none. The planet's surface gravity is only about three-tenths that of Earth, and any gases readily escape into space. Mercury's rate of rotation is very slow, and a day there lasts for 59 Earth-days. Consequently, on the side of Mercury which happens to be facing the Sun, intense and prolonged heating bakes the surface. It is estimated that, in perihelion, surface temperatures on Mercury rise to perhaps 790° F (420° C), a value exceeding the melting points of tin and lead. In contrast, temperatures on the shadowed side of Mercury may fall nearly to absolute zero. No other planet has so vast a temperature range on its surface. Mercury's cratered surface can be assumed to be completely devoid of any forms of life.

Because Mercury is so close to the Sun, we can see it only on special occasions. Most of the time, the Sun's blinding rays conceal the planet. It can be seen as either a morning star or an evening star for the short period of time that the Sun is below the horizon while Mercury is above the horizon, and then only when the planet is not too nearly in line with the Sun.

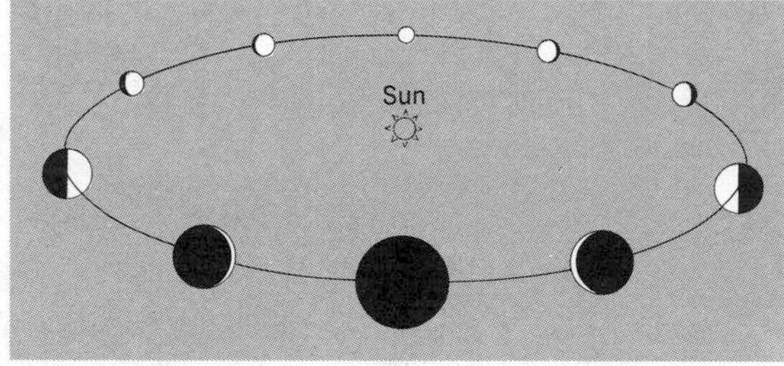

Figure 24.9 The orbit of Venus is seen here in perspective, as if viewed from a point above the north pole of Earth. In full phase (*opposite side of orbit*) Venus appears only one-sixth as large as when closest to Earth.

VENUS

Venus is the most brilliant object in the sky except for the Sun and Moon. Venus approaches closer to Earth than any other planet; a distance of some 26 million mi (42 million km) separates the two bodies at the minimum separation. The maximum distance is about 160 million mi (260 million km). As a result of this sixfold difference in separating distances, Venus seems to change greatly in diameter throughout its orbit (Figure 24.9). Moreover, the changing positions of Venus relative to Earth and Sun result in a series of phases of illumination ranging from a full disk to a thin crescent.

From the standpoint of diameter, mass, density, and length of year, Venus more closely resembles Earth than any other planet does. Moreover, Venus has a dense atmosphere, held by a gravitational force almost as strong as that of Earth. Atmospheric pressure at the surface of Venus is about 100 times as great as that on Earth.

Carbon dioxide constitutes 90% to 95% of the atmosphere of Venus. Oxygen has been found in substantial quantities, but water vapor and nitrogen have not been detected. Water was probably present on Venus at an early stage in the planet's history, but hydrogen atoms could not be held to the planet and most escaped into space.

Because of the presence of some kind of fine, suspended particles in its outer atmosphere, Venus reflects sunlight brilliantly, and nothing can be seen of its surface. The suspended particles have been identified as an aerosol of sulfuric acid.

Temperatures on the surface of Venus have been obtained from temperature analysis of radio waves. Because of the dense atmosphere, acting as a blanket to hold heat, surface temperatures are about 900° F (480° C). The complete absence of water excludes the pos-

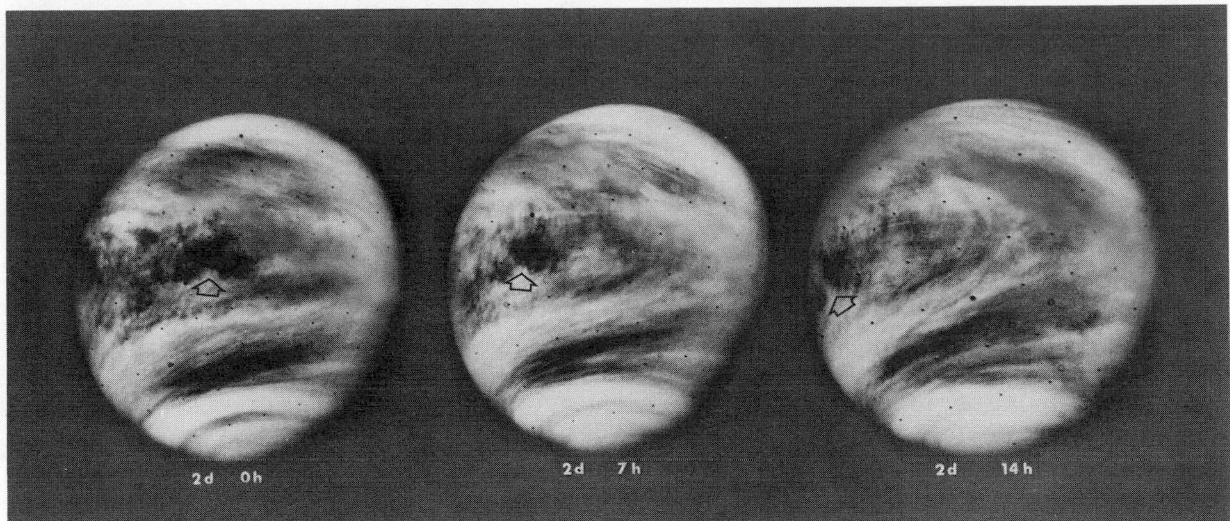

Figure 24.10 Spiral cloud bands change form rapidly on Venus, revealing a vigorous atmospheric circulation. The arrow points to the Venusian Eye, a huge turbulent convection cell in which heated gas is rising. (NASA)

sibility of life forms such as those found on Earth. Venus has a hostile environment, in which human beings, if located on the dark side and suitably protected from the very high atmospheric pressure and possible strong winds, might be able to survive for limited periods.

Until radar was developed, it was practically impossible to determine the rotation period of Venus. Reflected radar signals have yielded a period of 243 days, which is somewhat longer than the planet's period of revolution of 225 days. Furthermore, the rotation of Venus is slowly clockwise, or retrograde, in contrast to the direct rotation of the other planets. One effect of the very slow rotation is to give Venus a very simple system of atmospheric circulation in which the gas rises over the intensely heated equatorial area exposed to the sun. The center of rising air is seen in a dark spot called the Venusian Eye (Figure 24.10). At upper levels the gas spreads north and south, but sinks to the surface over the cold polar zones. We described this two-cell system in Chapter 21 (see Figure 21.11). Photographs of *Mariner 10* show this atmospheric circulation through the changing shapes of spiral cloud bands (Figure 24.10). Rotation is strong enough, however, to produce a Coriolis effect, turning the flow into a west-to-east pattern with speeds up to 220 mi (360 km) per hour. In this way the atmosphere spirals toward the poles.

MARS

Passing over planet Earth, we come to Mars, the first of the planets to be found in an orbit larger than that of Earth (Figure 24.11). Little more than half as large in diameter as Earth, the mass of Mars is only one-tenth that of Earth and the surface gravity only about one-third (Table 24.1).

Reddish in hue, Mars has definite surface features. Study of these features enabled astronomers to measure with precision the period of rotation of Mars, which is about 24½ hr, only a little longer than that of Earth. Also, the plane of Mars' equator is inclined about 25° with respect to its orbital plane, a value very close to Earth's inclination of 23½°.

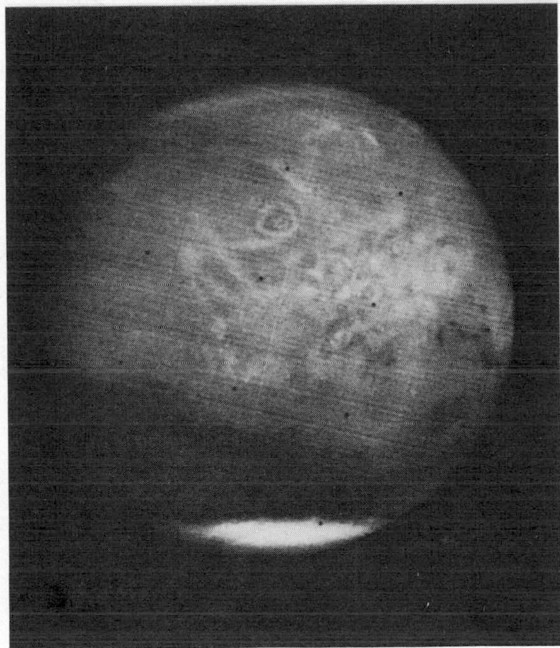

Figure 24.11 Far-encounter photograph of Mars taken from *Mariner 7* spacecraft at a distance of 267,000 mi (430,000 km). The dark spot at upper left is Nix Olympica, a great volcano. (NASA)

As seen from Earth, markings on the surface of Mars consist of dark and light areas in a permanent pattern. Much speculation arose over the significance of recurrent dark markings, which seemed to form an intersecting network of narrow bands. These were early interpreted as canals; their apparent straightness suggested to some astronomers that they were artificially produced by Martians, possibly to serve as irrigation canals.

Now that *Mariner* 9 has circled Mars at close range, we know a great deal about the varied Martian surface, with its craters, volcanoes, troughs, dunes, and channels. These features were described in Chapter 10. Of great interest in earth science is the seasonal growth and disappearance of the white polar caps on Mars, described in Chapter 10 and illustrated in Figure 10.21. Like Earth, Mars has a winter season in one hemisphere while it is summer in the opposite hemisphere. The polar cap grows during the autumn season of that hemisphere, spreading equatorward to a maximum in mid-winter, then receding with the approach of spring. (Note that the Martian year lasts 687 Earth-days, or nearly twice as long as an Earth-year.) Carbon dioxide in solid form (dry ice) is now regarded as the most probable substance of the polar caps.

The colors and color changes on Mars are remarkable. The dark areas are green, blue-green, or gray, and show a seasonal browning to earth-red colors (brick-red to ocher); they may represent barren areas of weathered rock or soil rich in hydroxides of iron. Note that the average density of Mars is 3.9 grams per cubic centimeter (gm/cc), a value well below that of Earth. This fact suggests that Mars may have a proportionally smaller core of iron and proportionally greater volume of rock.

The atmosphere of Mars is very thin. Atmospheric pressure at the surface is about one two-hundredth that at the Earth's surface. Carbon dioxide is the principal component of the Martian atmosphere. Nitrogen is scarcely measurable, and the amount of free oxygen is very small. Water vapor has been measured in extremely small quantity. Perhaps there was originally much more oxygen, some or most of which may have been taken up by chemical rock weathering. Although there is now little water in any form on Mars, there was probably once much more. The braided channels suggests the possibility of brief floods of water (Chapter 10). Some of the original water may have combined with minerals to be held there permanently. Escape of hydrogen would have been favored by the small gravity of Mars (only one-third that of Earth).

Surface air temperatures on Mars average much lower than on Earth because of greater distance from the Sun. Data from Mariner spacecraft gave surface temperature readings at middle latitudes of −171° F (−113° C) for a winter day, −36° F (−38° C) for a summer night. However, in equatorial latitudes under direct rays of the Sun, surface temperatures may rise over 85° F (30° C). Temperature measurements based on the infrared radiation of Mars substantiate these observations and suggest a total daily temperature range of 200 F° (112 C°) in Mars's equatorial region.

Of all the planets Mars would seem to offer the greatest possibility of harboring life, but that life would have to be adapted to a very scanty supply of oxygen and water and to the low density of atmosphere and small gravitational attraction. Mars would also seem the most favorable of the planets for human survival, but a visitor from Earth would have to provide his own life-support system.

THE ASTEROIDS

As early as the seventeenth century, astronomers had recognized the possibility that there might be a small planet between Mars and Jupiter, but it was not until 1801 that a small planet was observed in this region and named Ceres. Shortly thereafter a second object, Pallas, was found, followed by Juno and Vesta. Forty years later a fifth object was found, then many more smaller ones. Now called **asteroids,** these bodies have also been referred to as **minor planets.** All follow the planetary laws and are true planets of the Sun in the mechanical sense, if not in size.

Diameters of the four largest asteroids, named above, follow almost in the order of discovery given at right. Most are very much smaller—a few miles in diameter or less—and show only as points of light rather than as disks. The total number of asteroids runs into the ten of thousands, some 40,000 of which can be detected on photographs. The great majority of asteroids follow orbits between Mars and Jupiter, but some cut inside the orbit of Venus; one is known to sweep outward almost to Saturn's orbit. Their combined mass is perhaps 1/1000 to 1/500 that of Earth.

Ceres	480 mi (770 km)
Pallas	300 mi (480 km)
Vesta	240 mi (385 km)
Juno	120 mi (190 km)

Of particular interest is Eros, an irregularly shaped asteroid about 15 mi (25 km) long. Its orbit is highly eccentric, at times bringing it as close as 13½ million mi (21.7 million km) to Earth. The asteroid Icarus came within about 4 million mi (6.5 million km) of Earth on June 14, 1968. Observations showed that Icarus is irregular in shape, less than 1 mi (1.6 km) in width, and may be composed of iron. The smallest asteroids we can observe are about 1 mi (1.6 km) in diameter, but many are probably much smaller than this. It is entirely reasonable that many of the meteorites (Chapter 10) which strike Earth should be regarded as simply very small asteroids.

THE GREAT PLANETS

Strikingly unlike the terrestrial planets are the four great planets: Jupiter, Saturn, Uranus, and Neptune. Even the smallest of these, Neptune, has almost 4 times the diameter and 15 times the mass of Earth; whereas the giant of the group, Jupiter, has 11 times the diameter and 300 times the mass of Earth. Apart from their size, a second striking difference in these two groups of planets is that of density (Table 24.1). The least dense is Saturn, with a density of 0.7, about one-eighth the density of Earth and less than three-fourths that of liquid water. The other three have densities of 1.3 to 1.7 gm/cc, values only one-fourth that of Earth.

A third striking difference in the two groups of planets is the extremely low prevailing temperatures on the surfaces of the great planets, ranging from −216° F (−138° C) on Jupiter to −330° F (−210° C) on Neptune. A fourth striking difference is in composition. Recall that the four terrestrial planets are probably all composed

Figure 24.12 Photographed from *Pioneer 10* spacecraft, Jupiter displays a strongly banded upper atmosphere. The Great Red Spot lies just below the center of the photograph. (NASA)

of a rock mantle surrounding an iron core and have either no atmosphere or atmospheres of almost insignificant mass. In contrast, the four great planets have massive atmospheres of hydrogen and helium, with some ammonia, methane, and water. These substances in the solid state also make up most of the mass of each planet.

Jupiter appears as a somewhat flattened disk with dark and light bands extending across the surface in rough parallelism with the planet's equator (Figure 24.12). The bands are made irregular by cloudlike patches that over a period of days or weeks show changing patterns. Apparently the bands are produced by systems of flow in Jupiter's atmosphere analogous to Earth's planetary wind systems.

By analysis of ultraviolet light reflected from Jupiter's atmosphere it has been determined that the outer layer consists of 84% hydrogen and 15% helium. The remainder is largely methane and ammonia. The abundance of hydrogen in the atmosphere suggests that the planet as a whole has hydrogen as the predominant constituent. Jupiter may have a core of hydrogen in an extremely dense, metallic state.

Saturn is well known to all through its distinctive rings, which are seen as concentric bands of light and dark color lying in a very thin zone in the plane of the planet's equator (Figure 24.13). The rings consist largely of individual solid fragments, each revolving about the planet in an orbit as if each were an independent satellite of the planet. The particles may be on the order of the size of gravel, with some much larger, since they are capable of reflecting radar signals. Altogether the rings constitute not more than one-millionth of Saturn's mass.

Although Saturn is believed to be similar to Jupiter in composition and structure, Saturn's proportion of hydrogen may be larger to yield its lower density.

Uranus and Neptune are nearly twins in diameter and mass. Because of their great distances from Earth, these planets are rather difficult to observe and show little or no surface marking. Under spectroscopic analysis both planets show methane to be the dominant

Figure 24.13 The planet Saturn and its rings, photographed with the 100-in. (250-cm) Hooker telescope on Mount Wilson. (The Hale Observatories.)

Figure 24.14 This meteor trail showed a sudden increase in brightness as it traveled toward the lower right. (Yerkes Observatory.)

atmospheric constituent, whereas ammonia appears only in a trace. In addition to hydrogen compounds and helium in the solid state, rock may be present in the cores of Uranus and Neptune, since their densities are somewhat greater than those of Jupiter and Saturn.

PLUTO

Pluto is in a class by itself, because it is on the same order of size as the terrestrial planets but is located in a highly eccentric orbit beyond the great planets. Although its existence was long suspected because of irregularities in the orbits of Uranus and Neptune, Pluto was discovered only in 1930. It appeared as a very faint object, found to have changed position among the stars on successive photographs taken six days apart. The mass of Pluto has been calculated, from its distortion of Neptune's orbit, to be 11% of Earth's mass. The planet is too small to permit its diameter to be measured, but an estimate places the figure at not over 4000 mi (6400 km). Pluto's density has been estimated at about 5 gm/cc. Its surface temperature is judged to be not far above absolute zero.

METEOROIDS AND METEORS

Meteoroids are tiny particles of matter traveling at high velocities in space and entering the Earth's outer atmosphere in vast numbers. When they enter Earth's atmosphere they become luminous trails, called **meteors.** The trails can be studied to give valuable information about the physical properties and motions of the upper atmosphere (Figure 24.14).

It is estimated that most meteoroids have a mass less than one ounce and range downward in size to perhaps one-thousandth of that mass. Thousands of millions of such specks of solid matter strike Earth's atmosphere daily.

If you watch the sky carefully at a time when a large number of meteoroids are striking Earth, and if you sketch the meteor trails on a map of that portion of the sky, you will find that most of these lines radiate outward from an apparent center point among the stars termed the **radiant** (Figure 24.15). Meteors forming such a pattern constitute a **meteor shower.** For each meteor shower, the radiant occupies a fixed place among the stars. Actually, all the meteoroids of a shower are moving in parallel paths, but they seem to diverge because of the effect of perspective.

From a study of meteor trails, individual meteoroid swarms have been identified as groups of particles following highly eccentric orbits about the Sun (Figure 24.16). The swarms are named from the celestial point where the radiant lies. An example is the Leonid swarm, which seems to come from a point in the constellation of Leo. Each year, Earth crosses the orbit of the Leonid meteoroid swarm in the period November 14–15. Most of the Leonid meteoroids move in a group that makes a complete circuit of the Sun about every 33 years. At least eight other important swarms are known. Some meteoroids will be found scattered around the entire orbit of a swarm, so that at least a few will be seen each time Earth passes through the orbit.

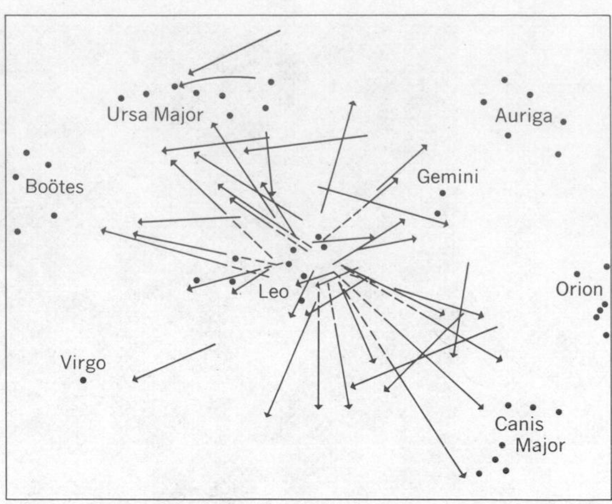

Figure 24.15 A sky-map plot of meteors observed during one night in mid-November shows that about two-thirds of them seem to radiate from a small area in the constellation of Leo. These are meteors of the Leonid swarm. (From *The Elements of Astronomy*, by E. A. Fath. Copyright 1934 by McGraw-Hill Book Company. Used by permission of the publisher.)

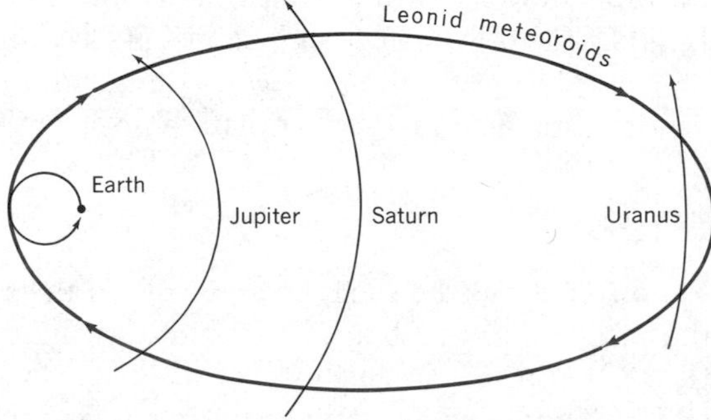

Figure 24.16 Orbit of the Leonid meteoroid swarm. (From *The Elements of Astronomy*, by E. A. Fath. Copyright 1934 by McGraw-Hill Book Company. Used by permission of the publisher.)

COMETS

Finally, we come to the comets, to many persons the most bizarre of the astronomical objects. The **comet** consists of a brightly luminous head, called the **coma,** from which a luminous **tail** streams off in a direction away from the Sun (Figure 24.17).

The apparent size of a comet varies greatly. The larger ones have tails 50 million mi (80 million km) long that cover a wide arc in the sky. Within the head of a large comet there is often a bright, starlike center, known as the **nucleus.** The material constituting the coma and tail is so diffuse that stars shine through it with undiminished intensity.

Most comets, like meteoroid swarms, follow highly eccentric elliptical orbits around the Sun (Figure 24.18). Because of their high velocity near perihelion, at which time they are visible from Earth, comets are seen for only a very short time in comparison with the total period of their revolution. At aphelion the velocities are greatly diminished, so that a comet spends most of its time among or beyond the outer planets. Most comets have orbital periods of tens of thousands of years. However some, among them Halley's comet, reappear at regular intervals ranging from three years to a few centuries. About 100 such periodic comets are known.

A particular comet group, known as the Jupiter family, about 40 in all, have orbits whose most distant points are close to Jupiter's orbit. It is believed that the gravitational attraction of Jupiter entrapped these comets into their relatively small orbits. Their periods range from 3.3 years (Encke's comet) to 8.6 years.

Comets have an extremely low density; the coma and tail consist entirely of fine dust particles or of gaseous matter driven off from the nucleus. The nucleus itself, however, is believed to be an aggregation of small particles of solid matter and frozen gases. Based upon spectroscopic analysis of its composition, this matter is largely methane, ammonia, carbon dioxide, and water.

During close approach to the Sun, some of the gas is vaporized and ionized under the intense heat of the Sun's rays, diffusing outward to form the comet tail, and under pressure of the solar wind the tail is pushed away from the Sun. As a result of passage close to the Sun, a comet may lose part of its mass or be completely disrupted.

It is thought that the diffuse dust particles remaining from disintegrated comets may constitute the meteoroid swarms. This possibility is strengthened by the fact that the highly eccentric orbits of meteoroid swarms and comets are much alike. Several cases are known in which the orbit of a meteoroid swarm is identical with the orbit of a previously known comet.

The source of matter for comets and meteoroids is a topic of speculation. The Dutch astronomer, Jan H. Öort, suggested that the comets originate in a vast cloud, or "reservoir," containing the substance of millions of comets and located far beyond the orbits of Neptune and Pluto. The matter in such a belt would be at a temperature close to absolute zero and would have been formed early in the history of the solar system.

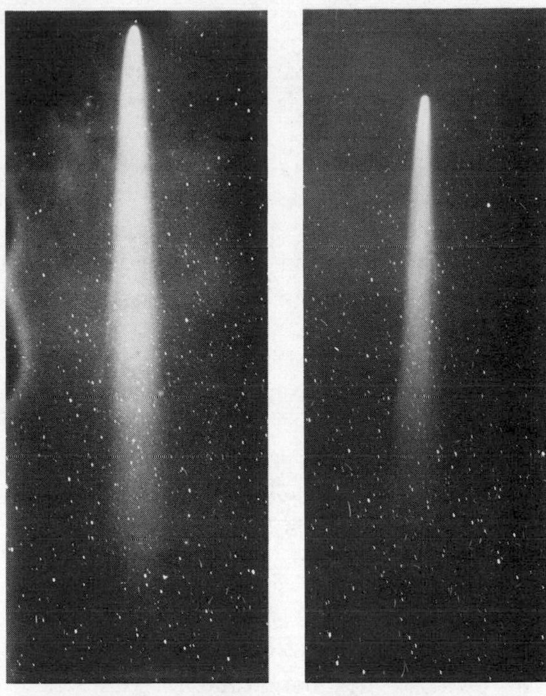

Figure 24.17 Halley's comet photographed on May 12 and 15, 1910, at Honolulu, Hawaii. The shorter tail (*right*) covers 30 degrees of arc. (The Hale Observatories.)

SATELLITES OF THE PLANETS

A **planetary satellite** is a solid object orbiting a planet. It is held in the planet's gravitational field. Altogether 32 satellites, or moons, have been identified in orbits around the nine planets, but the distribution is highly varied in terms of numbers per planet (Table 24.1). The two innermost planets, Mercury and Venus, have no moons. Mars has two: Deimos and Phobos. Both are smaller than Earth's single moon, and both orbit much closer to the parent body.

Jupiter's twelve moons consist of four large ones and eight small ones. Galileo, using the first astronomical telescope, discovered the four large moons in 1610, and they have since been designated the Galilean satellites. Their names are Io, Europa, Ganymede, and Callisto, stated in order outward. Inside the orbit of Io is another very small satellite. The remaining seven lie far beyond the orbit of Callisto and are very small objects.

The apparent positions of the Galilean moons change greatly from night to night, but all revolve in about the same plane and in the same direction (Figure 24.19). Galileo's observation of these satellites was a powerful point in favor of the Copernican theory of the solar system, because it provided a small-scale model of planets orbiting a Sun. Io, Europa, and Ganymede have densities that suggest they are of rock composition, whereas Callisto has a very low density and may consist of frozen water or ammonia.

Three of Jupiter's small moons have retrograde orbits. Saturn has ten known moons. The tenth moon, Janus, was discovered only in 1966. Saturn's farthest distant moon is Phoebe, traveling in a retrograde orbit of 8 million mi (13 million km) with a period of over 100 days. Little is known of the five moons of Uranus and the two of Neptune.

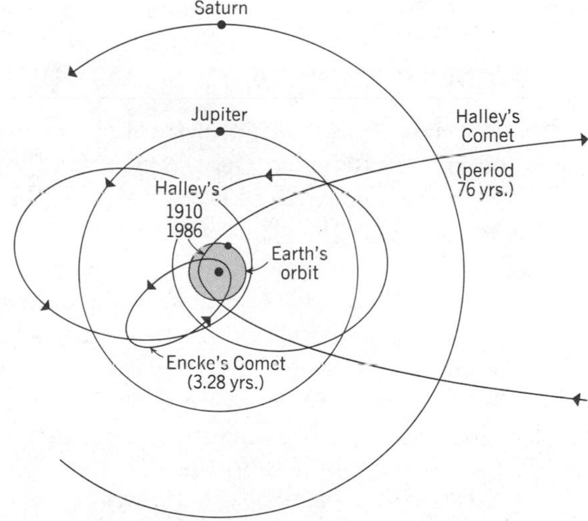

Figure 24.18 The orbits of Halley's comet, and some comets of the Jupiter family.

THE EARTH'S MOON

Of all the satellites, Earth's Moon is unique in being very large in comparison with the planet which it orbits. Its diameter is 2160 mi (3476 km), compared with an Earth diameter of about 8000 mi (13,000 km); the ratio of diameters is about 1 to 4. Our Moon has a mass of about one eighty-first that of Earth. Because of the Moon's relatively large size, astronomers have commented that the Earth-Moon system can be regarded as comprising a **binary planet.**

The average radius of the Moon's orbit is about 239,000 mi (384,000 km) (Figure 24.20). The Moon's orbit is quite strongly eccentric. When closest to Earth, the Moon is said to be in **perigee;** when most distant, in **apogee.** These terms apply to corresponding positions of any satellite, including the man-made orbiting satellites. Notice that the Moon orbits about the Earth in a direction that can be described as counterclockwise (direct motion), when we imagine ourselves to be viewing the system from a point above the Earth's north pole. Rotation of both Earth and Moon are also direct in this sense.

The period of the Moon's revolution about Earth, calculated in terms of 360° of angle with reference to the fixed stars, is 27.32 days. This period is the **sidereal month.** However, when we measure the revolution in reference to the Sun's position in the sky (from one new moon to the next) the period averages 29½ days. This period is the **synodic month.** We shall see that a monthly rhythm of the tides follows the synodic month.

The Moon always shows the same face to observers on Earth. This fact requires that the Moon's period of rotation upon its axis be exactly the same as its period of revolution with respect to the stars. So, the Moon rotates once in 27.32 days. Such a coincidence could scarcely be ascribed to mere chance. The Moon has a concentration of denser basaltic bodies beneath the maria on the side facing the Earth. Gravitational attraction for this excess mass slowed the Moon's rate of rotation, eventually locking the rotation into a period equal to its revolution.

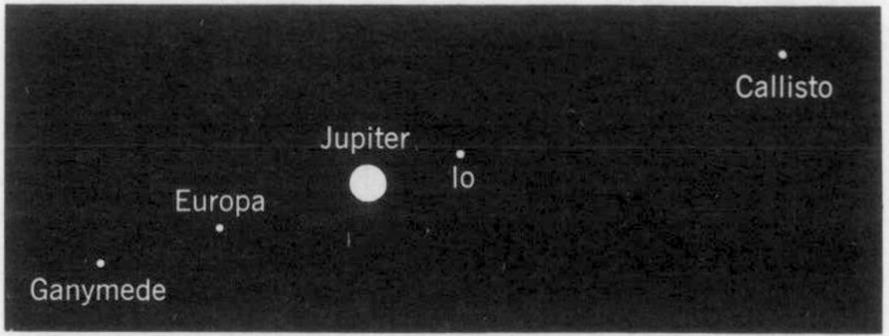

Figure 24.19 The four Galilean satellites of Jupiter as they might appear through binoculars or a small telescope. Distances from Jupiter are shown to correct scale, as if all four satellites were in a line at right angles to the observer. Diameters are not to scale.

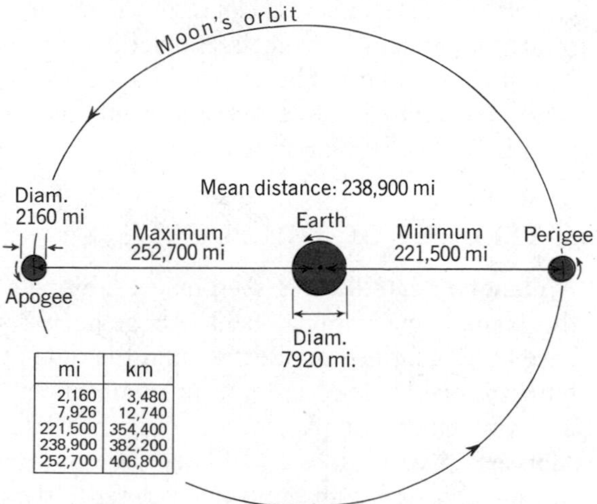

Figure 24.20 Dimensions of the Moon's orbit.

CONJUNCTION, OPPOSITION, AND QUADRATURE

We must know the simple geometrical relations of the Earth, Moon, and Sun in order to understand the phases of the Moon, the eclipses, and the tide (Figure 24.21). Assume for the moment that the Moon's orbit lies in the plane of the Earth's orbit, and consider only the alignment of the three bodies in a single plane.

When all three bodies lie along a single straight line, with Moon and Sun on the same side of Earth, the Moon and Sun are said to be in **conjunction;** when aligned with the Earth located between them, Moon and Sun are in **opposition.** A contrasting relation, called **quadrature,** exists when the line from Earth to Moon makes a right angle with respect to the line from Earth to Sun (Figure 24.21). In each synodic month the Moon will be found twice in quadrature.

The series of changes in appearance of the Moon throughout the synodic month are known as the **phases of the Moon** (Figure

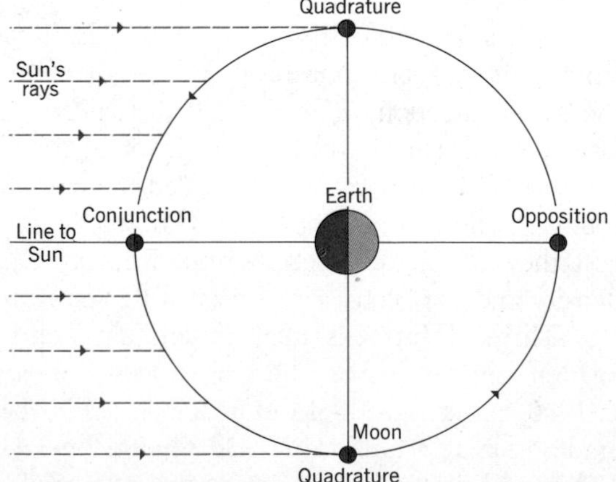

Figure 24.21 Relative position of Moon, Earth, and Sun determine the Moon's phases and the occurrences of eclipses.

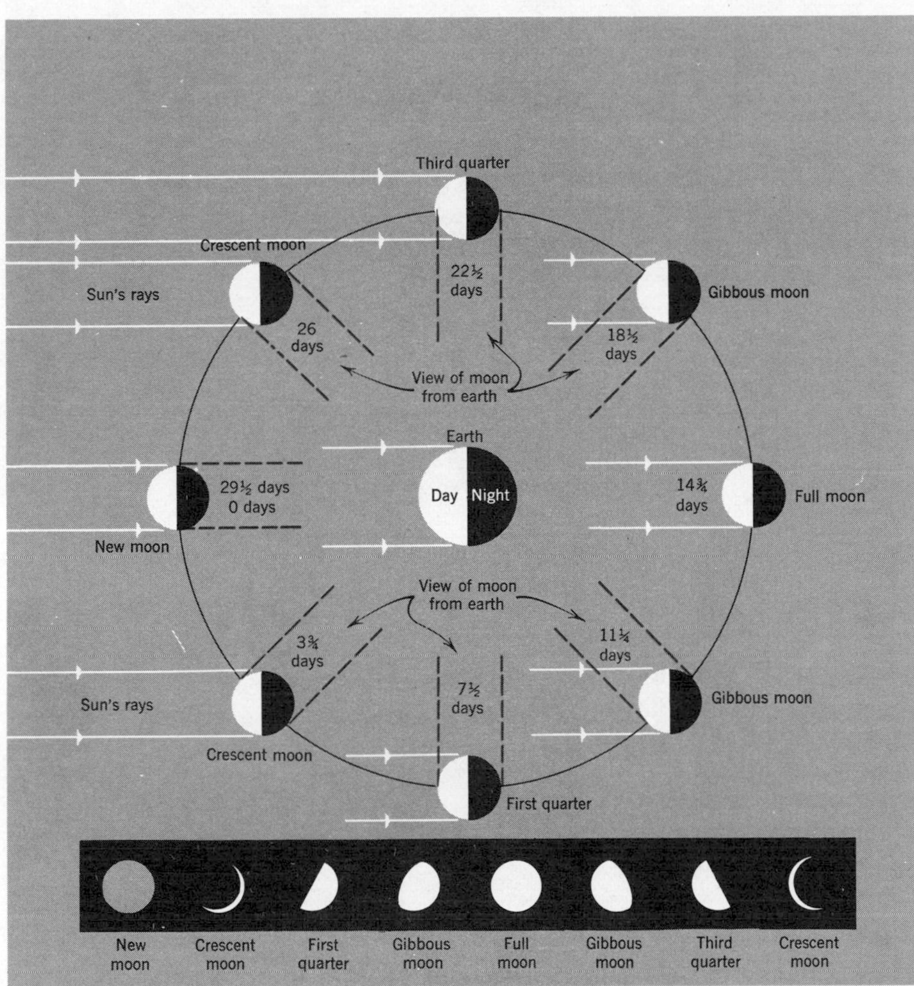

Figure 24.22 Phases of the Moon. The bottom diagram shows the Moon's outline as it appears in the southern half of the sky. (© 1969, John Wiley & Sons, New York.)

24.22). Moonlight consists of sunlight falling upon the Moon's surface and reflecting to Earth. The Moon is at all times divided into a sunlit hemisphere and a shadowed hemisphere. The phases of the Moon are simply the varying proportions of the sunlit and shadowed halves that we see from the Earth.

LUNAR ECLIPSE

When Sun and Moon are in opposition, the Moon may cross Earth's shadow, which is relatively large in proportion to the Moon's diameter. The result is an eclipse of the Moon, or **lunar eclipse.** If the Earth's shadow merely passes across one edge of the Moon's disk without entirely enveloping it, a **partial eclipse** is said to occur, whereas if Earth's shadow completely covers the Moon, as shown in Figure 24.23, a **total eclipse** results.

In a lunar eclipse two shadow zones are encountered (Figure 24.24). Because of the Sun's relatively large size there is formed an inner cone of complete shadow, the **umbra,** surrounded by a zone of partial shadow, the **penumbra.** As the Moon enters the penumbra, it goes gradually from full illumination into a region of increasingly dim light; then it crosses an abrupt boundary into the nearly total

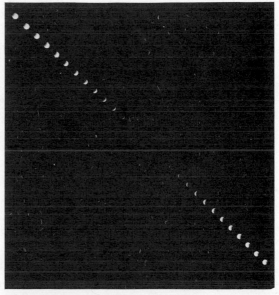

Figure 24.23 An eclipse of the Moon, recorded by multiple exposures at equal time intervals. The photograph has been condensed in the region of totality. (The American Museum—Hayden Planetarium.)

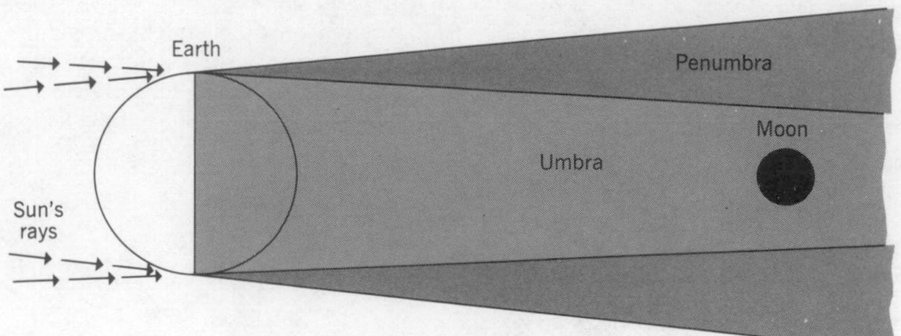

Figure 24.24 Diagram of a lunar eclipse.

darkness of the umbra. As Figure 24.24 shows, the diameter of the Earth's umbra, where it is crossed by the Moon, is about three times the Moon's diameter. In a total eclipse the Moon lies entirely within the umbra for almost 2 hr; the total elapsed time from its first contact with the umbra to the time it is entirely free of the umbra may last about 3¾ hr.

If the Moon's orbit lay exactly in the plane of the Earth's orbit, a total lunar eclipse would result once each synodic month, at the time of opposition (full moon). However, because the Moon's orbital plane is inclined by about 5° from the plane of the Earth's orbit, most of the times that the Moon is in opposition it will pass above or below the Earth's shadow. As a result, there are only two periods per year within which an eclipse can occur.

SOLAR ECLIPSE

An eclipse of the Sun, or **solar eclipse,** occurs when the Moon is in conjunction with the Sun and casts its shadow upon the Earth (Figure 24.25). For persons in the shadow zone on Earth, the Moon's disk seems to pass across the Sun's disk, partially or totally obscuring the Sun.

The Moon's core of total shadow, the umbra, forms only a very narrow track across the Earth—up to 168 mi (270 km) wide (Figure 24.25). For this reason a total eclipse is rarely seen by an individual unless he makes a special effort to travel to the predicted track. The shadow of totality travels across the Earth's surface at

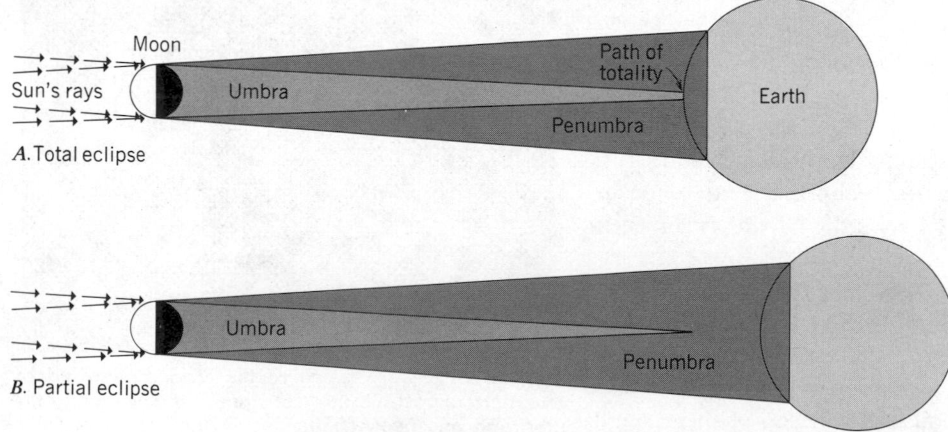

Figure 24.25 Diagrams of solar eclipses. (*A*) Long umbra cone and minimum separating distance allow a total eclipse to occur. (*B*) Short umbra cone and maximum separating distance allow partial eclipse only.

some 1000 to 4000 mi (1600 to 6400 km) per hour. The total phase lasts, at most, only 7½ min. The region of partial eclipse, in contrast, is a very broad zone, up to several thousand miles wide.

On some occasions, when the separating distance between Sun and Earth is greater than average, the umbra fails to reach the Earth. This situation is shown in the lower diagram of Figure 24.25. Then only the penumbra reaches the Earth and only a partial eclipse occurs. As in the case of the lunar eclipse, there are only two periods in each year when conditions are favorable for a solar eclipse.

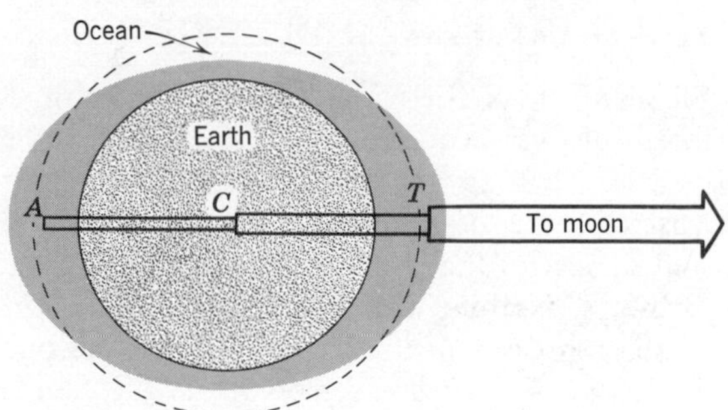

Figure 24.26 The Moon's attractive force tends to distort the Earth into an ellipsoid. (© 1969, John Wiley & Sons, New York.)

THE OCEAN TIDE

The periodic rise and fall of ocean level, or **ocean tide,** was known for centuries to be related to the Moon's path in the sky, but it was not until Newton published his law of gravitation in 1686 that the physical explanation became understood. Figure 24.26 shows the principle. The broad arrow represents the attraction which the Moon exerts upon the Earth. On the side nearest the Moon, point *T*, the attraction is stronger than at the Earth's center, point *C*, because gravitational force decreases with an increase in the separating distance between two masses. For the same reason the attractive force is even less at point A. These differences in force tend to distort the spherical shape of the Earth into a **prolate ellipsoid,** shaped somewhat like an American football.

Although the solid lithosphere makes only a very small response to this earth-stretching force, the oceans respond freely. As Figure 24.27 shows, the ocean water tends to move along the lines indicated by the surface arrows. The water moves toward two centers, one at A and one at *T*, but moves away from a belt girdling the globe on line passing about through the poles (*N* and *S*).

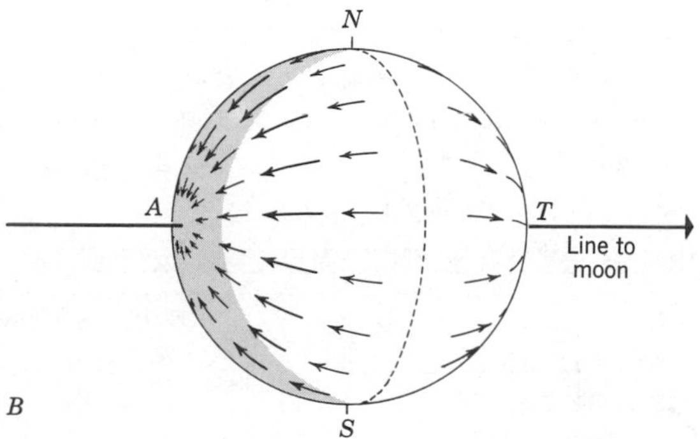

Figure 24.27 Tidal forces tend to move the ocean water toward two centers, on opposite sides of the globe, along a line connecting Earth and Moon.

Next, we realize that the Earth is rotating on its axis. This means that the two centers of tidal accumulation, or "tidal bulges," will be traveling continually around the Earth. For a fixed point on the globe, the tidal bulges will pass by twice daily. Each bulge produces a rise of ocean level to a maximum value, **high water.** The actual interval is close to 12½ hr between high waters. The belt of depressed surface will also pass by twice daily, so that we will have a minimum water level, or **low water,** 6¼ hr following a high water. This is the common tidal cycle found along most coastlines of the world.

Figure 24.28 shows a typical **tide curve** recorded throughout a 24-hr period. The difference in level between high water and low water, or **tide range,** was about 9 ft (3 m) in this particular case.

The Sun also produces a tide-raising force upon the Earth, but it is not as strong as the Moon's force, because of the much greater distance between Earth and Sun. However, the Sun's tide-raising force is added to the Moon's force during both conjunction and opposition, when all three bodies are in a single line. At such times, the tide range is increased, an event called the **spring tide.** When Sun and Moon are in quadrature, pulling at right angles, the forces are diminished, and tides of low range occur; these are called **neap tides.**

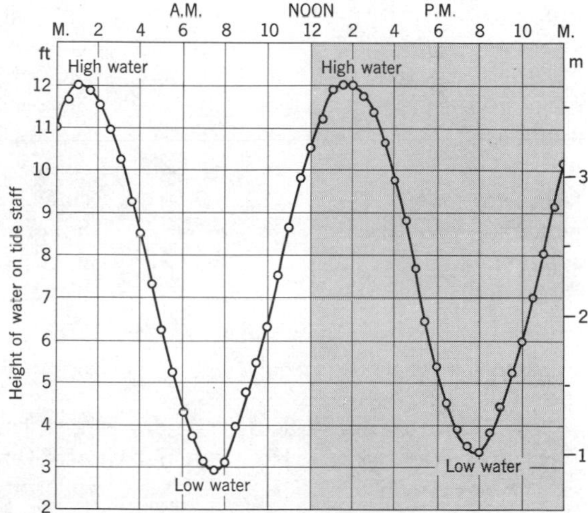

Figure 24.28 A graph of the rise and fall of tide at Boston Harbor for a 24-hour period. Dots show water level at half-hour intervals. (© 1969, John Wiley & Sons, New York.)

THE SOLAR SYSTEM IN REVIEW

All members of the solar system, whether they be giant planets or specks of dust in meteoroid swarms, orbit the Sun in obedience to Newton's laws of gravitation and motion. The satellites orbit their master planets under these same laws. Outer space is so nearly devoid of matter that these motions are frictionless along the orbital pathways. Nothing within the experience of man is so completely reliable, so exact in its schedule, and so beyond the power of man to alter, as the apparent motions of the Sun, Moon, and planets.

Yet the motions of the solar system do undergo very slow changes. These arise from the phenomenon of tides, which affects all pairs of objects in the solar system. The invisible force of gravitational attraction deforms the planets and satellites as they revolve and rotate. This flexing by tidal forces is a form of friction, and it robs the system of energy. So the orbits must change with time. Ultimately, the original energy of motion imparted to the solar system at the time of its formation will be converted into heat and dissipated into interstellar space. Perhaps, then, the ultimate demise of our solar system will take place as its members, one by one, fall into the Sun.

YOUR GEOSCIENCE VOCABULARY

heliocentric theory
geocentric theory
Ptolemaic system
Copernican system
solar system
planet
major planets
inner planets
terrestrial planets
outer planets
great planets
orbit
ellipse
focus
major axis
minor axis
focus
radius vector
perihelion
aphelion
eccentricity
revolution
rotation
year
day
direct motion
retrograde motion
gravitation
law of gravitation
first law of motion
centripetal force
asteroid
minor planets
meteoroid
meteor
meteor shower
radiant
comet
coma
tail of comet
nucleus of comet
planetary satellite
binary planet
perigee
apogee
sidereal month
synodic month
conjunction
opposition
quadrature
phases of the Moon
lunar eclipse; partial, total
umbra
penumbra
solar eclipse
ocean tide
prolate ellipsoid
high water
low water
tide curve
tide range
spring tide
neap tide

SELF-TESTING QUESTIONS

1. Compare the Ptolemaic system of astronomy with the Copernican system. How can the Copernican system be proved correct?

2. Name the members of the solar system. On what basis are the planets classified into groups? Compare the groups in terms of planetary size and density, distance from the sun, and periods of revolution.

3. Describe the planetary orbits in geometrical terms. State Kepler's first and second laws. Explain each. How are planetary revolution periods related to size of orbit? Describe the prevailing directions of planetary rotation and revolution.

4. Apply the principles of gravitation and centripetal force to the orbits of planets and their satellites.

5. Describe Mercury as a planet. What environmental conditions prevail on the surface of Mercury? When can Mercury be observed in the sky?

6. Describe Venus as a planet. What phases does Venus exhibit? How does atmosphere of Venus circulate? What environmental problems would space travelers from Earth encounter in making a visit to the surface of Venus?

7. Compare the surface environment of Mars with that of Venus and Mercury. What difficulty would visitors from Earth face in attempting to live on the surface of Mars?

8. Describe the asteroids. When were they discovered? How close do they come to Earth?

9. Describe Jupiter and Saturn in terms of size, surface temperature, and composition. Of what material are Saturn's rings composed? How was Pluto discovered? How does Pluto compare in size and density with the inner planets?

10. Describe a meteor. What can we learn by plotting their radiants on a sky map? How do meteoroid swarms travel?

11. Describe the form, structure, and orbit of a typical comet. What causes the tail? Which way does it point in the sky? Where do comets originate?

12. Which planets have satellites? Which have none? Describe the Galilean satellites. What part did they play in the debate over the validity of the Copernican theory of the solar system?

13. Describe the Moon's orbit in terms of dimensions and shape. Why does the same side of the Moon always face the Earth?

14. Describe in order the phases of the Moon throughout the synodic month.

15. Explain how a lunar eclipse occurs. When it is a partial eclipse and when a total eclipse? How long does an eclipse last?

16. Explain how a solar eclipse occurs. When is this eclipse total? Why is a total solar eclipse a rare event to witness?

17. Explain how ocean tides are generated. What controls the interval between successive high waters? What influence has the Sun in causing variations in the range of tide?

The Crab Nebula in the constellation of Taurus. It represents the gaseous remains of the supernova of A.D. 1054, spreading rapidly outward into space. (The Hale Observatories.)

25 STARS, GALAXIES, AND THE UNIVERSE

GUEST STARS AND BLACK HOLES

The Chinese called it the Guest Star, and it was uninvited as well. The Guest Star appeared on July 4 in the year A.D. 1054 as an object in the sky so brilliant that it could be seen in broad daylight for about 3 weeks. The Chinese carefully recorded its position on their sky charts; it was in the constellation of Taurus. Gradually the Guest Star faded into an ordinary star, and in 2 years it had disappeared from view. For reasons hard to understand, there is no record of the Guest Star of A.D. 1054 in any European writings. Surely Westerners must have seen it, for it was a supernova, an exploding star shining with a radiant energy over a billion times greater than the output of our Sun.

Looking into the same spot in the heavens today, astronomers find the Crab Nebula, a blot of luminescent gas that reminds one of the circular splash made by a diver entering a pool from the high board. Nine centuries have elapsed since the Guest Star supernova made its brilliant appearance. In those centuries the gaseous remains of the exploding star have spread outward in all directions. The present speed of expansion of the Crab Nebula is some 700 miles per second, and it occupies an area in space about 36 million-million miles across.

Supernovae occur about once in a century in our own Milky Way galaxy, in which there are about 100 billion stars. European history records only a few exploding stars like that seen in China and Japan in A.D. 1054. Not surprisingly, two of these were described by great astronomers of their day and subsequently named after them. One was Tycho's Nova, observed in the year 1572, by Tycho Brahe, the Danish astronomer. This guest star was visible for weeks in broad daylight. Kepler's Nova appeared in 1604, when Johannes Kepler, the German astronomer, was 35 years old. He described it as being brighter than the planet Jupiter. Observing these same points in the sky today, we find that each is surrounded by an expanding nebula of gases, the remains of the stellar explosion that was the supernova.

But now we have a new discovery to complete the story. Squarely in the center of the Crab Nebula astronomers have identified a point from which there comes a radio signal pulsating "on" and "off" at a steady rate of 30 pulses per second. The signal represents an enormous output of energy emanating from a rapidly rotating mass of extremely dense matter. It is all that remains from the explosion of A.D. 1054. Called a pulsar, because of its pulsing signal, this object is a neutron star, nearing the close of its long life cycle. In a single terminal outburst, this star threw off a great part of its mass, then contracted to an object perhaps only a few miles in diameter. Matter in this tiny ball is a million-billion times denser than water, for it consists of the nuclei of elements closely packed together.

As if the neutron star idea is not difficult enough for us laymen to swallow, astronomers have thrown at us the ultimate in incomprehensibility—the black hole. They tell us that a neutron star can continue to contract under its own gravitation, becoming smaller and denser without limit. When the star has shrunken to a diameter of about 4 miles, packing into that volume a mass equal to the mass of our Sun, its surface gravity is so strong that it pulls back into itself all light it might produce. In fact, no energy or matter can escape from its grip. With no radiation able to leave, the object becomes invisible—a veritable black hole in space. Physicists actually believe in black holes and are looking for evidence of their existence.

Much as we may talk about these incredible celestial configurations, there is no way we can grasp the enormity of things throughout the universe. Time, distance, and mass are scaled in quantities far beyond any meaningful measuring stick we Earthbound humans can apply.

ASTRONOMY AND EARTH SCIENCE

The Sun supplies Earth with energy of light and heat, as well as other forms of matter and energy. Earth also receives from all points in space a barrage of energy in the form of light, radio waves, X-ray emissions, and highly energetic nuclear particles. To understand planet Earth, we must know a great deal about the Sun and something of the stars beyond the Sun.

Astronomy on even the largest scale contributes knowledge of fundamental importance to an understanding of planet Earth. Astronomers and geologists have often worked as partners in developing scientific hypotheses. This cooperation comes about because evidence obtained from matter on Earth can in some instances be applied outward to problems in astronomy. For instance, an early calculation of the age of the universe proved incompatible with ages of rocks on Earth, and the astronomical theory was accordingly revised.

In this chapter we will try to place Earth in its proper perspective in the universe. With that objective in mind we first investigate our Sun as a star.

THE SUN

Our Sun is a true star. It is a huge sphere of incandescent gas more than 100 times the diameter of Earth, with a mass more than 330,000 times that of Earth and a volume of 1.3 million times that of Earth. The Sun's surface gravity is 34 times as great as that of Earth.

The Sun's diameter is about 900,000 mi (1,400,000 km). The Sun lies at an average distance from Earth of about 93 million mi (150 million km). Traveling at the speed of light, which is roughly 186,000 mi (300,000 km) per second, solar radiation takes about 8⅓ min to reach the Earth.

Like our Earth the Sun rotates upon an axis, but with an important difference: Earth is solid and has a uniform rate of rotation at its surface, whereas the Sun is a gaseous body and does not have the same rate of rotation from one part of its surface to another. From a study of the movements of sunspots we know that the equatorial region of the Sun rotates with a period of about 27 days, whereas at progressively higher latitudes the rotation is slower.

The visible surface layer of the Sun is called the **photosphere.** The outer limit of the photosphere constitutes the edge of the Sun's visible disk. Gases in the photosphere are at a density less than that of Earth's atmosphere at sea level.

Temperature at the base of the photosphere is about 11,000° F (6000° K) but decreases to about 7700° F (4300° K) at the outer photosphere boundary. Light production is extremely intense within the photosphere. Beneath the photosphere, temperatures and pressures increase to enormously high values in the Sun's interior, or **nucleus.** Here temperatures are between 22 and 32 million °F (13 to 18 million °K).

Above the photosphere lies a low solar atmosphere, the **chromosphere.** It is a region which includes rosy, spikelike clouds of hydro-

Figure 25.1 This photograph of the Sun's outer corona was taken during a total eclipse. The Moon's disk completely covers the Sun, permitting this pearly-white, tenuous outer layer of gases to be seen. (The Hale Observatories.)

gen gas called **solar prominences.** Still farther above the Sun's surface is the **corona,** a region of pearly-white streamers of light. The corona constitutes the Sun's outer atmosphere (Figure 25.1). At times the solar prominences reach far out into the corona as luminous archlike bodies (Figures 25.2 and 25.3) rising to heights of over one million mi (1.6 million km). Temperatures increase outward through the chromosphere and the corona until values as high as 4 million °F (2 million °K) are reached. Surprisingly, the photosphere, or Sun's surface, is its coolest layer.

The corona extends far out through the solar system and envelops the planets. It is known as the solar wind in the region surrounding the inner planets (Chapter 19). This wind consists of charged particles—electrons and protons—derived from the breakup of solar hydrogen atoms.

Although almost all the known elements can be detected by analysis in the Sun's rays, hydrogen is the predominant constituent of the Sun, with helium also abundant. It is estimated that hydrogen constitutes at least 90% of the Sun, and hydrogen and helium together total about 98%.

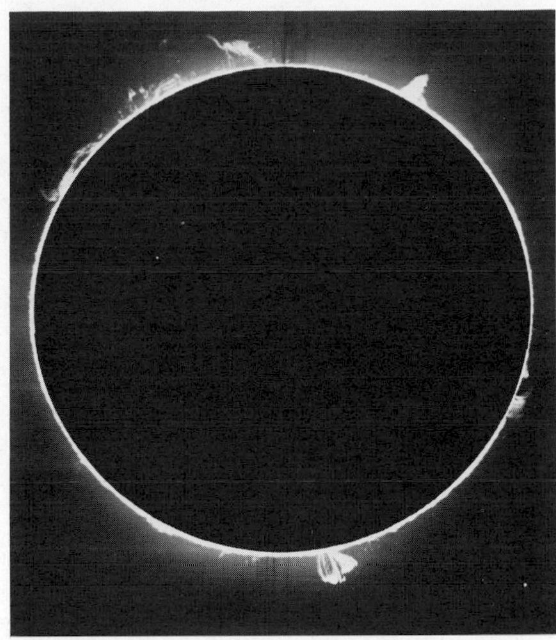

Figure 25.2 A photograph of the entire edge of the Sun shows several prominences. (The Hale Observatories.)

THE SUN'S INTERIOR

At temperatures over 4 million °K within the interior of a star, there occur several forms of reactions in which helium is produced. The source of the Sun's energy is the conversion of hydrogen into helium by nuclear fusion within the Sun's interior. During the fusion process, mass is converted into energy. The quantity of energy produced by conversion of matter into energy is enormous. At its present rate of energy production, the mass of the Sun will diminish only one-millionth part of its mass in 15 million years (m.y.).

Prior to the discovery of nuclear reactions, the Sun's energy was attributed entirely to the mechanical process of contraction under its own gravitation. The process is known as **Helmholtz contraction.** It depends on the principle that a gas forced to occupy a smaller volume undergoes a rise of temperature. Calculations made over 100 years ago by the physicist, Hermann von Helmholtz (1821–1894), demonstrated that the amount of energy produced by the Sun in 1 year could be derived through a reduction of about 280 ft (85 m) in its diameter.

Assuming that the Sun was formed from a highly dispersed body of gases, gravitational contraction to its present diameter was calculated to require 50 m.y. As you learned in Chapter 5, the radiometric dating of the age of rocks and meteorites points to a vastly longer span of time in which the solar system has endured in essentially the complete form we find it today. While the Helmholtz contraction process does not account for the Sun's present production of energy, it remains a valid principle when applied to the early stages of contraction of dispersed matter to produce a star.

Heat produced in the Sun's innermost core region moves outward by a process of radiation through the extremely dense gas of the interior. In a zone nearer the Sun's exterior, a process of convection (mixing) transports the heat to the surface.

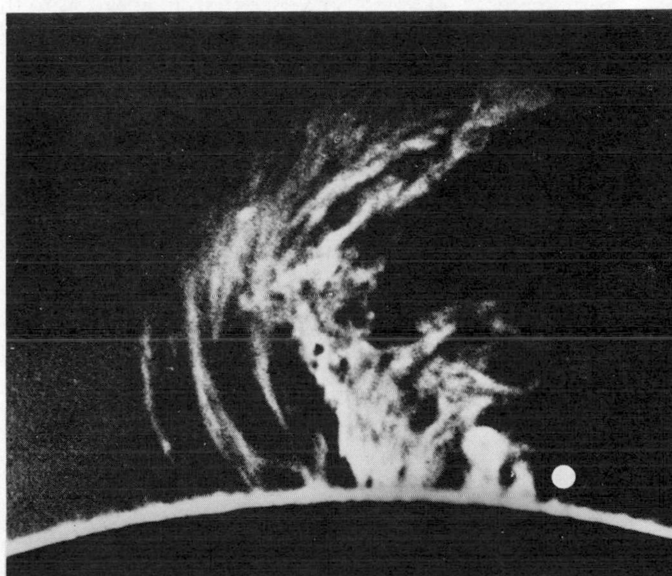

Figure 25.3 This great solar prominence rose to a height of 140,000 mi (225,000 km). The white dot at lower right represents the Earth at the same scale. (The Hale Observatories.)

SUNSPOTS AND SOLAR FLARES

A **sunspot** is a dark spot on the Sun's photosphere. The spot normally consists of a darker central region, surrounded by a somewhat lighter border (Figure 25.4). A single sunspot may be from 500 to 50,000 mi (800 to 80,000 km) across and represents a strong disturbance extending far down into the Sun's interior. The spot has a somewhat lower temperature than the surrounding photosphere. Sunspots form and disappear over a time span of several days to several weeks, during which time they can be seen to move with the Sun's rotation. The frequency of sunspots follows an average cycle of about 11 years.

It has been found that the sunspots have powerful magnetic fields associated with them—several thousand times as great in intensity as the magnetic field at the Earth's surface. This magnetism takes the form of strong poles associated with the sunspots. Adjacent spots of a pair in the same hemisphere have opposite polarity.

The same intense magnetic fields that are associated with sunspots also produce **solar flares,** which are emissions of ionized hydrogen gas from the vicinity of the sunspots. It is from such flares that X rays are sent out, followed by streams of charged particles. These particles reach the Earth about a day later. The emissions received from solar flares cause the Van Allen radiation belt to become intensely radioactive. At such times the magnetic field at the Earth's surface is severely disrupted. This phenomenon is a magnetic storm; it interferes seriously with radio communication (Chapter 19).

Solar flares occur in much greater numbers than sunspots. As many as 2000 to 4000 flares occur per year during times of maximum sunspot activity. Flares are thus about 20 times more frequent events than sunspots, but their duration is correspondingly much shorter. A single sunspot group in the course of its duration will produce as many as 40 flares. So we see that, in addition to the steady solar wind, the Earth intercepts intense bursts of X rays and ionized particles at irregular intervals.

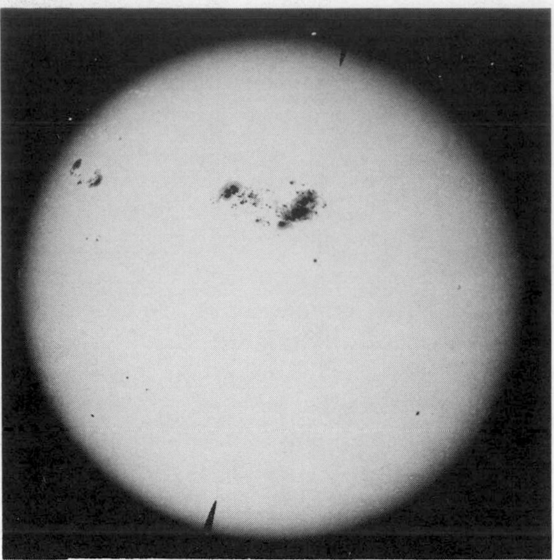

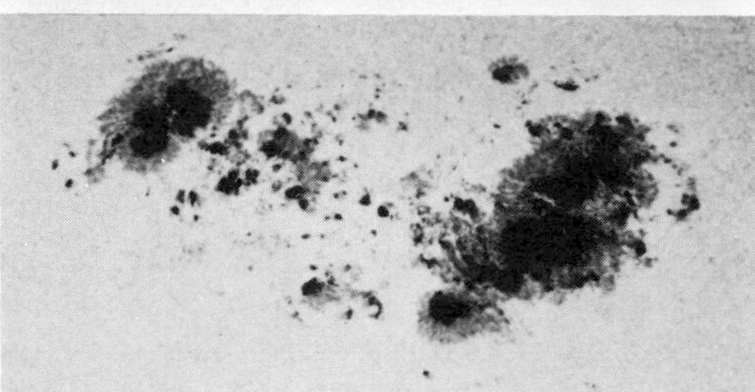

Figure 25.4 The whole disk of the Sun (*above*) shows a large sunspot group. Below is an enlargement of the group of spots, showing the darker and lighter regions. (The Hale Observatories.)

UNITS OF INTERSTELLAR DISTANCE

The vastness of interstellar space requires us to use units of length quite different from those applicable to the solar system. Consider the fact that the nearest star to our Sun, Alpha Centauri, is about 300,000 times more distant from the Sun than the Sun is from Earth. A convenient unit of interstellar distance is the **light-year;** it is the distance traveled by light in 1 year's time. Multiplying the speed of light, 186,000 mi/sec, by the number of seconds in the year gives a value of approximately 6 million million mi as the distance equal to 1 light-year. Alpha Centauri is about 4.3 light-years distant from the Sun.

Astronomers also make use of another measure of distance. This measure is based on the principle of **stellar parallax:** As the Earth moves across its great orbital distance each year, the nearer stars should seem to change their apparent positions in relation to the more distant ones. **Parallax** is a word used in optics; it means a difference in the apparent relative positions of objects when viewed

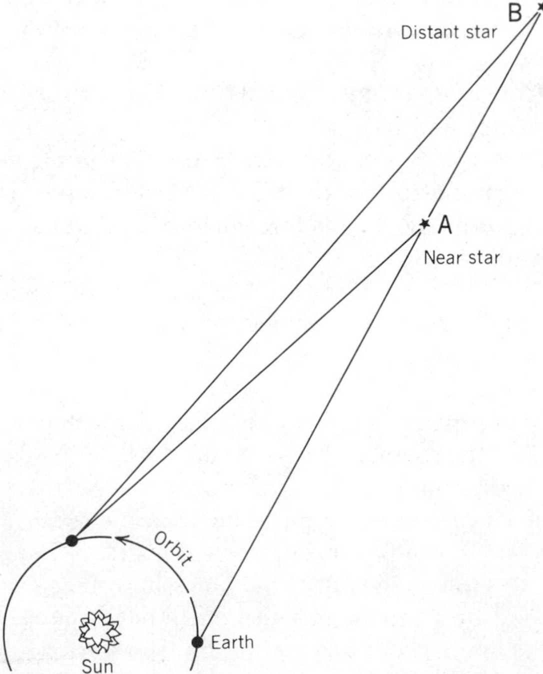

Figure 25.5 The principle of the parallax of stars.

from different points. For example, as you walk along a road; a tree close by may appear in line with a distant tree. As you move farther along, the two trees will seem to become separated by a widening distance.

The principle of parallax is illustrated in Figure 25.5 (the angles are greatly exaggerated). A near star, A, may appear very close to a distant star, *B*, when viewed in the spring. But as the Earth moves in its orbit, star A will seem to shift its location in the sky, so that in the autumn star A may be separated from star *B* by a very small angle. The closer star A is to us, the greater the parallax effect.

The parallax of the stars was first measured in 1838 by the Prussian astronomer Bessel, who discovered that a faint star in the constellation of Cygnus was displaced annually by a very small amount. Since then, the parallaxes of about 10,000 stars have been measured.

A star having a parallax of exactly 1 second of arc would lie at a distance of about 20 million million mi from the Sun. This distance is 1 **parsec,** a term coined from the words parallax and second. Alpha Centauri lies at a distance of 1.3 parsecs from the Sun. One parsec is equal to 3.26 light-years.

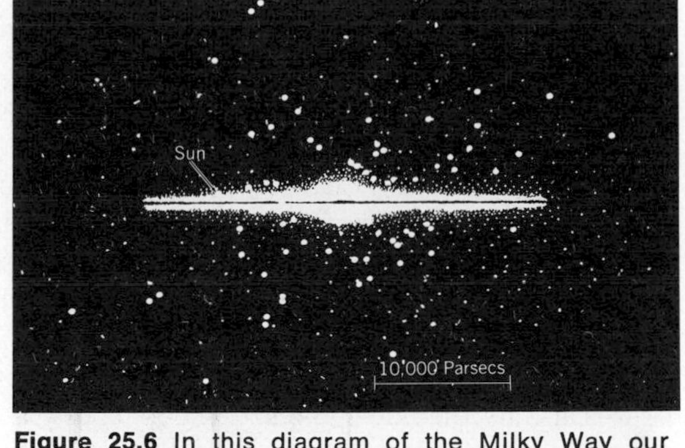

Figure 25.6 In this diagram of the Milky Way our galaxy is viewed from a point in the plane of the spiral. The large spots represent star clusters; the small spots represent stars. (Reprinted from *The Universe* by Otto Struve by permission of The M.I.T. Press, Cambridge, Massachusetts. Copyright © 1962 by The Massachusetts Institute of Technology.)

THE SUN IN OUR GALAXY

In its larger setting, our Sun is but one star among some billions of stars grouped into an assemblage called a **galaxy.** Our galaxy, in turn, is but one of a vast number of widely separated galaxies constituting the **universe,** which is the sum total of all matter and energy that exists.

Our galaxy has the form of a great disk, or wheel, with a marked central thickening at the hub (Figure 25.6). If it could be seen from an outside vantage point, our galaxy would probably be quite similar to the Whirlpool nebula and to the Great Spiral galaxy located in the constellation of Andromeda (Figure 25.7).

Figure 25.7 Three spiral galaxies photographed with the 200-in. (500 cm) telescope. (*Left*) The Great Spiral Galaxy, M 31, in the constellation Andromeda. (*Center*) Whirlpool Nebula, spiral galaxy M 51. (*Right*) Spiral galaxy NGC 4565, seen edge on. (The Hale Observatories.)

Our Sun occupies a position more than halfway out from the center toward the rim of the galaxy (Figure 25.6). As we look out into the plane of the disk, we see the stars of the galaxy massed in a great band, the Milky Way, which completely encircles the sky. For this reason our galaxy is named the Milky Way galaxy.

The Milky Way galaxy rotates about its hub, the center part turning more rapidly than the more distant outer regions. At the position occupied by our Sun, a full cycle of rotation requires about 200 m.y. The velocity of the solar system in this circuit is about half a million mi (800,000 km) per hour.

Thickness of the Milky Way galaxy is from 5000 to 15,000 light-years, its diameter about 100,000 light-years. The galaxy has a system of **spiral arms,** comparable to those in the Andromeda spiral (Figure 25.7). Each arm consists of individual aggregations of stars, known as **star clouds.** Each cloud has a dimension of 5000 to 20,000 light-years. Altogether, about 100 billion stars are contained in the galaxy.

Figure 25.8 This globular star cluster is in the constellation of Hercules. (The Hale Observatories.)

The Milky Way galaxy also contains gas clouds and clouds of cosmic dust. Concentration of these clouds is particularly heavy in the plane of the galactic disk (Figure 25.6). Surrounding the disk is a vast **halo** of widely scattered stars and **globular star clusters** (Figure 25.8).

PROPERTIES OF STARS

To understand our Sun we must compare it with other stars. We use the word **star** to refer to discrete concentrations of matter in our galaxy, bound into single units by gravitation. In this way a star stands distinct from highly dispersed matter in the form of gas clouds and dust clouds.

Measurable properties that distinguish one star from another and enable classification to be made are mass, size, (volume, radius, or surface area), density, luminosity, and temperature. Temperature in turn determines the type of radiation emitted by the star.

Mass of a star, which refers to the quantity of matter present, varies over a wide range. Taking the mass of our Sun as unity (1.0), the masses of stars range from as small as about one-tenth that of the Sun to about 20 times greater than the Sun. Stars also have a great range in diameter. For example, a small companion star to Sirius has a diameter only one-thirtieth that of the Sun, whereas the diameter of Antares is almost 500 times greater than that of the Sun.

Density of a star refers to the degree of concentration of mass within a given volume of space. The average density of the Sun is only slightly more than the density of water at the Earth's surface. Stars show a truly enormous range in density, from less than one-millionth that of the Sun to more than 100 million times as great. The companion star to Sirius, referred to above as a small star, has a mass almost equal to that of the Sun and, consequently, a density 35,000 times that of water on the Earth's surface.

Luminosity of a star is the measure of its total radiant energy output as if measured at the star itself. The luminosity of the Sun is taken as unity (1.0). The range of luminosity among stars is from as low as one-millionth that of the Sun to as high as half a million times as great. However, for most stars the luminosity ranges between one ten-thousandth and 10,000 times that of the Sun.

Star temperature, always given in degrees Kelvin (°K), refers to the surface temperature. Temperatures range from below 3500° K to 80,000° K. A star's color is closely related to its surface temperature: The hottest stars are blue; those only a little cooler are white; at progressively lower temperatures star color ranges from yellow through orange to red.

STELLAR DISTANCES AND BRIGHTNESS

To the observer on Earth the great range in brightness of the stars has long been recognized by designations of **star magnitude.** These designations are used for purposes of navigation and general descriptive astronomy. Many of you are familiar with a system used on star charts in which the brightest stars are classed as of the "first magnitude," those of lesser brightness as "second magnitude," and so on, down to the sixth magnitude.

When placed on an exact basis, the **apparent visual magnitude** of celestial objects resolves itself into a scale of numbers. In this magnitude scale each integer value represents an increase in light intensity by a factor of 2.5 over the next larger integer. Thus a star of magnitude 1 is 2.5 times as bright as one of magnitude 2, but 6.25 (2.5 × 2.5) times as bright as one of magnitude 3. The magnitude scale, which is a logarithmic (constant ratio) scale, extends through zero into negative numbers. According to this scale, the Sun's apparent visual magnitude is −26.7, the Moon when full, −12.7, and Venus in brightest phase, −4.5.

Table 25.1 gives the apparent visual magnitudes of the 15 brightest stars, together with information on luminosity and distance. Apparent visual magnitude is measured by sensitive photoelectric meters attached to telescopes. Magnitudes as faint as +24 can be measured. Consider next this concept: Apparent visual magnitude depends upon two factors, (1) luminosity of the star and (2) its

Table 25.1. **The fifteen brightest stars**

Name	Constellation	Apparent visual magnitude	Luminosity (Sun = 1)	Distance (light-years)
Sirius	Canis Major	−1.44	23	8.7
Canopus	Carina	−0.72	1,500	180
Alpha Centauri	Centaur	−0.27	1.5	4.3
Arcturus	Boötes	−0.05	110	36
Vega	Lyra	0.03	55	26.5
Capella	Auriga	0.09	170	47
Rigel	Orion	0.11	40,000	800
Procyon	Canis Minor	0.36	7.3	11.3
Betelgeuse	Orion	0.40	17,000	500
Achernar	Eridanus	0.49	200	65
Beta Centauri	Centaur	0.63	5,000	300
Altair	Aquila	0.77	11	16.5
Aldebaran	Taurus	0.80	100	53
Alpha Crucis	Southern Cross	0.83	4,000	400
Antares	Scorpius	0.94	5,000	400

distance from Earth. Light emitted from a point source diminishes very rapidly with increasing distance. For two stars of equal distance from Earth, the one with the greater luminosity will appear to be the brighter. On the other hand, a near star of low luminosity might appear just as bright as a distant star of high luminosity. For example, Betelgeuse and Achernar have nearly the same apparent visual magnitude, but Betelgeuse is vastly more luminous and is much farther away (Table 25.1).

To reduce the actual stellar luminosities to a scale that correlates with the scale of magnitudes, a system of absolute magnitudes is used. The **absolute magnitude** of a star is the apparent visual magnitude it would have if it were located at a distance of 10 parsecs from the Sun. In Figure 25.9, absolute magnitude is scaled on the left-hand side of the graph in numbers ranging from under −4 to over +16. By reading across to the right-hand side of the graph, you can read the corresponding value of luminosity.

STAR MASS AND LUMINOSITY

You might well reason that the larger a star, the greater will be its luminosity, since the area of radiating spherical surface increases greatly as the diameter increases. Your reasoning ignores one fact. A hot star may be radiating 10,000 times more strongly per square centimeter of its surface than a cool star.

On the other hand, there is a sound scientific reason to associate increased mass with increased luminosity. The more massive the star, the greater will be the gravitational pressure tending to cause contraction and, consequently, the higher will be the star's internal temperature. As internal temperature increases, the rate of production of energy by the nuclear fusion processes also increases. For this reason, as a general rule, the larger the mass of a star, the greater will be its output of radiant energy.

Figure 25.9 is a graph in which luminosity and corresponding absolute magnitude are plotted against mass for a number of stars whose mass and luminosity have both been independently measured.

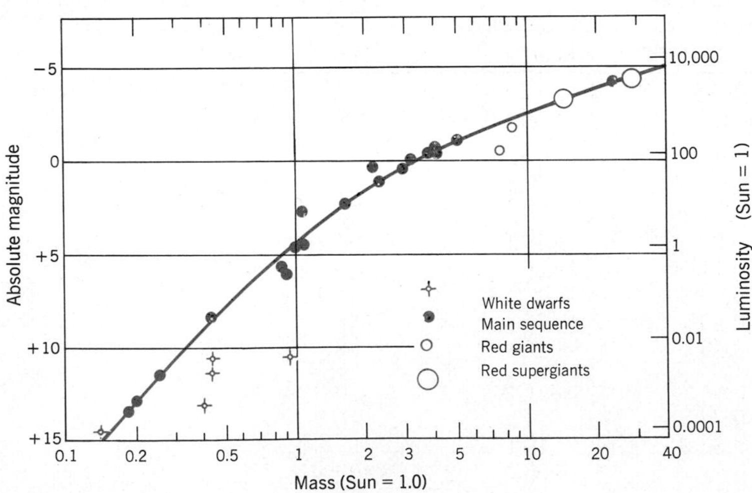

Figure 25.9 Mass-luminosity diagram. (After T. G. Mehlin, 1959, *Astronomy*, New York, John Wiley & Sons, p. 45, Figure 2-3.)

For the most part, the stars fall on or close to a broadly curved line. At the upper right are stars of enormous size known as **red supergiants,** below them and to the left are **red giants.** In the middle of the graph are stars of the **main sequence,** ranging from 100 times to about 1/500 the Sun's mass. However, you see on the graph a group of stars known as the **white dwarfs,** whose plotted positions lie far off the typical curve. These are very small stars of extremely high density. They produce far less heat from thermonuclear processes than do stars of the main sequence having equivalent masses. Apparently, the white dwarfs have largely exhausted their supplies of hydrogen and have contracted into an abnormally dense state.

The mass-luminosity curve is useful to the astronomer because it enables him to estimate the mass of a star of the main sequence when its luminosity is known, or vice versa.

SPECTRAL CLASSES OF STARS

The radiation spectrum produced on the photosphere of a star consists of the full sequence of wavelengths appropriate to the temperature of the radiating surface. We discussed these concepts in Chapter 20 under the subject of the Sun's radiation spectrum. As this radiation passes through the star's atmosphere (chromosphere) the various elements that make up the atmospheric gas absorb certain wavelengths. Where absorption occurs, black lines show on the color spectrum.

Each element creates its particular set of absorption lines on the spectrum, a fact which allows elements in a star to be identified with certainty. Moreover, it is possible to determine the physical state of the absorbing element, whether it exists as neutral atoms or in the ionized state. From the spectral lines the temperature of the star's atmosphere and the proportions in which each element is present can also be determined.

Stars are classified according to the **spectral class** to which each belongs. Arranged according to temperature, from hottest to coolest, the six major classes are designated *B, A, F, G, K,* and *M.*

SPECTRUM-LUMINOSITY RELATIONSHIPS

About 1910 two astronomers, Hertzsprung and Russell, working independently, plotted star luminosity against position in the spectral sequence. They found that a distinct and meaningful relationship exists. Figure 25.10 is the **Hertzsprung-Russell diagram** (or simply **H-R diagram**), in which each point represents a star. Luminosity is scaled on the vertical axis, and a corresponding scale in terms of absolute magnitude is given as well. On the horizontal axis, spectral classes are arranged in sequence from highest temperature, on the left, to lowest temperature, on the right. A scale of temperatures in degrees Kelvin is shown across the bottom.

It is obvious that most of the stars plotted on the H-R diagram lie in a diagonal band commencing with high temperature and great luminosity at the upper left and ending with low temperature and small luminosity at the lower right. This band is the main sequence.

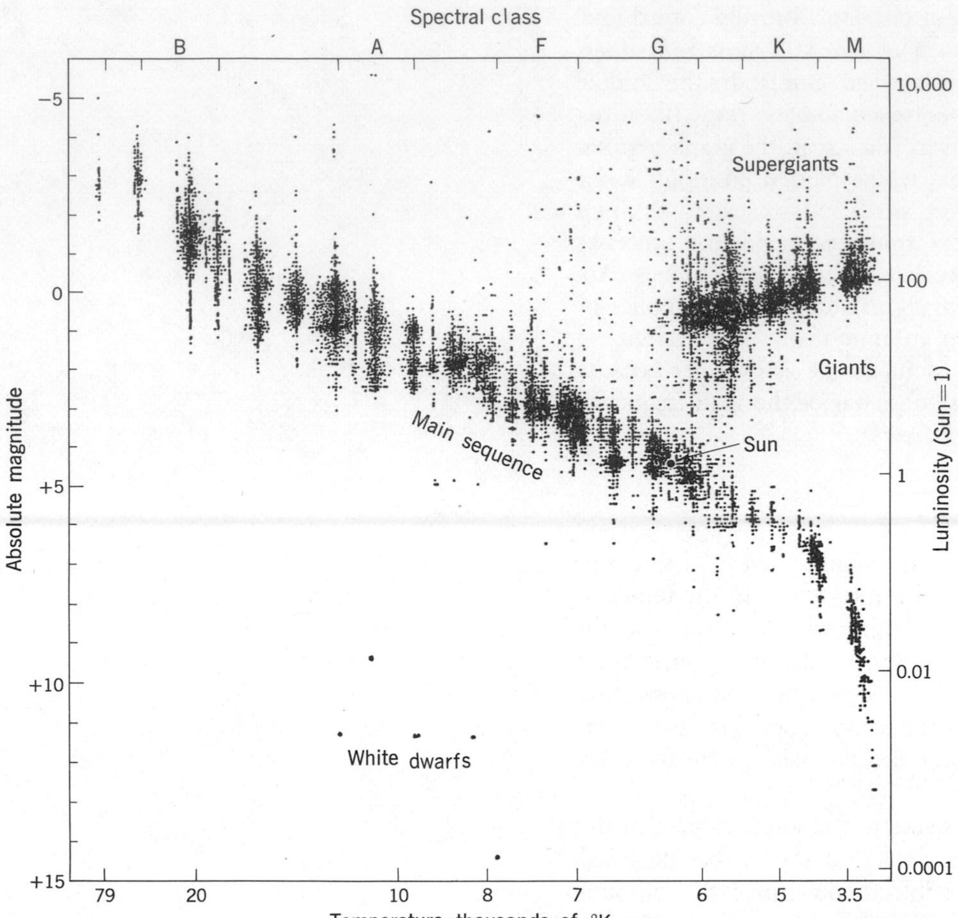

Figure 25.10 The Hertzsprung-Russell spectrum-luminosity diagram. Each dot represents a star. Altogether a sample of 6700 stars is recorded on the diagram. (Yerkes Observatory.)

Our Sun lies about two-thirds of the way down this main sequence. A large, isolated cluster of points above and to the right of the main sequence consists of the red supergiants and red giants. These enormous stars have great luminosity despite their relatively cool temperatures. They fall into the spectral classes *K* and *M*. In the lower part of the diagram are a very few stars, the white dwarfs. We have already noted that these stars are very small but of extremely great density. They are relatively hot stars.

THE LIFE HISTORY OF A STAR

Information we have reviewed thus far can be organized to describe the life history of a star.

Within our Milky Way galaxy there are clouds of cold gas and dust whose temperature is close to absolute zero. Certain of these clouds appear as dark globules on astronomical photographs, because the gas effectively absorbs most or all of the starlight that would otherwise pass through from distant stars on the far side. Diameters of the dark globules are on the order of 100 to 1000 times the diameter of the solar system.

As a working hypothesis, we will assume that the cloud of cold gas making up a dark globule represents the initial stage in the life history of a star. Through the gravitational attraction which all particles of the gas cloud exert upon all other particles, the cloud begins to contract, occupying a smaller volume. Through the Helmholtz principle, already explained, the temperature of the contracting body of gas increases, and particularly so near the center of the mass, where pressures are greatest.

Eventually, contraction forms a star with an interior temperature exceeding 1 million °K. At this point the first of a series of nuclear reactions begins to take place, converting matter into energy and causing the star to begin emitting large amounts of electromagnetic radiation. As contraction continues and interior temperatures rise, other forms of nuclear reactions develop and sustain a high level of energy production. A fully developed star such as our Sun has now come into existence.

As the Hertzsprung-Russell diagram shows, stars of the main sequence span a very great range in both temperature and mass. Those of small mass can attain only comparatively low temperatures and pressures. These small bodies therefore produce energy at a relatively slow rate, resulting in stars of faint luminosity. Such small stars will have an extremely long life because the utilization of the hydrogen supply takes place so very slowly.

On the other hand, stars at the high-temperature and large-mass end of the sequence are converting their hydrogen supply into energy at an extremely fast rate. Their life expectancies will be short. For example, a star of mass 10 times that of the Sun will radiate energy about 10,000 times as rapidly as the Sun. The life of such a large star must therefore be on the order of 1% of the life of our Sun, or as short as 100 m.y. The small stars will correspondingly have lives vastly longer than the Sun, life spans as long as thousands of billions of years.

Figure 25.11 is a graph with essentially the same field as the H-R diagram, but it does not show the plots of the individual stars. The diagonal band shows the position of the main sequence. The chain of arrows represents the evolution of a single star of about the size of our Sun. The time-path of the star's evolution enters from the right and moves horizontally toward the line of the main sequence. This horizontal time-path is covered comparatively rapidly and represents the stage of contraction of the gas cloud and its rise in temperature. When the star begins to consume its hydrogen by thermonuclear processes it is located on the line of the main sequence. As appreciable amounts of the star's hydrogen are transformed into helium, the star may brighten slightly, moving slowly to a position slightly above its original main sequence location.

The next stage in the life history of a star comes when its hydrogen supply is seriously depleted. Nuclear activity ceases first in the central region of the star, which then contracts. Nuclear activity continues in a surrounding zone that gradually moves outward from the center toward the surface. As this happens, the star may expand greatly. Although the luminosity remains high, the surface temperature falls, and the star spectrum changes toward the red region.

On the H-R diagram (Figure 25.10) this change requires that the plotted position of the star depart from the main sequence and

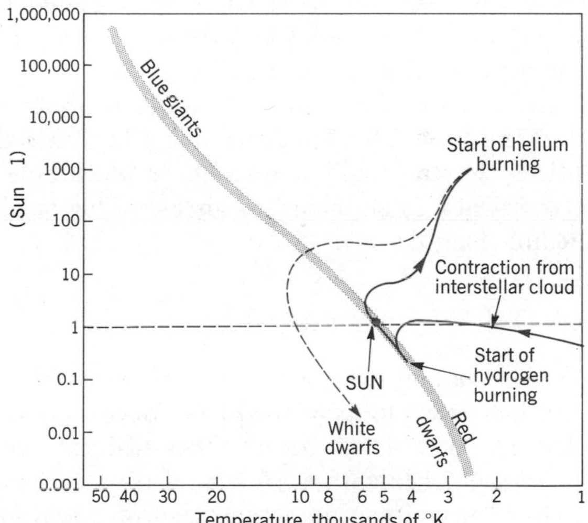

Figure 25.11 A simplified H-R diagram showing inferred evolution of an average star. (Reprinted from *The Universe* by Otto Struve by permission of The M.I.T. Press, Cambridge, Massachusetts. Copyright © 1962 by The Massachusetts Institute of Technology.)

move toward the upper right, occupying a position among the red giants and perhaps reaching the position of the supergiants. This time-path is shown on Figure 25.11.

The final hypothetical stage in the life of the star is suggested by a time-path that moves downward and to the left, then sharply downward to the region of the white dwarfs. The change into a white dwarf may be quite rapid. The star is now "burned out" and has only a faint luminosity despite its high temperature.

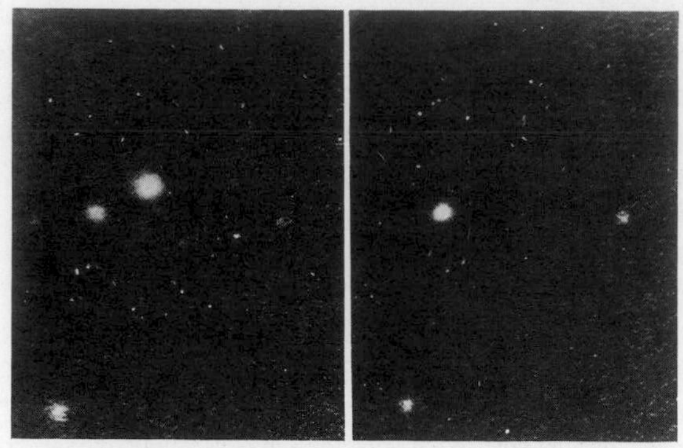

Figure 25.12 Comparison photographs of a pulsar in "on" (*left*) and "off" (*right*) phases. (Lick Observatory photograph, University of California.)

PULSARS AND BLACK HOLES

A recently discovered class of stars are the **pulsars.** These stars flash "on" and "off" rapidly, emitting both light waves and radio waves in the same rhythm. Light pulses range in frequency from about 1 pulse per 4 sec in the slowest rhythm to as high as 30 pulses per sec. In the case of the high rate of pulsation, the star appears to the eye and on photographs to be continuously bright, but special techniques can reveal the flashing on and off (Figure 25.12).

To explain the periodic emission of pulsars it has been suggested that they are extremely small, dense, dwarf stars, of a kind called a **neutron star.** The star is rotating rapidly on an axis. The emitting source is situated at one spot on the star and thus gives forth a single turning ray of light or of radio waves, as does a lighthouse or rotating beacon light.

The magnetic field of a pulsating neutron star is thought to be enormously strong—some thousand billion times as strong as that of our Sun. A measurable slowing of pulse rate in these stars suggests that energy is being dissipated at a rapid rate.

The continued gravity-caused collapse of a neutron star might conceivably lead to one of the strangest of all possible galactic objects. According to Einstein's theory, gravitational attraction acts upon light. Accordingly, it has been suggested that a neutron star might, if it became sufficiently dense, develop such an intense gravitational force that it would pull back into itself all light that it produces, and for that matter, all forms of electromagnetic radiation by means of which its presence might be detected. Under such conditions, the star would be invisible—a **black hole** in space. The star is envisioned as shrinking to an incredibly small size with an incredibly high density.

NOVAE AND SUPERNOVAE

On occasion, an extremely faint star bursts into intense brightness, then fades back to its original level. Such stars are known as **novae,** meaning "new stars," because they had not been noticed prior to the episode of brightness. A typical nova increases in brightness by 10 to 12 magnitudes in a time span of a few hours to a few days. Immediately after the outburst, the brightness falls off rapidly for a time and then tends to level off and to diminish gradually over many weeks. Within years the brightness returns to the original level.

Increase in brightness of the nova is associated with an explosive increase in its size, which may be a diameter increase of from 100 to 200 times. This expansion takes place in the photosphere of the star and is not an explosive enlargement of the entire star interior. The expanded layer of gases gradually dissipates and is lost into space, revealing the main body of the star intact. Novae are interpreted as being stars in the white-dwarf stage, near the end of the stellar life cycle. The explosion represents a short period of instability during the final stages of contraction into an extremely dense small star.

A very rare type of nova is one that attains sudden brightness in the range between a quarter-billion and 1½ billion times the luminosity of the Sun. If this kind of star is in our own galaxy, its apparent magnitude may exceed that of the brightest planets. Such phenomena are known as **supernovae.** They occur within our galaxy about once every several hundred years. Following the outburst, an expanding cloud of gas and dust, or **nebula,** has been observed surrounding the site of the supernova. The Ring Nebula is an example of this feature (Figure 25.13).

The supernovae probably originate from extremely massive stars that have transformed so much of their hydrogen to heavier elements that they become explosively unstable. Unlike the typical nova, the outburst of a supernova blows off the major part of the material of the star, changing the nature of the star drastically.

In 1969, the central star of the Crab Nebula was shown to be a pulsar, or rapidly rotating neutron star, only a few miles in diameter, but having a density on the order of 1 million billion times the density of water. It emits extremely regular radio pulses at the rate of 30 per sec. Thus, evidence is accumulating to show that a neutron star is the dense mass remaining after the outburst of a supernova.

GALAXIES AND THE UNIVERSE

Our observable universe consists of widely spaced galaxies, of which an estimated 10 billion can now be viewed telescopically, but no outer limit to the universe can be recognized. Within the limits of observation there may be as many as 100 billion galaxies.

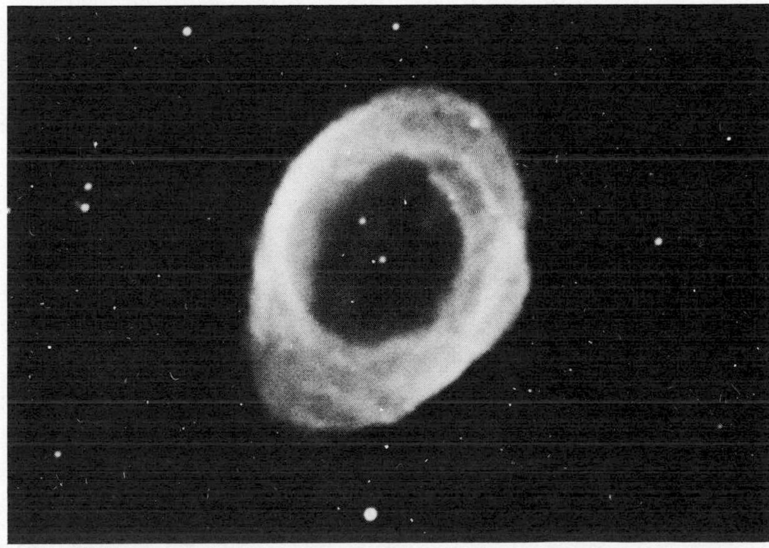

Figure 25.13 The Ring Nebula is the gaseous remains of a supernova, spreading rapidly outward in all directions. A compact dwarf star, seen at the center, is what remains of the body which exploded to produce the nebula. (The Hale Observatories.)

Figure 25.14 A barred spiral galaxy in the constellation of Pegasus. (The Hale Observatories.)

Galaxies fall into several classes, according to their shapes. **Spiral galaxies,** such as our Milky Way, are illustrated by the Andromeda spiral, which is the closest galaxy to our own (Figure 25.7). Its distance is about 2 billion light-years, and its diameter is a bit larger than our Milky Way galaxy. Another class of galaxies is the **barred spirals,** in which the two arms uncoil from a central bar (Figure 25.14). Equally important are galaxies of the **elliptical** group (Figure 25.15). These are ellipsoidal or spherical masses having a high degree of symmetry. Their form suggests that they, like the spirals, are rotating. In addition, there are galaxies of highly irregular shape, but these are relatively few.

Within the nearer galaxies individual stars and star clusters can be recognized. Clouds of dust and gas that are typical of the spiral galaxies seem to be absent from the elliptical types. Attempts have been made to arrange the several forms of galaxies into an evolutionary series. Edwin P. Hubble, the astronomer who did much of the pioneering work in galactic investigation, suggested a classification which begins with the almost spherical elliptical galaxies and then progresses to the more flattened systems. The series then branches into two parallel arms, one for the spirals and the other for barred spirals, and perhaps ends with the irregular galaxies (Figure 25.16).

When more was known about the galaxies and the ages of the stars in them, Harlow Shapley suggested a reversed evolutionary sequence. It might well begin with the irregular galaxies, developing into spiral systems in which the nucleus would move tightly into the arms with increasing age. Then, as the stars aged and the interstellar clouds of gas and dust were eliminated, the spirals might

Figure 25.15 An elliptical galaxy in the constellation of Andromeda. (The Hale Observatories.)

evolve into elliptical systems of varying degrees of flattening. It is still not understood why some spirals take the normal form and some become barred spirals.

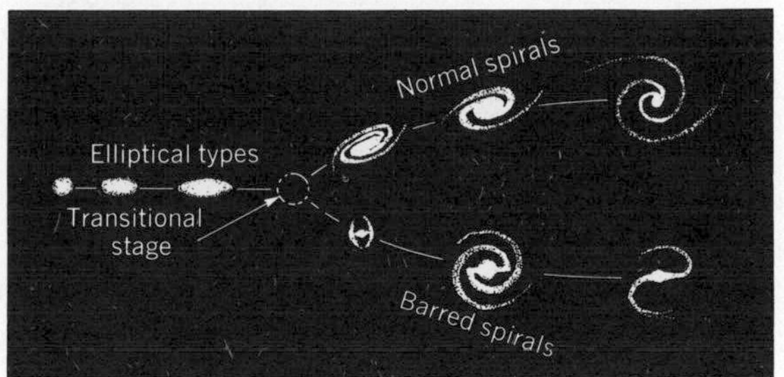

Figure 25.16 This diagram shows the sequence of nebular types as arranged by the astronomer Edwin Hubble. No nebulae have been recognized in the transitional stage, which is hypothetical. (From E. Hubble, 1936, *The Realm of the Nebulae,* New Haven, Yale Univ. Press, p. 45.)

THE DOPPLER EFFECT

One of the most remarkable findings about galaxies is evidence that they are rapidly moving away from us, and that the farther away they are, the faster they are receding. To understand the evidence for such a statement we must investigate a principle of science applying to the light spectrum and other forms of wave motion, such as sound waves.

The principle involved is the **Doppler effect,** familiar in its manifestation in sound waves. The pitch of a sound of fixed vibration period seems higher as the emitting source is brought rapidly toward us, and lower as it recedes from us. The sudden drop in pitch of a locomotive horn as it passes by at close range is a good example.

A very simple analogy may help to illustrate the Doppler principle. Suppose that we stand beside a long horizontal conveyer belt and place pebbles on the belt at uniform intervals of time. If the belt speed is constant, the pebbles will be uniformly spaced. Now, if as we place the pebbles we also walk slowly in the direction in which the belt is moving, the pebbles will be spaced closer together; whereas if we walk in a direction opposite to the belt motion, again placing pebbles at the same intervals of time, they will be spaced farther apart on the belt. In a similar way the frequency of light rays is changed as the emitting source moves.

Take now the case of a star emitting a given light spectrum. When the star is moving earthward, it appears to us that the frequencies of vibration constituting the light rays are all increased slightly. This increase in frequency results in a slight change in the color, since the color we observe is determined by the frequencies of light, and these have been increased by the motion. When the star is moving away from the Earth a reverse effect occurs: The frequencies of vibration are reduced.

The Doppler effect has been an essential tool of astronomy. It has enabled astronomers to measure the speed of a star in the line of sight. This speed is called the **radial velocity.** Using this method, the speed of rotation of the Milky Way galaxy was determined.

THE RED SHIFT

In the case of a galaxy, its color spectrum is shifted toward the red end of the spectrum when the galaxy is moving away from us. For short, this effect is called the **red shift.** A particular line in the spectrum identified with a particular element, such as calcium, is found to be displaced toward the right in the spectrum. The amount of the red shift can then be interpreted in terms of radial velocity.

The red shift of the spectra of galaxies increases in direct proportion to their distance from our point of observation in the solar system. Assuming that the red shift is a true Doppler effect, all galaxies must be in radial motion, receding outward in straight lines from our galaxy. Moreover, the speed of recession increases proportionately with increasing distance. This principle was discovered by Edwin Hubble and is known as **Hubble's law.**

THE ORIGIN OF THE UNIVERSE

The geometry of apparent radial outward motion of the galaxies can be visualized in terms of a universe that is expanding uniformly in volume. From any single vantage point in this system all other objects will appear to be moving radially outward. Hubble's discovery of a law of increase of radial velocity proportionate with distance quickly led to a new theory of origin of the universe.

Cosmology as a science concerns itself with the nature and origin of the universe. Among the first to propose a cosmological theory based on Hubble's discovery was Canon Lemaître, a Belgian. He referred to the concept as a "fireworks theory." The theory requires an initial point in time at which all matter was concentrated into a small space. From this center it expanded explosively outward in all directions. The elements were created during this explosion and were later formed into the galaxies. Although now commonly referred to as the "big-bang" theory of the universe, the title of **evolutionary theory** is perhaps more fitting.

Using Hubble's first derived estimates of the rate of velocity increase with distance of separation, it could be calculated that all matter of the universe was concentrated into a small space about 2 billion years (b.y.) ago. This point in time was designated the **age of the universe.** Age of the universe in years would be equal to distance in light-years of the most rapidly traveling galaxy.

As we explained in Chapter 10, age determination of meteorites and rocks based upon analysis of radioactive decay of certain elements gives ages much greater than 2 b.y. Ages of meteorites and certain lunar rock fragments are found to be 4½ b.y. Theoretical investigations along a number of lines—such as the probable ages of the stars and the stability of star clusters in our galaxy—suggest an age at least twice this great for the universe.

In 1952, data obtained from the 200-in. reflector at the Mount Palomar Observatory required Hubble's calculations to be modified, and, consequently, to increase the calculated age of the universe to nearly 5 b.y. This modification placed the evolutionary theory in accord with the established ages of meteorites and rocks. More recently, the figures have again been revised to increase the age of the universe to as great as 10 b.y., or even greater.

The evolutionary theory conforms to the principle that the distribution of galaxies is uniform in all directions throughout space. Under this concept, to an observer from any galaxy the average composition of the universe would appear the same. It is interesting to consider that under Hubble's principle, the radial velocity of extremely distant galaxies, with respect to our observation point in the solar system, must reach and finally equal the speed of light. This distance would constitute the observable limit of the universe, beyond which we could receive no light or radio waves from the emitting sources.

The hypothesis of a **pulsating universe** has also been suggested, as a modification of the "big-bang" hypothesis. Immediately after the initial explosion, all of the matter would be moving outward with high velocities, but the mutual gravitational attraction between all of the parts would tend to slow the outward motion, perhaps finally stopping it and causing the entire system to contract. All of the material would eventually come back to a central point in an implosion that would annihilate all forms of matter—stars, galaxies, and even individual atoms. The result would be another "cosmic bomb," which would explode and start the whole process over again. The interval for one complete cycle has been estimated to be something less than 100 b.y.

A major rival theory of the universe holds that there was no single point in time at which matter was concentrated in one place. Instead, the production of matter has gone on throughout intergalactic space at a constant rate during all time. Rate of production of matter in the form of hydrogen atoms has been equaled by the rate at which matter is dispersed by the expansion of the universe. This **steady-state theory** of cosmology was proposed in 1948 by the astronomers H. Bondi, T. Gold, and F. Hoyle.

The most recent evaluations of information concerning galaxies and other distant objects seem to place the evolutionary theory of the universe in a stronger scientific position than the steady-state theory. However, we can anticipate modified and new cosmological theories of the universe to be brought forward from time to time as new information is gained from the development and use of newer tools of astronomy.

RADIO ASTRONOMY AND QUASARS

Part of the electromagnetic radiation spectrum, that in the long-wave region, consists of radio waves (see Figure 19.8). Radio waves with wavelengths between about 1 cm and about 20 m can pass through our atmosphere and be received by **radio telescopes.** These instruments use a huge, concave, bowl-shaped (parabolic) antenna that can be aimed at a distant emitting source (Figure 25.17).

Figure 25.17 The world's largest radio telescope antenna is this parabolic dish, 1000 ft (300 m) in diameter, nestled in a natural bowl-shaped valley. It belongs to the Arecibo Observatory in Puerto Rico. (Courtesy of the National Astronomy and Ionosphere Center, operated by Cornell University under contract with the National Science Foundation.)

Thousands of radio-emitting sources have been discovered and their positions plotted, but only a few can be identified with stellar objects that appear on photographs. Some sources of radio emission lie within our Milky Way galaxy; others are in distant galaxies, referred to as **radio galaxies.** These radio galaxies, of which about 150 have been identified, are the most powerful of all radio-emission sources.

Hydrogen gas clouds within our galaxy are also radio-wave emitting sources. Our Sun shows strong radio-wave emission at those times when a solar flare is in progress. A number of stars are known to have flares of similar nature; at such times their brilliance is greatly increased. For this reason radio emissions received from these stars are believed to be associated with flares.

Among the most important of astronomical discoveries in recent years has been the finding of extremely small sources of intensely powerful radio emission not related to any surrounding galaxy. They were named **quasistellar radio sources,** but the term has since been reduced to **quasars.** These emission sources appear only as pinpoints of light. The distribution of the 100 or so quasars identified is quite random with respect to direction from the Earth.

A particularly striking feature of quasars is that the lines in their spectra show a very great shift toward the red. Although its use here may be questioned, if the same red shift-distance relationship developed for galaxies is applied to quasars, the extremely large red shift would lead to the conclusion that they are on the order of 1 to 10 billion light-years away. If so, they are the most distant known objects in the universe. The luminosities and energy outputs of the quasars must be truly enormous. One hypothesis explains the quasars as formed from gas clouds sent outward from the center of an exploding universe at a speed up to 80% that of light.

X-RAY ASTRONOMY

The Earth's atmosphere effectively blocks nearly all X-ray, gamma ray, and ultraviolet radiation from terrestrial telescopes. Much of the infrared radiation is also blocked. Until orbiting satellites were available, astronomers had to depend largely upon the visible light spectrum and upon radio waves for their knowledge of stars, galaxies, and other energy-emitting objects of the universe. Now, telescopes mounted on spacecraft orbit high above the atmosphere. They are highly specialized in function to receive specified bands of the electromagnetic spectrum over its entire range.

Launched in 1970, a satellite named Uhuru began to pick up sources of X-ray emission from many celestial sources. In 1972 another orbiting telescope package, named Copernicus, began its observations in the X-ray and ultraviolet bands. Uhuru identified more than 100 X-ray emission sources. Many of these X-ray sources lie in our galaxy, but some are in other galaxies. Intense sources have been picked up from quasars and from pulsars in the remnants of supernovae. Other sources are suspected of being black holes, into which X-ray emitting gases are being drawn from nearby stars.

COSMIC PARTICLES

The Earth's atmosphere is continually bombarded with elementary particles traveling at speeds approaching the speed of light and having enormous energy and penetrating power. This form of radiation from outer space is the **cosmic particle,** often called the "cosmic ray." It is an entirely independent phenomenon from the electromagnetic radiation spectrum of a star.

Cosmic particles are protons, that is, parts of the atomic nucleus. Approximately 90% are hydrogen nuclei, 9% are helium nuclei, and 1% are heavier nuclei. The energy of cosmic particles is enormous.

Cosmic particles approach our Earth from all directions. Their space paths seem to be quite at random, and they can be visualized as constituting a kind of cosmic "gas" in which particles undergo random collisions and can take an infinite variety of paths and a wide range of speeds. Sources of cosmic particles are considered to be varied. They are produced in solar flares, but most come from other sources, believed to be the explosions of supernova and other forms of explosive activity in the central parts of our own and other galaxies. It has been suggested that galaxies which emit radio waves are also sources of important amounts of cosmic radiation.

Cosmic particles are important in the environment of life on the Earth's surface. The extremely high energy of cosmic particles enables them to penetrate deep into the lower atmosphere and to reach the Earth's surface. This penetration is accompanied by an elaborate series of secondary nuclear reactions making up a shower of particles and secondary forms of radiation. When a high-speed cosmic particle impacts the nucleus of an atom within the atmosphere, there are produced neutrons and protons, mesons, and gamma radiations. The effect of such secondary radiation upon life forms is to induce genetic changes (mutations in genes) which are important in the process of organic evolution.

MAN'S PLACE IN THE UNIVERSE

Seen in its relative position among the other stars of the Milky Way galaxy, our Sun is a fairly typical star in most respects. It lies somewhat below the midpoint of the main sequence of stars, belonging to a spectral class which has moderate surface temperatures in terms of the total temperature range. Luminosity and mass are about midway on the scale of those values. Extreme constancy of energy output over vast spans of geologic time characterizes the Sun, a behavior in strong contrast to the violent energy emissions of the novae.

Our Sun represents one of the basic forms of energy systems, that of conversion of matter to energy in nuclear reactions occurring within a gaseous medium under enormously high pressures and temperatures. The life span of our Sun is neither very short nor very long in comparison with the range found among stars, but it is long enough to assure that our terrestrial environment can continue with little change for a span of time vastly longer than that which has already elapsed as geologic time.

In reference to the total size of the Milky Way galaxy, our Sun is no more than an insignificant particle of matter, while in the context of the universe of galaxies, it comes close to being nothing at all. Within the universe there must be a very large number of stars quite similar to our Sun, and many of these must have planets resembling our own.

Reason leads us to suppose that spontaneous development of organic life and its evolution to highly complex states must have been replicated a great number of times on unknown planets. But we also realize that the vastness of interstellar and intergalactic space reduces almost to zero the possibilities of identifying and communicating with even the closest of such organic complexes. Despite such odds, the possibility of a discovery that Man on planet Earth is not alone in the universe continues to fire the popular imagination.

YOUR GEOSCIENCE VOCABULARY

photosphere
nucleus of Sun
chromosphere
solar prominence
corona
Helmholtz contraction
sunspot
solar flare
light-year
stellar parallax
parallax
parsec
galaxy
universe
spiral arms
star clouds
halo
globular star cluster

star
luminosity
star magnitude
apparent visual magnitude
absolute magnitude
red supergiants
red giants
main sequence
white dwarfs
spectral class
Hertzsprung-Russell diagram (H-R diagram)
pulsars
neutron star
black hole
nova (novae)
supernova
nebula

spiral galaxy
barred spiral galaxy
elliptical galaxy
Doppler effect
radial velocity
red shift
Hubble's law
cosmology
evolutionary theory
age of the universe
pulsating universe
steady-state theory
radio telescope
radio galaxy
quasistellar radio source (quasar)
cosmic particle

SELF-TESTING QUESTIONS

1. In what respect is the Sun's rotation different from that of Earth? Describe the various layers, or shells, comprising the Sun. Where are temperatures highest?

2. What process operates in the Sun's interior to produce energy from matter? Explain the principle of Helmholtz contraction. Does this principle apply to the Sun's source of energy? Of what elements is the Sun composed?

3. Describe sunspots. How are they related to magnetic fields? How are they related to solar flares?

4. What units are used in the measurement of interstellar distances? How far away is the nearest star? Describe the phenomenon of stellar parallax, and show how it is used to measure distance to a star.

5. Describe the Milky Way galaxy and its motions. Where are spiral arms, star clouds, the halo, and globular star clusters located in this galaxy? Where in our galaxy is the Sun located?

6. What important physical properties of stars are subject to measurement? Compare our Sun with other stars in terms of mass, density, luminosity, and surface temperature.

7. What range of apparent visual magnitudes is found among celestial objects? What relation does luminosity bear to apparent visual magnitude? How is the absolute magnitude of a star determined? Why is it used?

8. Relate star mass to luminosity. Describe the mass-luminosity curve. Where on this curve are the red supergiants? the red giants? the main sequence? Where are the white dwarfs located with respect to the mass-luminosity curve? Explain their position.

9. What use is made of the spectrum of a star? What is the significance of the main spectral classes?

10. What spectrum-luminosity relationships emerge from the Hertzsprung-Russell diagram? Describe this diagram. How do the major groups of stars fit into it?

11. Trace the evolution of a star, beginning with a cloud of cold dust and gas. What causes interior heating of a star? At what point do nuclear reactions begin to occur? Track the stages of development of a star on the H-R diagram. In what way is the life expectancy of a star related to its mass? What is the end stage in the life of a star?

12. Describe the emission of pulsars. What hypothesis explains the pulsation? To what class of stars do the pulsars belong?

13. Describe the occurrence of a nova and explain what happens to the star. What event does a supernova represent in the life history of a star? What kind of star remains after a supernova has dispersed?

14. About how many galaxies can be observed with existing telescopes? How many galaxies are estimated to exist? Classify galaxies according to their forms. What sequence of development of galaxies was proposed by E. P. Hubble? by Harlow Shapley?

15. Explain the Doppler effect upon the spectrum of a star. How is the Doppler effect used to measure radial velocity of a star?

16. What is cosmology? Describe the red shift of the galaxies and relate it to their radial velocities. What is Hubble's law? Describe the geometry of a universe conforming to Hubble's law.

17. Describe the evolutionary ("big-bang") theory of the universe. How, according to this theory, can the age of the universe be estimated? Compare earlier and recent estimates of age of the universe. Do these estimates agree with radiometric ages of meteories?

18. Describe the hypothesis of a pulsating universe. In what way does the steady-state theory differ from the evolutionary theory?

19. Describe the radio-emitting source in space beyond the Earth. How are radio emissions received? Does the Sun emit radio waves? What are radio galaxies? What is remarkable about the emission spectra of quasars? How are quasars interpreted?

20. What are cosmic particles? Of what forms of matter are they composed? What happens when a cosmic particle enters the Earth's atmosphere? Where do cosmic particles come from?

21. Summarize the characteristics of our Sun as compared with other stars, and comment upon the importance of those characteristics in determining our planetary environment through geologic time.

22. What do you estimate to be the probability of existence of planets with life systems similar to those of planet Earth in other solar systems in our own and other galaxies? Would communication with advanced life forms on such distant planets be possible?

GLOSSARY

This Glossary contains definitions of all terms printed in boldface type in the text and listed at the end of each chapter under the heading Your Geoscience Vocabulary. Glossary terms are listed in alphabetical order, as in the Index. Italicized terms will be found as individual entries elsewhere in the Glossary.

The Glossary can be a valuable aid in reviewing for your final examination. Be sure you have a good grasp of the meaning of each term. If further explanation or an illustration is needed, refer to the Index, where the same term is listed with a boldface numeral giving the text page on which the term is introduced and explained.

ablation Wastage of glacial ice by both melting and evaporation.

abrasion Erosion of *bedrock* of a *stream channel* by impact of particles carried in a *stream* and by rolling of larger rock fragments over the stream bed. Abrasion is also an activity of glacial ice, waves, and wind.

abrasion platform Sloping, nearly flat *bedrock* surface extending out from the foot of a *marine cliff* under the shallow water of a breaker zone.

absolute magnitude *Apparent visual magnitude* that a *star* would possess if it were located at a distance of 10 *parsecs* from the sun.

absolute zero Value of 0° K (*degrees Kelvin*); –273° C.

absorption of radiation Transfer of energy of *electromagnetic radiation* into heat energy within a *gas* or *liquid* through which the radiation is passing.

abyssal plain Large expanse of very smooth, flat ocean floor found at depths of 15,000 to 18,000 ft (4600 to 5500 m).

accelerated erosion *Soil erosion* occurring at a rate much faster than *soil* horizons can be formed from the parent *regolith*.

accretion Process of coming together under *gravitation* and adhering of particles of the *solar nebula* to form *planetesimals*.

acidic lava *Lava* of *felsic mineral* composition, usually *rhyolite*.

active agents Fluid agents, including running water, glacial ice, waves and currents, and winds that carry out the action of the *external earth processes* through *erosion*.

adiabatic lapse rate (See *dry adiabatic lapse rate*, *wet adiabatic lapse rate*.)

adiabatic process Change of temperature within a *gas* because of compression or expansion, without gain or loss of heat from the outside.

advection fog *Fog* produced by *condensation* within the moist basal air layer moving over cold land or water.

age of the universe Elapsed time since the primordial explosion that gave rise to the *universe*, according to the *evolutionary theory*.

aggradation Raising of *stream channel* altitude by continued deposition of *bed load*.

air mass Extensive body of air within which upward gradients of temperature and moisture are fairly uniform over a large area.

air pollutants Foreign matter, causing *air pollution* and consisting of *particulates* and *chemical pollutants*.

air pollution Introduction of foreign matter into the lower *atmosphere* by man's activities through combustion of fuels and other activities.

air pressure (See *atmospheric pressure*.)

air temperature (See *annual temperature range*, *mean annual temperature*, *mean monthly temperature*.)

alluvial fan Low, gently sloping, conical accumulation of coarse *alluvium* deposited by a *braided stream* undergoing *aggradation* below the point of emergence of the channel from a narrow *canyon*.

alluvial meanders Sinuous bends of a *graded stream* flowing in the alluvial deposit of a *floodplain*.

alluvial river River (*stream*) of low gradient flowing upon thick deposits of *alluvium* and experiencing approximately annual overbank flooding of the adjacent *floodplain*.

alluvial terrace *Terrace* carved in *alluvium* by a *stream* during *degradation*.

alluvium Any stream-laid *sediment* deposit found in a *stream channel* and in low parts of a stream valley subject to flooding. (See *reworked alluvium*.)

alpine glacier Long narrow mountain *glacier* on a steep downgrade, occupying the floor of a troughlike valley.

alpine system Narrow *tectonic* belt, severely deformed by *folding* and *overthrust faulting* in recent geologic time, making high mountains like those of the European Alps.

aluminosilicates *Silicate minerals* containing aluminum as an essential *element*.

amphibole group Complex *aluminosilicate minerals* rich in calcium, magnesium, and iron, dark in color, high in *density*, and classed as *mafic minerals*. (See *hornblende*.)

andesite *Extrusive igneous rock* of *diorite* composition; occurs as *lava*.

andesite line Line around margins of the Pacific Ocean basin, marking the division between oceanic *basalt lavas* and *andesite lavas* of the *circum-Pacific belt*.

angular unconformity Variety of *unconformity* in which the younger layered *rocks* (*strata*) above the unconformity lie upon the truncated layers of the older rock beneath with an angular discordance. (See *unconformity, disconformity.*)

anhydrite *Evaporite mineral,* composition calcium sulfate.

annual temperature range Difference between the *mean monthly temperature* of the warmest and coldest months of the year.

annular streams *Streams* forming a pattern of incomplete concentric circles, as in the case of *subsequent streams* in valleys encircling an eroded *sedimentary dome.*

anorthosite *Igneous rock* formed of *silicate minerals,* largely *plagioclase feldspar;* a *felsic igneous rock.*

antarctic circumpolar current *Ocean current* system flowing eastward in the zone of *prevailing westerlies* of the Southern Hemisphere.

anthracite Grade of *coal* very high in fixed carbon content, with little volatile matter, and of metamorphic development, called hard coal.

anticlinal mountain Long, narrow ridge or mountain formed on an *anticline.*

anticlinal valley Valley eroded in weak *strata* on the central line or axis of an eroded *anticline.*

anticline Upfold of *strata* or other layered *rock* in an archlike structure; a class of *folds.* (See *syncline.*)

anticyclone Center of high *atmospheric pressure.*

aphelion Point on the earth's elliptical *orbit* at which the earth is farthest from the sun.

apogee Point in the moon's *orbit* that is farthest from the earth.

apparent visual magnitude *Star magnitude* stated in a logarithmic scale of numbers in which each integer decrease represents an increase in light intensity by a factor of 2.5 over the next higher integer.

aquiclude *Rock* mass or layer that impedes or prevents the movement of *ground water;* it is *impermeable,* or nearly so.

aquifer *Rock* mass or layer that readily transmits and holds *ground water;* possesses both high *porosity* and high *permeability.*

artesian well Drilled well in which water rises under hydraulic pressure above the level of the surrounding *water table* and may reach the surface.

assimilation Incorporation of *country rock* into a rising *magma* by means of melting or chemical reaction, thereby changing the composition of the magma.

asteroids Solid bodies, numbering in the tens of thousands, orbiting the sun between the orbits of Mars and Jupiter, ranging in diameter downward from 480 mi (770 km); also called *minor planets.*

asthenosphere Soft layer of the *mantle,* beneath the rigid *lithosphere.*

astrogeology Branch of *geology* applying principles and methods of geology to all condensed matter and gases of the *solar system* outside the earth.

atmosphere Envelope of gases surrounding the earth, held by *gravity.*

atmospheric pressure Pressure exerted by the *atmosphere* because of the force of *gravity* acting upon the overlying column of air. (See *standard sea level pressure.*)

atoll Circular or closed-loop *coral reef* enclosing an open lagoon with no island inside.

augite Mineral of the *pyroxene group,* usually black, greenish-black, or dark green.

aurora borealis, aurora australis Light glow in the *magnetosphere* over high latitudes, taking the form of bands, rays, or draperies. (See *northern lights.*)

axial rift Narrow, trenchlike depression situated along the center line of the Mid-Oceanic Ridge and identified with active *sea-floor spreading.*

backwash Return flow of *swash* water under the influence of *gravity.*

banding Layered arrangement of strongly knit *mineral* crystals into bands of differing mineral composition in certain *metamorphic rocks,* especially *gneiss.*

bank caving Incorporation of masses of *alluvium* or other bank materials into a *stream channel* because of undermining, usually in high flow stages.

banks Rising ground slopes comprising the sides of a *stream channel.*

barchan dune *Sand dune* of crescentic base outline with sharp crest and steep lee *slip face,* with crescent points (horns) pointing downwind.

barometer Instrument for measurement of *atmospheric pressure.*

barometric pressure (See *atmospheric pressure.*)

barred spiral galaxy *Spiral galaxy* having bar-shaped central region to which *spiral arms* are attached.

barrier-island coast *Coast* with a broad zone of shallow water offshore, (a lagoon) shut off from the ocean by a barrier island.

barrier reef *Coral reef* separated from the mainland *shoreline* by a lagoon.

basal slip Blocklike down-valley motion of an entire *glacier,* or large segment of a glacier, slipping over the bed.

basalt *Extrusive igneous rock* of *gabbro* composition; occurs as *lava.*

basaltic cinder cone Small conical hill built of basaltic *tephra,* without *lava* flows.

basaltic rock General term for rock of lower layer of *continental crust* and for *oceanic crust,* composed of *mafic igneous rock; rock* of composition of *basalt.*

baselevel Lower limiting surface or level that can ultimately be attained by a *stream* under conditions of stability of the *crust* and sea level; an imaginary surface equivalent to sea level projected inland.

basic lava *Lava* of *mafic mineral* composition, usually *basalt.*

batholith Large, deep *pluton* (body of *intrusive igneous rock*), usually with an area of surface exposure greater than 40 sq mi (100 sq km).

bauxite Mixture of several *clay minerals,* consisting largely of aluminum oxide and water with impurities; a principal *ore* of aluminum.

beach Thick, wedge-shaped accumulation of *sand, gravel,* or *cobbles* in the zone of breaking waves. (See *shingle beach, pocket beach.*)

beach drifting Transport of *sand* on a *beach* parallel with a shoreline by a succession of landward and seaward water movements at times when *swash* approaches obliquely.

bed load That fraction of the total *load* of a *stream* being

bedding planes (See *stratification planes.*)

moved in *traction.*

bedrock Solid *rock* in place with respect to the surrounding and underlying rock and relatively unchanged by *weathering* processes.

binary planet Two linked planetary bodies of relatively comparable *mass,* each in orbit about their common center of gravity; used in reference to the earth-moon pair.

biosphere Sum total of all living organic forms of the earth, together with the environment with which they interact.

biotite Mineral of the *mica group* containing magnesium and iron, usually black; one of the *mafic minerals.*

bituminous coal Grade of *coal* with substantial content of volatiles; called soft coal.

black hole *Neutron star* that has attained a sufficiently high *density* to draw all *electromagnetic radiation* into itself through its enormously strong *gravitation,* and thus to lack any direct means of detection.

block faulting Faulting accompanying *crustal spreading* within the continents, leading to the dislocation of crustal blocks along *normal faults* and the occurrence of *grabens, horsts,* and *fault block mountains.*

blowout (1) Break-out of *petroleum* under pressure through the *rock* mass surrounding an oil well, usually occurring on the sea floor. (2) Shallow surface depression produced by continued *deflation.*

bottom currents Currents sweeping over the floor of the ocean at great depth and capable of transporting coarse *sediment,* often along the contour of the bottom and, in some cases, up the grade of the bottom.

boulders *Sediment* particles larger than 256 mm in diameter.

braided stream Stream with shallow channel in coarse *alluvium* carrying multiple threads of fast flow, subdividing and rejoining repeatedly and continually shifting in position.

breccia General term for *sediment* consisting of angular *rock* fragments in a matrix of finer sediment particles. (See *sedimentary breccia, volcanic breccia.*)

breeding (nuclear) Process of inducing *nuclear fission* in *isotopes* of uranium, plutonium, and thorium, which would not otherwise undergo spontaneous fission.

brine Concentrated solution of salts in water.

brown clay Ocean-floor *pelagic sediment* that is inorganic; a *detrital sediment,* consisting mostly of *clay minerals.*

butte Prominent, steep-sided hill or peak, often representing the final remnant of a resistant *rock* layer in a region of *horizontal strata.*

calcite *Mineral* having the composition calcium carbonate.

caldera Very large, steep-sided circular depression resulting from the explosion and subsidence of a large *composite volcano.*

canyon (See *gorge.*)

capillary water Water clinging to a solid surface by means of the force of capillary film tension.

cap rock Impervious *rock* layer overlying a *reservoir rock* and blocking the upward escape of *petroleum* and *natural gas.*

capture hypothesis Hypothesis that the moon was drawn into the earth's orbit by entrapment in the earth's *gravity* field.

carbonates (carbonate minerals, carbonate rocks) *Minerals* that are carbonate compounds of calcium, magnesium or both, i.e., calcium carbonate or magnesium carbonate. (See *calcite, dolomite.*)

carbonation Reaction of carbonic acid in rainwater, soil water, and *ground water* with *minerals;* most strongly affects *carbonates* (carbonate minerals and rocks); an activity of *chemical weathering.*

catastrophists Naturalists of the late eighteenth century, led by Baron Cuvier, who explained all disruption of *strata* and extinction of organisms as occurring in a single great catastrophe. (See *uniformitarianism.*)

cell of thunderstorm Central, rapidly rising air column or air bubble within the core of a *thunderstorm.*

Celsius scale Temperature scale in which the freezing point of water is 0°, the boiling point 100°.

cementation *Lithification* of *sediment* by *mineral* deposition in the interstices between grains or in other pore spaces of *rock.*

central depression Steep-walled, flat-floored depression at the summit of a *shield volcano;* a result of subsidence.

central eye Cloud-free central vortex of a *tropical cyclone.*

centripetal force Tension force acting to keep an object in circular motion about a fixed center; same as *gravitation* force in the case of motion of a *planet* in *orbit* about the sun.

chalcedony *Mineral,* composition *silica* (silicon dioxide), lacking in visible crystalline structure.

chalk Variety of *limestone;* soft, earthy, white; formed of hard parts of various marine organisms (foraminifera, algae).

channel (See *graded channel, stream channel.*)

channel flow (stream flow) Movement of *runoff* to a lower level in a *stream channel.*

chatter marks Curved fractures produced by ice pressure on the surface of *bedrock* subjected to *grinding* action of moving glacial ice; usually associated with *glacial striations.*

chemical pollutants *Gases* other than the normal gaseous components of the *pure dry air* introduced by man's activities into the *atmosphere.*

chemical precipitate *Sediment* consisting of *mineral* matter chemically precipitated from a water solution in which the matter has been transported in the dissolved state as *ions.*

chemical weathering Chemical change in rock-forming *minerals* through exposure to atmospheric conditions in the presence of water, mainly chemical reactions of *oxidation, hydrolysis,* and *carbonation,* or direct solution.

chert *Sedimentary rock,* composed largely of *chalcedony* and various impurities, in the form of nodules and layers, often with *limestone* layers.

chromosphere Low layer of the sun's atmosphere, immediately above the *photosphere;* contains *solar prominences.*

circle of illumination Complete circle on earth's surface, dividing the globe into a sunlit hemisphere and a shadowed hemisphere.

circulation cell Any closed flow circuit in a *gas* or *liquid.* (See *Hadley cell.*)

circum-Pacific belt Chains of *andesite volcanoes* making up mountain belts and *island arcs* surrounding the Pacific Ocean basin.

cirque Bowl-shaped depression holding the collecting ground and *firn* of an *alpine glacier.*

clastic rock *Rock* formed from *clastic sediments* through compaction and/or cementation.

clastic sediment *Sediment* consisting of particles broken away physically from a parent *rock* source.

clay *Sediment* particles smaller than 0.004 mm in diameter.

clay minerals Class of *minerals* produced by alteration of *silicate minerals,* having plastic properties when moist.

claystone *Sedimentary rock* formed by *lithification* of *clay.*

clear air turbulence (CAT) Small-scale, often intense air turbulence found in the *jet stream.*

cleavage Property of a *mineral* to split readily in a set of parallel planes or along two or three sets of parallel planes.

cliff Sheer, near-vertical rock wall formed from flat-lying resistant layered rocks, usually *sandstone, limestone,* or *lava flows.* Cliff may refer to any near-vertical rock wall. (See *marine cliff.*)

cloud families Groups of cloud varieties defined in terms of height range or degree of vertical development.

clouds Dense concentrations of suspended water or ice particles in diameter ranges 20 to 50 *microns.* (See *cumuliform clouds, stratiform clouds.*)

cloud seeding Fall of ice crystals from the anvil top of a *cumulonimbus cloud,* serving as *nuclei* of *condensation* at lower levels. (Seeding may also be carried out artificially.)

coal *Hydrocarbon compounds* comprising a *rock* formed of compacted, lithified, and altered accumulations of plant remains (*peat*).

coal measures Accumulations of *coal* seams with interbedded *shale, sandstone,* and *limestone.*

coastal blowout dune High *sand dune* of the *parabolic dune* class formed adjacent to a *beach,* usually with deep *deflation* hollow enclosed within the dune ridge.

coastal plain Coastal belt, emerged from beneath the sea as a former *continental shelf,* underlain by *strata* with gentle *dip* seaward.

coastline (coast) Zone in which coastal processes operate or have a strong influence. (See *barrier-island coast, coral-reef coast, delta coast, fault coast, fiord coast, ria coast, volcano coast.*)

cobbles *Sediment* particles between 64 mm and 256 mm in diameter.

cold front Moving weather *front* along which a cold *air mass* is forcing itself beneath a warm air mass, causing the latter to be lifted.

colloids (mineral) *Mineral* particles of extremely small size, capable of remaining indefinitely in *suspension* in water; typically in the form of thin plates or scales.

colluvium Deposit of *mineral* particles accumulating from *overland flow* at the base of a slope and originating from higher slopes where *sheet erosion* is in progress. (See *alluvium.*)

columnar jointing System of flat-faced columnlike masses of fine-textured *igneous rock* of a *lava flow, sill,* or *dike,* typically with 5 or 6 sides per column, produced by volume shrinkage accompanying cooling.

coma Brightly luminous head of a *comet.*

comet Member of the *solar system* consisting of highly diffuse matter in the form of a brightly luminous *coma* and a *tail,* seen when passing close to the sun.

composite volcano *Volcano* constructed of alternate layers of *lava* and *tephra* (*volcanic ash*).

compound (chemical) Substance consisting of two or more *elements,* always occurring in the same combination with respect to kinds of atoms and their proportions.

condensation Process of change of matter in the *gaseous state* (*water vapor*) to the *liquid state* (liquid water) or *solid state* (ice).

cone of depression Conical configuration of the lowered *water table* around a well from which water is being rapidly withdrawn.

conglomerate *Clastic rock* consisting of *pebbles* or *cobbles,* usually well-rounded, in a matrix of *sand* or *silt.*

conjunction Alignment of the sun, earth, and moon, with the sun and moon on same side of the earth.

consequent stream *Stream* that takes its course down the slope of an *initial landform,* such as a newly emerged *coastal plain* or a *volcano.*

contact metamorphism Rock metamorphism localized in *country rock* adjacent to *intrusive igneous rock* (*magma*), resulting from intense heating or infusion of chemical solutions.

continental collision Event in *plate tectonics* in which *subduction* brings two segments of the *continental crust* into contact, leading to *suturing* and formation of a *suture.*

continental crust *Crust* beneath the continents; thicker and less dense than *oceanic crust.*

continental drift Hypothesis, introduced by Alfred Wegener and others early in the 1900s, of the breakup of a parent continent, *Pangaea,* starting near the close of the Mesozoic Era, and resulting in the present arrangement of *continental shields* and intervening *ocean basin floors*.

continental margins One of three major divisions of the *ocean basins,* being the zones directly adjacent to the continent and including the *continental shelf, continental slope,* and *continental rise*.

continental nuclei Oldest masses of the *continental shields,* generally older than about 2½ billion years.

continental rise Gently sloping sea floor lying at the foot of the *continental slope* and leading gradually into the *abyssal plain*.

continental rupture *Crustal spreading* affecting the *continental crust,* so as to cause a *rift valley* to appear and to widen, eventually creating a new belt of *oceanic crust*.

continental shelf Shallow, gently sloping belt of sea floor adjacent to the continental *shoreline* and terminating at its outer edge in the *continental slope*.

continental shields Ancient crustal *rock masses* of the continents, largely *igneous rock* and *metamorphic rock,* and mostly of *Precambrian* age.

continental slope Steeply descending belt of sea floor between the *continental shelf* and the *continental rise*.

continuity (See *principle of continuity.*)

convectional precipitation *Precipitation* formed within a spontaneously rising air column or air bubble, usually within a *cumulonimbus cloud*.

convection currents General term for any overturning motions within fluids. Referring to the solid earth, convection currents are slow rising and sinking motions within the *mantle*.

convergence Inflow of air along converging flow lines toward a line or center of lower *atmospheric pressure*.

Copernican system *Heliocentric theory* of the *solar system* devised by Nikolaus Copernicus in the early 1500s, replacing the *Ptolemaic system*.

coral reef Rocklike accumulation of carbonate *mineral* secreted by corals and algae in shallow water along a marine *shoreline*.

coral-reef coast *Coast* built out by accumulations of *limestone* in coral reefs.

core of earth Spherical central mass of the earth composed largely of iron and consisting of an outer liquid zone and an interior solid zone.

core of sediment Long, narrow cylindrical sample of ocean bottom *sediment* or *rock,* obtained by penetrating the sea floor with a length of open pipe, or by rotary drilling methods.

coring Activity of obtaining submarine *cores* from the ocean bottom.

Coriolis force Fictitious force, tending to deflect any object in motion toward the right of its direction of motion in the Northern Hemisphere; toward the left in the Southern Hemisphere.

corona Sun's outer atmosphere, above the *chromosphere;* a region of pearly white streamers of light reaching far out into the *solar system*.

corrosion Erosion of *bedrock* of a *stream channel* (or other *rock* surfaces) by chemical reactions between solutions in stream water and *mineral surfaces*.

cosmic particle Radiation arriving from outer space in the form of a high-energy particle, which is a *proton* traveling at extremely high velocity; most are *nuclei* of hydrogen atoms.

cosmology Science dealing with the nature and origin of the *universe*.

counter-radiation *Longwave radiation* of *atmosphere* directed downward to the earth's surface.

country rock *Rock* surrounding a mass of an *intrusive igneous rock* and predating the *intrusion*.

crater Central summit depression associated with the principal *vent* of a *volcano*.

crevasse Gaping crack in the brittle surface ice of a *glacier*.

cross-bedding System of sloping laminations within a massive *sandstone* layer, indicating deposition in a *sand dune, delta,* or sand bar in a stream bed or *beach*.

crust of earth Outermost solid shell or layer of the earth, composed largely of *silicate minerals*.

crustal roots Those portions of the *continental crust* extending deeply into the *mantle* beneath high-standing mountain and plateau areas.

crystalline solid Matter in the *solid state* consisting of atoms locked into a regularly repeating three-dimensional space-lattice pattern.

crystallization Process in which matter becomes a *crystalline solid* from another state, the *gaseous state* or *liquid state*.

cuesta *Erosional landform* developed on *resistant strata* having low to moderate *dip* and taking the form of an asymmetrical low ridge or hill belt with one side a steep scarp and the other a gentle slope; usually associated with a *coastal plain*.

cumuliform clouds *Clouds* of globular shape, often with extended vertical development.

cumulonimbus clouds Large, dense *cumuliform clouds* yielding *precipitation*.

Curie point Critical temperature during cooling of *magma,* below which crystallized iron and titanium minerals become magnetized by lines of force of the earth's magnetic field. (See *hard magnetism, soft magnetization.*)

cutoff The cutting-through of a *meander neck,* so as to bypass the stream flow in an *alluvial meander* and cause it to be abandoned.

cyclone Center of low *atmospheric pressure.* (See *tropical cyclone, wave cyclone.*)

cyclone family Succession of *wave cyclones* tracking eastward along the *polar front* while developing from open stage to occluded stage.

daughter product *Element* produced by *radioactive decay* of another element.

day Period of time required for the earth, or any *planet* or *planetary satellite*, to complete one full circle of *rotation* on its axis.

debris flood Streamlike flow of muddy water heavily charged with *sediment* of a wide range of size grades, including *boulders*, generated by sporadic torrential rains upon steep mountain watersheds.

deep environment Environment of high pressure and high temperature to which *rock* is subjected deep within the earth's *crust*.

deep-focus earthquake *Earthquake* with *focus* at a depth between 185 and 400 mi (300 and 650 km).

deflation Lifting and transport in suspension by wind of loose particles of *soil* or *regolith* from dry ground surfaces.

degradation Downcutting of a *stream*, causing the *stream channel* to be lowered in altitude.

degrees Kelvin Temperature scale in which *absolute zero* is equivalent to −273° C on the Celsius scale.

delta *Sediment* deposit built by a *stream* entering a body of standing water and formed of the *load* carried by the stream.

delta coast *Coast* bordered by a *delta*.

delta kame Flat-topped hill of *stratified drift* representing a glacial *delta* constructed adjacent to an *ice sheet* in a *marginal lake*.

dendritic drainage pattern *Drainage pattern* of treelike branched form, in which the smaller *streams* take a wide variety of directions and show no parallelism or dominant trend.

density Quantity of *mass* per unit of volume, stated in gm/cc.

denudation Total action of all processes whereby the exposed *rocks* of the continents are worn down and the resulting *sediments* are transported to the sea by the *active agents;* includes *weathering* and *mass wasting*.

deposition (See *stream deposition*.)

depth recorder (See *precision depth recorder*.)

desert pavement Surface layer of closely fitted *pebbles* or coarse *sand* from which finer particles have been removed by *deflation*.

detrital sediment *Sediment* consisting of *mineral* fragments derived by *weathering* of preexisting *rock* and transported to places of accumulation by currents of air, water, or ice.

dew-point lapse rate Rate at which *dew-point temperature* falls within a rising air mass; about 1 F°/1000 ft (0.2 C°/100m).

dew-point temperature Temperature of *saturated air*.

diffuse reflection Form of *scattering* in which solar rays are deflected or reflected by minute dust particles or cloud particles.

diffuse sky radiation *Shortwave radiation* sent earthward by *scattering* and by *diffuse reflection*.

dike Thin layer of *intrusive igneous rock,* often near-vertical or with steep *dip*, occupying a widened fracture in the *country rock*, and typically cutting across older rock planes.

diorite *Intrusive igneous rock* consisting dominantly of intermediate *plagioclase feldspar* and *pyroxene*, with some *amphibole* and *biotite;* a *felsic igneous rock,* occurs as a *pluton*.

dip Acute angle between an inclined natural *rock* plane or surface and an imaginary horizontal plane of reference; always measured perpendicular to the *strike*. Also a verb, meaning to incline toward.

direct motion Direction of *revolution* or *rotation* of a *planet* or a *planetary satellite* similar to that taken by the earth and moon, which is counterclockwise as viewed from a point in space above the earth's north pole. (See *retrograde motion*.)

discharge Volume of flow moving through a given cross section of a *stream* in a given unit of time; commonly given in cubic feet per second.

disconformity Surface of separation between two *formations* of parallel *strata*, representing a large time gap, usually including an episode of erosional removal of part of the lower formation; a variety of *unconformity*.

disseminated deposit *Ore* deposit consisting of metallic *minerals* widely disseminated throughout a large body of *rock*. Example: *porphyry copper*.

divergence Outflow of air along diverging flow lines away from a line or center of higher *atmospheric pressure*.

doldrums Belt of calms and variable *winds* occurring at times along the *equatorial trough*.

dolomite *Mineral* or *sedimentary rock* having the composition calcium-magnesium carbonate.

Doppler effect Change in pitch of sound waves as sound-emitting source approaches and recedes from an observer.

drainage pattern Geometrical pattern formed by a total *stream channel* network as depicted on a map. (See *dendritic drainage pattern, rectangular drainage pattern, trellis drainage pattern*.)

drawdown Difference in height between the base of the *cone of depression* and the original *water table* surface.

drift (See *glacial drift, stratified drift*.)

drift current Slow *ocean current* induced by prevailing *winds*.

drifts (in mines) Horizontal tunnels or passageways in mines.

dripstone *Travertine* accumulations produced at points in *limestone caverns* where water is dripping from the ceiling and upon the floor.

drizzle Form of *precipitation* consisting of slowly falling droplets smaller than 0.5 mm in diameter.

drumlin Hill of glacial *till*, oval or elliptical in basal outline and with smoothly rounded summit, formed by plastering of till beneath moving, debris-ladened glacial ice.

dry adiabatic lapse rate Rate at which rising air is cooled by expansion when no *condensation* is occurring; 5.5 F°/1000 ft (1.0 C°/100 m).

dry stage Stage in rising air mass in which *dry adiabatic lapse rate* holds, *condensation* not occurring.

dune (See *sand dune*.)

dunite *Ultramafic igneous rock* consisting almost entirely of *olivine*.

dust Particles of *silt* and *clay* grade sizes carried in suspension by wind. Any finely divided solid particles suspended in the *atmosphere*.

dust storm Heavy concentration of *dust* in a turbulent *air mass*, often associated with a *cold front*.

dynamo theory *Earth magnetism* explained by dynamo action within a rotating liquid *core*.

dynamothermal metamorphism *Rock metamorphism* taking place at great depth through large-scale shearing under high pressure and high temperature during *tectonic activity*.

earthflow Moderately rapid downhill flowage of masses of water-saturated *soil*, *regolith*, or weak *shale*, typically forming a steplike terrace at the top and a bulging toe at the base.

earth magnetism Magnetic phenomenon of the earth as a *planet*, including the internal and external magnetic fields believed to be generated by motions within the liquid *core*.

earthquake A trembling or shaking of the ground produced by the passage of *seismic waves*. (See *deep-focus earthquake*, *intermediate-focus earthquake*, *shallow-focus earthquake*.)

easterlies (See *tropical easterlies*.)

eccentricity Degree of flattening of an *ellipse* or of an elliptical planetary *orbit*.

economic geology Branch of *geology* dealing with *minerals* and *rocks* of economic value, including *metalliferous deposits*, *nonmetallic deposits*, and *fossil fuels*.

elastic limit The limit of strength of an elastic solid when strained, beyond which bending gives way to sudden rupture on fractures or undergoes some other permanent change in shape.

elastic-rebound theory Theory that *earthquakes* are generated by sudden release of elastic *strain* accumulated in *rock* across a *fault plane*, allowing the strained rock masses to rebound to their original shapes.

electromagnetic radiation Wavelike form of energy radiated by any substance possessing heat; travels through space at the speed of light.

electromagnetic spectrum Total collection of *frequencies* and *wavelengths* of *electromagnetic radiation* capable of being radiated by matter.

element (chemical) Natural subdivision of matter consisting of like atoms, i.e., atoms having a fixed combination of neutrons, protons, and electrons.

elevated shoreline Inactive (defunct) *shoreline* raised above the level of wave action by a sudden crustal rise.

ellipse Geometrical figure formed by a line lying in a plane and connecting all points located so that the sum of the two *radius vectors* is a constant.

elliptical galaxy *Galaxy* of elliptical outline lacking in clouds of dust and *gas*.

emergence Exposure of submarine *landforms* by a lowering of sea level or a rise of the *crust*, or both.

energy balance Condition of balance between incoming and outgoing forms of radiant energy, synonymous with *radiation balance*.

energy deficit Condition in which rate of outgoing radiant energy exceeds rate of incoming radiant energy at a given time and place.

energy surplus Condition in which rate of incoming radiant energy exceeds the rate of outgoing radiant energy at a given time and place.

entrenched meanders Winding, sinuous valley bends produced by *degradation* of a *stream* with trenching of the *bedrock* by downcutting.

environmental temperature lapse rate Rate of temperature decrease upward through the *troposphere*; standard value is 3½ F°/1000 ft (6.4 C°/km).

eon Largest segment of geologic time, the two eons being the Cryptozoic and the Phanerozoic. (*Eon* is also used to denote a time span of 1 billion years.)

epicenter Ground surface point directly above the *focus* of an *earthquake*.

epoch Geologic time unit, a subdivision of a *period*.

epoch (paleomagnetic) (See *normal epoch*, *reversed epoch*.)

equatorial current West-flowing *ocean current* in the belt of *trade winds*.

equatorial trough Low-pressure trough centered more or less over the equator and situated between the two belts of *trade winds*.

equinox Instant in time when the *subsolar point* falls on the earth's equator and the *circle of illumination* passes through both poles. Vernal equinox occurs on March 20 or 21; autumnal equinox on September 22 or 23.

era Subdivision of the *eon* of geologic time; the Phanerozoic Eon is subdivided into Paleozoic, Mesozoic, and Cenozoic Eras.

erosion (See *stream erosion*.)

erosional landforms Class of the *sequential landforms* shaped by the removal of *regolith* or *bedrock* by *active agents* of erosion. Examples: *canyon*, glacial *cirque*, *marine cliff*.

erratic boulder *Boulder* carried by glacial ice far from the place of its origin in a *bedrock outcrop*.

esker Narrow, often sinuous embankment of coarse *gravel* and *boulders* deposited in the bed of a meltwater *stream* enclosed in a tunnel within stagnant ice of an *ice sheet*.

Eurasian-Melanesian belt Major *primary arc* system extending from southern Europe across southern Asia and Indonesia.

evaporation Process in which water in the *liquid state* or *solid state* passes into the *vapor state*.

evaporites Class of *chemical precipitate sediment* and *sedimentary rock* composed of soluble salts deposited from salt water bodies. (See *anhydrite*, *gypsum*, *halite*.)

evolution (See *organic evolution*.)

evolutionary theory Theory of *cosmology* holding that the *universe* originated with an explosion of a dense concentration of matter and is continually expanding.

exfoliation Development of sheets, shells, or scales of *rock* upon exposed *bedrock* or *boulders*, usually in concentric, spheroidal arrangements. *Spontaneous expansion* is one cause of exfoliation.

exfoliation dome Smoothly rounded *rock* knob or hilltop bearing rock sheets or shells produced by *spontaneous expansion.*

external earth processes Processes acting upon *rocks* and *minerals* exposed to the *atmosphere* and *hydrosphere* and powered by *solar energy;* processes involved in *denudation.*

extrusive igneous rock *Rock* produced by the solidification of *lava* or ejected fragmental *igneous rock* (*tephra*).

Fahrenheit scale Temperature scale in which the freezing point of water is 32°, the boiling point 212°.

fall (of meteorite) Collection of a *meteorite* whose arrival was witnessed.

fallout *Gravity* fall of atmospheric *particulates,* reaching the ground.

fault Sharp break in *rock* with displacement (slippage) of block on one side with respect to adjacent block. (See *left-lateral fault, normal fault, overthrust fault, right-lateral fault, transcurrent fault, transform fault.*)

fault block Blocklike *rock* mass lying between two parallel *normal faults.* (See *graben, horst*).

fault block mountain Mountain consisting of an uplifted *fault block,* which may be a *horst* or a tilted block.

fault coast *Coast* formed when the *shoreline* comes to rest against a *fault scarp.*

faulting Process of formation of a *fault;* a form of *tectonic activity.* (See *block faulting.*)

fault plane Surface of slippage between two earth blocks moving relative to each other during *faulting.*

fault scarp Clifflike surface feature produced by *faulting* and exposing the *fault plane;* commonly associated with a *normal fault.*

fauna A natural assemblage of several animal species, used as a group to identify the age of a *formation.* (See *succession of faunas.*)

feldspar *Aluminosilicate mineral* group consisting of silicate of aluminum and one or more of the metals potassium, sodium, or calcium. (See *plagioclase feldspar, potash feldspar.*)

felsic igneous rock *Igneous rock* dominantly composed of *felsic minerals.*

felsic minerals, felsic mineral group *Quartz* and *feldspars* treated as a *mineral* group of light color and relatively low *density.* (See *mafic minerals.*)

ferro-alloy metals Metals used principally as alloys with iron to create steels with special properties. Examples: vanadium, chromium.

find (of meteorite) Collection of a *meteorite* whose fall was not observed.

fiord Narrow, deep ocean embayment partially filling a *glacial trough.*

fiord coast Deeply embayed, rugged *coast* formed by partial *submergence* of *glacial troughs.*

fireball Brilliant flash of light produced by the friction of a large *meteoroid* passing through the earth's *atmosphere.*

firn Granular old *snow* forming a surface layer in the *zone of accumulation* of a *glacier.*

firn field Field of *glacier* in which *firn* accumulates.

first law of motion Every body continues in a state of rest or of uniform motion in a straight line unless made to change that state by some external force.

fissile Adjective describing a *rock,* usually a *shale,* that readily splits up into small flakes or scales.

fission hypothesis Hypothesis that the moon was formed out of the earth's *mantle,* torn from the earth by the centrifugal force of rapid earth *rotation.*

fissure Cracklike vent from which *lava flows* or *tephra* may issue.

flatiron Triangular *erosional landform* formed between adjacent *canyons* cut through a *hogback.*

flood basalts Large-scale outpourings of *basalt lava* to produce thick accumulations of *basalt* over large areas.

floodplain Belt of low, flat ground, present on one or both sides of a *stream channel,* subject to inundation by flood about once annually and underlain by *alluvium.*

flowstone *Travertine* deposits formed on *limestone cavern* floors where water is flowing over the surface in *streams* or held in pools.

fluid Substance that flows readily when subjected to unbalanced stresses; may exist as a *gas* or a *liquid.*

fluid agents (See *active agents.*)

fluvial denudation *Denudation* occurring largely through the action of running water as *runoff* in *overland flow* and *stream flow,* together with *weathering* and *mass wasting.*

focus (pl. **foci**) Point within the earth at which the energy of an *earthquake* is released and from which the *seismic waves* emanate.

focus of ellipse One of two points (foci) lying on the *major axis* of an *ellipse,* from which a *radius vector* extends to a point on the ellipse.

fog *Cloud* layer in contact with land or sea surface, or very close to that surface. (See *advection fog, radiation fog.*)

folding Process of formation of *folds;* a form of *tectonic activity.*

folds Wavelike corrugations of *strata* (or other layered *rock* masses) as a result of crustal compression; a form of *tectonic activity.*

foliation Crude layering structure or *cleavage* imposed on *rock* by *dynamothermal metamorphism.*

foredunes Ridge of irregular *sand dunes* typically found adjacent to beaches on low-lying coasts and bearing a partial cover of plants.

formation Distinctive layer or sequence of similar *strata* of *sedimentary rock* deposited in a small span of geologic time and identified by a proper name.

fossil Ancient plant and animal remains or their traces or impressions preserved in *sedimentary rocks.* (See *index fossil.*)

fossil fuels Collective term for *coal, petroleum,* and *natural gas* capable of being utilized by combustion as energy sources.

fractionation of magma Change in *magma* composition

during cooling as crystallized *minerals* are physically removed from the magma body.

freezing Change from *liquid state* to *solid state* accompanied by release of *latent heat of fusion* becoming *sensible heat.*

frequency Number of wave crests moving past a fixed point in a given unit of time; expressed as the number of wave cycles per second.

fringing reef *Coral reef* directly attached to land with no intervening lagoon of open water.

front Surface of contact between two unlike *air masses.* (See *cold front, occluded front, polar front, warm front.*)

frontal precipitation *Precipitation* occurring within an *air mass* forced to rise along a *front.*

frost action *Rock* breakup by forces accompanying the *freezing* of water.

frost heaving Lifting of *soil, regolith,* or individual *pebbles* and *cobbles* by growth of needlelike ice crystals formed during freezing of soil water.

fumarole Vent in *rock* surface emitting volcanic gases, mostly steam, at high temperature.

gabbro *Intrusive igneous rock* consisting largely of *pyroxene* and calcic *plagioclase feldspar,* with variable amounts of *olivine;* a *mafic igneous rock,* occurs as a *pluton.*

galaxy Assemblage of *stars,* numbering several billions, widely separated from other galaxies and distributed throughout space of the *universe.* (See *barred spiral galaxy, elliptical galaxy, radio galaxy, spiral galaxy.*)

gamma rays High-energy form of radiation at the extreme short *wavelength* (high *frequency*) end of the *electromagnetic spectrum.*

gas *Fluid* of very low *density* (as compared with a *liquid* of the same chemical composition) that expands to fill uniformly any small container and is readily compressed.

gaseous state *Fluid* state of matter having the properties of a *gas.*

geocentric theory Theory of *solar system* astronomy holding that the planets, moon, and sun revolve around the earth.

geochronometry Measurement of absolute ages of *rocks,* usually by *radiometric age* determinations.

geode Spherical mass of rock containing a cavity lined with *mineral* crystals having freely developed faces and ends.

geologic map Map showing the surface extent and boundaries of *rock* units of different ages or rock varieties.

geologic norm Stable natural condition in a humid climate in which slow *soil erosion* is paced by maintenance of *soil* horizons bearing a plant communuity in an equilibrium state.

geology Science of the solid earth, including the earth's origin and history, materials comprising the earth, and processes acting within the earth and upon its surface.

geomagnetic axis Central axis of the earth's magnetic field, defining the *magnetic poles.*

geomorphology Science of *landforms,* including their history and processes of origin.

geophysics Branch of *geoscience* that applies principles and methods of physics to the study of the earth; includes specialties of *seismology, earth magnetism,* and *gravity.*

geoscience Synonym for *geology,* but often used to include atmospheric, oceanic, and space sciences with traditional geology.

geostrophic wind *Wind* at high levels above the earth's surface blowing parallel with a system of straight, parallel *isobars.*

geosyncline Thick, lens-shaped accumulation of *sedimentary strata* deposited in a long, narrow belt of crustal subsidence.

geothermal energy Heat energy of igneous origin drawn from steam or hot water beneath the earth's surface.

geothermal gradient The rate at which temperature rises with increasing depth beneath the earth's solid surface.

geysers Periodic jetlike emissions of hot water and steam from narrow vents.

glacial drift General term for all varieties and forms of *rock* debris deposited in close association with Pleistocene *ice sheets.*

glacial grooves Long, troughlike depressions on *bedrock* surfaces worn by the *grinding* action of glacial ice.

glacial plucking Removal of masses of *bedrock* from beneath an *alpine glacier* or *ice sheet* as ice moves forward suddenly.

glacial striations Scratches or scorings on *bedrock* or on glacial *boulders* (*erratics*), produced by the grinding action of moving glacial ice.

glacial trough Deep, steep-sided rock trench of a U-shaped cross section formed by *alpine glacier* erosion.

glaciated rock knob Prominent knob of hard *bedrock* shaped by *grinding* and *plucking* of moving glacial ice.

glaciation (1) General term for the total process of growth and landform modification of *glaciers.* (2) Single episode or time period in which *ice sheets* formed, spread, and disappeared, as contrasted with *interglaciation.*

glacier Large natural accumulation of land ice affected by present or past flowage. (See *alpine glacier, outlet glacier.*)

glacier advance Forward movement of *glacier terminus* resulting from an increase in the rate of *glacier* nourishment or a decrease in the rate of *ablation,* or both.

glacier equilibrium State of balance in the activity of a *glacier* in which the rate of nourishment equals the rate of *ablation* and the *terminus* holds a fixed position.

glacier recession Up-valley recession of a *glacier terminus* resulting from an increase in the rate of *ablation* or a decrease in the rate of nourishment, or both.

glacier terminus Lower end of an *alpine glacier.*

glaciologist Scientist who investigates *glaciers* and glacial activity.

glaze Ice layer accumulated upon solid surfaces by the freezing of falling *rain* or *drizzle.*

globular star cluster Closely clustered *stars* found in the *halo* surrounding a *spiral galaxy*.

gneiss Variety of *metamorphic rock* showing *banding* and commonly rich in quartz and feldspar. (See *granite gneiss, injection gneiss.*)

Gondwana Hypothetical Southern Hemisphere continent existing throughout Paleozoic and early Mesozoic Eras, consisting of *continental shields* of South America, Africa, India, Madagascar, Australia, and Antarctica; part of *Pangaea*.

gorge (canyon) Steep-sided *bedrock* valley with a narrow floor limited to the width of a *stream channel*.

graben Trenchlike depression representing the surface of a *fault block* dropped down between two opposed, infacing *normal faults*. (See *rift valley*.)

graded bedding Variety of lamination within *turbidites* in which each bed represents the deposit of one *turbidity current* flow.

graded channel (graded stream) *Stream* with its *gradient* adjusted to achieve a balanced state in which the average *bed load* transport rate is matched to the averate bed load input rate.

graded profile Smoothly descending *longitudinal profile* displayed by a *graded stream*.

graded stream (See *graded channel*.)

granite *Intrusive igneous rock* consisting largely of *quartz*, *potash feldspar*, and sodic *plagioclase feldspar*, with minor amounts of *biotite* and *hornblende*; a *felsic igneous rock*, occurs as a *pluton*.

granite-gabbro series *Igneous rock* series ranging in composition from *felsic igneous rock* (*granite*) to *mafic igneous rock* (*gabbro*).

granite gneiss Form of *gneiss* having the composition of *granite*.

granitic rock General term for *rock* of the upper layer of the *continental crust*, composed largely of *felsic igneous rock*; *rock* of composition similar to that of *granite*.

granitization Transformation of *country rock* into *granite* by a metamorphic process through infusion of chemical solutions; contrasted with the *magmatic theory* of the origin of granite.

granular disintegration Grain-by-grain breakup of the outer surface of coarse-grained *rock*, yielding *gravel* and leaving behind rounded *boulders*.

gravitation Mutual attraction between any two masses.

gravity Gravitational attraction of the earth upon any small *mass* near the earth's surface; also applies to the moon. (See *gravitation*.)

gravity well Drilled or dug well in which the standing water level coincides with the level of the surrounding *water table*.

great planets The four large *outer planets:* Jupiter, Saturn, Uranus, and Neptune.

greenhouse effect Accumulation of heat in the lower *atmosphere* through the absorption of *longwave radiation* from the earth's surface.

greenstone Variety of *schist* derived by *metamorphism* from *basalt lavas*.

grinding *Abrasion* of *bedrock* performed by a *glacier* in motion.

groin Man-made wall or embankment built out into the water at right angles to the *shoreline*.

ground moraine *Moraine* distributed beneath a large expanse of land surface covered at one time by an *ice sheet*.

ground radiation *Electromagnetic radiation* emitted by the earth's land or ocean surfaces; it is *longwave radiation*, mostly *infrared radiation*.

ground water *Subsurface water* occupying the *saturated zone* and moving under the force of *gravity*.

gullies Deep, V-shaped trenches carved by newly formed *streams* in rapid headward growth during advanced stages of *accelerated soil erosion*.

gypsum *Evaporite mineral*, composition calcium sulfate with water.

gyres Large circular *ocean current* systems centered upon the oceanic *subtropical high-pressure cells*.

Hadley cell Atmospheric *circulation cell* in low latitudes involving rising air over an *equatorial trough* and sinking air over *subtropical high-pressure belts*.

hail Form of *precipitation* consisting of pellets or spheres of ice with a concentric layered structure.

half-life Time span required for a given radioactive *isotope* to be reduced to one-half its initial quantity through spontaneous *radioactive decay*.

halite *Evaporite mineral*, composition sodium chloride; rock salt.

halo Region of widely scattered *stars* surrounding the main disk of a *spiral galaxy*.

hanging trough Tributary *glacial trough* with floor high above the floor of the main trough that it joins.

hard magnetism Final, permanent state of magnetization of magnetic *minerals* during cooling and crystallization of *magma*.

haze Light concentration of *particulates* suspended in the lower *atmosphere* and causing reduced visibility.

heat engine Mechanical system in which motion is powered by heat energy.

heat island Persistent region of higher air temperatures centered over a city.

heavy minerals Group of *minerals* having exceptionally high *density*, usually 4 gm/cc and greater, typically occurring in *detrital sediments*. Example: *magnetite*.

heavy oil *Petroleum* enclosed in *sand* or *sandstone* layers but not capable of rising spontaneously to the surface through drilled wells.

heliocentric theory Theory of *solar system* astronomy holding that the planets revolve around the sun.

Helmholtz contraction Contraction of a large mass of matter under its own *gravitation*, accompanied by a rise in internal temperature.

hematite Mineral, composition iron oxide; a principal *ore* of iron.

Hertzsprung-Russell diagram (H-R diagram) Graphical representation of stars plotted on a field in which the vertical axis represents *luminosity,* the horizontal axis *spectral class* and star temperature.

high water Point in the cycle of the *ocean tide* when the ocean water reaches its highest point.

historical geology Branch of *geology* dealing with the earth's history as interpreted from the stratigraphic record and its *fossil* remains, and from any other form of evidence.

hogback Sharp-crested, often sawtooth ridge formed of the upturned edge of a resistant *rock* layer of *sandstone, limestone,* or *lava.*

homoclinal ridge Narrow-crested ridge formed by the erosion of a resistant bed of steep *dip* in a region of *folds;* essentially identical in form and structure to a *hogback.*

homogeneous structure Uniform physical and chemical properties in all directions throughout the *bedrock* of a *landmass.*

horizontal strata Strata lying in approximately horizontal attitude.

horn Sharp, steep-sided mountain peak formed by the intersection of headwalls of glacial *cirques.*

hornblende Mineral of the *amphibole group,* commonly black or dark green.

horst *Fault block* uplifted between two *normal faults.*

hot springs *Springs* discharging heated water approaching the boiling point in temperature; usually related to *igneous rock intrusion* at depth.

Hubble's law Principle recognized by the astronomer Edwin Hubble that the *radial velocity* of recession of a *galaxy* increases proportionately with increasing distance.

humidity General term for the amount of water vapor present in the air. (See *relative humidity, specific humidity.*)

hurricane *Tropical cyclone* of the western North Atlantic and Caribbean Sea.

hydraulic action *Stream erosion* by impact force of the flowing water upon the bed and banks of the *stream channel.*

hydraulic head Difference in level of the *water table* between one point and another, setting up a pressure difference and causing the flow of *ground water* from higher to lower points.

hydrocarbon compounds Naturally occurring organic *compounds* consisting of carbon, hydrogen, and oxygen. (See *coal, natural gas, peat, petroleum.*)

hydrogeologist *Geologist* specializing in the geologic aspects of *ground water* problems; also called a groundwater geologist or geohydrologist.

hydrologic cycle Total plan of movement, exchange, and storage of the earth's free water in *gaseous state, liquid state,* and *solid state.*

hydrology Science of the earth's water and its motions through the *hydrologic cycle.*

hydrolysis Chemical union of water molecules with *minerals* to form different, more stable mineral *compounds.*

hydropower Power source using the kinetic energy of water moving to lower levels in *streams* under the influence of *gravity.*

hydrosphere Total water realm of the earth's surface zone, including the *oceans,* surface waters of the lands, *ground water,* and water held in the *atmosphere.*

hydrothermal solutions Highly heated, watery solutions exuded from a crystallizing *magma* and permeating the overlying *country rock.*

hygroscopic nuclei *Nuclei* of *condensation* having a high affinity for water; usually salt particles.

ice fall Abrupt steepening of the down-valley gradient of a *valley glacier,* causing the surface ice to break apart in deep *crevasses.*

ice lobes (glacial lobes) Broad tonguelike extensions of an *ice sheet* resulting from more rapid ice motion where terrain was more favorable.

ice sheet Large thick plate of glacial ice moving outward in all directions from a central region of accumulation.

ice shelf Thick plate of floating glacial ice attached to an *ice sheet* and fed by the ice sheet and by *snow* accumulation.

ice storm Occurrence of heavy *glaze* of ice on solid surfaces.

ice wedge Vertical, wall-like body of ground ice, often tapering downward, occupying a shrinkage crack in *silt* of *permafrost* areas.

igneous rock *Rock* solidified from a high-temperature molten state; rock formed by cooling of *magma.* (See *extrusive igneous rock, felsic igneous rock, intrusive igneous rock, mafic igneous rock, ultramafic igneous rock.*)

illite *Clay mineral* derived by *chemical weathering* from such *silicate minerals* as *feldspar* and *muscovite mica.*

ilmenite Mineral of iron-titanium oxide composition; black, high in *density,* classed as a *mafic mineral.*

impact metamorphism Formation of *lunar breccia* by fragmentation and *lithification* of lunar *rock* during impacts that formed *lunar craters.*

impermeable rock *Rock* of very low *permeability.* Example: *shale.*

index fossil *Fossil* species particularly well suited, because of its limited range in time, to establish the geologic age of a *formation.*

inertia Tendency of any mass to resist a change in a state of rest or of uniform motion in a straight line.

infiltration Absorption and downward movement of *precipitation* into the *soil* and *regolith.*

infrared radiation *Electromagnetic radiation* in the *wavelength* range of 0.7 to about 200 *microns.*

initial landforms *Landforms* produced directly by *internal earth processes* of *volcanism* and *tectonic activity.* Examples: *volcano, fault scarp.*

injection gneiss Form of *gneiss* containing layers of igneouslike *rock,* suggesting intrusion of *granite magma* between layers of older rock.

inner planets The four *planets* lying closest to the sun, also called *terrestrial planets:* Mercury, Venus, Earth, and Mars.

intensity scale Any scale of numbers describing the intensity of earth shaking felt at a given point during an *earthquake.* (See *Mercalli scale.*)

interface Contact layer or zone of interaction between any two of the earth realms: *atmosphere, hydrosphere,* or *lithosphere.*

interglaciation Time period of mild climate in which *ice sheets* wasted back and largely disappeared, between *glaciations.*

interlobate moraine *Moraine* formed between two adjacent *ice lobes.*

intermediate-focus earthquake *Earthquake* with *focus* at a depth between 35 and 150 mi (55 and 240 km).

intermediate lava *Lava* of composition intermediate between *felsic igneous rock* and *mafic igneous rock,* usually *rhyolite.*

internal earth processes Geologic processes acting within the earth's *crust,* powered by internal earth forces and deriving energy from internal heat sources, such as *radioactivity* (*radiogenic heat*). *Volcanism* and *tectonic activity* are the manifestations.

intertropical convergence zone (ITC) Zone of convergence of *air masses* of *tropical easterlies* along the axis of the *equatorial trough.*

intrusion Entry of *magma* into space previously occupied by older, solid *rock* (*country rock*) below the surface.

intrusive igneous rock *Igneous rock* body produced by solidification of *magma* beneath the surface, surrounded by preexisting rock.

inversion lid Top surface of a *low-level temperature inversion,* resisting mixing of colder air below with warmer air above.

ion Electrically charged atom or molecule, produced when an atom or molecule loses or gains an electron.

ionization Process of formation of *ions.*

ionizing radiation Very short *wavelength electromagnetic radiation,* such as that of *X rays* and *radioactivity,* capable of causing *ionization* of atoms exposed to the radiation.

ionosphere Atmospheric layer in the altitude range 50 to 250 mi (80 to 400 km) in which a large number of gas molecules undergo *ionization* by action of shortwave solar radiation.

irons *Meteorites* composed almost entirely of nickel-iron alloy.

island arc Chain of volcanic islands, part of a *primary arc,* paralleling a *subduction zone.*

isobar Line drawn on a map passing through all points having the same *atmospheric pressure.*

isobaric surfaces Surfaces of equivalent *atmospheric pressure.*

isostasy Equilibrium state, resembling flotation, in which crustal masses stand at levels determined by their thickness and *density;* equilibrium being achieved by flowage of denser *mantle rock* of the underlying *asthenosphere.*

isotherm Line drawn on a map passing through all points having the same air temperature.

isotope A variety of a given *element,* differing in *mass number* from other varieties of the same element.

jet stream High-speed air flow in narrow bands within the *upper-air westerlies* and along certain other global latitude zones at high levels.

joints Fractures within *bedrock,* usually occurring in parallel and intersecting sets of planes.

kame Hill composed of sorted coarse water-laid *glacial drift,* largely *sand* and *gravel,* built into an impounded water body within stagnant ice or against the margin of an *ice sheet.*

kame terrace *Kame* taking the form of a flat-topped *terrace* built between a body of stagnant glacial ice and a rising valley wall.

kaolinite *Clay mineral* typically formed by *hydrolysis* from *potash feldspar* (also from micas).

Kennelly-Heaviside layer Layer in the lower part of the *ionosphere,* capable of reflecting radio waves.

Kepler's laws Three laws of planetary motion discovered by Johannes Kepler. First law: The orbit of every *planet* is an *ellipse.* Second law: A planet moves in its *orbit* about the sun at a varying speed, such that the *radius vector* from the sun sweeps over equal areas in equal times. (Third law not covered.)

kerogen Waxy substance of *hydrocarbon compounds* held in *oil shale* and capable of yielding *petroleum* upon distillation.

kettle Deep, more-or-less circular depression in *glacial drift* produced by melting of an ice block and collapse of the enclosing drift.

killing frost Below-freezing air temperatures occurring near the ground during the growing season of sensitive plants.

knob and kettle Terrain of numerous small knobs of *glacial drift* and deep depressions usually situated along the *moraine* belt of a former *ice sheet.*

laccolith *Intrusive igneous rock body,* lenslike and circular, with flat floor and domed upper surface, intruded into *sedimentary rocks* so as to lift the overlying beds.

land breeze Local wind blowing from land to sea during the night.

landforms Configurations of the land surface taking distinctive forms and produced by natural processes. Examples: hill, valley, plateau. (See *depositional landforms, erosional landforms, initial landforms, sequential landforms.*)

landmass Large area of *continental crust* lying above sea level (*baselevel*) and thus available for removal by *denudation.*

landslide Very rapid sliding of large masses of *bedrock* on steep mountain slopes or from high cliffs.

land subsidence Sinking or settling of the ground surface over areas of *rock* removal (such as *coal* mines) or of fluid removal (such as *ground water* or *petroleum*).

langley (ly) Unit of intensity of solar radiation equal to one gram-calorie per square centimeter.

lapse rate (See *environmental temperature lapse rate, dry adiabatic lapse rate, wet adiabatic lapse rate.*)

latent heat Heat absorbed and held in storage in a *liquid* or a *solid* during the processes of *condensation* or *freezing*, respectively; distinguished from *sensible heat.*

latent heat of fusion *Latent heat* released during *melting* or absorbed during *freezing.*

latent heat of vaporization *Latent heat* released during change from *liquid state* to *gaseous state* or absorbed in the reverse change.

lateral moraine *Moraine* forming an embankment between the ice of an *alpine glacier* and the adjacent valley wall.

lateral planation Sidewise cutting of a shifting *stream channel* usually occurring on the outside of a bend.

laterite Rocklike *ore* deposit formed as a layer beneath the *soil* consisting of *minerals* rich in iron, aluminum, or manganese, such as *limonite* and *bauxite.*

Laurasia Hypothetical Northern Hemisphere continent existing throughout Paleozoic and early Mesozoic Eras, consisting of what are now *continental shields* of North America and Eurasia; part of *Pangaea.*

lava *Magma* emerging upon the earth's solid surface, exposed to air or water. (See *acidic lava, basic lava, intermediate lava.*)

lava flow Outpouring of molten *lava* (*magma*) upon the earth's surface or sea floor, congealing to form *extrusive igneous rock.*

law of gravitation Any two masses attract each other with a force directly proportional to the product of their masses and inversely proportional to the square of the distance between them.

law of motion (See *first law of motion.*)

left-lateral fault Variety of *transcurrent fault* in which the motion of block on the opposite side from observer is to the left.

level of condensation Altitude at which the *dry stage* of rising air ends and above which *condensation* sets in.

lightning Electric arc generated between parts of a *cumulonimbus cloud* or between cloud and ground.

light-year Unit of interstellar distance equal to the distance traveled by light in one year's time.

lignite Low grade of *coal,* intermediate in development between *peat* and coal; called brown coal.

limestone Nonclastic *sedimentary rock* in which *calcite* is the predominant *mineral,* and with varying minor amounts of magnesium carbonate, *silica,* or other minerals, and *clay.*

limestone caverns Interconnected subterranean cavities formed in *limestone* by *carbonation* occurring in slowly moving *ground water.*

limonite A *mineral* or group of minerals consisting largely of iron oxide and water, produced by *chemical weathering* of other iron-bearing minerals.

liquid *Fluid* that maintains a free upper surface and is only very slightly compressible, as compared with a *gas.*

liquid state *Fluid* state of matter having the properties of a *liquid.*

lithification Process of hardening of *sediment* to produce *sedimentary rock.*

lithosphere General term for the entire solid earth realm; in *plate tectonics,* it refers to the strong, brittle outermost *rock* layer lying above the *asthenosphere.*

lithospheric plate Segment of *lithosphere* moving as a unit, in contact with adjacent *lithospheric plates* along *subduction zones,* zones of *crustal spreading,* or *transform faults.*

littoral drift Transport of *sediment* parallel with the *shoreline* by the combined action of *beach drifting* and *longshore current* transport.

load Solid material carried by a *stream.* (See *bed load, suspended load.*)

lode Dense concentration of *ores* in numerous, closely spaced *veins.*

loess Accumulation of yellowish to buff-colored, fine-grained *sediment,* largely of *silt* grade, upon upland surfaces after transport in suspension (i.e., carried in a *dust storm*).

longitudinal profile Graphic representation of the descending course of a *stream* or *stream channel* from upper end to lower end; altitude is plotted on the vertical scale, downstream distance on the horizontal scale.

longitudinal wave Form of *seismic wave* motion found in the *primary wave.*

longshore current Current in the breaker zone and running parallel with the *shoreline,* set up by the oblique approach of waves.

longwave radiation *Electromagnetic radiation* emitted by the earth, largely in the range from 3 to 50 *microns.*

lowland Broad, open valley between *cuestas* of a *coastal plain. Lowland* may refer to any low area of land surface.

low-level temperature inversion Reversal of normal *environmental temperature lapse rate* in an air layer near the ground.

low water Point in cycle of *ocean tide* when ocean water reaches its lowest point.

luminosity Measure of the total radiant energy output of a *star.*

lunar breccia Lunar *rock* consisting of angular rock fragments enclosed in smaller rock fragments, formed by *impact metamorphism* during formation of *lunar craters.*

lunar craters Circular, rimmed depressions on the moon's surface, largely formed by the impact of *asteroids.*

lunar eclipse Apparent darkening of the full moon, occurring at *opposition* when the moon crosses the earth's shadow. (See *partial eclipse of moon, total eclipse of moon.*)

lunar highlands (terrae) Light-colored, higher upland regions of the moon's surface.

lunar maria Dark-colored plains of the moon's surface underlain by *basalt.* (Latin *mare:* sea.)

lunar regolith Particles of finely divided *rock* in a loose state forming a surface layer over most of the moon.

Lystrosaurus Genus of mammal-like reptiles living in the Triassic Era and inhabiting all parts of *Gondwana.*

mafic igneous rock *Igneous rock* dominantly composed of *mafic minerals.*

mafic minerals, mafic mineral group *Minerals,* largely *silicates,* rich in magnesium and iron, dark in color, and of relatively great *density.*

magma Mobile, high-temperature molten state of *rock,* usually of *silicate mineral* composition and with dissolved gases (*volatiles*).

magmatic differentiation Concept that a single parent *magma* can give rise to two or more magmas of different composition.

magmatic segregation Segregation of certain components of a *magma* by *crystallization* of *minerals,* which then sink because of greater *density* to form a discrete *rock* or *mineral* layer.

magmatic theory Concept holding that *granite* consists of solidified *magma* brought up from great depth; opposed to the concept of *granitization.*

magnetic anomaly Any departure of magnetic intensity or inclination from a constant, normal value, as observed by use of the *magnetometer.*

magnetic equator Imaginary circle in a plane at right angles to the *geomagnetic axis,* marking the place where lines of force parallel that axis.

magnetic event A relatively short period within either a *normal epoch* or a *reversed epoch,* during which magnetic polarity was reversed from that of the epoch.

magnetic poles Points where the *geomagnetic axis* emerges from the earth's surface; designated north and south magnetic poles.

magnetic storm Disturbance of the earth's magnetic field under the impact of strong bursts of energetic particles (electrons and *protons*) of the *solar wind* entering the *magnetosphere.*

magnetism, magnetization (See *Curie point, earth magnetism, hard magnetism, soft magnetization.*)

magnetite *Mineral* of iron-oxide composition, black, high in *density,* strongly magnetic, classed as a *mafic mineral.*

magnetometer Instrument that detects and records the intensity of a magnetic field.

magnetopause Outer boundary surface of the *magnetosphere.*

magnetosphere External portion of the earth's magnetic field, shaped by pressure of the *solar wind* and contained within the *magnetopause.*

magnitude (See *star magnitude.*)

main sequence Class of *stars* of intermediate *mass* and *luminosity* forming a linear band on the mass-luminosity diagram, ranging in temperature from about 5000 to 20,000° K.

major axis Longest diameter of an *ellipse.*

major planets The nine large *planets:* Mercury, Venus, Earth, Mars, Jupiter, Saturn, Uranus, Neptune, and Pluto.

manganese nodules *Mineral* masses, resembling *pebbles* or *cobbles,* found on the deep ocean floor and consisting of manganese oxides and iron oxides, or of coatings of those minerals around nuclei of volcanic rock.

mantle *Rock* layer or shell of the earth beneath the *crust* and surrounding the *core,* composed of *ultramafic rock* of *silicate mineral* composition.

marble Variety of *metamorphic rock* derived from *limestone* or *dolomite* by recrystallization under pressure.

marginal lake Lake impounded between an ice front and rising ground slopes.

maria (See *lunar maria.*)

marine cliff *Rock* cliff shaped and maintained by the undermining action of breaking waves.

marine placer *Placer deposit* occurring in *sand* and *gravel* of an ocean *beach* or shallow-water bar.

marine terrace Former *abrasion platform* elevated to become a steplike coastal *landform.*

mass Quantity of matter.

mass number Total number of *neutrons* and *protons* in the *nucleus* of the atom.

mass wasting Spontaneous downward movement of *soil, regolith,* and *bedrock* under the influence of *gravity;* does not include the action of *active agents.*

maximum-minimum thermometer Pair of *thermometers* recording the maximum and minimum air temperatures since the last set.

mean annual temperature Mean of daily air temperature means for a given year or succession of years.

mean daily temperature Sum of daily maximum and minimum air temperature readings divided by two.

meander neck Narrow strip of *floodplain* lying between two segments of *stream channel* comprising the entering and exiting parts of a strongly compressed *alluvial meander* bend. (See *cutoff.*)

meanders (See *alluvial meanders.*)

mean monthly temperature Mean of daily air temperature means for a given calendar month.

medial moraine *Moraine* riding on the surface of an *alpine glacier* and composed of debris carried down-valley from the point of junction of two ice streams.

megahertz Wave *frequency* unit; one megahertz is a frequency of one million cycles per second.

melting Change from *solid state* to *liquid state,* accompanied by absorption of *sensible heat* to become *latent heat.*

Mercalli scale (modified) An *earthquake intensity scale* using 12 levels designated by Roman numerals I through XII.

mercurial barometer *Barometer* using the Torricelli principle, in which *atmospheric pressure* counterbalances a column of mercury in a tube.

meridional winds *Winds* moving across the parallels of latitude, in a north-south, or south-north direction, along the meridians.

mesa Table-topped *plateau* of comparatively small extent bounded by *cliffs* and occurring in a region of *horizontal strata.*

mesopause Upper limit of the *mesosphere.*

mesosphere Atmospheric layer of upwardly diminishing temperature, situated above the *stratopause* and below the *mesopause.*

metalliferous deposits *Ores* of metallic *elements.* Examples: iron ore, copper ore.

metals recycling Ratio of *old scrap* metal used to the total use of primary metal plus *new scrap.*

metamorphic rock *Rock* altered in physical structure and/or chemical (*mineral*) composition in the *solid state* by action of heat, pressure, shearing stress, or infusion of elements, all taking place at substantial depth beneath the surface.

metamorphism Process of formation of *metamorphic rock.* (See *contact metamorphism, dynamothermal metamorphism, impact metamorphism.*)

metaquartzite Variety of *metamorphic rock* consisting almost wholly of *silica,* formed by *dynamic metamorphism* of *quartz sandstone.*

metasomatism Emplacement of *minerals* in preexisting *rock* or in older minerals by solutions emanating from an intruding *magma.*

meteor Light trail in upper *atmosphere* produced by infall of a *meteoroid.*

meteorite Meteor that has landed on earth, becoming a variety of *rock.*

meteoritic hypothesis Interpretation of the *lunar craters* as impact features produced by the infall of large *meteoroids* or *asteroids.*

meteoroid Fragment of solid matter entering the earth's *atmosphere* from outer space; most have a mass less than one ounce.

meteorology Science of the *atmosphere;* particularly the physics of the lower or inner atmosphere.

meteor shower Occurrence of a large number of *meteors* in a short time period, usually associated with a particular *meteoroid* swarm.

mica group *Aluminosilicate* mineral group of complex chemical formula having perfect *cleavage* into thin sheets. (See *biotite, muscovite.*)

micron Length unit; one micron equals 0.0001 cm.

microwaves *Electromagnetic radiation* in the *wavelength* range of about 0.03 to 1 cm.

Mid-Oceanic Ridge One of three major divisions of the *ocean basins,* being the central belt of submarine mountain topography with a characteristic *axial rift.*

millibar Unit of *atmospheric pressure;* one-thousandth of a bar. Bar is a force of one million dynes per square centimeter.

mineral A naturally occurring inorganic substance, usually having a definite chemical composition and a characteristic atomic structure. (See *felsic minerals, mafic minerals, silicate minerals.*)

mineral replacement Action of slowly moving *ground water* in which molecule-by-molecule replacement of one *mineral* by another takes place.

minor axis Shortest diameter of an *ellipse.*

minor planets Synonym for *asteroids.*

Moho Contact surface between the earth's *crust* and *mantle;* a contraction of Mohorovicić, the seismologist who discovered this feature.

monsoon system System of low-level *winds* blowing into a continent in summer and out of it in winter, controlled by *atmospheric pressure* systems developed seasonally over the continent.

montmorillonite *Clay mineral* derived by the chemical alteration of *silicate minerals* in various *igneous rocks.*

moraine Accumulation of rock debris carried by an *alpine glacier* or an *ice sheet* and deposited by the ice to become a *depositional landform.* (See *ground moraine, interlobate moraine, lateral moraine, medial moraine, recessional moraine, terminal moraine.*)

mud *Sediment* consisting of a mixture of *clay* and *silt* with water, often with minor amounts of *sand* and sometimes with organic matter.

mudcracks Vertical cracks, forming intersecting fracture networks, produced by shrinkage of *sediment* rich in *clay* when drying out under exposure to the *atmosphere.*

mudflow Rapid flowage of *mud* stream down a *canyon* floor and spreading out upon a plain at the foot of mountains; often contributes to the building of an *alluvial fan.*

mudstone *Sedimentary rock* formed by the *lithification* of *mud.*

muscovite Mineral of the *mica group,* light in color or colorless, often called white mica.

native metals Metals occurring as chemical *elements* in the natural state. Examples: native gold, native copper.

natural gas Naturally occurring mixture of *hydrocarbon compounds* (principally methane) in the *gaseous state* held within certain porous *rocks.*

natural levee Belt of higher ground parallelling a meandering *alluvial river* on both sides of the *stream channel* and built up by deposition of fine *sediment* during periods of *overbank flooding.*

neap tide Unusually small *tide range* occurring when the sun and moon are in *quadrature.*

nebula Diffuse interstellar cloud of *gas* and dust; can be produced by an explosion of a *supernova.* (See *solar nebula.*)

neutron One of the two forms of particles comprising the *nucleus* of the atom, the other being the *proton.*

neutron star Extremely small, extremely dense, dwarf *star* produced by *gravititation* collapse following the exhaustion of nuclear fuel.

new scrap Scrap metal derived as cuttings or other waste during manufacturing processes.

nonmetallic deposits *Minerals* and *rocks* useful to man as structural materials or for extraction and synthesis of chemicals, but not used for their metals.

nonrenewable earth resources Earth materials, largely *minerals,* that cannot be renewed by natural processes at a rate comparable to the rate of withdrawal and consumption by man.

normal epoch Time span in geologic record in which the

polarity of the earth's magnetic field was like that of the present time. (See *reversed epoch.*)

normal fault Variety of *fault* in which the *fault plane* inclines (*dips*) toward the downthrown block and a major component of the motion is vertical.

northern lights *Aurora borealis.*

northwesterlies *Westerly winds* of the Southern Hemisphere.

nova (novae) Event in life of a *star* characterized by a sudden great increase in brightness occurring within a span of hours or days, followed by a gradual decrease in brightness; associated with an explosive increase in diameter.

nuclear fission Breakdown of the *nucleus* of an *element,* producing a different element and releasing a large quantity of energy. Most commonly used *isotope* for fission energy is uranium-235.

nuclear fuels Radioactive *isotopes,* principally of uranium, used to furnish heat by controlled *nuclear fission.*

nuclear fusion Fusing of the nuclei of two atoms to yield a different *element,* with the release of a large quantity of energy. Example: fusion of two hydrogen nuclei to yield one helium *nucleus.*

nuclei (atmospheric) Minute particles of solid matter suspended in the *atmosphere* and serving as cores for *condensation* of water or ice.

nucleus of atom Dense central portion of an atom consisting of *neutrons* and *protons.*

nucleus of comet Starlike center sometimes seen within the *coma* of a *comet.*

nucleus of sun Sun's interior region, lying beneath the *photosphere.*

nuée ardente Glowing volcanic avalanche; a cloud of incandescent *dust* and *gas;* travels rapidly down the steep side of a *volcano.*

obsidian Black or dark-colored *volcanic glass* of *rhyolite* composition.

occluded front Weather *front,* along which a moving *cold front* has overtaken a *warm front,* forcing the warm *air mass* aloft.

ocean-basin floors One of the major divisions of the *ocean basins,* comprising the deep portions consisting of *abyssal plains, seamounts,* and low hills.

ocean basins Deep, flat-floored depressions of subglobal dimension, underlain largely by *oceanic crust* and holding the water of the *oceans.*

ocean current Persistent, dominantly horizontal flow of ocean water. (See *drift current.*)

oceanic crust *Crust* beneath the *ocean basins;* thinner and denser than *continental crust.*

oceanic trench Narrow, deep depression in the sea floor representing the *subduction* of an oceanic *lithospheric plate* beneath the margin of a continental lithospheric plate; often associated with an *island arc.*

oceans Standing bodies of salt water occupying the *ocean basins* of the earth. (See *world ocean.*)

ocean tide Periodic rise and fall of the ocean level induced by gravitational attraction between the earth and moon in combination with earth *rotation.*

oil pool Accumulation of *petroleum* saturating a *reservoir rock.*

oil shale *Shale* containing dispersed *hydrocarbon compounds,* capable of yielding *petroleum* or *natural gas* by distillation when heated.

oil spill Release of a large quantity of *petroleum,* usually by accident, from an offshore oil well, oil transport vessel, or pipeline.

old scrap Scrap metal salvaged from discarded metallic products.

olivine *Silicate mineral* with magnesium and iron but no aluminum, usually olive-green or grayish-green; a *mafic mineral.*

ooze Ocean-floor *pelagic sediment* rich in *tests* of *plankton;* may be calcareous or siliceous.

opposition Alignment of the sun, earth, and moon, with the sun and moon on opposite sides of the earth.

orbit Path followed by a *planet* revolving around the sun, or by a *planetary satellite* revolving around a planet.

ore *Mineral* deposit sufficiently concentrated to be extracted at a profit for refinement and industrial use.

organically derived sediment *Sediment* consisting of the remains of nonliving plants or animals, or of *mineral* matter produced by the activities of plants or animals.

organic evolution Continual change in the form and structure of a species of organisms with the passage of time, as a result of changes within genetic materials.

orogeny Term in *stratigraphy* for a major episode of *tectonic activity* resulting in *strata* being deformed by *folding* and *faulting.*

orographic precipitation *Precipitation* induced by the forced rise of moist air over a mountain barrier.

orthoclase Mineral name for a variety of *potash feldspar.*

outcrop Surface exposure of *bedrock.*

outer planets The five *planets* whose *orbits* lie beyond the orbit of Mars: Jupiter, Saturn, Uranus, Neptune, and Pluto.

outgassing Process of exudation of water and other gases from the earth's *crust* through *volcanoes,* to become a permanent part of the earth's *hydrosphere* and *atmosphere.*

outlet glacier Tonguelike ice stream, resembling an *alpine glacier,* fed from an *ice sheet.*

outwash plain Flat, gently sloping plain built up of *sand* and *gravel* by the *aggradation* of meltwater *streams* in front of the margin of an *ice sheet.*

overbank deposits *Alluvium* deposited during *overbank flooding.*

overbank flooding Rise of flood waters in an *alluvial river* so as to spill over the *banks* and inundate the *floodplain.*

overburden *Rock* or *regolith* of no commercial value overlying an *ore* deposit.

overland flow Motion of the surface layer of water over a sloping ground surface at times when the *infiltration* rate is exceeded by the *precipitation* rate; a form of *runoff*.

overthrust fault *Fault* characterized by the overriding of one *fault block* (or *thrust sheet*) over another along a gently inclined *fault plane;* associated with crustal compression.

overthrust faulting Process of formation of an *overthrust fault*.

overturned fold *Anticline* tightly compressed and turned over upon an adjacent *syncline*.

oxbow lake, oxbow swamp Crescent-shaped lake or swamp representing the abandoned channel left by the *cutoff* of an *alluvial meander*.

oxidation Chemical union of free oxygen with metallic *elements* in *minerals*.

ozone Gas with a molecule consisting of three atoms of oxygen, O_3.

ozone layer Layer in the *stratosphere*, mostly in the altitude range 12 to 31 mi (20 to 35 km), in which a concentration of *ozone* is produced by the action of solar *ultraviolet rays*.

paleomagnetism Relict magnetism within magnetic *minerals* in *rocks*, representing the state of the earth's magnetic field at the time the rock was formed.

paleontologist Scientist who pursues *paleontology*.

paleontology Study of ancient life based on *fossil* remains of plants and animals.

Pangaea Hypothetical parent continent, enduring until near the close of the Mesozoic Era, consisting of the *continental shields* of *Laurasia* and *Gondwana* joined into a single unit. (See *continental drift*.)

parabolic dunes Isolated low *sand dunes* of parabolic outline, with points directed into the prevailing wind.

parallax Optical phenomenon of an apparent moving apart or coming together of two distant objects because of changing alignment as viewed by a moving observer; applies to *stars* as seen from opposite points in the earth's *orbit*.

parsec Unit of interstellar distance equal to the distance from the *solar system* to a *star* having a *stellar parallax* of one second of arc.

partial eclipse of moon *Lunar eclipse* in which the earth's shadow passes across only a part of the moon's disk.

particulates Solid and liquid particles capable of being suspended in air for long periods of time.

peat Partially decomposed, compacted accumulation of plant remains occurring in a bog environment.

pebbles *Sediment* particles between 2 mm and 64 mm in diameter.

pegmatite *Veins, dikes,* and irregular pockets of very large *mineral* crystals formed in the final stages of *magma crystallization*.

pelagic sediment *Sediment* of the deep ocean floor settling from suspension in the surface water layer; may be organic in origin (*tests* of *plankton*) or *detrital sediment*. (See *ooze, brown clay*.)

peneplain Land surface of low elevation and slight relief produced in the late stages of *denudation* of a *landmass*.

penumbra Zone of partial shadow from the sun's rays, surrounding the *umbra* of the earth and moon.

perched water table *Ground water* layer or lens formed over an *aquiclude* and occupying a position above the main *water table*.

peridotite *Intrusive igneous rock* consisting largely of *olivine* and *pyroxene;* an *ultramafic igneous rock,* occurs as a *pluton*.

perigee Point in the moon's *orbit* that is closest to the earth.

perihelion Point on the earth's elliptical *orbit* at which the earth is nearest to the sun.

period Time subdivision of the *era*, each ranging between about 35 and 70 million years in duration.

permafrost Condition of permanently frozen water in the *soil, regolith,* and *bedrock* in cold climates of subarctic and arctic regions.

permeability Property of relative ease of movement of *ground water* through *rock* when unequal pressure (*hydraulic head*) exists.

petroleum (crude oil) Natural liquid mixture of many complex *hydrocarbon compounds* of organic origin, found in accumulations (*oil pools*) within certain *sedimentary rocks*.

petroleum trap Any one of a variety of arrangements of rock layers and structures capable of trapping and holding *petroleum* and *natural gas*. Examples: *anticline, dome*.

phases of moon Changing appearance of the moon through *synodic month,* revealing first increasing, then decreasing portions of the illuminated hemisphere of the moon.

photochemical reactions Action of sunlight upon pollutant *gases* to synthesize new, often toxic compounds or gases in polluted air.

photosphere Visible outer surface of the sun.

physical oceanography Physical science of the *oceans*.

physical weathering Breakup of massive *rock* (*bedrock*) into small particles through the action of physical forces acting at or near the earth's surface. (See *weathering*.)

pit crater Deep, narrow, steep-walled depression within central depression or elsewhere on *shield volcano,* typically containing fluid *lava* on the floor.

placer deposit *Ore* deposit consisting of metallic *mineral* grains forming part of a *gravel* deposit in a *stream* or *beach*. (See *marine placer*.)

plagioclase feldspar *Aluminosilicate mineral* with sodium or calcium or both.

plain (See *abyssal plain*.)

planetary satellite Solid object revolving in *orbit* around a *planet;* often called a moon.

planetesimals Aggregations of matter formed into discrete solid bodies during *accretion* of the *solar nebula*.

planets Solid objects, mostly of large mass, revolving around the sun in an elliptical orbit. (See *binary planet,* *great planets, inner planets, major planets, minor planets, outer planets, terrestrial planets.*)

plankton Collective term for small, often microscopic, floating organisms living in the surface layer of the ocean; some species are plants, others are animals.

plate tectonics Branch of *tectonics* or theory of *tectonic activity* dealing with *lithospheric plates* and their activity.

plateau Upland surface, more or less flat and horizontal, upheld by a resistant bed or *formation* of *sedimentary rock* or *lava flows* and bounded by a steep *cliff*.

plunge of fold Descent in altitude of the axis of a *fold*.

pluton Large body of coarse-grained *intrusive igneous rock*, typically of *felsic minerals*. Example: *batholith* of *granite*.

pocket beach *Beach* of crescentic outline located at a bay head.

polar easterlies System of easterly surface *winds* at high latitude, best developed in the Southern Hemisphere, over Antarctica.

polar front *Front* lying between cold polar and warm tropical *air masses*, often situated along a *jet stream* within the *upper-air westerlies*.

polar low Persistent center of low *atmospheric pressure* over high latitudes in the upper atmosphere.

polar wandering Slow rotation of the entire *lithosphere* as a global shell over the plastic *asthenosphere*, so that the continents seem to wander with respect to the rotational axis of the earth, which remains fixed in space.

porosity Total volume of pore space within a given volume of *rock*; a ratio, expressed as a percent.

porphyry Texture class of *igneous rock* consisting of widely separated large crystals enclosed in a fine-grained rock mass.

porphyry copper Copper *ore* in the form of a *disseminated deposit* occurring within a body of *porphyry*.

potash feldspar *Aluminosilicate mineral* with potassium the dominant metal.

pothole Cylindrical cavity in hard *bedrock* of a *stream channel* produced by *abrasion* of a large spherical or discus-shaped mass of *rock* rotating within the cavity.

Precambrian time All of geologic time older than the beginning of the Cambrian Period, i.e., older than 600 million years.

precipitation Particles of liquid water or ice that fall from the *atmosphere* and may reach the ground. (See *convectional precipitation, frontal precipitation, orographic precipitation.*)

precision depth recorder Instrument used to measure water depth below a vessel, using reflected sound waves.

pressure cell Center of high or low *atmospheric pressure*, identified with a *cyclone* or an *anticyclone*.

pressure gradient Change of *atmospheric pressure* measured along a line at right angles to the *isobars*.

pressure-gradient force Force acting horizontally, tending to move air in the direction of lower *atmospheric pressure*.

prevailing westerlies *Westerly winds* near the earth's surface. (See *northwesterlies, southwesterlies.*)

primary arcs Narrow, curving belts of *volcanoes, oceanic trenches*, or recent *alpine system* mountains representing recent or current *tectonic activity* and *volcanism* along subduction zones.

primary wave (P-wave) Variety of *seismic wave* in which motion is forward and backward (longitudinal) in the path of wave travel, alternately causing compression and expansion of the rock; also called a *longitudinal wave*.

principle of continuity Principle of *stratigraphy* that *strata* at one locality can be matched in age with strata at a distant locality by tracing the unbroken layer sequence across the intervening country.

principle of superposition Principle of *stratigraphy* that *strata* are arranged in order of decreasing age upward within any sequence of strata that has not been subsequently overturned.

prolate ellipsoid Solid geometrical figure having the cross section of an *ellipse* through all possible planes passed through the *major axis*, or longest axis, and a circular cross section in the plane of the shortest diameter.

proton One of the two forms of particles comprising the *nucleus* of the atom, the other being the *neutron*.

Ptolemaic system *Geocentric theory* of the *solar system* put forth by Claudius Ptolemy in the second century A.D.

pulsars Class of small *neutron stars* emitting regular pulses of light and radio waves in the same rhythm.

pulsating universe Theory of *cosmology* that the *universe* undergoes alternate expansion and contraction in a cycle on the order of every 100 billion years.

pumice Light-colored, glassy *scoria* of very low *density*, resembling solidified foam.

pure dry air Mixture of *gases* of constant proportions comprising the nonvarying part of the lower *atmosphere* and excluding *water vapor*.

pyroclastic sediment *Sediment* consisting of particles thrown into the air by volcanic explosion in the form of *tephra* or *volcanic ash*.

pyroxene group Complex *aluminosilicate minerals* rich in calcium, magnesium, and iron, dark in color, high in *density*, classed as *mafic minerals*. (See *augite*.)

quadrature Configuration in which the line from the moon to the earth is at right angles to the line from the moon to the sun, occurring twice each *synodic month*.

quartz Mineral of silicon dioxide (*silica*) composition.

quasar (See *quasistellar radio source*.)

quasistellar radio source (quasar) Extremely distant, pinpoint source of extremely powerful radio wave emission, showing a very strong *red shift* in its light spectrum.

radial streams *Streams* flowing radially outward from a central peak or highland such as a *sedimentary dome* or *volcano*.

radial velocity Speed of motion of a *star* or *galaxy* in a direction along the line of sight to that object.

radiant Point among stars from which *meteors* of a *meteor shower* appear to emanate.

radiation balance Condition of balance between incoming energy of solar *shortwave radiation* and outgoing *longwave radiation* emitted by the earth into space.

radiation fog *Fog* produced by radiational cooling of the basal air layer.

radioactive decay Spontaneous breakdown of a radioactive *isotope*, occurring by the flying off of a small part of the *nucleus*, and leading to transformation into another *element*, called a *daughter product*.

radioactivity Spontaneous emission of particles and energy from unstable *isotopes* of certain *elements* during the process of *radioactive decay*.

radio galaxy *Galaxy* that emits strong radio waves.

radiogenic heat Heat energy generated by *radioactive decay*.

radiometric age Age of *rock* determined by the measurement of ratios of radioactive *isotopes* to their *daughter products*.

radio telescope Form of telescope adapted to receive radio waves from distant emitting sources in space.

radius vector Straight line connecting one *focus* of an *ellipse* with a point on the ellipse.

rain Form of *precipitation* consisting of falling water drops, usually 0.5 mm or larger in diameter.

rainshadow desert Belt of arid climate to lee of a mountain barrier, produced as a result of *adiabatic* warming of descending air.

rays (lunar) Streaks of lighter-colored surface material radiating from certain *lunar craters*, interpreted as fragments thrown out during the impact that produced the crater.

reaction in magma Changes in composition of crystallized *silicate minerals* in a *magma* as cooling proceeds.

recessional moraine *Moraine* produced at the ice margin during a temporary halt in the recessional phase of *glaciation*.

recharge Addition of water to the *saturated zone* by percolation from the *unsaturated zone* above.

rectangular drainage pattern *Drainage pattern* in which *subsequent streams* develop in two parallel sets at about right angles to each other, expressing control by faults.

recumbent fold Form of *overturned fold* in which the plane of the fold axis is nearly horizontal.

red beds Accumulations of reddish *sedimentary rock*, largely *shale* with interbedded *sandstone* or *siltstone*, colored by red oxides of iron.

red giants Class of large red *stars* of *mass* intermediate between the *red supergiants* and most stars of the *main sequence*, and of comparatively great *luminosity* and low temperature.

red shift Shift in lines of the spectrum of a *star*, *galaxy*, or *quasar* toward the red end of the spectrum as a result of its *radial velocity* away from the observer.

red supergiants Class of enormously large *stars*, red in color, with very high *luminosities*, but comparatively cool, and with *masses* from 10 to 40 times greater than the sun.

reef limestone *Limestone* consisting of skeletal structures secreted by corals and algae in *coral reefs*, and accumulations of detrital fragments of reefs.

regolith Layer of *mineral* particles overlying the *bedrock*; may be derived by *weathering* of underlying bedrock or transported from other locations by *active agents*. (See also *lunar regolith*.)

relative humidity Ratio of *water vapor* present in the air to the maximum quantity possible for *saturated air* at the same temperature.

reservoir rock A porous *rock* mass capable of holding a large concentration of *petroleum*.

residual regolith *Regolith* formed in place by *weathering* of bedrock.

retrograde motion Motion opposite in direction to *direct motion*.

reversed epoch Time span in geologic record in which the polarity of the earth's magnetic field was reversed from that of the present time. (See *normal epoch*.)

revolution Motion of a *planet* in its *orbit* around the sun, or of a *planetary satellite* around a planet.

reworked alluvium *Alluvium* repeatedly set in motion and deposited in a *stream channel* in alternating rising and falling stages of flow.

rhyolite *Extrusive igneous rock* of *granite* composition; occurs as *lava*.

ria coast Deeply embayed *coast* formed by partial *submergence* of a *landmass* previously shaped by *fluvial denudation*.

Richter scale Scale of magnitude numbers describing the quantity of energy released by an *earthquake*.

rift Long, narrow, trenchlike surface feature produced during the pulling apart of the *crust* (as in the case of *sea floor spreading*), or along a *transcurrent fault*. (See *rift valley*.)

rift valley Trenchlike valley with steep, parallel sides; essentially a *graben* between two *normal faults*; associated with *crustal spreading*. (See *axial rift*.)

right-lateral fault Variety of *transcurrent fault* in which the motion of block on the opposite side from the observer is to the right.

rilles Narrow, canyonlike features indenting into the moon's surface; some are straight, others sinuous.

rock A natural aggregate of minerals in the *solid state*; usually hard and consisting of one, two, or more *mineral varieties*.

rock basin Overdeepened section of *bedrock* floor of a *cirque* or *glacial trough* forming a depression, often holding a *tarn*.

rockfall Free fall of particles or masses of *bedrock* from a steep *cliff* face, typically accumulating at the cliff base in the form of *talus*.

rock flour Finely ground *rock* of *clay* and fine *silt* grades, carried in suspension by meltwater issuing from a *glacier*.

rock step Abrupt down-step in the rock floor of a *glacial trough*.

rock terrace *Terrace* carved in *bedrock* during the *degradation* of a *stream channel* induced by the crustal rise or a fall of the sea level.

rock transformation cycle Total cycle of changes in which *rock* of any one of the three major rock classes—*igneous rock, sedimentary rock, metamorphic rock*—is transformed into rock of one of the other classes.

Rossby waves (See *upper-air waves.*)

rotation Spinning of a spherical object around an axis.

runoff Flow of water from continents to oceans by way of *stream flow* and *ground water* flow; a term in the water balance of the *hydrologic cycle*. In a more restricted sense, runoff refers to surface flow by *overland flow* and *channel flow*.

salinity Degree of concentration of salts in a water solution; ratio of weight of dissolved salts to weight of water (gm/kg).

saltation Leaping, impacting, and rebounding of spherical *sand* grains transported over a sand or *pebble* surface by wind.

salt-crystal growth A form of *weathering* in which *rock* is disintegrated by the growth of salt crystals during dry weather periods.

salt marsh Peat-covered expanse of *sediment* built up to the level of high tide over a previously formed *tidal flat*.

sand *Sediment* particles between 0.06 mm and 2 mm in diameter.

sand-blast action Wind erosion of hard bedrock carried out by the impact of *sand grains* traveling in *saltation*.

sand drift Accumulation of wind-transported *sand* in the lee of an obstacle.

sand dune Hill or ridge of loose, well-sorted *sand* shaped by wind and usually capable of downwind motion.

sand sea Field of *transverse dunes*.

sandspit Narrow, fingerlike embankment of *sand* constructed by *littoral drift* into the open water of a bay.

sand storm Dense, low layer of *sand* grains traveling in *saltation* over a *sand dune* or *beach* surface.

saprolite Layer of deeply decayed *igneous rock* or *metamorphic rock*, rich in *clay minerals*, typically formed in warm, humid climates; a form of *residual regolith*.

saturated air Air holding the maximum possible quantity of *water vapor* at a given temperature and pressure.

saturated zone Zone beneath the land surface in which all pores of the *bedrock* or *regolith* are filled with *ground water*.

scattering Turning aside by reflection of solar *shortwave radiation* by gas molecules of the *atmosphere*.

schist Foliated *metamorphic rock* in which *mica* flakes are typically found oriented parallel with *foliation* surfaces.

scoria *Lava* or *tephra* fragments containing numerous cavities produced by expanding gases during cooling.

scrap metal (See *new scrap, old scrap.*)

sea breeze Local wind blowing from sea to land during the day.

sea-floor spreading Pulling apart of the *crust* along a *rift*, typically the *axial rift* of the *Mid-Oceanic Ridge*, and representing the separation of two *lithospheric plates* bearing *oceanic crust*.

sea level pressure (See *standard sea level pressure.*)

seamount Prominent, isolated conical or peaklike structure rising above the *abyssal plain;* usually identified as an extinct submarine *volcano*.

secondary ores *Ores* produced and accumulated by non-igneous processes through *chemical weathering* with or without transport by *subsurface water*.

secondary wave (S-wave) Variety of *seismic wave* in which motion is back and forth at right angles (sidewise) to the direction of wave travel; also called a *transverse wave*.

sediment Finely divided *mineral* matter and organic matter derived directly or indirectly from preexisting *rock* and from life processes. (See *chemical precipitate, detrital sediment, organically derived sediment, pelagic sediment.*)

sedimentary breccia *Detrital sediment* consisting of angular fragments in a matrix of finer fragments.

sedimentary dome Up-arched *strata* forming a circular structure with domed summit and flanks with moderate to steep outward *dip*.

sedimentary ore *Ore* contained in *sedimentary rock* and formed as part of the original *sediment* of that deposit.

sedimentary rock *Rock* formed from accumulations of *sediment*.

seismic sea wave (tsunami) Train of sea waves set off by an *earthquake* (or other sea floor disturbance) traveling over the ocean surface with a velocity proportional to the square root of the ocean depth.

seismic waves Waves sent out during an *earthquake* by *faulting* or other crustal disturbance from an earthquake *focus* and propagated through the solid earth.

seismogram Record of *seismic waves* as produced by a *seismograph*.

seismograph Instrument for detecting and recording *seismic waves*.

seismology Study of *earthquakes* and the interpretation of *seismic waves* in reference to the physical properties of earth layers through which the waves pass; a branch of *geophysics*.

sensible heat Heat measurable by a *thermometer;* an indication of the intensity of kinetic energy of molecular motion within a substance.

sequential landforms *Landforms* produced by *external earth processes* in the total activity of *denudation*. Examples: canyon, alluvial fan, floodplain.

shale *Fissile, sedimentary rock* of *mud* or *clay* composition, showing lamination.

shallow-focus earthquake *Earthquake* with *focus* less than 35 mi (55 km) beneath the earth's surface.

shearing A type of internal motion, or deformation, in which thin layers of molecules glide over one another.

sheet erosion Phase of *accelerated soil erosion* in which thin layers of *soil* are removed without formation of *shoestring rills* or *gullies*.

shield volcano Low, often large, domelike accumulation

of *basalt lava flows* emerging from long radial fissures on flanks.

shingle beach *Beach* composed of well-rounded *cobbles.*

shoestring rills Long, narrow channels of small depth incised into *soil, regolith,* or land slopes undergoing *accelerated erosion.*

shoreline Shifting line of contact between water and land.

shortwave radiation *Electromagnetic radiation* in the range from 0.2 to 3 *microns,* including most of the energy spectrum of solar radiation.

sidereal month Period of time required for the moon to complete one earth-orbit of 360° with reference to the fixed stars (27.32 days).

silica *Mineral* form of silicon dioxide, as in *quartz* and *chalcedony.*

silicate magma *Magma* from which *silicate minerals* are formed.

silicates, silicate minerals *Minerals* (*compounds*) containing silicon and oxygen atoms, linked in the crystal space lattice in units of four oxygen atoms to each silicon atom.

siliceous sinter Mineral matter, largely *silica,* forming encrustations and terraces around and near *hot springs* and *geysers.*

sill *Intrusive igneous rock* in the form of a plate where *magma* was forced into a natural parting in the *country rock,* such as a bedding surface in a sequence of *sedimentary rocks.*

silt *Sediment* particles between 0.004 mm and 0.06 mm in diameter.

siltstone *Sedimentary rock* consisting of particles largely of *silt* grade.

sinkhole Surface depression in *limestone,* leading down into *limestone caverns.*

slate Compact, fine-grained variety of *metamorphic rock,* derived from *shale,* showing well-developed *cleavage.*

sleet Form of *precipitation* consisting of ice pellets, which may be frozen raindrops.

slip face Steep face of an active *sand dune,* receiving sand by saltation over the dune crest and repeatedly sliding because of oversteepening.

slump Form of *landslide* in which a single large block of *bedrock* moves downward with backward rotation upon an upwardly concave fracture surface.

smog Mixture of *particulate* matter and *chemical pollutants* in a visible and toxic low air layer over urban areas.

snow Form of *precipitation* consisting of ice particles.

snow line Altitude above which snowbanks remain throughout the entire year on high mountains.

soft layer of mantle A layer within the *mantle* in which the temperature is close to the melting point, causing the mantle *rock* to be weak; synonym for *asthenosphere.*

soft magnetization First stage, not permanent, of magnetization of *minerals* below the *Curie point.*

soil Surface layer over the lands, formed of both inorganic (*mineral*) matter and organic matter (humus) and possessing a set of physical, chemical, and organic properties favorable to the support of plants.

soil creep Extremely slow downhill movement of *soil* and *regolith* as a result of continued agitation and disturbance of the particles by such activities as *frost action,* temperature changes, or wetting and drying of the soil.

soil erosion Removal of particles of *soil* or *regolith* by the eroding action of water moving down a hill slope in *overland flow.*

solar constant Intensity of solar radiation falling upon a unit area of surface held at right angles to the sun's rays at a point outside the earth's *atmosphere;* equal to about 2 gram-calories per square centimeter per minute (2 cal/cm^2/min), or 2 *langleys* per minute (2 ly/min).

solar eclipse Apparent darkening of all or part of the sun's disk, occurring at *conjunction,* when the earth passes across the moon's shadow.

solar energy Energy arriving as *electromagnetic radiation* from the sun, including such energy stored as heat in air, soil, or water.

solar flare Emission of ionized hydrogen *gas* from the vicinity of a *sunspot,* traveling outward through the *solar system* as a body of intense *X rays* and ionized particles.

solar nebula Primordial body of diffuse *gas* and dust that through condensation gave rise to the *solar system.*

solar prominences Rosy, spikelike clouds of hydrogen gas rising into the sun's *chromosphere* and often into the *corona.*

solar wind Flow of electrons and protons emanating from the sun and traveling outward in all directions through the *solar system.*

solar system Collective term for the sun, the planets and their satellites, the asteroids, meteoroids, and comets, all orbiting the sun.

solid state State of matter that is dense, has strength to resist flowage, shows elastic properties, and may be a *crystalline solid* or an amorphous (noncrystalline) substance. Examples: *mineral* crystal, *rock, volcanic glass.*

solifluction Tundra (arctic) variety of *earthflow* in which the saturated thawed layer over *permafrost* flows slowly downhill to produce multiple terraces and *solifluction lobes.*

solifluction lobe Bulging mass of material with steep curved front moved downhill by *solifluction.*

solstice Instant in time when the subsolar point is on either the tropic of cancer, 23½° N latitude (June 21 or 22), or the tropic of capricorn, 23½° S latitude (December 21 or 22).

sorting Separation of one grade size of *sediment* particles from another by the action of currents of air or water.

source region Extensive land or ocean surface over which an *air mass* derives its temperature and moisture characteristics.

southwesterlies *Westerly winds* of the Northern Hemisphere.

specific humidity Mass of *water vapor* contained in a unit mass of air.

spectral class Subdivisions of the total range of spectral types of *stars,* with six major classes arranged from hottest to coolest star temperatures.

spheroidal weathering Production of thin, soft concentric shells of decomposed *rock* as *chemical weathering* penetrates *joints* in *bedrock* under a protective cover of *regolith.*

spiral arms Bands of *stars* extending outward from the central region of a *galaxy* in the form of spirals partly wrapped around the central region.

spiral galaxy *Galaxy* of circular platelike outline with *spiral arms.*

spontaneous expansion Volume expansion of *bedrock* as *denudation* uncovers the rock, relieving the overlying confining pressure.

springs Discharges of *ground water* from points on the ground surface or from the floor of a *stream* or *lake,* or at a *shoreline.*

spring tide Unusually large *tide range* produced by combined tidal forces of the sun and moon when in *conjunction* or *opposition.*

standard sea level pressure Average height of the mercury column in a *mercurial barometer:* 76 cm, 29.92 in., 1013.2 mb.

star Large, discrete concentration of matter within a *galaxy,* bound into a single unit by *gravitation;* contrasted with highly dispersed matter in the form of gas clouds and dust clouds.

star clouds Individual aggregations of *stars* within the *spiral arms* of a *galaxy.*

star magnitude Brightness (light intensity) of a *star* as compared with that of other stars and other celestial objects. (See *absolute magnitude, apparent visual magnitude.*)

steady-state theory Theory of *cosmology* holding that the production of matter in the *universe* has gone on continually throughout intergalactic space through all time.

stellar parallax Changing arc of separation between two *stars,* one near and one distant, as observed at opposite seasons, because of *parallax.*

stock Small *pluton,* area of surface exposure less than 40 sq mi (100 sq km); a projecting portion of a *batholith.*

stone rings Linked ringlike ridges of *cobbles* or *boulders* lying at the surface of the ground in arctic and alpine tundra regions.

stones *Meteorites* composed largely of *silicate minerals,* mostly *olivine* and *pyroxene,* with only 20% or less nickel-iron.

stony irons *Meteorites* of composition intermediate between *irons* and *stones.*

stoping Upward *intrusion,* incorporating masses of *country rock* into the rising *magma.*

storm surge Rapid rise of coastal water accompanying the arrival of an intense traveling *cyclone.*

storm track Path of a traveling *cyclone,* as shown on a weather map.

strain General term for deformation of a solid by bending or volume change when stress is applied.

strata Layers of *sediment* or *sedimentary rock* in which individual beds are separated from one another along *stratification planes.*

stratification planes (bedding planes) Planes of separation between individual *strata* in a sequence of *sedimentary rock.*

stratified drift *Glacial drift* made up of sorted and layered *clay, silt, sand,* or *gravel* deposited from meltwater in *stream channels* or *marginal lakes* close to the ice front.

stratiform clouds *Clouds* of layered, blanketlike form.

stratigrapher Scientist who pursues *stratigraphy.*

stratigraphy Branch of *historical geology* dealing with the sequence of events in the earth's history as interpreted from the evidence found in *sedimentary rocks.*

stratopause Upper limit of the *stratosphere,* transitional upward into the *mesosphere.*

stratosphere Layer of *atmosphere* lying directly above the *troposphere.*

stream Long, narrow body of flowing water occupying a *stream channel* and moving to lower levels under the force of *gravity.* (See *annular stream, consequent stream, graded stream, radial stream, subsequent stream.*)

stream capacity Maximum *load* of solid matter that can be carried by a *stream* for a given *discharge.*

stream channel Long, narrow, troughlike depression occupied and shaped by a *stream* moving to progressively lower levels.

stream deposition Accumulation of transported particles on a *stream* bed, upon the adjacent *floodplain,* or in a body of standing water.

stream erosion Progressive removal of *mineral* particles from the floor or sides of a *stream channel* by drag force of the moving water, or by *abrasion,* or by *corrosion.*

stream flow (See *channel flow.*)

stream gradient Rate of descent to lower elevations along the length of a *stream channel,* stated in ft/mi, degrees, or percent.

streamline flow Fluid flow in which all particles of the fluid move in parallel paths conforming with the solid boundaries confining the flow.

stream transportation Down-valley movement of eroded particles in a *stream channel* in solution, in *suspension,* or in *traction* as *bed load.*

stream velocity Speed of flow of water in a *stream,* measured in the downstream direction at a given point above the bed or stated as an average value for the stream as a whole.

strike Compass direction of the line of intersection of an inclined *rock* plane and a horizontal plane of reference. (See *dip.*)

strip mining Mining method in which *overburden* is first removed from a seam of *coal,* or a *sedimentary ore,* allowing the coal or ore to be extracted.

structure section Cross section in the vertical plane depicting *rock* varieties and their structures from the surface down to a given depth.

sustained-yield energy source Source of energy that is constantly being replaced or regenerated by natural processes. Examples: *solar energy, hydropower, tidal power.*

subduction Descent of the downbent edge of a *lithospheric plate* into the *asthenosphere* so as to pass beneath the edge of the adjoining plate.

sublimation Process of change of *water vapor* (*gaseous state*) to ice (*solid state*) or vice versa.

submarine canyon Narrow, V-shaped trench cut into the *continental slope,* usually attributed to *erosion* by *turbidity currents.*

submergence Inundation or partial drowning of a former land surface by a rise of sea level or a sinking of the *crust* or both.

subsequent stream *Stream* that develops its course by *stream erosion* along a band or belt of weaker *rock.*

subsidence theory Hypothesis, advanced by Charles Darwin, explaining *atolls* by subsidence of the *oceanic crust* and continued upbuilding of coral reefs.

subsolar point Point at which solar rays are perpendicular to the earth's surface.

subsurface water Water of the lands held in *soil, regolith,* or *bedrock* below the surface.

subtropical high-pressure belts Belts of persistent high *atmospheric pressure* trending east-west and centered about on latitudes 30° N and S.

subtropical high-pressure cells *Pressure cells* within the *subtropical high-pressure belts.*

succession of faunas *Faunas* differing one from the next within a succession of *strata* or *formations.*

sunspot Dark spot on the sun's surface, forming and disappearing in a time span of several days to several weeks, associated with a powerful magnetic field.

supercooled water Water films or droplets remaining in a *liquid state* at air temperatures well below the freezing point.

supergiants (See *red supergiants.*)

supernova Variety of *nova* in which brightness increase is extremely great, representing the explosive demolition of a small *star* and leaving only a *neutron star.*

superposition (See *principle of superposition.*)

surface environment Environment of low pressure and low temperature to which *rock* is exposed near the earth's surface.

surface water Water of the lands flowing exposed (as *streams*) or impounded (as ponds, lakes, or marshes).

surface waves Variety of *seismic wave,* traveling at the earth's solid surface, in which principal motion is circular in a vertical plane paralleling the direction of wave travel.

suspended load That part of the stream *load* carried in *suspension.*

suspension Form of *stream transportation* in which particles of *clay, silt,* or fine *sand* are held in upward currents in *turbulent flow.*

suture Term in *plate tectonics* referring to the narrow zone of crustal deformation and alpine structure produced by *continental collision.* Example: Himalaya range.

suturing Process of formation of a *suture.*

swash Surge of water up the *beach* slope (landward) following the collapse of a breaker.

synclinal mountain Ridge or elongate mountain developed by erosion on a *syncline.*

synclinal valley Valley eroded in weak *strata* along the central trough or axis of a *syncline.*

syncline Downfold of *strata* (or other layered *rock*) in a troughlike structure; a class of *folds.* (See *anticline.*)

synodic month Period of time required for the moon to complete one earth-orbit with reference to the sun's position in the sky, i.e., from one new moon to the next.

synoptic map Weather map showing the state of the *atmosphere* at a given time over a large area.

tail of comet Diffuse, luminous projection extending from the *coma* of a *comet* in a direction away from the sun.

talus Accumulation of loose rock fragments derived by *rockfall* from a *cliff.*

talus cone Accumulation of *talus* in the form of a partial cone with apex at the top, heading in a ravine or *gully.*

tarn Small lake occupying a *rock basin* in a *cirque* or *glacial trough.*

tectonic Relating to *tectonic activity* and *tectonics.*

tectonic activity Crustal processes of bending (*folding*) and breaking (*faulting*), concentrated on or near active *lithospheric plate* boundaries. (See *tectonics.*)

tectonics Branch of *geology* specializing in the study of structural features of the earth's *crust* and their origin. (See *plate tectonics, tectonic activity.*)

temperature (See *air temperature.*)

temperature inversion (See *low-level temperature inversion.*)

tephra Collective term for all size grades of solid *igneous rock* blown under gas pressure from a volcanic *vent.*

terminal moraine *Moraine* deposited as an embankment at the *terminus* of an *alpine glacier* or at the leading edge of an *ice sheet.*

terrace Any steplike *landform* having a flattened tread bounded by a rising steeper slope on one side and a descending steeper slope on the other. (See *alluvial terrace, rock terrace, marine terrace.*)

terrae (See *lunar highlands.*)

terrestrial planets The four *inner planets,* all having *rock* outer shells and iron cores: Mercury, Venus, Earth, and Mars.

tests Hard parts, skeletal structures, of *plankton.*

texture (of rock) Physical property of *rock* pertaining to the sizes, shapes, and arrangements of *mineral* grains it contains, i.e., coarse-grained or fine-grained.

thermal pollution Introduction of large amounts of waste heat into water bodies or into the *atmosphere,* usually from electricity-generating plants using water as a coolant.

thermocline Water layer in which temperature changes rapidly in the vertical direction.

thermometer Instrument measuring temperature. (See *maximum minimum thermometer, recording thermometer.*)

thermosphere Atmospheric layer of upwardly rising temperature, lying above the *mesopause.*

thrust sheet Sheetlike mass of *rock* moving forward over a low-angle *overthrust fault.*

thunder Sound waves sent out from a *lightning* stroke into the surrounding air.

thunderstorm Intense, local convectional storm associated with a *cumulonimbus cloud* and yielding heavy *precipitation,* along with *lightning* and *thunder,* and sometimes the fall of *hail.*

tidal flat Accumulation of fine *sediment,* including organic matter, built up to tide level in bays and estuaries.

tidal power Power derived from tidal currents moving through constricted coastal passages, usually modified by dam structures.

tide curve Graphical presentation of the rhythmic rise and fall of ocean water because of *ocean tides.*

tide range Difference in height between one *high water* and the following or preceding *low water* in *ocean tides.*

till Heterogeneous mixture of *rock* fragments ranging in size from *clay* to *boulders,* deposited beneath moving glacial ice or directly from the melting in place of stagnant glacial ice.

tillite Lithified glacial *till,* one of the forms of *sedimentary rock.*

tornado Small, very intense wind vortex with extremely low *air pressure* in center, formed beneath a dense *cumulonimbus cloud* in proximity to a *cold front.*

total eclipse of moon *Lunar eclipse* in which the earth's dark shadow, or *umbra,* completely covers the moon's disk.

traction Dragging action in which particles of *sand, gravel,* or *cobbles* are moved along the floor of a *stream channel* by force of the flowing water.

trade winds Surface *winds* in low latitudes, representing the low-level air flow within the *tropical easterlies.*

transcurrent fault Variety of *fault* on which the motion is dominantly horizontal along a near-vertical *fault plane.*

transform fault Special case of a *transcurrent fault* making up the boundary of two moving *lithospheric plates,* usually found along an offset of the *Mid-Oceanic Ridge* where *crustal spreading* is in progress.

transportation (See *stream transportation.*)

transported regolith *Regolith* formed of *mineral* matter carried by fluid agents from a distant source and deposited upon the *bedrock* or upon older regolith. Examples: *floodplain silt,* lake *clay, beach sand.*

transverse dunes Wavelike field of *sand dunes* with crests running at right angles to the direction of the prevailing wind.

transverse wave Form of *seismic wave* motion found in the *secondary wave.*

travertine Carbonate *mineral* matter, usually *calcite,* accumulating upon *limestone cavern* surfaces situated in the *unsaturated zone.*

trellis drainage pattern *Drainage pattern* characterized by a dominant parallel set of major *subsequent streams,* joined at right angles by numerous short tributaries; typical of *coastal plains* and belts of *eroded folds.*

triangular facets Steeply inclined *bedrock* surfaces of triangular outline occurring between *canyons* carved into the *fault scarp* near the base of a *fault block mountain.*

tropical cyclone Intense traveling *cyclone* of tropical latitudes, accompanied by high winds and heavy rainfall.

tropical easterlies Low-latitude *wind* system of persistent air flow from east to west between the two *subtropical high-pressure belts.*

tropopause Boundary between *troposphere* and *stratosphere.*

troposphere Lowermost layer of the *atmosphere* in which air temperature falls steadily with increasing altitude.

trough (See *glacial trough, hanging trough.*)

tsunami (See *seismic sea wave.*)

tuff Lithified deposit of *volcanic ash;* a form of *pyroclastic sediment.*

turbidites Class of ocean-bottom *sediments* consisting of the accumulations of deposits from *turbidity currents.*

turbidity current Rapid downslope streamlike flow of turbid (muddy) sea water close to the sea bed, and often confined within a *submarine canyon* on the *continental shelf,* or flowing down the side of an *oceanic trench.*

turbulent flow Fluid flow in which individual water particles move in complex eddies, superimposed on the average downstream flow path.

typhoon *Tropical cyclone* of the western North Pacific and coastal waters of Southeast Asia.

ultramafic igneous rock *Igneous rock* composed almost entirely of *mafic minerals,* usually *olivine* or *pyroxene.*

ultraviolet rays *Electromagnetic radiation* in the *wavelength* range of 0.2 to 0.4 *microns.*

umbra Long, conical zone of complete shadow from the sun's rays, extending from the earth and moon.

unconformity Discordant relationship between *strata* of one geologic age and older *rocks* lying beneath, the surface of separation representing denudation through a large gap in time, usually following *orogeny* affecting the older rock mass. (See *angular unconformity, disconformity.*)

uniformitarianism Concept, introduced by James Hutton in the late 1700s, that geologic processes acting in the past are essentially the same as those seen in action today; opposed to the view of *catastrophists.*

universe Sum total of all matter and energy that exists.

unsaturated zone *Subsurface water* zone in which pores are not fully saturated, except at times when *infiltration* is very rapid; lies above the *saturated zone.*

unstable air mass *Air mass* with substantial content of *water vapor,* capable of breaking into spontaneous convectional activity leading to the development of heavy showers and *thunderstorms.*

upper-air waves Large-scale horizontal undulations in the flow path of *upper-air westerlies;* also called *Rossby waves.*

upper-air westerlies System of *westerly winds* in the upper *atmosphere* over middle and high latitudes.

valley sedimentation Process of accumulation of *alluvium* on a valley floor.

Van Allen radiation belt Doughnut-shaped belt of intense *ionizing radiation* surrounding the earth, within the inner *magnetosphere*.

vein Thin, sheetlike layer of *minerals* or *igneous rock*, intruded or precipitated in fractures in the *country rock*; often comprising complex intersecting systems; a form of occurrence of *ores*.

vent Place of emergence of *lava*, *tephra*, and gases from a *volcano*.

volcanic neck Isolated, narrow steep-sided peak formed by erosion of *igneous rock* previously solidified in the feeder pipe of an extinct *volcano*.

ventifact *Pebble* or *cobble* shaped by *sand-blast action*, often with three curved faces joining in three sharp edges and two points.

visible light *Electromagnetic radiation* in the *wavelength* range of 0.4 to 0.7 *microns*.

volatiles *Elements* and *compounds* normally existing in the gaseous state under atmospheric conditions, dissolved in *magma*.

volcanic ash Finely divided *extrusive igneous rock* blown under gas pressure from a volcanic *vent*.

volcanic bomb *Tephra* of *cobble* or *boulder* sizes, plastic when thrown from a volcanic vent; often formed into spheroidal or spindle shapes.

volcanic breccia *Breccia* consisting of *pyroclastic sediment*.

volcanic cone Conical landform identified with a *volcano*, usually with steeply rising side slopes culminating in a central *crater*.

volcanic dike *Dike* associated with a *volcano* and serving as a passageway for *magma* that erupted at the surface during volcano building.

volcanic glass *Lava* of glassy (noncrystalline) *texture*, resulting from rapid cooling.

volcanism General term for *volcano* building and related forms of extrusive igneous activity.

volcano Conical, circular structure built by accumulation of *lava flows* and *tephra*, including *volcanic ash*. (See *composite volcano, shield volcano*.)

volcano coast *Coast* formed by *volcanoes* and *lava flows* built partly below and partly above sea level.

walls (lunar) Steep, straight *cliffs* crossing the lunar surface; probably produced by *faulting*. (See *fault scarp*.)

warm front Moving weather *front* along which a warm *air mass* is sliding up over a cold air mass, leading to production of *stratiform clouds* and *precipitation*.

washout Downsweeping of atmospheric *particulates* by *precipitation*.

waterfall Abrupt descent of a *stream* over a *bedrock* downstep in the *stream channel*.

water gap Narrow transverse gorge cut across a narrow ridge by a *stream*, usually in a region of *folds*.

water table Upper boundary surface of the *saturated zone*; the upper limit of the *ground water* body. (See *perched water table*.)

water vapor *Gaseous state* of water.

wave amplitude In wave motion, the width of space between crests and troughs, measured at right angles to the direction of wave travel.

wave-cut notch Recess at the base of a *marine cliff* where wave impact is concentrated.

wave cyclone Traveling, vortexlike *cyclone* involving interaction of cold and warm *air masses* along sharply defined *fronts*.

wave frequency Number of waves passing a fixed point in a given unit of time.

wavelength Distance separating one wave crest from the next in any uniform succession of traveling waves.

wave refraction Bending of a wave front as it travels in shallow water, caused by changes in depth of the bottom.

weathering Total of all processes acting at or near the earth's surface to cause physical disruption and chemical decomposition of *rock*. (See *chemical weathering, physical weathering*.)

well (See *artesian well, gravity well*.)

Wentworth scale Scale of particle sizes, applied to *detrital sediments*, in which size grades are named and defined.

westerly winds *Wind* system in which flow is from west to east. (See *northwesterlies, prevailing westerlies, southwesterlies, upper-air westerlies*.)

west-wind drift *Drift current* moving eastward in zone of *prevailing westerlies*.

wet (saturation) adiabatic lapse rate Reduced *adiabatic lapse rate* when *condensation* is taking place in rising air; value ranges from 2 to 3 F°/1000 ft (0.3 to 0.6 C°/100 m).

white dwarfs Class of very small *stars* of extremely high *density* and very low *luminosity* positioned far below the *main sequence* of the *Hertzsprung-Russell* diagram.

wind Air motion, dominantly horizontal relative to the earth's surface. (See *geostrophic wind, meridional winds, westerly winds*.)

world ocean Collective term for combined *oceans* of the globe.

xenolith Fragment of *country rock* enclosed by *magma* during *stoping* and retaining its identity after solidification of the *magma* into *igneous rock*.

X rays High-energy form of radiation at the extreme short *wavelength* (high *frequency*) end of the *electromagnetic spectrum*.

year Period of time required for one complete *revolution* of a planet in its *orbit* about the sun.

zigzag ridges Ridges formed in a belt of eroded *folds* having regional *plunge* of their axes, so that the ridge crests double back and forth sharply across the area.

zone of ablation Lower portion of *glacier* in which *ablation* exceeds gain of mass by snowfall; zone of wastage of a glacier.

zone of accumulation Upper portion of a *glacier* in which the *firn* becomes transformed into glacial ice; the zone of nourishment of a glacier.

INDEX

The number in boldface is the page in the text on which the term is found in boldface with a definition or explanation.

76 77 78 9 8 7 6 5 4 3 2 1